Reformulation: Nonsmooth, Piecewise Smooth, Semismooth and Smoothing Methods

Applied Optimization

Volume 22

Series Editors:

The titles published in this series are listed at the end of this volume.

Reformulation: Nonsmooth, Piecewise Smooth, Semismooth and Smoothing Methods

Edited by

Masao Fukushima

Kyoto University,
Kyoto, Japan

and

Liqun Qi

The University of New South Wales,
Sydney, Australia

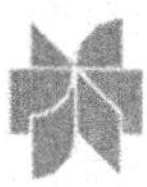

KLUWER ACADEMIC PUBLISHERS

DORDRECHT / BOSTON / LONDON

A C.I.P. Catalogue record for this book is available from the Library of Congress.

ISBN 978-1-4419-4805-2

Published by Kluwer Academic Publishers,
P.O. Box 17, 3300 AA Dordrecht, The Netherlands.

Sold and distributed in the North, Central and South America
by Kluwer Academic Publishers,
101 Philip Drive, Norwell, MA 02061, U.S.A.

In all other countries, sold and distributed
by Kluwer Academic Publishers,
P.O. Box 322, 3300 AH Dordrecht, The Netherlands.

Printed on acid-free paper

Contents

Preface vii

Solving Complementarity Problems by Means of a New Smooth Constrained Nonlinear Solver 1
Roberto Andreani and José Mario Martínez

ε-Enlargements of Maximal Monotone Operators: Theory and Applications 25
Regina S. Burachik, Claudia A. Sagastizábal and B. F. Svaiter

A Non-Interior Predictor-Corrector Path-Following Method for LCP 45
James V. Burke and Song Xu

Smoothing Newton Methods for Nonsmooth Dirichlet Problems 65
Xiaojun Chen, Nami Matsunaga and Tetsuro Yamamoto

Frictional Contact Algorithms Based on Semismooth Newton Methods 81
Peter W. Christensen and Jong-Shi Pang

Well-Posed Problems and Error Bounds in Optimization 117
Sien Deng

Modeling and Solution Environments for MPEC: GAMS & MATLAB 127
Steven P. Dirkse and Michael C. Ferris

Merit Functions and Stability for Complementarity Problems 149
Andreas Fischer

Minimax and Triality Theory in Nonsmooth Variational Problems 161
David Yang Gao

Global and Local Superlinear Convergence Analysis of Newton-Type Methods for Semismooth Equations with Smooth Least Squares 181
Houyuan Jiang and Daniel Ralph

Inexact Trust-Region Methods for Nonlinear Complementarity Problems 211
Christian Kanzow and Martin Zupke

Regularized Newton Methods for Minimization of Convex Quadratic Splines with Singular Hessians 235
Wu Li and John Swetits

Regularized Linear Programs with Equilibrium Constraints 259
Olvi L. Mangasarian

Reformulations of a Bicriterion Equilibrium Model 269
Patrice Marcotte

A Smoothing Function and its Applications 293
Ji-Ming Peng

On the Local Super–Linear Convergence of a Matrix Secant Implementation of the Variable Metric Proximal Point Algorithm for Monotone Operators 317
Maijian Qian and James V. Burke

Reformulation of a Problem of Economic Equilibrium 335
Alexander M. Rubinov and Bevil M. Glover

A Globally Convergent Inexact Newton Method for Systems of Monotone Equations 355
Michael V. Solodov and Benar F. Svaiter

On the Limiting Behavior of the Trajectory of Regularized Solutions of a P_0-Complementarity Problem 371
Roman Sznajder and M. Seetharama Gowda

Analysis of a Non-Interior Continuation Method Based on Chen-Mangasarian Smoothing Functions for Complementarity Problems 381
Paul Tseng

A New Merit Function and a Descent Method for Semidefinite Complementarity Problems 405
Nobuo Yamashita and Masao Fukushima

Numerical Experiments for a Class of Squared Smoothing Newton Methods for Box Constrained Variational Inequality Problems 421
Guanglu Zhou, Defeng Sun and Liqun Qi

Preface

The concept of "reformulation" has long been playing an important role in mathematical programming. A classical example is the penalization technique in constrained optimization that transforms the constraints into the objective function via a penalty function thereby reformulating a constrained problem as an equivalent or approximately equivalent unconstrained problem. More recent trends consist of the reformulation of various mathematical programming problems, including variational inequalities and complementarity problems, into equivalent systems of possibly nonsmooth, piecewise smooth or semismooth nonlinear equations, or equivalent unconstrained optimization problems that are usually differentiable, but in general not twice differentiable. Because of the recent advent of various tools in nonsmooth analysis, the reformulation approach has become increasingly profound and diversified.

In view of growing interests in this active field, we planned to organize a cluster of sessions entitled "Reformulation – Nonsmooth, Piecewise Smooth, Semismooth and Smoothing Methods" in the 16th International Symposium on Mathematical Programming (ismp97) held at Lausanne EPFL, Switzerland on August 24–29, 1997. Responding to our invitation, thirty-eight people agreed to give a talk within the cluster, which enabled us to organize thirteen sessions in total. We think that it was one of the largest and most exciting clusters in the symposium. Thanks to the earnest support by the speakers and the chairpersons, the sessions attracted much attention of the participants and were filled with great enthusiasm of the audience.

To take advantage of this opportunity and to popularize this interesting research field further, we decided to edit a book that primarily consists of papers based on the talks given in the above-mentioned cluster of sessions. After more than a year from the first call-for-papers, we are now in the position of publishing this book containing twenty-two refereed papers. In particular, these papers cover such diverse and important topics as

- Linear and nonlinear complementarity problems
- Variational inequality problems
- Nonsmooth equations and nonsmooth optimization problems
- Economic and network equilibrium problems
- Semidefinite programming problems
- Maximal monotone operator problems
- Mathematical programs with equilibrium constraints.

The reader will be convinced that the concept of "reformulation" provides extremely useful tools in advancing the study of mathematical programming from both theoretical and practical aspects. We hope that this book helps the current state of the art in mathematical programming to go one step further.

We wish to mention that Olvi Mangasarian is one of the pioneers of reformulation methods. He also contributes a paper to this book. We would like to dedicate this book to him on the occasion of his 65th birthday in January 1999.

We are grateful to all authors who contributed to this book. We would like to mention that most of them also served as anonymous referees for submitted papers. Moreover we are indebted to the following colleagues who kindly helped us by reviewing the papers submitted to this book: Bintong Chen, Francisco Facchinei, Alfredo Iusem, Helmut Kleinmichel, Donghui Li, Tom Luo, Jiri Outrata, Houduo Qi, Stefan Scholtes, Jie Sun, Kouichi Taji, Francis Tin-Loi, and Akiko Yoshise. Last but not least, we would like to express our special thanks to Nobuo Yamashita and Guanglu Zhou for their patient assistance and to Houyuan Jiang for his helpful suggestion during the compilation process.

May 1998

Masao Fukushima, Kyoto University
Liqun Qi, University of New South Wales

Reformulation: Nonsmooth, Piecewise Smooth, Semismooth and Smoothing Methods, pp. 1–24
Edited by M. Fukushima and L. Qi

Solving Complementarity Problems by Means of a New Smooth Constrained Nonlinear Solver

Roberto Andreani* and José Mario Martínez†

Abstract Given $F : \mathbb{R}^n \to \mathbb{R}^m$ and Ω a closed and convex set, the problem of finding $x \in \mathbb{R}^n$ such that $x \in \Omega$ and $F(x) = 0$ is considered. For solving this problem an algorithm of Inexact-Newton type is defined. Global and local convergence proofs are presented. As a practical application, the Horizontal Nonlinear Complementarity Problem is introduced. It is shown that the Inexact-Newton algorithm can be applied to this problem. Numerical experiments are performed and commented.

Key Words nonlinear systems, inexact–Newton method, global convergence, convex constraints, box constraints, complementarity.

1 INTRODUCTION

The problem considered in this paper is to find $x \in \Omega \subset \mathbb{R}^n$ such that

$$F(x) = 0, \tag{1.1}$$

where Ω is closed and convex, $F : \Omega \to \mathbb{R}^m$, and first derivatives of F exist and are continuous on an open set that contains Ω, except, perhaps, at the solutions of (1.1).

When $m = n$ this is the constrained nonlinear system problem, considered recently in [17]. By means of the introduction of slack variables, any nonlinear

*Department of Applied Mathematics, IMECC-UNICAMP, University of Campinas, CP 6065, 13081-970 Campinas SP, Brazil . This author was supported by FAPESP (Grant 90-3724-6, 93/02479-6), FINEP and FAEP-UNICAMP.
†Department of Mathematics, IMECC-UNICAMP, University of Campinas, CP 6065, 13081-970 Campinas SP, Brazil. This author was supported by FAPESP (Grant 90-3724-6), FINEP and FAEP-UNICAMP. email: martinez@ime.unicamp.br

feasibility problem given by a set of equations and inequalities can be reduced to the form (1.1). See [21] and references therein.

We introduce an Inexact-Newton-like algorithm (see [6]) for solving (1.1). At each iteration of this algorithm, a search direction d is computed such that

$$\|F'(x_k)d + F(x_k)\| \le \theta_k \|F(x_k)\|, \quad x_k + d \in \Omega, \quad |d| \le \Delta,$$

where $\|\cdot\| = \|\cdot\|_2$, $|\cdot|$ is an arbitrary norm, Δ is large and $\theta_k \in [0, 1)$. If the Inexact-Newton step does not exist, the tolerance θ_k is increased. Otherwise, we try to find a new point in the direction of the computed step. We prove that global convergence holds in the sense that, under some conditions, a solution of (1.1) is met and, in general, a stationary point of the problem can be found. We establish conditions under which local convergence results can also be proved. A practical implementation of the algorithm is given for the case in which Ω is an $n-$dimensional box.

As an application of the Inexact-Newton algorithm we define the Horizontal Nonlinear Complementarity Problem (HNCP). This is a generalization of the Horizontal Linear Complementarity Problem (HLCP) (see [2]) for which several interesting applications exist. The HNCP also generalizes the well-known Nonlinear Complementarity Problem. See [5], [9], [14], [15], [20] and [22]. We prove that, under some conditions on the HNCP, stationary points of this problem coincide with their global solutions and we also analyze the conditions under which local convergence results hold. Other applications of problem (1.1) an an interior-point algorithm for solving it can be found in [21]. Finally, we comment some numerical experiments.

The Inexact-Newton method for solving (1.1) is given in Section 2 of the paper. In Section 3 we prove the convergence results. In Section 4 we discuss the application to HNCP. In Section 5 we show the numerical experiments and in Section 6 we state some conclusions.

2 THE INEXACT-NEWTON ALGORITHM

Let $\Omega \subset I\!R^n$ be closed and convex, $\Omega \subset \mathcal{A}$, $\mathcal{A}$ an open set of $I\!R^n$. We assume that $F : \mathcal{A} \to I\!R^m$ is continuous and that the Jacobian matrix $F'(x) \in I\!R^{m\times n}$ exists and is continuous for all $x \in \mathcal{A}$ such that $F(x) \neq 0$.

We say that x_* is a *stationary point* of problem (1.1) if it is a minimizer of $\|F(x_*) + F'(x_*)(x - x_*)\|$ subject to $x \in \Omega$. Clearly, this is equivalent to say that x_* is a minimizer of $q(x) \equiv \|F(x_*) + F'(x_*)(x - x_*)\|^2$ subject to $x \in \Omega$. If the convex set Ω satisfies a constraint qualification (as it is the case of the application of Section 4, where Ω is the positive orthant) this is equivalent to say that the first order optimality conditions (Karush-Kuhn-Tucker) of the problem of minimizing the convex quadratic $q(x)$ on Ω are satisfied at x_*. But, since $\nabla q(x_*) = \nabla f(x_*)$ (where $f(x) \equiv \|F(x)\|^2$) it turns out that the stationarity of x_* corresponds exactly to the satisfaction of the Karush-Kuhn-Tucker conditions for minimizers of f on Ω.

Below we describe the main model algorithm of this paper. The algorithm computes a sequence of points $x_k \in \Omega$, starting from an arbitrary x_0. The following parameters are given independently of the iteration index k: $\sigma \in (0,1)$ (generally $\sigma \in [10^{-4}, 0.1]$) is associated to a sufficient decrease condition for the norm of F. The reduction of the stepsize α_k is related to two backtracking parameters $\kappa_1, \kappa_2 \in (0,1)$ (generally $\kappa_1 = 0.1, \kappa_2 = 0.9$). The scale-dependent large number $\Delta > 0$ is the maximum size admitted, initially, for the steplength. The parameter $\bar{\rho} \geq 1$ is an upper bound for the extrapolation factor related to the pure Inexact-Newton iteration. Finally $\theta_0 \in (0,1)$ is the initial value for the Inexact-Newton parameter θ_k. In Algorithm 2.1 it is explained how to obtain x_{k+1}, given the current iterate x_k, for all $k = 0, 1, 2, \ldots$ We initialize $\alpha_0 = 1$, which reflects our original intention of performing full Inexact-Newton steps.

Algorithm 2.1.

Given $x_k \in \Omega$, $\alpha_k \in (0,1]$, $\theta_k \in [0,1)$, the steps for obtaining x_{k+1}, α_{k+1}, and θ_{k+1} or for declaring finite convergence, are the following:

Step 0. *Stop if x_k is a solution.*
If $F(x_k) = 0$, stop. (A solution has been found.)

Step 1. *Find the Inexact-Newton direction.*
Try to find $d_k \in I\!R^n$ such that

$$x_k + d_k \in \Omega, \quad |d_k| \leq \Delta \tag{2.2}$$

and

$$\|F'(x_k)d_k + F(x_k)\| \leq \theta_k \|F(x_k)\|. \tag{2.3}$$

If a direction d_k satisfying (2.2)–(2.3) does not exist, define $x_{k+1} = x_k$, $\alpha_{k+1} = \alpha_k$, $\theta_{k+1} = (1 + \theta_k)/2$, and finish the iteration. If a direction d_k satisfying (2.2)–(2.3) is found, define $\theta_{k+1} = \theta_k$.

Step 2. *Reject the trial point if the norm increased.*
If

$$\|F(x_k + \alpha_k d_k)\| \geq \|F(x_k)\|, \tag{2.4}$$

define $x_{k+1} = x_k$, $\alpha_{k+1} \in [\kappa_1\alpha_k, \kappa_2\alpha_k]$ and finish the iteration.

Step 3. *Test sufficient decrease and try to extrapolate.*
Define $\gamma_k = 1 - \theta_k^2$. If

$$\|F(x_k + \alpha_k d_k)\| \leq (1 - \frac{\sigma\gamma_k\alpha_k}{2})\|F(x_k)\|, \tag{2.5}$$

find $\rho_k \in [1, \bar{\rho}]$ such that $x_k + \rho_k\alpha_k d_k \in \Omega$ and

$$\|F(x_k + \alpha_k \rho_k d_k)\| \leq \|F(x_k + \alpha_k d_k)\|, \tag{2.6}$$

define $x_{k+1} = x_k + \alpha_k \rho_k d_k$, $\alpha_{k+1} = 1$ and finish iteration k.

If (2.5) does not hold, define $x_{k+1} = x_k + \alpha_k d_k$, choose

$$\alpha_{k+1} \in [\kappa_1 \alpha_k, \kappa_2 \alpha_k]. \tag{2.7}$$

and finish the iteration.

Some algorithms for solving $n \times n$ nonlinear systems accept trial points generated by Newton-like procedures only when some sufficient decrease condition holds. See, for example, [7]. Here we adopt the point of view that a trial point deserves to be accepted if the norm of F decreases, since this feature tends to produce larger steps far from the solution, as it is generally desired in minimization algorithms. Moreover, using the extrapolation step we try to find a point along the Inexact-Newton direction, the distance of which to x_k is larger than the distance of the pure Inexact-Newton trial point to the current point. In this way, we try to alleviate the tendency of backtracking algorithms to produce short steps.

Let us now describe a particular version of Algorithm 2.1, which corresponds to the case in which Ω is an n−dimensional box, perhaps with some infinite bounds. So,

$$\Omega = \{x \in I\!R^n \mid x_i \geq \ell_i \text{ for } i \in I \text{ and } x_i \leq u_i \text{ for } i \in J\},$$

where I and J are subsets of $\{1, \ldots, n\}$. In this case we use $|\cdot| = \|\cdot\|_\infty$, so that the constraints (2.2) also define a box. The key point is that the step d_k is computed by means of the consideration of the following box-constrained quadratic minimization problem:

$$\text{Minimize } \Phi(d) \text{ subject to } x_k + d \in \Omega \text{ and } \|d\|_\infty \leq \Delta, \tag{2.8}$$

where $\Phi(d) = \|F'(x_k)d + F(x_k)\|^2$. A necessary and sufficient condition for a global minimizer of (2.8) is that $\|P(d - \nabla\Phi(d)) - d\| = 0$, where P is the orthogonal projection on the feasible region of (2.8). Moreover, if $0 \in I\!R^n$ is a solution of (2.8) it turns out that x_k is a stationary point of (1.1). The particular case of the model algorithm that we are going to describe also uses the scale-dependent parameter $\Delta > 0$, the sufficient descent constant $\sigma \in (0, 1)$, the maximum extrapolation parameter $\bar{\rho} \geq 1$ and the initial Inexact-Newton tolerance $\theta_0 \in [0, 1)$. We initialize, as before, $\alpha_0 = 1$. Moreover, we use an additional tolerance $\eta_k \in [0, 1)$ for deciding termination of the method used for solving the subproblem (2.8). Roughly speaking, η_k is the degree of accuracy used in the approximate solution of (2.8). Different strategies for the choice of η_k at each iteration will be discussed in Section 5.

Algorithm 2.2.
Given $x_k \in \Omega$, $\alpha_k \in (0,1]$, $\theta_k \in [0,1)$, $\eta_k \in [0,1)$, the steps for obtaining x_{k+1}, α_{k+1}, and θ_{k+1} or for declaring finite convergence, are the following:

Step 0. *Stop the execution if x_k is a solution.*
If $F(x_k) = 0$, stop. (A solution has been found.)

Step 1. *Stop the execution if x_k is stationary.*
If $\|P(-\nabla\Phi(0))\| = 0$, stop. (A stationary point has been found.)

Step 2. *Compute the step d_k.*
If $k > 0$ and $x_k = x_{k-1}$ take $d_k = d_{k-1}$. Otherwise, choose $d_k \in \mathbb{R}^n$ such that

$$x_k + d_k \in \Omega, \quad \|F'(x_k)d_k + F(x_k)\| \leq \theta_k \|F(x_k)\| \quad \text{and} \quad \|d_k\|_\infty \leq \Delta. \tag{2.9}$$

If such a choice is not possible, replace $\theta_k \leftarrow (1 + \theta_k)/2$ and repeat Step 2.

Step 3. *Repeat the current iterate if the norm did not decrease.*
Define $\theta_{k+1} = \theta_k$. If (2.4) takes place, define $x_{k+1} = x_k, \alpha_{k+1} = \alpha_k/2$ and finish the iteration.

Step 4. *Test sufficient decrease.*
Define $\gamma_k = 1 - \theta_k^2$. If (2.5) does not hold, define $x_{k+1} = x_k + \alpha_k d_k$, $\alpha_{k+1} = \alpha_k/2$ and finish iteration k.

Step 5. *Try extrapolation step.*
Step 5.1. Set $\underline{\rho} \leftarrow 1, \rho \leftarrow 2$.
Step 5.2. If

$$\rho > \bar{\rho} \quad \text{or} \quad \|F(x_k + \alpha_k \rho d_k)\| \geq \|F(x_k + \alpha_k \underline{\rho} d_k)\|$$

set

$$\rho_k = \underline{\rho}, \quad x_{k+1} = x_k + \alpha_k \rho_k d_k, \quad \alpha_{k+1} = 1$$

and finish iteration k.
Step 5.3. Set $\underline{\rho} \leftarrow \rho$, $\rho \leftarrow 2\rho$ and go to Step 5.2.

As we mentioned above, for computing a direction d_k satisfying (2.9), we apply an iterative method to (2.8), stopping when an increment $\bar{d}$ is obtained such that $x_k + \bar{d} \in \Omega$, $\|\bar{d}\|_\infty \leq \Delta$ and

$$\|P(\bar{d} - \nabla\Phi(\bar{d})) - \bar{d}\| \leq \eta_k \|P(-\nabla\Phi(0))\|. \tag{2.10}$$

If the step $\bar{d}$ obtained in this way satisfies (2.3), we define $d_k = \bar{d}$. Otherwise, we continue the execution of the quadratic solver stopping only when an iterate $\bar{d}$ satisfies $\|F'(x_k)\bar{d} + F(x_k)\| \leq \theta_k \|F(x_k)\|$ or when a global minimizer of (2.8) is found.

If a minimizer $\bar{d}$ of (2.8) is found but $\|F'(x_k)\bar{d}+F(x_k)\| > \theta_k\|F(x_k)\|$, then the condition (2.9) cannot be fulfilled, so we increase θ_k. In Algorithm 2.1 we changed the iteration index in this case while in Algorithm 2.2 we preferred to redefine θ_k without increasing the iteration number. Clearly, this is only a formal modification that does not affect the convergence properties of the algorithm. However, from the description adopted in Algorithm 2.2 it is clear that no additional work is needed when one increases θ_k. In fact, if $\bar{d}$ is a minimizer of (2.8) and the null vector is not a solution of (2.8), the condition $\|F'(x_k)\bar{d} + F(x_k)\| \leq \theta\|F(x_k)\|$ is necessarily satisfied for large enough $\theta < 1$. Therefore, this rejection involves only a redefinition of θ_k.

3 CONVERGENCE RESULTS

The first theorem in this section proves that, if θ_k is increased a finite number of times then Algorithm 2.1 finds a solution of (1.1).

Theorem 3.1. *Let $\{x_k\}$ be a sequence generated by Algorithm 2.1, such that $\theta_k = \bar{\theta} < 1$ for all $k \geq k_0$. Then, every limit point of $\{x_k\}$ is solution of (1.1).*

Proof. Without loss of generality, assume that $\theta_k = \bar{\theta}$ for all $k = 0, 1, 2, \ldots$. A straightforward calculation shows that (2.3) implies that

$$\langle F'(x_k)d_k, F(x_k)\rangle \leq -\frac{\gamma_k}{2}\|F(x_k)\|^2.$$

Define $\varphi(x) = \frac{1}{2}\|F(x)\|^2$ and $\bar{\gamma} = 1 - \bar{\theta}^2$. Since $\nabla\varphi(x) = F'(x)^T F(x)$, we have that

$$\langle \nabla\varphi(x_k), d_k\rangle \leq -\frac{\bar{\gamma}}{2}\|F(x_k)\|^2, \tag{3.1}$$

for all $k = 0, 1, 2, \ldots$.

Let x_* be a limit point of $\{x_k\}$ and $K_0 \subset I\!N$, such that

$$\lim_{k\in K_0} x_k = x_*.$$

Let us call

$$K_1 = \{k \in K_0 \text{ such that (2.5) holds}\}.$$

Assume first that K_1 is infinite and $\limsup_{k\in K_1}\alpha_k > 0$. Let K_2 be an infinite subset of K_1 such that

$$\alpha_k \geq \bar{\alpha} > 0$$

for all $k \in K_2$. Therefore we obtain

$$1 - \frac{\sigma\bar{\gamma}\alpha_k}{2} \leq 1 - \frac{\sigma\bar{\gamma}\bar{\alpha}}{2} \equiv r < 1$$

for all $k \in K_2$. Therefore $\{\|F(x_k)\|\}$ is a nonincreasing sequence such that $\|F(x_{k+1})\| \leq r\|F(x_k)\|$ for all $k \in K_2$. This implies that $\|F(x_k)\| \to 0$. Therefore $F(x_*) = 0$.

The following possibilities remain to be considered:

(i)

$$K_1 \text{ is infinite but } \lim_{k\in K_1} \alpha_k = 0. \tag{3.2}$$

(ii)

$$K_1 \text{ is finite;} \tag{3.3}$$

Let us consider first (i). Assume, by contradiction, that $F(x_*) \neq 0$. Without loss of generality, assume that $\alpha_k < 1$ for all $k \in K_1$. So, by (2.5) and (2.7), we have that, for all $k \in K_1$,

$$\alpha_k \in [\kappa_1\alpha_{k-1}, \kappa_2\alpha_{k-1}] \tag{3.4}$$

and

$$\|F(x_{k-1} + \alpha_{k-1}d_{k-1})\|^2 > (1 - \frac{\sigma\bar{\gamma}\alpha_{k-1}}{2})^2\|F(x_{k-1})\|^2. \tag{3.5}$$

By (3.2) and (3.4) we have that

$$\lim_{k\in K_1} \alpha_{k-1} = 0.$$

So, since $|d_k|$ is bounded, we see that

$$\lim_{k\in K_1} x_{k-1} = x_*.$$

Now, by (3.5),

$$\frac{\varphi(x_{k-1} + \alpha_{k-1}d_{k-1}) - \varphi(x_{k-1})}{\alpha_{k-1}} > -\frac{\sigma\bar{\gamma}\|F(x_{k-1})\|^2}{2} + \frac{\sigma^2\bar{\gamma}^2\alpha_{k-1}\|F(x_{k-1})\|^2}{8}.$$

for all $k \in K_1$. Since $\alpha_{k-1} \to 0$ and $|d_{k-1}|$ is bounded, we have that $x_{k-1} + \alpha_{k-1}d_{k-1}$ tends to x_* for $k \in K_1$. But $F(x_*) \neq 0$ so $F'(x)$ exists and is continuous in a neighborhood of x_*. For large enough $k \in K_1$ we have that both x_{k-1} and $x_{k-1}+\alpha_{k-1}d_{k-1}$ belong to that neighborhood. So, we can apply the Mean Value Theorem, which implies that for large enough $k \in K_1$, there exists $\xi_{k-1} \in [0,1]$ such that

$$\langle\nabla\varphi(x_{k-1} + \xi_{k-1}\alpha_{k-1}d_{k-1}), d_{k-1}\rangle > -\frac{\sigma\bar{\gamma}\|F(x_{k-1})\|^2}{2} + \frac{\sigma^2\bar{\gamma}^2\alpha_{k-1}\|F(x_{k-1})\|^2}{8}. \tag{3.6}$$

Since $|d_k| \leq \Delta$ for all k, there exists K_3, an infinite subset of K_1, such that

$$\lim_{k\in K_3} d_{k-1} = d.$$

Taking limits for $k \in K_3$ on both sides of (3.6), we obtain

$$\langle\nabla\varphi(x_*), d\rangle \geq -\frac{\sigma\bar{\gamma}\|F(x_*)\|^2}{2}.$$

So, for large enough $k \in K_3$, defining $\sigma' = \frac{\sigma+1}{2}$, we have that

$$\langle \nabla\varphi(x_{k-1}), d_{k-1}\rangle \geq -\frac{\sigma'\bar{\gamma}\|F(x_{k-1})\|^2}{2} > -\frac{\bar{\gamma}\|F(x_{k-1})\|^2}{2} \tag{3.7}$$

Therefore, (3.7) contradicts (3.1). This proves that $F(x_*) = 0$.

Let us now assume (ii). Since K_1 is finite, there exists $k_3 \in K_0$ such that (2.5) does not hold for all $k \in K_0$ and $k \geq k_3$. Therefore

$$\lim_{k \in K_0} \alpha_k = 0,$$

and we can repeat the former proof, with minor modifications, for proving that $F(x_*) = 0$. ∎

In Theorem 3.1 we proved that if θ_k needs to be increased only a finite number of times, then any acumulation point of a sequence generated by Algorithm 2.1 is a solution of problem. In the next theorem we show what happens if θ_k needs to be increased infinitely many times. Essentially, we show that, in that case, we find a stationary point of the problem.

Theorem 3.2. *Suppose that, in Algorithm 2.1, θ_k is increased infinitely many times and define*

$$K_4 = \{k \in \{0,1,2,\ldots\} \mid \theta_{k+1} > \theta_k\}.$$

Then, every limit point of the sequence $\{x_k\}_{k\in K_4}$ is stationary.

Proof. Let $x_* \in \Omega$ be a limit point of $\{x_k\}_{k\in K_4}$. If $F(x_*) = 0$ we are done, so let us assume that $\|F(x_*)\| > 0$. This implies that $F'(x)$ exists and is continuous at x_*. Suppose, by contradiction, that x_* is not stationary. Therefore, there exists $d \in I\!R^n$ such that $|d| \leq \Delta/2$, $x_* + d \in \Omega$ and

$$\|F'(x_*)d + F(x_*)\| < \|F(x_*)\|.$$

Define

$$\frac{\|F'(x_*)d + F(x_*)\|}{\|F(x_*)\|} = r < 1.$$

Choose $r' \in (r,1)$. By continuity of F and F', we have that

$$\frac{\|F'(x_k)(x_* + d - x_k) + F(x_k)\|}{\|F(x_k)\|} \leq r' \tag{3.8}$$

for large enough $k \in K_4$. But, since $|d| \leq \Delta/2$ for large enough $k \in K_4$, $|x_* + d - x_k| \leq \Delta$. So, (3.8) contradicts the fact that (2.3) cannot be achieved for $k \in K_4$ and $\theta_k \to 1$. ∎

Remark. The hypotheses of Theorems 3.1 and 3.2 have an interesting interpretation from a non strictly algorithmic point of view. In fact, the operator

$L(x) = F(\bar{x}) + F'(\bar{x})(x - \bar{x})$ is the first-order affine model of $F(x)$ that uses information available at $\bar{x}$. In Theorem 3.1 we assume that the inequality $\|L(x)\| \leq \bar{\theta}\|L(\bar{x})\|$, $\bar{\theta} < 1$, holds for some $x \in \Omega$, $\|x - \bar{x}\| \leq \Delta$ for all algorithmic iterates. Its conclusion implies that whenever Ω is bounded or $\|F(x)\|$ has bounded level sets, a solution of the system *exists*. In other words, Theorem 3.1 can be interpreted as an existence theorem that says that solutions of a bounded problem exist when it is possible a sufficient decrease of the model of the system on a neighborhood of fixed (not necessarily small) arbitrary size.

Now, many times one applies nonlinear system solvers to problems where a solution is not known to exist and sometimes the solution does not exist at all. In this case, according to Theorem 3.1, sufficient reduction of the model cannot be possible at all points. However, Theorem 3.2 ensures that cluster points of iterates where sufficient reduction is not possible are stationary.

Now we prove two auxiliary lemmas which, in turn, will be useful in the local convergence analysis.

Lemma 3.3. *If $F(x_k) \neq 0$ and (2.3) holds, then $\|F'(x_k)\| \neq 0$ and*

$$\|d_k\| \geq \frac{1 - \theta_k}{\|F'(x_k)\|}\|F(x_k)\|. \tag{3.9}$$

Proof. Since $\theta_k < 1$ and $F(x_k) \neq 0$, we obtain that $F'(x_k) \neq 0$. Now, by (2.3),

$$\|F(x_k)\| - \|F'(x_k)\|\|d_k\| \leq \|F'(x_k)d_k + F(x_k)\| \leq \theta_k\|F(x_k)\|.$$

So,

$$\|F'(x_k)\|\|d_k\| \geq (1 - \theta_k)\|F(x_k)\|$$

and the thesis follows from this inequality. ∎

Lemma 3.4. *Let $\{w_k, k = 0, 1, 2, \ldots\}$ be a sequence in $\mathbb{R}^n$ and assume that*

$$\lim_{k\to\infty} \frac{\|w_{k+1} - w_k\|}{\|w_k - w_{k-1}\|} = 0. \tag{3.10}$$

Then, $\{w_k\}$ converges R-superlinearly to some $w_* \in \mathbb{R}^n$.

Proof. By (3.10), there exists $k_0 \in \mathbb{N}$ such that

$$\frac{\|w_{k+1} - w_k\|}{\|w_k - w_{k-1}\|} \leq \frac{1}{2} \quad \text{for all} \quad k \geq k_0. \tag{3.11}$$

So,

$$\|w_{k+1} - w_k\| \leq \frac{1}{2^{k-k_0}}\|w_{k_0+1} - w_{k_0}\| \tag{3.12}$$

for all $k \geq k_0$.

Now, by (3.12), given $\varepsilon > 0$, there exists $k_1 \in \mathbb{N}$ such that

$$\|w_{k+1} - w_k\| \le \frac{\varepsilon}{2} \text{ for all } \; k \ge k_1.$$

Define $k_2 = \max\{k_1, k_0\}$. For all $l, j \ge k_2$ we have that

$$\|w_j - w_l\| \le \sum_{i=l}^{j-1} \|w_{i+1} - w_i\| \le (\sum_{i=l}^{j-1} \frac{1}{2^{i-l}})\|w_{l+1} - w_l\| \le (\sum_{i=l}^{j-1} \frac{1}{2^{i-l}})\frac{\varepsilon}{2} \le \varepsilon$$

Then, $\{w_k\}$, is a Cauchy sequence and so, it converges to some $w_* \in \mathbb{R}^n$.

Now, if $k \ge k_2$, we have that

$$\|w_k - w_*\| \le \sum_{i=k}^{\infty} \|w_{i+1} - w_i\| \le 2\|w_{k+1} - w_k\|. \tag{3.13}$$

Therefore, the sequence $\{\|w_k - w_*\|\}$ is bounded by the Q-superlinear sequence $\{2\|w_{k+1} - w_k\|\}$. This implies that the convergence of $\{w_k\}$ is R-superlinear. ∎

The following assumption states that x_* is a solution of the problem such that, in some neighborhood of this point, the norm of $F(x)$ is bounded below by a quantity of the order of the distance between x and x_*. As a consequence, x_* is the unique solution of the system on that neighborhood.

Assumption 3.1. *Strong Local Unicity (SLU)*

There exist $x_* \in \Omega$, $\varepsilon, c > 0$ such that $F(x_*) = 0$ and $\|x - x_*\| \le c\|F(x)\|$ for all $x \in \Omega$ such that $\|x - x_*\| \le \varepsilon$.

In the following theorem we prove that, under some conditions on the direction d_k, Algorithm 2.1 converges superlinearly to a solution of the problem. The assumptions include a Lipschitz condition on the Jacobian, the fact that the Inexact-Newton direction tends to be a Newton direction (the direction is an approximate solution of $F'(x_k)d + F(x_k) = 0$ with accuracy tending to zero) and, finally, that the norm of d_k has the same order as $\|F(x_k)\|$. In this way R-superlinear convergence to some solution x_* is obtained. If, in addition, the SLU assumption holds, the convergence is Q-superlinear.

Theorem 3.5. *Assume that*

$$\{x \in \Omega \text{ such that } \|F(x)\| \le \|F(x_0)\|\} \text{ is bounded} \tag{3.14}$$

and the following Lipschitz condition on $F'(x)$ holds:

$$\|F'(x) - F'(y)\| \le L\|x - y\| \tag{3.15}$$

for all $x, y \in \Omega$.

Assume that the sequence $\{x_k\}$ is generated by Algorithm 2.1 and that there exists $t_k \to 0$, $\beta > 0$ such that

$$\|F'(x_k)d_k + F(x_k)\| \le t_k \|F(x_k)\| \tag{3.16}$$

and

$$\|d_k\| \le \beta \|F(x_k)\| \tag{3.17}$$

for all $k = 0, 1, 2, \ldots$. Then, there exists $k_0 \in \{0, 1, 2, \ldots\}$ such that $\alpha_k = 1$ for all $k \ge k_0$. Moreover,

$$\lim_{k \to \infty} \frac{\|F(x_{k+1})\|}{\|F(x_k)\|} = 0, \tag{3.18}$$

and the sequence is R-superlinearly convergent to some $x_* \in \Omega$ such that $F(x_*) = 0$. Finally, if x_* satisfies the SLU Assumption, the convergence is Q-superlinear.

Proof. By (3.15), (3.16) and (3.17), for all $k = 0, 1, 2, \ldots$, we have that

$$\begin{aligned}
\|F(x_k + \alpha_k d_k)\| &\le \|F(x_k) + \alpha_k F'(x_k) d_k\| + \frac{L}{2}\alpha_k^2 \|d_k\|^2 \\
&\le \|\alpha_k [F(x_k) + F'(x_k) d_k]\| + (1 - \alpha_k)\|F(x_k)\| + \frac{L}{2}\alpha_k^2 \|d_k\|^2 \\
&\le \alpha_k t_k \|F(x_k)\| + (1 - \alpha_k)\|F(x_k)\| + \frac{L}{2}\alpha_k^2 \|d_k\|^2 \\
&\le \alpha_k t_k \|F(x_k)\| + (1 - \alpha_k)\|F(x_k)\| + \frac{L}{2}\alpha_k^2 \beta^2 \|F(x_k)\|^2 \\
&= (\alpha_k t_k + 1 - \alpha_k + \frac{L}{2}\alpha_k^2 \beta^2 \|F(x_k)\|)\|F(x_k)\| \\
&= [1 - \alpha_k(1 - t_k - \frac{L}{2}\alpha_k \beta^2 \|F(x_k)\|)]\|F(x_k)\|.
\end{aligned} \tag{3.19}$$

By (3.14), (3.16) and Theorem 3.1 we have that $\|F(x_k)\| \to 0$. Since $t_k \to 0$ and $\sigma\gamma_k/2 \in (0,1)$, (3.19) implies that there exists $k_1 \in \{0, 1, 2, \ldots\}$ such that

$$\|F(x_k + \alpha_k d_k)\| \le (1 - \frac{\sigma\gamma_k\alpha_k}{2})\|F(x_k)\|$$

if $k \ge k_1$. By the definition of Algorithm 2.1, this implies that $\alpha_k = 1$ for all $k \ge k_0 \equiv k_1 + 1$.

Now, by (3.19) with $\alpha_k = 1$ for $k \ge k_0$, we obtain

$$\|F(x_{k+1})\| \le \|F(x_k + d_k)\| \le [1 - (1 - t_k - \frac{L}{2}\beta^2 \|F(x_k)\|)]\|F(x_k)\|.$$

Therefore, (3.18) holds.

By the compactness assumption and Lipschitz condition there exists $c_1 \in \mathbb{R}$ such that $\|F'(x_k)\| \le c_1$ for all $k = 0, 1, 2, \ldots$, then , by Lemma 3.3, there exists $\beta_1 > 0$ such that

$$\|d_k\| \geq \beta_1 \|F(x_k)\|. \tag{3.20}$$

Since $\rho_k \leq \bar{\rho}$ for all k, by (3.17), (3.18) and (3.20), we obtain

$$\|x_{k+1} - x_k\| = \rho_k \|d_k\| \leq \bar{\rho}\beta \|F(x_k)\| \leq \bar{\rho}\beta (\frac{\|F(x_k)\|}{\|F(x_{k-1})\|}) \|F(x_{k-1})\|$$

$$\leq \frac{\bar{\rho}\beta}{\beta_1} u_k \|d_{k-1}\| = \frac{\bar{\rho}\beta}{\beta_1 \rho_{k-1}} u_k \|x_k - x_{k-1}\| \leq M u_k \|x_k - x_{k-1}\|$$

for all $k \geq k_1$, where $M = \frac{\bar{\rho}\beta}{\beta_1}$ and $u_k = \frac{\|F(x_k)\|}{\|F(x_{k-1})\|} \to 0$. Therefore

$$\frac{\|x_{k+1} - x_k\|}{\|x_k - x_{k-1}\|} \to 0$$

So, by Lemma 3.4, $\{x_k\}$ converges R-superlinearly to some solution $x_* \in \Omega$.

If Assumption 3.1 holds, there exists $k_2 \in \mathbb{N}$ such that, for all $k \geq k_2$,

$$\|x_{k+1} - x_*\| \leq c \|F(x_{k+1})\|. \tag{3.21}$$

But, by (3.15) there exists $L_1 > 0$ such that

$$\|x_k - x_*\| \geq \frac{1}{L_1} \|F(x_k)\|. \tag{3.22}$$

By (3.21) and (3.22) we have that

$$\frac{\|x_{k+1} - x_*\|}{\|x_k - x_*\|} \leq c L_1 \frac{\|F(x_{k+1})\|}{\|F(x_k)\|}$$

Therefore, by (3.18), $\{x_k\}$ converges Q-superlinearly to x_*. ∎

The following auxiliary lemma will be useful to prove a local convergence result in Section 4. Its proof follows from elementary analysis considerations and will be omitted here.

Lemma 3.6. *Assume that $F \in C^1$, Ω convex, $x_* \in \Omega$ and let $\mathcal{V}_1(\Omega, x_*)$ be the set of directions $d \in \mathbb{R}^n$ such that $d = \lim_{k\to\infty} \frac{x_k - x_*}{\|x_k - x_*\|}$ for some sequence $\{x_k\} \subset \Omega$ that converges to x_*. Then the following propositions are equivalent:*

(i)

$$\inf_{v \in \mathcal{V}_1(\Omega, x_*)} \|F'(x_*)v\| = \zeta > 0.$$

(ii) *There exists $c, \varepsilon > 0$, such that*

$$\|x - x_*\| \leq c \|F(x) - F(x_*)\|,$$

for all $x \in \Omega$ such that $\|x - x_*\| \leq \varepsilon$.

4 APPLICATION TO COMPLEMENTARITY PROBLEMS

Assume that $G, H : \mathbb{R}^q \to \mathbb{R}^q$ represent two transformation processes that depend on nonnegative sequential inputs $y_1, \ldots, y_q$ (inputs of G) and $z_1, \ldots, z_q$ (inputs of H). However, at each time stage $i \in \{1, \ldots, q\}$ only one of the two controls $\{y_i, z_i\}$ can be positive. In other words, the very essence of the work does not admit performing two jobs simultaneously. Suppose that, as a result of the transformations, the following synchronization equation must hold:

$$G(y) + H(z) = 0. \tag{4.1}$$

Clearly, the conditions stated above on the controls impose that

$$y^T z = 0 \quad \text{and} \quad y, z \geq 0. \tag{4.2}$$

The system (4.1)-(4.2) defines the Horizontal Nonlinear Complementarity Problem (HNCP). If G and H are affine functions, we have the Horizontal Linear Complementarity Problem (HLCP) (see [2, 12]). It can be shown that the optimality conditions of quadratic programming problems form an HLCP and that the optimality conditions of linearly constrained minimization problems form an HNCP. If $H(z) = -z$ for all $z \in \mathbb{R}^q$, the HNCP reduces to the classical Nonlinear Complementarity Problem (NCP).

Let us define $n = 2q, m = q + 1, x = (y, z), \Omega = \{x \in \mathbb{R}^n \mid x \geq 0\}$, and

$$F(x) = F(y, z) = \begin{pmatrix} G(y) + H(z) \\ y^T z \end{pmatrix}$$

Then, the HNCP consists on finding a solution of

$$F(y, z) = 0, \quad y, z \in \Omega. \tag{4.3}$$

We seek to solve problem (4.3) using the algorithms defined in Section 2. With this purpose, it will be useful to characterize the conditions under which stationary points of (4.3) are, in fact, global solutions. An answer to this question is given in Theorem 4.1. Let us recall first that, since Ω satisfies a constraint qualification, the points that satisfy Karush-Kuhn-Tucker conditions for $\|F(y, z)\|^2$ on Ω are exactly the stationary points defined at the beginning of Section 2.

Theorem 4.1. *Assume that (y_*, z_*) is stationary point of problem (4.3) and $H'(z_*), G'(y_*)$ are such that for all $u \in \mathbb{R}^q$, $u \neq 0$, there exists $D(u) \in \mathbb{R}^{q \times q}$, a diagonal matrix with strictly positive diagonal elements, such that*

$$u^T H'(z_*) D(u) G'(y_*)^T u < 0.$$

Then, (y_*, z_*) is solution of (4.1)-(4.2).

Proof. The solutions of (4.3) are minimizers of the problem

$$\text{Minimize } \|G(y) + H(z)\|^2 + (y^T z)^2 \quad \text{subject to} \quad y, z \geq 0. \tag{4.4}$$

Let us call

$$u_* = G(y_*) + H(z_*). \tag{4.5}$$

The first-order optimality conditions of (4.4) are:

$$2G'(y_*)^T u_* + 2(y_*^T z_*) z_* - \mu_1 = 0, \tag{4.6}$$

$$2H'(z_*)^T u_* + 2(y_*^T z_*) y_* - \mu_2 = 0, \tag{4.7}$$

$$y_*^T \mu_1 = 0, \tag{4.8}$$

$$z_*^T \mu_2 = 0, \tag{4.9}$$

$$y_* \geq 0, \quad z_* \geq 0, \quad \mu_1 \geq 0, \quad \mu_2 \geq 0. \tag{4.10}$$

By (4.6)- (4.10), we obtain

$$4[H'(z_*)^T u_*]_i [G'(y_*]^T u_*]_i = 4(y_*^T z_*)^2 [y_*]_i [z_*]_i + [\mu_1]_i [\mu_2]_i \geq 0, \tag{4.11}$$

for all $i = 1, \ldots, q$.

Suppose, for a moment, that $u_* \neq 0$. Then, by the hypothesis of the theorem, there exists a diagonal matrix $D(u_*) \in \mathbb{R}^{q \times q}$ with strictly positive diagonal entries, such that

$$\sum_{i=1}^{q} [H'(z_*)^T u_*]_i [D(u_*)]_{i,i} [G'(y_*)^T u_*]_i < 0. \tag{4.12}$$

Since $[D(u_*)]_{i,i} > 0$, the equality (4.11) cannot be true. This contradiction came from the assumption $u_* \neq 0$. Therefore, $G(y_*) + H(z_*) = 0$. Thus, by (4.6) and (4.7), we have that

$$2(y_*^T z_*) z_* - \mu_1 = 0 \tag{4.13}$$

and

$$2(y_*^T z_*) y_* - \mu_2 = 0. \tag{4.14}$$

Multiplying (4.13) and (4.14), we obtain, using (4.8) and (4.9):

$$4(y_*^T z_*)^3 + \mu_1^T \mu_2 = 0.$$

So, by (4.10),

$$y_*^T z_* = 0,$$

as we wanted to prove. ∎

Remark. In the Nonlinear Complementarity Problem $NCP(G)$ $(H(z) = -z)$, the hypothesis of Theorem 4.1 reads: For all $u_* \in \mathbb{R}^q, u_* \neq 0$, there exists a diagonal matrix $D(u_*)$ with strictly positive diagonal entries such that $u_*^T D(u_*) G'(y_*)^T u_* > 0$. Friedler and Pták [13] proved that this hypothesis holds if $G'(y_*)$ is a $P-$matrix (all principal minors of $G'(y_*)$ are positive).

Let us recall (see [4]) that a matrix $A \in \mathbb{R}^{n\times n}$ is called an S-matrix if there exists $x \in \mathbb{R}^n, x \geq 0$ such that $Ax > 0$. It can be proved (see [13]) that A is an S-matrix if, and only if, $\{y \in \mathbb{R}^n \mid y \geq 0, y \neq 0, A^T y \leq 0\}$ is empty.

A matrix $A \in R^{n\times n}$ is said to be column-sufficient (see [4]) if for all $u \in R^n$ such that

$$[u]_i[Au]_i \geq 0 \text{ for all } i = 1, \ldots, n$$

one necessarily has that

$$[u]_i[Au]_i = 0 \text{ for all } i = 1, \ldots, n.$$

A matrix A is row-sufficient if A^T is column-sufficient and it is called sufficient if it is both column-sufficient and row-sufficient.

Although any P-matrix is necessarily a sufficient S-matrix, the reciprocal is not true. For example, the sufficient S-matrix

$$\begin{bmatrix} 0 & 1 \\ -1 & 1 \end{bmatrix}$$

is not a P-matrix.

Using the definitions above we can prove a new sufficient condition that guarantees that a stationary point of $NCP(G)$ is a solution of the problem.

Theorem 4.2. *Assume that (y_*, z_*) is a stationary point of system (4.3) associated to NCP(G). If $G'(y_*)$ is a row-sufficient S-matrix then y_* is solution of $NCP(G)$.*

Proof. In this case $H(z) = -z$ and, in Theorem 4.1, $u_* = G(y_*) - z_*$. Then, by (4.11), we obtain

$$[u_*]_i[G'(y_*)^T u_*]_i \leq 0 \tag{4.15}$$

for all $i = 1, \ldots, q$. Since $G'(y_*)$ is row-sufficient, we have that

$$[u_*]_i[G'(y_*)^T u_*]_i = 0 \tag{4.16}$$

for all $i = 1, \ldots, q$. By (4.11), we obtain $[y_*]_i[z_*]_i = 0$ for all $i = 1, \ldots, q$, so

$$y_*^T z_* = 0. \tag{4.17}$$

Thus, by (4.6) and (4.7),

$$2G'(y_*)^T(-u_*) = -\mu_1 \leq 0, \tag{4.18}$$

$$-2u_* = \mu_2 \geq 0. \tag{4.19}$$

Suppose that $u_* \neq 0$, then by (4.18) and (4.19) $G'(y_*)$ could not be an S-matrix. Therefore $u_* = 0$. So, by (4.17), y_* is solution of $NCP(G)$. ∎

As we mentioned above, the Horizontal Linear Complementarity Problem (HLCP) is the particular case of HNCP in which G and H are affine functions. So, it can be stated in the following way:

Find $y, z \in \mathbb{R}^q, y, z \geq 0$ such that

$$Qy + Rz = b, \quad y^T z = 0. \tag{4.20}$$

We are going to prove that the stationary points of (4.20) are solutions of the problem under different hypotheses than those stated for the general HNCP. This result generalizes the Equivalence Theorem proved in [12]. As in [12], we say that the HLCP defined by (4.20) is feasible if there exist $y, z \geq 0$ such that

$$Qy + Rz = b.$$

An additional definition concerning pairs of square matrices is necessary. (See [19].) Given $M, N \in \mathbb{R}^{q\times q}$, $A = (M, N) \in \mathbb{R}^{q\times 2q}$, we say that the pair (M, N) is row-sufficient if

$$\left[\begin{pmatrix} r \\ s \end{pmatrix} \in \text{range } A^T, \quad r \circ s \leq 0\right]$$

implies that

$$r \circ s = 0.$$

Here $\circ$ denotes the Hadamard product; i.e., $x \circ y$ is the vector whose components are the products of the corresponding components of x and y.

Theorem 4.3. *Assume that the HLCP (4.20) is feasible, (y_*, z_*) is a stationary point and the pair $(Q, -R)$ is row-sufficient. Then, (y_*, z_*) is solution of the Horizontal Linear Complementarity Problem (4.20).*

Proof. As in Theorem 4.1, the minimization problem associated to (4.20) is:

$$\text{Minimize } \|Qy + Rz - b\|^2 + (y^T z)^2 \quad \text{subject to} \quad y, z \geq 0. \tag{4.21}$$

Define $G(y) = Qy - b$ and $H(z) = Rz$, then $G'(y) = Q$ and $H'(z) = R$. Calling

$$u_* = Qy_* + Rz_* - b, \tag{4.22}$$

the inequality (4.11) of Theorem 4.1 remains

$$4[R^T u_*]_i[Q^T u_*]_i = 4(y_*^T z_*)^2[y_*]_i[z_*]_i + [\mu_1]_i[\mu_2]_i \geq 0 \quad \text{for all } i = 1, \ldots, q. \tag{4.23}$$

Since the pair $(Q, -R)$ is row-sufficient, whenever

$$[R^T u_*]_i[Q^T u_*]_i \geq 0 \text{ for all } \; i = 1, \ldots, q$$

one necessarily has that

$$[R^T u_*]_i[Q^T u_*]_i = 0 \text{ for all } \; i = 1, \ldots, q.$$

Therefore, by (4.23) we obtain:

$$[y_*]_i[z_*]_i = 0 \text{ for all } i = 1, \ldots, q. \tag{4.24}$$

Now, considering (4.6)-(4.10) in Theorem 4.1, with the obvious specifications relative to the linear case, the first-order conditions of (4.21) turn out to be

$$2Q^T u_* - \mu_1 = 0, \tag{4.25}$$

$$2R^T u_* - \mu_2 = 0, \tag{4.26}$$

$$y_*^T \mu_1 = 0, \tag{4.27}$$

$$z_*^T \mu_2 = 0, \tag{4.28}$$

$$y_* \geq 0, \quad z_* \geq 0, \quad \mu_1 \geq 0, \quad \mu_2 \geq 0. \tag{4.29}$$

Now, (4.25)-(4.29), are necessary and sufficient conditions for global minimizers of the following convex quadratic minimization problem:

$$\text{Minimize } \|Qy + Rz - b\|^2 \quad \text{subject to} \quad y \geq 0, \ z \geq 0. \tag{4.30}$$

So, by the feasibility of the HLCP, it turns out that (y_*, z_*) is global solution of (4.30) with minimun value zero, that is

$$Qy_* + Rz_* - b = 0. \tag{4.31}$$

Therefore, by (4.24) and (4.31), (y_*, z_*) is a solution of the Horizontal Linear Complementarity Problem. ∎

In Section 3 we showed that the key condition that allows us to prove superlinear convergence of some variations of Algorithm 2.1 is the Strong Local

Unicity Assumption (SLCU). Therefore, it is interesting to study sufficient conditions under which this assumption holds in the case of the Horizontal Nonlinear Complementarity Problem. Probably, when these sufficient conditions are fulfilled, any reasonable method will converge quickly, if started from a good enough initial approximation. On the other hand, difficult problems will be probably characterized by situations where those sufficient conditions are not true. When applied to the Nonlinear Complementarity Problem, the condition introduced in Theorem 4.4 corresponds to conditions given by Mangasarian [18] for the uniqueness of the solution of NCP and Moré [20] for superlinear convergence of a specific NCP algorithm.

Given the set of indices $\mathcal{L} \subset \{1, \ldots, q\}$ and $\mathcal{K} \subset \{1, \ldots, q\}$ and the matrix $A \in I\!R^{q \times q}$, we denote $A_{\mathcal{L},\mathcal{K}}$ the submatrix of A such that $A_{\mathcal{L},\mathcal{K}} \equiv (a_{i,j})$ with $i \in \mathcal{L}$ and $j \in \mathcal{K}$.

Theorem 4.4. *Assume that (y_*, z_*) is a solution of the HNCP(G,H) such that*

$$[y_*]_i + [z_*]_i > 0 \quad \textit{for all} \quad i = 1, \ldots, q. \tag{4.32}$$

Moreover, suppose that the matrix formed by the columns of $G'(y_*)$ that correspond to $[y_*]_i > 0$ and the columns of $H'(z_*)$ that correspond to $[z_*]_i > 0$ is nonsingular. Then (y_*, z_*) satisfies Assumption 3.1.

Proof. Let us call

$$\mathcal{J} = \{i \in \{1, \ldots, q\} \mid [y_*]_i > 0\}, \quad \mathcal{M} = \{i \in \{1, \ldots, q\} \mid [z_*]_i > 0\}.$$

Since (y_*, z_*) is solution of $HNCP(G, H)$, by (4.32) we obtain

$$\begin{gathered} [y_*]_{\mathcal{J}} > 0, \quad [y_*]_{\mathcal{M}} = 0, \quad [z_*]_{\mathcal{J}} = 0, \quad [z_*]_{\mathcal{M}} > 0, \\ \mathcal{I} = \{1, \ldots, q\} = \mathcal{J} \cup \mathcal{M} \ \text{ and } \ \mathcal{J} \cap \mathcal{M} = \emptyset. \end{gathered} \tag{4.33}$$

Assume that $F(y, z)$ does not verify the strong unicity condition at (y_*, z_*), then by Lemma 3.6 there exists $\tau = (\tilde{y}, \tilde{z}) \in I\!R^{2q}$, $\|\tau\| = 1$ and $\{(\tilde{y}_k, \tilde{z}_k)\} \subset \Omega = I\!R^{2q}_+$ such that

$$(\tilde{y}_k, \tilde{z}_k) \to (y_*, z_*),$$

$$\tau = \begin{pmatrix} \tilde{y} \\ \tilde{z} \end{pmatrix} = \lim_{k \to \infty} \frac{\begin{pmatrix} \tilde{y}_k \\ \tilde{z}_k \end{pmatrix} - \begin{pmatrix} y_* \\ z_* \end{pmatrix}}{\left\| \begin{pmatrix} \tilde{y}_k \\ \tilde{z}_k \end{pmatrix} - \begin{pmatrix} y_* \\ z_* \end{pmatrix} \right\|} \tag{4.34}$$

and

$$F'(y_*, z_*)\tau =$$

$$\begin{bmatrix} [G'(y_*)]_{\mathcal{I},\mathcal{J}} & [G'(y_*)]_{\mathcal{I},\mathcal{M}} & [H'(z_*)]_{\mathcal{I},\mathcal{J}} & [H'(z_*)]_{\mathcal{I},\mathcal{M}} \\ [z_*]^T_{\mathcal{J}} & [z_*]^T_{\mathcal{M}} & [y_*]^T_{\mathcal{J}} & [y_*]^T_{\mathcal{M}} \end{bmatrix} \begin{bmatrix} \tilde{y}_{\mathcal{J}} \\ \tilde{y}_{\mathcal{M}} \\ \tilde{z}_{\mathcal{J}} \\ \tilde{z}_{\mathcal{M}} \end{bmatrix} = 0. \tag{4.35}$$

By (4.33) and (4.35), we have that

$$[z_*]^T_{\mathcal{M}} \tilde{y}_{\mathcal{M}} + [y_*]^T_{\mathcal{J}} \tilde{z}_{\mathcal{J}} = 0. \tag{4.36}$$

Now, by (4.33) and (4.34), calling

$$\pi_k = \left\| \begin{pmatrix} \tilde{y}_k \\ \tilde{z}_k \end{pmatrix} - \begin{pmatrix} y_* \\ z_* \end{pmatrix} \right\|,$$

we obtain

$$\tilde{y}_{\mathcal{M}} = \lim_{k\to\infty} \frac{[\tilde{y}_k]_{\mathcal{M}}}{\pi_k} \geq 0 \text{ and } \tilde{z}_{\mathcal{J}} = \lim_{k\to\infty} \frac{[\tilde{z}_k]_{\mathcal{J}}}{\pi_k} \geq 0. \tag{4.37}$$

Then, by (4.33), (4.36) and (4.37), we have that

$$\tilde{y}_{\mathcal{M}} = 0 \text{ and } \tilde{z}_{\mathcal{J}} = 0. \tag{4.38}$$

Therefore, by (4.35),

$$\left[\ [G'(y_*)]_{\mathcal{I},\mathcal{J}} \quad [H'(z_*)]_{\mathcal{I},\mathcal{M}}\ \right] \left[\begin{array}{c} \tilde{y}_{\mathcal{J}} \\ \tilde{z}_{\mathcal{M}} \end{array} \right] = 0. \tag{4.39}$$

By the hypothesis of nonsingularity, we obtain

$$\tilde{y}_{\mathcal{J}} = 0 \text{ and } \tilde{z}_{\mathcal{M}} = 0. \tag{4.40}$$

By (4.38) and (4.40), we have that $\tau = 0$, which is a contradiction. ∎

5 NUMERICAL IMPLEMENTATION

In this section we describe a computer implementation of Algorithm 2.2. The computation of the direction d_k satisfying (2.9) takes into account, as mentioned in Section 2, the box-constrained quadratic subproblem (2.8), with a first stopping criterion given by (2.10). The second stopping criterion used in the computer implementation was

$$\begin{array}{c} \|F'(x_k)\bar{d} + F(x_k)\| \leq \theta_k \|F(x_k)\| \\ \text{or} \\ \|P(\bar{d} - \nabla\Phi(\bar{d})) - \bar{d}\| \leq 10^{-12} \|P(-\nabla\Phi(0))\| \end{array} \tag{5.1}$$

This means that at Step 2 of Algorithm 2.2 we run the quadratic solver with the stopping criterion (2.10), obtaining the increment $\bar{d}$. If $\|F'(x_k)\bar{d}+F(x_k)\| \leq \theta_k\|F(x_k)\|$ we accept $d_k = \bar{d}$. Otherwise, we continue the execution of the quadratic solver with the stopping criterion (5.1).

We wish to deal with large-scale problems. So, we used, for solving (2.8), the algorithm introduced in [10] and improved in [1]. This algorithm does not use matrix factorizations at all. In fact, conjugate gradient iterations are used within the faces of the box and, when the current face needs to be abandoned, orthogonal chopped-gradient directions are employed (see [1, 10, 11]).

The algorithmic parameters used in the experiments were: $\Delta = 100$, $\bar{\rho} = 8$, $\sigma = 10^{-4}$ and $\theta_0 = 0.9999$. Practical convergence to a solution was declared when $\|F(x_k)\| \leq 10^{-8}$. A current iterate was declared stationary if the choice (2.9) is not possible with $\theta_k = 0.999995$.

We wish to test the efficiency of different strategies for the choice of η_k. The tested strategies were:

(1) $\eta_k = 0.1$ for all $k = 0, 1, 2, \ldots$

(2) $\eta_k = 10^{-3}$ for all $k = 0, 1, 2, \ldots$

(3) $\eta_k = 10^{-8}$ for all $k = 0, 1, 2, \ldots$

(4) $\eta_k = 1/(k+2)$ for all $k = 0, 1, 2, \ldots$

(5) An adaptive strategy given by

$$\eta_0 = 0.1, \ \ \eta_k = (\|F(x_k)\|/\|F(x_{k-1}\|)^2 \ \text{ if } \ \eta_{k-1}^2 \leq 0.1,$$

$$\text{and} \ \ \eta_k = \eta_{k-1}^2 \ \text{ if } \ \eta_{k-1}^2 > 0.1.$$

The strategy (5) was inspired on a choice suggested in [8] for the forcing term of Inexact-Newton methods for solving nonlinear systems.

We generated a set of test problems with known solution defining

$$[y_*]_i = 1 \ \text{ if } \ i \ \text{ is odd, and } \ [y_*]_i = 0 \ \text{ otherwise,} \tag{5.2}$$

and

$$[z_*]_i = 1 \ \text{ if } \ i \ \text{ is even, and } \ [z_*]_i = 0 \ \text{ otherwise.} \tag{5.3}$$

The following nonlinear mappings $T_i : \mathbb{R}^q \to \mathbb{R}^q$, $i \in \{1, 2, 3, -1, -2, -3\}$, were considered:

$$[T_1(x)]_i = 10(x_{i+1}^2 - x_i) \ \text{ if } \ i \ \text{ is odd,}$$

$$[T_1(x)]_i = x_{i-1} - 1 \ \text{ if } \ i \ \text{ is even,}$$

$$[T_2(x)]_1 = 2x_1 + e^{x_1} + x_2,$$

$$[T_2(x)]_i = -x_{i-1} + 2x_i + e^{x_1} + x_{i+1}, i = 2, \ldots, q-1,$$

$$[T_2(x)]_q = -x_{q-1} + 2x_q + e^{x_q},$$

$$T_3(x) = x,$$

$$T_{-i}(x) = -T_i(x), i = 1, 2, 3.$$

We generated 12 problems $P_{ij}, (i,j) \in \mathcal{P}$, where

$$\mathcal{P} = \{(i,j) \ | \ i = 1, 2, 3, \ \ |j| \leq 3, |j| \leq |i|\}.$$

in the following way:

For each $(i,j) \in \mathcal{P}$, we computed $b = T_i(y_*) + T_j(z_*)$ and we defined $G(y) = T_i(y) - b/2$ and $H(y) = T_j(y) - b/2$. So, (y_*, z_*) is a solution of the Horizontal Nonlinear Complementarity Problem P_{ij}, defined by G and H.

We tried to solve the problems using three initial points:

(1) $x_0 = x_* + (\underline{2} - x_*)$;

(2) $x_0 = x_* + 10(\underline{2} - x_*)$;

(3) $x_0 = x_* + 100(\underline{2} - x_*)$;

where $\underline{2} = (2, \ldots, 2)^T$.

Problems P_{ij} with $|i| = 2$ or $|j| = 2$ cannot be run for the third initial point because an overflow is produced in the computation of the nonlinear system at x_0. Therefore, 30 different problems were generated. We tested the five strategies for η_k described above for each of these problems, using $q = 500$ (so $n = 1000$, $m = 501$).

The solution points of these problems are nondegenerate in the sense that $[y_*]_i + [z_*]_i > 0$ for all $i = 1, \ldots q$. With the aim of investigating the local behavior of the algorithm in degenerate problems we defined an additional set of problems $P'_{ij}, (i,j) \in \mathcal{P}$ using

$$[y_*]_i = 1 \quad \text{if} \quad i \le q/2, \ i \text{ odd, and} \quad [y_*]_i = 0 \quad \text{otherwise,} \tag{5.4}$$

and

$$[z_*]_i = 1 \quad \text{if} \quad i \le q/2, \ i \text{ even, and} \quad [z_*]_i = 0 \quad \text{otherwise.} \tag{5.5}$$

Since our theoretical results show that, in this case, problems may arise with the local convergence, we used only the initial approximation $x_0 = x_* + 0.1(\underline{2} - x_*)$.

Our codes were written in Fortran 77 with double precision, we used the DOS-Microsoft compiler, and the tests were run in a Pentium-166 MHz. The complete numerical results, in six different tables, and the source codes are available from the authors. Here we are only going to state the conclusions of the numerical study.

Surprisingly, the behavior of different accuracy strategies in the 30 nondegenerate problems is remarkably similar, with one exception: the performance of Strategy 3 in problem (1, −2) with the second initial point is much poorer than the performance of the other strategies. In this problem, Strategy 5 was more efficient than the others. Roughly speaking, with all the strategies we obtained solutions of the problems, except in three cases, in which we obtained stationary points. In most cases, the computer time was very moderate, in spite of the modest computer environment used.

As expected, problems with degenerate solution were, in general, more difficult than the nondegenerate ones. Although no meaningful differences were detected regarding the performance of the three strategies tested, we verified that in two problems ((1, −2) and (1, −3)) 100 iterations were not enough for achieving convergence. In four problems convergence was achieved but the distance between the approximate solution and the true one was greater than 10^{-6}. Finally, in general, the computer time used for solving these degenerate problems with initial point close to the solution is of the same order (if not greater) than the computer time used to solve nondegenerate problems starting far from the solution.

6 CONCLUSIONS

The method presented in this paper is able to deal with large scale box constrained, non-necessarily square, nonlinear systems of equations. Usually, these problems are treated as ordinary large-scale minimization problems in practical optimization. See, for example, [3, 11] and the CUTE collection of problems of Conn, Gould and Toint. We think that to consider the specific nonlinear-system structure can represent a practical advantage. On one hand, the sufficient decrease condition (2.3) on the model imposes a residual decrease proportional to the operator norm, and not only to some projected gradient measure, as a minimization algorithm would do. This probably implies that the nonlinear-system algorithm tends to be more active trying to find roots of the system, and less prone to converge to local nonglobal minimizers. On the other hand, high order convergence can be obtained using only first order information. In [17] a similar idea for square systems was applied successfully to turning point problems, with a different concept of near-stationarity.

Inexact-Newton methods using conditions similar to (2.3) are well developed and largely used for solving large-scale (unconstrained) nonlinear systems. The contribution of this paper is a natural extension of those methods to constrained problems.

The essential tool for computing the search direction at each iteration is a box-constrained quadratic solver that uses conjugate gradient directions inside the faces of the box and theoretically justified procedures for leaving the faces. Most of the computer time used by the algorithm is spent in the computation of these directions, at least in the numerical examples tested.

We used the new method for solving a set of Horizontal Nonlinear Complementarity Problems (HNCP), which are generalizations of Nonlinear Complementarity Problems (NCP) and Horizontal Linear Complementarity Problems (HLCP). We proved that, in many cases, stationary points of the HNCP are global solutions. These results generalize well-known existing results for the HLCP and the NCP.

We tried five different strategies related to the accuracy required in the solution of the quadratic subproblem at each iteration. Theoretical results concerning order of convergence seem to suggest that degenerate problems are more difficult than problems with nondegenerate solution, at least from the local convergence point of view. This theoretical prediction was confirmed by the experiments.

References

[1] R. H. Bielschowsky, A. Friedlander, F. M. Gomes, J. M. Martínez and M. Raydan [1995], An adaptive algorithm for bound constrained quadratic minimization, to appear in *Investigación Operativa.*

[2] J. F. Bonnans and C. C. Gonzaga [1996], Convergence of interior point algorithms for the monotone linear complementarity problem, *Mathematics of Operations Research* 21, pp. 1-25

[3] Conn, A. R.; Gould, N. I. M.; Toint, Ph. L. [1988], Global convergence of a class of trust region algorithms for optimization with simple bounds, *SIAM Journal on Numerical Analysis* 25, pp. 433 - 460. See also *SIAM Journal on Numerical Analysis* 26 [1989] pp. 764 - 767.

[4] R. W. Cottle, J-S. Pang and R. E. Stone [1992], The linear complementarity problem, Academic Press, Boston.

[5] T. De Luca, F. Facchinei and C. Kanzow [1994], A semismooth equation approach to the solution of nonlinear complementarity problem, *Mathematical Programming* 75, pp. 407-439.

[6] R. S. Dembo, S. C. Eisenstat and T. Steihaug [1982], Inexact Newton methods, *SIAM Journal on Numerical Analysis* 14, pp. 400-408.

[7] S. C. Eisenstat and H. F. Walker [1994], Globally convergent inexact Newton methods, *SIAM Journal on Optimization* 4, pp. 393-422.

[8] S. C. Eisenstat and H. F. Walker [1996], Choosing the forcing terms in an Inexact Newton method, *SIAM Journal on Scientific Computing* 17, pp. 16-32.

[9] A. Fischer [1994], An NCP-function and its use for the solution of complementarity problems, Technical Report MATH-NM-12-1994, Technische Universität Dresden, Dresden, Germany.

[10] A. Friedlander and J. M. Martínez [1994], On the maximization of a concave quadratic function with box constraints, *SIAM Journal on Optimization* 4, pp. 177-192.

[11] A. Friedlander, J. M. Martínez and S. A. Santos [1994], A new trust region algorithm for bound constrained minimization, *Applied Mathematics and Optimization* 30, pp. 235-266.

[12] A. Friedlander, J. M. Martínez and S. A. Santos [1995], Solution of linear complementarity problems using minimization with simple bounds, *Journal of Global Optimization* 6, pp. 1-15.

[13] M. Friedler and V. Pták [1966], Some generalizations of positive definiteness and monotonicity. *Numerische Mathematik* 9, pp. 163-172.

[14] S. A. Gabriel and J-S. Pang [1992], An inexact NE/SQP method for solving the nonlinear complementarity problem, *Computational Optimization and Applications* 1, pp. 67-91.

[15] H. Jiang and L. Qi [1994], A new nonsmooth equations approach to nonlinear complementarities, *Applied Mathematics Report, AMR94/31*, School of Mathematics, The University of South Wales, Sydney Australia.

[16] J. J. Júdice [1994], Algorithms for linear complementarity problems, in *Algorithms for Continuous Optimization*, edited by E. Spedicato, Kluwer, pp. 435-474.

[17] D. N. Kozakevich, J. M. Martínez and S. A. Santos [1996], Solving nonlinear systems of equations with simple constraints, to appear in *Computational and Applied Mathematics.*

[18] O. L. Mangasarian [1980], Locally unique solutions of quadratic programs, linear and nonlinear complementarity problems, *Mathematical Programming* 19, pp. 200-212.

[19] O. L. Mangasarian and J-S. Pang[1995], The extended linear complementarity problem, *SIAM Journal of Matrix Analysis and Applications*, 16, pp. 359-368.

[20] J. J. Moré [1996], Global methods for nonlinear complementarity problems, *Mathematics of Operations Research*, 21, pp. 589-614.

[21] T. Wang, R. D. C. Monteiro and J-S. Pang [1996], An interior point potential reduction method for constrained equations, *Mathematical Programming* 74, pp. 159-195.

[22] N. Yamashita and M. Fukushima [1995], On stationary points of the implicit Lagrangian for the nonlinear complementarity problems, *Journal of Optimization Theory and Applications* 84, pp. 653-663.

Reformulation: Nonsmooth, Piecewise Smooth, Semismooth and Smoothing Methods, pp. 25–43
Edited by M. Fukushima and L. Qi

ε-Enlargements of Maximal Monotone Operators: Theory and Applications

Regina S. Burachik*, Claudia A. Sagastizábal† and B. F. Svaiter‡

Abstract Given a maximal monotone operator T, we consider a certain ε-enlargement T^ε, playing the role of the ε-subdifferential in nonsmooth optimization. We establish some theoretical properties of T^ε, including a transportation formula, its Lipschitz continuity, and a result generalizing Brønsted & Rockafellar's theorem. Then we make use of the ε-enlargement to define an algorithm for finding a zero of T.

Key Words maximal monotone operators, enlargement of an operator, Brønsted & Rockafellar's theorem, transportation formula, algorithmic scheme.

1 INTRODUCTION AND MOTIVATION

Given a convex function $f: \mathbb{R}^N \to \mathbb{R} \cup \{\infty\}$, the *subdifferential* of f at x, i.e. the set of *subgradients* of f at x, denoted by $\partial f(x)$, is defined as

$$\partial f(x) = \{u \in \mathbb{R}^N : f(y) - f(x) - \langle u, y - x\rangle \geq 0 \text{ for all } y \in \mathbb{R}^N\}.$$

This concept has been extended in [BR65], where the *ε-subdifferential* of f at x was defined for any $\varepsilon \geq 0$ as follows:

$$\partial_\varepsilon f(x) := \{u \in \mathbb{R}^N : f(y) - f(x) - \langle u, y - x\rangle \geq -\varepsilon \text{ for all } y \in \mathbb{R}^N\}.$$

The introduction of a "smearing" parameter ε gives an enlargement of $\partial f(x)$ with good continuity properties (see § 2.3 below for a definition of continuous

*Engenharia de Sistemas de Computação, COPPE–UFRJ, CP 68511, Rio de Janeiro–RJ, 21945–970, Brazil. email: regi@cos.ufrj.br

†INRIA, BP 105, 78153 Le Chesnay, France. Also Depto. Engenharia Elétrica, PUC-Rio. Caixa Postal 38063, 22452-970 Rio de Janeiro, Brazil. email: Claudia.Sagastizabal@inria.fr or sagastiz@ele.puc-rio.br

‡IMPA, Estrada Dona Castorina, 110. Rio de Janeiro, RJ, CEP 22460-320, Brazil. email: benar@impa.br

multifunctions). Actually, the ε-subdifferential calculus appears as a powerful tool in connexion with nonsmooth optimization: the initial ideas presented by [BM73] and [Nur86] resulted in the so-called methods of ε-descent, [HUL93, Chapter XIII], which are the antecessors of bundle methods, [Kiw90], [SZ92], [LNN95], [BGLS97]. The convergence of such methods is based on some crucial continuity properties of the multifunction $x \hookrightarrow \partial_\varepsilon f(x)$, such as its upper and Lipschitz continuity and also a transportation formula. We refer to [HUL93, Chapter XI] for a deep study of all these classical properties of the ε-subdifferential.

In this paper we address some continuity issues of the ε-enlargement for maximal monotone operators T^ε, introduced in [BIS97] and defined as follows:

Definition 1.1 *Given $T: H \to \mathcal{P}(H)$, a maximal monotone operator on a Hilbert space H, and $\varepsilon \geq 0$, the* ε-enlargement *of T at x is defined by*

$$\begin{array}{rccl} T^\varepsilon: & H & \to & \mathcal{P}(H) \\ & x & \hookrightarrow & \{u \in H : \langle v - u, y - x\rangle \geq -\varepsilon\,, \forall y \in H\,, v \in T(y)\}\,. \end{array} \tag{1.1}$$

□

Observe that the definition above is similar to the definition of $\partial_\varepsilon f(\cdot)$. When $T = \partial f$, $\partial_\varepsilon f(x) \subseteq T^\varepsilon(x)$; this result, as well as examples showing that the inclusion can be strict, can be found in [BIS97].

First, we prove that the multifunction $x \hookrightarrow T^\varepsilon(x)$ shares with $x \hookrightarrow \partial_\varepsilon f(x)$ some properties that make ε-descent methods work, specifically, a theorem generalizing Brønsted & Rockafellar's, a transportation formula, and the Lipschitz-continuity.

Second, we develop an ε-descent-like method to find a zero of T using the ε-enlargements $T^\varepsilon(x)$. More precisely, consider the unconstrained convex program

$$\text{Find } x \text{ such that } 0 \in \partial f(x)\,.$$

It is well-known that $\partial f(\cdot)$ is a maximal monotone operator (see [Mor65]). In this sense, the problem above can be generalized to

$$\text{Find } x \text{ such that } 0 \in T(x)\,, \tag{1.2}$$

which is the classical problem of finding a zero of a maximal monotone operator T. We denote the solution set of (1.2) by $\mathcal{S}$.

The algorithm we introduce to solve (1.2) can be outlined as follows. Given an arbitrary y and $v \in T(y)$, because T is monotone, the set $\mathcal{S}$ is contained in the halfspace

$$H_{y,v} := \{z \in H : \langle z - y, v\rangle \leq 0\}\,. \tag{1.3}$$

Let x^k be a current (non optimal) iterate, i.e., $0 \notin T(x^k)$, and let $(y^k, v^k \in T(y^k))$ be such that $x^k \notin H_{y^k,v^k}$. Then project x^k onto H_{y^k,v^k} to obtain a new iterate:

$$x^{k+1} := \mathrm{P}_{H_{y^k,v^k}}(x^k) = x^k - \frac{\langle x^k - y^k, v^k\rangle}{\|v^k\|^2} v^k\,. \tag{1.4}$$

Because x^{k+1} is a projection, it is closer to the solution set than x^k. This simple idea exploiting projection properties has also been used successfully in other contexts, for example to solve variational inequality problems ([Kor76], [Ius94], [SS96], [IS97], [Kon97], [SS97]).

The use of extrapolation points y^k for computing the direction v^k comes back to [Kor76] and her *extragradient* method for solving continuous variational inequality problems. When solving (1.2), in order to have convergence and good performances, appropriate y^k and v^k should be found. For instance, since $0 \notin T(x^k)$, it may seem natural to simply find $s^k = \mathrm{P}_{T(x^k)}(0)$ and pick $y^k = x^k - t^k s^k$, for some positive stepsize t^k. Nevertheless, because $x \hookrightarrow T(x)$ is not continuous, (in the sense defined in §2.3) the method can converge to a point which is not a solution, as an example in [Luc97, page 4], shows. In addition, such a method may behave like a steepest-descent algorithm for nonsmooth minimization, whose numerical instabilities are well known, [HUL93, Chapter VII.2.2].

When T is the subdifferential of f, a better direction s^k can be obtained by projecting 0 onto a set bigger than $\partial f(x^k)$, namely the $\partial_\varepsilon f(x^k)$. Accordingly, when T in (1.2) is an arbitrary maximal monotone operator, it seems a good idea to parallel this behaviour when generating y^k in (1.4). Here is where the ε-enlargements $T^\varepsilon(x)$ come to help: the directions s^k will be generated by computing $\mathrm{P}_{T^\varepsilon(x^k)}(0)$, for appropriate values of ε.

The paper is organized as follows. In Section 2 we establish a generalization of the Brønsted & Rockafellar's theorem, a "transportation formula" and the Lipschitz-continuity for T^ε. Then, in § 3, we define an algorithm combining the projections and ε-descent techniques outlined above. This conceptual algorithm is proved to be convergent in § 3.2 and is the basis to develop the implementable bundle-like method described in the companion paper [BSS97]. We finish in § 4 with some concluding remarks.

2 PROPERTIES OF T^ε

We gather in this section some results which are important not only for theoretical purposes, but also in view of designing the implementable algorithm of [BSS97]. These properties are related to the continuity of the ε-enlargement as a multifunction and also to "transportation formulæ" relating elements from T^ε with those from T.

We need first some notation. Given a set $A \subseteq H$ and a multifunction $S: H \to \mathcal{P}(H)$

- the closure of A is denoted by $\overline{A}$,
- we define the set $S(A) := \bigcup_{a \in A} S(a)$.
- The domain, image and graph of S are respectively denoted by

$$\begin{aligned} D(S) &:= \{x \in H : S(x) \neq \emptyset\}, \\ R(S) &:= S(H) \text{ and} \\ G(S) &:= \{(x,v) : x \in D(S) \text{ and } v \in S(x)\}. \end{aligned}$$

- S is *locally bounded* at x if there exists a neighbourhood U of x such that the set $S(U)$ is bounded.
- S is *monotone* if $\langle u - v, x - y\rangle \geq 0$ for all $u \in S(x)$ and $v \in S(y)$, for all $x, y \in H$.
- S is *maximal monotone* if it is monotone and, additionally, its graph is not properly contained in the graph of any other monotone operator. □

Recall that any maximal monotone operator is locally bounded in the interior of its domain ([Roc69, Theorem 1]).

2.1 Extending Brønsted & Rockafellar's theorem

For a closed proper convex function f, the theorem of Brønsted & Rockafellar, see for instance [BR65], states that any ε-subgradient of f at a point x_ε can be approximated by some *exact* subgradient, computed at some x, possibly different from x_ε.

The ε-enlargement of Definition 1.1 also satisfies this property:

Theorem 2.1 *Let $T : H \to \mathcal{P}(H)$ be maximal monotone, $\varepsilon > 0$ and $(x_\varepsilon, v_\varepsilon) \in G(T^\varepsilon)$. Then for all $\eta > 0$ there exists $(x, v) \in G(T)$ such that*

$$\|v - v_\varepsilon\| \leq \frac{\varepsilon}{\eta} \quad \textit{and} \quad \|x - x_\varepsilon\| \leq \eta\,. \tag{2.1}$$

Proof. For an arbitrary positive coefficient β define the multifunction

$$\begin{array}{rccl} G_\beta \colon & H & \to & \mathcal{P}(H) \\ & y & \hookrightarrow & \beta T(y) + \{y\}\,. \end{array}$$

Since βT is maximal monotone, by Minty's theorem [Min62], G_β is a surjection:

$$\exists\ (x, v) \in G(T) \quad \text{such that} \quad G_\beta(x) = \beta v + x = \beta v_\varepsilon + x_\varepsilon\,.$$

This, together with Definition 1.1, yields

$$\begin{aligned} \langle v_\varepsilon - v, x_\varepsilon - x\rangle &= -\beta\|v - v_\varepsilon\|^2 \\ &= -\frac{1}{\beta}\|x - x_\varepsilon\|^2 \quad \geq -\varepsilon\,. \end{aligned}$$

Choosing $\beta := \eta^2/\varepsilon$, the result follows. ■

Observe that the proof above only uses the ε-inequality characterizing elements in T^ε. Accordingly, the same result holds for any other enlargement of T, as long as it is contained in T^ε.

The result above can also be expressed in a set-formulation:

$$T^\varepsilon(x) \subset \bigcap_{\eta>0} \ \bigcup_{y \in B(x,\eta)} \{T(y) + B(0, \frac{\varepsilon}{\eta})\}\,, \tag{2.2}$$

where $B(x,\rho)$ denotes the unit ball centered in x with radius ρ. This formula makes clear that the value $\eta = \sqrt{\varepsilon}$ is a compromise between the distance to x and the degree of approximation. The value of η which makes such quantities equal gives the following expression

$$T^\varepsilon(x) \subset \bigcup_{y\in B(x,\sqrt{\varepsilon})} \{T(y) + B(0,\sqrt{\varepsilon})\}.$$

It follows that T^ε is locally bounded together with T. The following result, which extends slightly Proposition 2 in [BIS97], gives further relations between the two multifunctions.

Corollary 2.1 *With the notations above, the following hold:*

(i) $R(T) \subset R(T^\varepsilon) \subset \overline{R(T)}$,

(ii) $D(T) \subset D(T^\varepsilon) \subset \overline{D(T)}$,

(iii) If $d(\cdot,\cdot)$ *denotes the point-to-set distance, then* $d((x_\varepsilon, v_\varepsilon); G(T)) \leq \sqrt{2\varepsilon}$

Proof. The leftmost inclusions in (i) and (ii) are straightforward from Definition 1.1. As for the right ones, they follow from Theorem 2.1, making $\eta \to +\infty$ and $\eta \to 0$ in (i) and (ii) respectively.

To prove (iii), take $\eta = \sqrt{\varepsilon}$ in (2.1), write

$$d((x_\varepsilon, v_\varepsilon); G(T))^2 \leq \|x - x_\varepsilon\|^2 + \|v - v_\varepsilon\|^2 \leq 2\varepsilon,$$

and take square roots. ∎

2.2 Transportation Formula

We already mentioned that the set $T^\varepsilon(x)$ approximates $T(x)$, but this fact is of no use as long as there is no way of *computing* elements of $T^\varepsilon(x)$. The question is then how to construct an element in $T^\varepsilon(x)$ with the help of some elements $(x^i, v^i) \in G(T)$. The answer is given by the "transportation formula" stated below. Therein we use the notation

$$\Delta_m := \{\alpha \in \mathbb{R}^m \mid \alpha_i \geq 0, \sum_{i=1}^m \alpha_i = 1\}$$

for the unit-simplex in $\mathbb{R}^m$.

Theorem 2.2 *Let T be a maximal monotone operator defined in a Hilbert space H. Consider a set of m triplets*

$$\{(\varepsilon_i, x^i, v^i \in T^{\varepsilon_i}(x^i))\}_{i=1,\ldots,m}.$$

For any $\alpha \in \Delta_m$ *define*

$$\begin{array}{rcl} \hat{x} &:=& \sum_1^m \alpha_i x^i \\ \hat{v} &:=& \sum_1^m \alpha_i v^i \\ \hat{\varepsilon} &:=& \sum_1^m \alpha_i \varepsilon_i + \sum_1^m \alpha_i \langle x^i - \hat{x}, v^i - \hat{v} \rangle . \end{array} \tag{2.3}$$

Then $\hat{\varepsilon} \geq 0$ *and* $\hat{v} \in T^{\hat{\varepsilon}}(\hat{x})$.

Proof. Recalling Definition 1.1, we need to show that the inequality in (1.1) holds for any $(y, v) \in G(T)$. Combine (2.3) and (1.1), with (ε, x, u) replaced by $(\varepsilon_i, x^i, v^i)$, to obtain

$$\begin{array}{rcl} \langle \hat{x} - y, \hat{v} - v \rangle &=& \sum_1^m \alpha_i \langle x^i - y, \hat{v} - v \rangle \\ &=& \sum_1^m \alpha_i \left[\langle x^i - y, \hat{v} - v^i \rangle + \langle x^i - y, v^i - v \rangle \right] \\ &\geq& \sum_1^m \alpha_i \langle x^i - y, \hat{v} - v^i \rangle - \sum_1^m \alpha_i \varepsilon_i . \end{array} \tag{2.4}$$

Since

$$\begin{array}{rcl} \sum_1^m \alpha_i \langle x^i - y, \hat{v} - v^i \rangle &=& \sum_1^m \alpha_i \left[\langle x^i - \hat{x}, \hat{v} - v^i \rangle + \langle \hat{x} - y, \hat{v} - v^i \rangle \right] \\ &=& -\sum_1^m \alpha_i \langle x^i - \hat{x}, v^i - \hat{v} \rangle + 0 \\ &=& -\sum_1^m \alpha_i \langle x^i - \hat{x}, v^i - \hat{v} \rangle , \end{array}$$

with (2.4) and (2.3) we get

$$\langle \hat{x} - y, \hat{v} - v \rangle \geq -\hat{\varepsilon} . \tag{2.5}$$

For contradiction purposes, suppose that $\hat{\varepsilon} < 0$. Then $\langle \hat{x} - y, \hat{v} - v \rangle > 0$ for any $(y, v) \in G(T)$ and the maximality of T implies that $(\hat{x}, \hat{v}) \in G(T)$. In particular, the pair $(y, v) = (\hat{x}, \hat{v})$ yields $0 > 0$!. Therefore $\hat{\varepsilon}$ must be nonnegative. Since (2.5) holds for any $(y, v) \in G(T)$, we conclude from (1.1) that $\hat{v} \in T^{\hat{\varepsilon}}(\hat{x})$. ∎

Observe that when $\varepsilon_i = 0$, for all $i = 1, \ldots, m$, this theorem shows how to construct $\hat{v} \in T^{\hat{\varepsilon}}(\hat{x})$, using $(x^i, v^i) \in G(T)$.

The formula above holds also when replacing T^ε by $\partial_\varepsilon f$, with f a proper closed and convex function. This is Proposition 1.2.10 in [Lem80], where an equivalent expression is given for $\hat{\varepsilon}$:

$$\hat{\varepsilon} = \sum_{i=1}^m \alpha_i \varepsilon_i + \frac{1}{2} \sum_{i,j=1}^m \alpha_i \alpha_j \langle x^i - x^j, v^i - v^j \rangle .$$

Observe also that, when compared to the standard transportation formula for ε-subdifferentials, Theorem 2.2 is a *weak* transportation formula, in the sense that it only allows to express some *selected* ε-subgradients in terms of subgradients.

The transportation formula can also be used for the ε-subdifferential to obtain the lower bound:

$$\text{If} \quad \{(v^i \in \partial_{\varepsilon_i} f(x^i))\}_{i=1,2} \quad \text{then} \quad \langle x^1 - x^2, v^1 - v^2 \rangle \geq -(\varepsilon_1 + \varepsilon_2)\,.$$

In the general case, we have a weaker bound:

Corollary 2.2 *Take* $v^1 \in T^{\varepsilon_1}(x^1)$ *and* $v^2 \in T^{\varepsilon_2}(x^2)$. *Then*

$$\langle x^1 - x^2, v^1 - v^2 \rangle \geq -(\sqrt{\varepsilon_1} + \sqrt{\varepsilon_2})^2 \tag{2.6}$$

Proof. If ε_1 or ε_2 are zero the result holds trivially. Otherwise choose $\alpha \in \Delta_2$ as follows

$$\alpha_1 := \frac{\sqrt{\varepsilon_2}}{\sqrt{\varepsilon_1} + \sqrt{\varepsilon_2}} \qquad \alpha_2 := 1 - \alpha_1 = \frac{\sqrt{\varepsilon_1}}{\sqrt{\varepsilon_1} + \sqrt{\varepsilon_2}} \tag{2.7}$$

and define the convex sums $\hat{x}$, $\hat{v}$ and $\hat{\varepsilon}$ as in (2.3). Because $\hat{\varepsilon} \geq 0$, we can write

$$\begin{aligned} 0 \leq \hat{\varepsilon} &= \alpha_1\varepsilon_1 + \alpha_2\varepsilon_2 + \alpha_1\langle x^1 - \hat{x}, v^1 - \hat{v}\rangle + \alpha_2\langle x^2 - \hat{x}, v^2 - \hat{v}\rangle \\ &= \alpha_1\varepsilon_1 + \alpha_2\varepsilon_2 + \alpha_1\alpha_2\langle x^2 - x^1, v^2 - v^1\rangle\,, \end{aligned}$$

where we have used the identities $x^1 - \hat{x} = \alpha_2(x^1 - x^2)$, $x^2 - \hat{x} = \alpha_1(x^2 - x^1)$, $v^1 - \hat{v} = \alpha_2(v^1 - v^2)$ and $v^2 - \hat{v} = \alpha_1(v^2 - v^1)$ first, and then $\alpha_1\alpha_2^2 + \alpha_1^2\alpha_2 = \alpha_1\alpha_2$. Now, combine the expression above with (2.3) and (2.7) to obtain

$$\sqrt{\varepsilon_1\varepsilon_2} + \frac{\sqrt{\varepsilon_1\varepsilon_2}}{(\sqrt{\varepsilon_1} + \sqrt{\varepsilon_2})^2}\langle x^1 - x^2, v^1 - v^2\rangle \geq 0$$

Rearranging terms and simplifying the resulting expression, (2.6) is proved. ∎

2.3 Lipschitz-Continuity of T^ε

All along this subsection, $H = \mathbb{R}^N$ and $D := D(T)^0$ is the interior of $D(T)$. The reason for taking $H = \mathbb{R}^N$ is that for the results below we need bounded sets to be compact.

A closed-valued locally bounded multifunction S is *continuous* at $\bar{x}$ if for any positive ϵ there exist $\delta > 0$ such that

$$\|x - \bar{x}\| \leq \delta \quad \Longrightarrow \quad \begin{cases} S(x) & \subset \quad S(\bar{x}) + B(0, \epsilon) \\ S(\bar{x}) & \subset \quad S(x) + B(0, \epsilon) \end{cases}$$

Furthermore, S is *Lipschitz continuous* if, given a (nonempty) compact set $K \subseteq D$, there exists a nonnegative constant L such that for any $y^1, y^2 \in K$ and $s^1 \in S(y^1)$ there exists $s^2 \in S(y^2)$ satisfying $\|s^1 - s^2\| \leq L\|y^1 - y^2\|$.

We will prove that the application $(\varepsilon, x) \hookrightarrow T^\varepsilon(x)$ is Lipschitz continuous in D. We start with a technical lemma.

Lemma 2.1 *Assume that D is nonempty. Let $K \subset D$ be a compact set and take $\rho > 0$ such that*

$$\tilde{K} := K + \overline{B(0,\rho)} \subset D. \tag{2.8}$$

Define $\tilde{M} := \sup\{\|u\| \mid u \in T(\tilde{K})\}$. Then, for all $\varepsilon \geq 0$, we have that

$$\sup\{\|u\| : u \in T^{\varepsilon}(K)\} \leq \frac{\varepsilon}{\rho} + \tilde{M}. \tag{2.9}$$

Proof. Because T is locally bounded and $\tilde{K}$ is compact, $T(\tilde{K})$ is bounded. Then $\tilde{M}$ is finite. To prove (2.9), take $x_\varepsilon \in K$ and $v_\varepsilon \in T^{\varepsilon}(x_\varepsilon)$. Apply Theorem 2.1 with $\eta := \rho$: there exists a pair $(x, v) \in G(T)$ such that

$$\|x - x_\varepsilon\| \leq \rho \quad \text{and} \quad \|v - v_\varepsilon\| \leq \frac{\varepsilon}{\rho}.$$

Then $x \in \tilde{K}$, $\|v\| \leq \tilde{M}$, and therefore

$$\|v_\varepsilon\| \leq \|v_\varepsilon - v\| + \|v\| \leq \frac{\varepsilon}{\rho} + \tilde{M},$$

so that (2.9) follows. ∎

Now we prove the Lipschitz-continuity of T^{ε}. Our result strengthens Theorem 1(ii) in [BIS97].

Theorem 2.3 *Assume that D is nonempty. Let $K \subset D$ be a compact set and $0 < \underline{\varepsilon} \leq \bar{\varepsilon} < +\infty$. Then there exist nonnegative constants A and B such that for any (ε_1, x^1), $(\varepsilon_2, x^2) \in [\underline{\varepsilon}, \bar{\varepsilon}] \times K$ and $v^1 \in T^{\varepsilon_1}(x^1)$, there exists $v^2 \in T^{\varepsilon_2}(x^2)$ satisfying*

$$\|v^1 - v^2\| \leq A\|x^1 - x^2\| + B|\varepsilon_1 - \varepsilon_2|. \tag{2.10}$$

Proof. With ρ, $\tilde{K}$ and $\tilde{M}$ as in Lemma 2.1, we claim that (2.10) holds for the following choice of A and B:

$$A := \left(\frac{1}{\rho} + \frac{2\tilde{M}}{\underline{\varepsilon}}\right)\left(\frac{\bar{\varepsilon}}{\rho} + 2\tilde{M}\right), \qquad B := \left(\frac{1}{\rho} + \frac{2\tilde{M}}{\underline{\varepsilon}}\right). \tag{2.11}$$

To see this, take x^1, x^2, ε_1, ε_2 and v^1 as above. Take $l := \|x^1 - x^2\|$ and let x^3 be in the line containing x^1 and x^2 such that

$$\|x^3 - x^2\| = \rho, \qquad \|x^3 - x^1\| = \rho + l, \tag{2.12}$$

as shown in Figure 1.

Then, $x^3 \in \tilde{K}$ and

$$x^2 = (1-\theta)x^1 + \theta x^3 \quad , \text{ with } \quad \theta = \frac{l}{\rho + l} \in [0,1) .$$

Now, take $u^3 \in T(x^3)$ and define

$$\tilde{v}^2 := (1-\theta)v^1 + \theta u^3 .$$

Theorem 2.2 yields $\tilde{v}^2 \in T^{\tilde{\varepsilon}_2}(x^2)$, with

$$\begin{aligned} \tilde{\varepsilon}_2 &= (1-\theta)\varepsilon_1 + (1-\theta)\langle x^1 - x^2, v^1 - \tilde{v}^2 \rangle + \theta \langle x^3 - x^2, u^3 - \tilde{v}^2 \rangle \\ &= (1-\theta)\varepsilon_1 + \theta(1-\theta)\langle x^1 - x^3, v^1 - u^3 \rangle . \end{aligned}$$

Use Lemma 2.1 with $v^1 \in T^{\varepsilon_1}(x^1)$, together with the definition of $\tilde{M}$, to obtain

$$\|v^1 - u^3\| \le \|v^1\| + \|u^3\| \le \tilde{M} + (\frac{\varepsilon_1}{\rho} + \tilde{M}) \le \frac{\varepsilon_1}{\rho} + 2\tilde{M} . \tag{2.13}$$

Using now Cauchy-Schwarz, (2.13), (2.12), and recalling the definition of θ, we get

$$\begin{aligned} \tilde{\varepsilon}_2 &\le (1-\theta)\varepsilon_1 + \theta(1-\theta)\|x^1 - x^3\| \, \|v^1 - u^3\| \\ &\le (1-\theta)\varepsilon_1 + \theta(1-\theta)(\rho + l)\left(\frac{\varepsilon_1}{\rho} + 2\tilde{M}\right) \\ &= \varepsilon_1 + \frac{\rho l}{\rho + l} 2\tilde{M}. \end{aligned} \tag{2.14}$$

The definition of $\tilde{v}^2$ combined with (2.13) yields

$$\|v^1 - \tilde{v}^2\| = \theta\|v^1 - u^3\| \le \theta\left(\frac{\varepsilon_1}{\rho} + 2\tilde{M}\right) , \tag{2.15}$$

as well as

$$\|v^1 - \tilde{v}^2\| \le \|x^1 - x^2\| \frac{1}{\rho}\left(\frac{\varepsilon_1}{\rho} + 2\tilde{M}\right) . \tag{2.16}$$

Now consider two cases:

(i) $\tilde{\varepsilon}_2 \le \varepsilon_2$,

(ii) $\tilde{\varepsilon}_2 > \varepsilon_2$.

If (i) holds, $\tilde{v}^2 \in T^{\tilde{\varepsilon}_2}(x^2) \subseteq T^{\varepsilon_2}(x^2)$. Then, choosing $v^2 := \tilde{v}^2$ and using (2.16) together with (2.11), (2.10) follows.

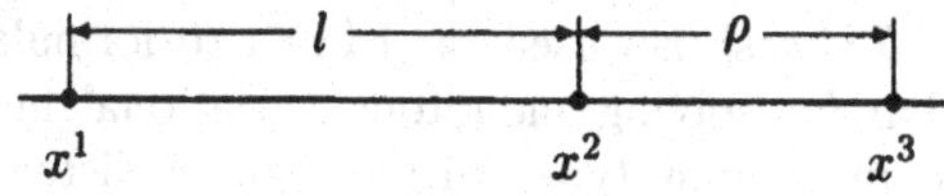

Figure 1.1

In case (ii), define $\beta := \frac{\varepsilon_2}{\tilde{\varepsilon}_2} < 1$ and $v^2 := (1-\beta)u^2 + \beta\tilde{v}^2$, with $u^2 \in T(x^2)$. Because of Theorem 2.2, $v^2 \in T^{\varepsilon_2}(x^2)$. Furthermore, (2.9) together with (2.15) lead to

$$\begin{aligned} \|v^2 - v^1\| &\leq (1-\beta)\|u^2 - v^1\| + \beta\|\tilde{v}^2 - v^1\| \\ &\leq (1-\beta)\left(\tfrac{\varepsilon_1}{\rho} + 2\tilde{M}\right) + \beta\theta\left(\tfrac{\varepsilon_1}{\rho} + 2\tilde{M}\right) \\ &= (1-\beta(1-\theta))\left(\tfrac{\varepsilon_1}{\rho} + 2\tilde{M}\right) \end{aligned} \tag{2.17}$$

Using (2.14) we have that $\beta \geq \dfrac{\varepsilon_2}{\varepsilon_1 + \frac{\rho l}{\rho + l} 2\tilde{M}}$.

Some elementary algebra, the inequality above and the definitions of θ and l, yield

$$\begin{aligned} 1-\beta(1-\theta) &\leq l\left(\frac{\varepsilon_1 + \rho 2\tilde{M}}{(\rho+l)\varepsilon_1 + \rho l 2\tilde{M}}\right) + \frac{\rho(\varepsilon_1 - \varepsilon_2)}{(\rho+l)\varepsilon_1 + \rho l 2\tilde{M}} \\ &\leq \|x^1 - x^2\|\left(\frac{1}{\rho} + \frac{2\tilde{M}}{\varepsilon_1}\right) + |\varepsilon_1 - \varepsilon_2|\frac{1}{\varepsilon_1}\,. \end{aligned} \tag{2.18}$$

Altogether, with (2.17), (2.18) and our assumptions on $\varepsilon_1, \varepsilon_2, \underline{\varepsilon}, \overline{\varepsilon}$, the conclusion follows. ∎

The continuity of $T^\varepsilon(x)$ as a multifunction is straightforward. In particular, it also has a closed graph:

For any sequence $\{(\varepsilon_i, x^i, v^i \in T^{\varepsilon_i}(x^i))\}_i$ such that $\varepsilon_i > 0$ for all i,

$$\lim_{i\to\infty} x^i = x, \quad \lim_{i\to\infty} \varepsilon_i = \varepsilon, \quad \lim_{i\to\infty} v^i = v \implies v \in T^\varepsilon(x)\,. \tag{2.19}$$

Let us mention that this result was already proved in [BIS97, Proposition 1(iv)].

3 CONCEPTUAL ALGORITHMIC PATTERNS

The way is now open to define iterative algorithms for solving (1.2). We have at hand an enlargement $T^\varepsilon(x)$, which is continuous on x and ε. We present in this section a conceptual algorithm, working directly with the convex sets $T^\varepsilon(x)$, together with its convergence proof. As already mentioned, this algorithm will be the basis of the implementable algorithm in [BSS97], where the sets $T^\varepsilon(x)$ are replaced by suitable polyhedral approximation by means of a bundle strategy which makes an extensive use of the transportation formula in Theorem 2.2.

The ε-enlargement T^ε was also used in [BIS97] to formulate and analyze a conceptual algorithm for solving monotone variational inequalities. The method is of the proximal point type, with Bregman distances and inexact resolution of the related subproblems.

We assume from now on that $D(T) = \mathbb{R}^N$, so that T maps bounded sets in bounded sets, and we have a nonempty solution set S.

3.1 The algorithm

Before entering into technicalities, we give an informal description on how the algorithm works. Following our remarks in § 1, iterates will be generated by projecting onto halfspaces $H_{y,v}$ from (1.3). We also said that to ensure convergence, the pair $(y, v \in T(y))$ has to be "good enough", in a sense to be precised below. Since $y = x^k - ts$, where $s = \mathrm{P}_{T^{\varepsilon_k}(x^k)}(0)$ for some $\varepsilon_k > 0$, this comes to say that ε_k has to be good enough. The control on $\varepsilon_k = \varepsilon 2^{-j_k}$ is achieved in Step 1 below. More precisely, Lemma 3.2 and Theorem 3.1 in§ 3.2 show how the decreasing sequence $\{\varepsilon_k\}$ converges to 0.

Now we describe the algorithm.

Conceptual Algorithmic Scheme (CAS):

Choose positive parameters τ, R, ε and σ, with $\sigma \in (0,1)$.

INITIALIZATION: Set $k := 0$ and take $x^0 \in \mathbb{R}^N$.

K-STEP:

Step 0 **(stopping test)**

(0.a) If $0 \in T(x^k)$, then STOP.

Step 1 **(computing search direction)**

(1.a) Set $j := 0$.

(1.b) Compute $s^{k,j} := \operatorname{argmin}\{\|v\|^2 \mid v \in T^{\varepsilon 2^{-j}}(x^k)\}$.

(1.c) If $\|s^{k,j}\| \leq \tau\, 2^{-j}$ then set $j := j+1$ and LOOP to (1.b).

(1.d) Else, define $j_k := j$ and $s^k := s^{k,j_k}$.

Step 2 **(line search)**

(2.a) Set $l := 0$.

(2.b) Define $y^{k,l} := x^k - R\,2^{-l} s^k$ and take $v^{k,l} \in T(y^{k,l})$.

(2.c) If $\langle v^{k,l}, s^k\rangle \leq \sigma \|s^k\|^2$, then set $l := l+1$ and LOOP to (2.b).

(2.d) Else, define $l_k := l$, $v^k := v^{k,l_k}$ and $y^k := y^{k,l_k}$.

Step 3 **(projection step)**

(3.a) Define $x^{k+1} := x^k - \langle v^k, x^k - y^k\rangle v^k / \|v^k\|^2$.

(3.b) Set $k := k+1$ and LOOP to k-step. □

A few comments are now in order.

In *Step 1* a search direction s^k is computed. It satisfies $s^k \in T^{\varepsilon 2^{-j_k}}(x^k)$ with $\|s^{k,j_k}\| > \tau\, 2^{-j_k}$. We show in Proposition 3.2(i) that an infinite loop cannot occur in this step.

In *Step 2*, using the direction s^k, a pair $(y^k, v^k \in T(y^k))$ is obtained, by iterating on l. This pair is such that not only $x^k \notin H_{y^k,v^k}$, but it is also "far enough" and gives a nonnegligible progress on $\|x^{k+1} - x^k\|$. This is shown in Proposition 3.3.

Variational inequality problems (VIP) can be considered as constrained versions of (1.2). Accordingly, a "constrained" variant of (CAS) could be devised in a straightforward manner by adding a subalgorithm ensuring feasibility of iterates. Rather than introducing this extra complication, we chose to focus our efforts in designing an algorithmic pattern oriented to future implementations. As a result, in [BSS97] we analyze a "Bundling Strategy", whose K-STEP gives an *implementable* version of the one in (CAS). As already mentioned, for this mechanism to work, transportation formulæ like the one in Theorem 2.2 are crucial, because they allow the iterative construction of polyhedral approximations of $T^{\varepsilon 2^{-j_k}}(x^k)$ while preserving convergence. Further on, in § 3.2, we analyze the convergence of (CAS). One of the key arguments for the proof in Lemma 3.2 below is the Lipschitz-continuity of T^{ε}, stated in Theorem 2.3. In turn, the proof of the latter result is based on the (useful!) transportation formula.

Along these lines, the related recent work [Luc97] introduces two algorithms to solve VIP with multivalued maximal monotone operators. These methods, named *Algorithms I and II*, ensure feasibility of iterates by making inexact orthogonal projections. For the unconstrained case, both algorithms first choose a direction in some enlargement $T^{\varepsilon_k}(x^k)$, like in (1.b) of (CAS), and perform afterwards a line search similar to *Step 2* in (CAS). The important difference between (CAS) and the methods in [Luc97] is how the parameter ε_k is taken along iterations:

- *Algorithm I* uses a constant $\varepsilon_k = \varepsilon$ for all k. As a result, see [Luc97, Theorem 3.4.1], $\{x^k\}$ is a bounded sequence whose cluster points $\bar{x}$ are ε-solutions of (1.2), i.e., points satisfying $0 \in T^{\varepsilon}(\bar{x})$.
- *Algorithm II* makes use of a dynamical ε_k, varying along iterations. In this case, if the sequence of iterates $\{x^k\}$ is infinite, it converges to a solution of (1.2). If finite termination occurs, the last generated point x^{k_f} is shown to be an ε_{k_f}-solution ([Luc97, Theorem 3.6.1]).

In (CAS), instead, a close and careful control on ε_k is made during the iterative process. This control is decisive to prove convergence to a solution, even for the finite termination case (see Theorem 3.1 below).

Summing up, we believe that both *Algorithm II* in [Luc97] and (CAS) are very important steps for developing implementable schemes for solving VIP, because they are on the road to extending bundle methods to a more general framework than nonsmooth convex optimization. However, when compared to the methods in [Luc97], we also think that (CAS) makes a more effective use of the enlargements $T^{\varepsilon_k}(x^k)$ and gives a deeper insight of the mechanism that makes algorithms of this type work, both in theory and in practice.

3.2 Convergence Analysis

For proving convergence of (CAS), we use the concept of Fejér-convergent or Fejér-monotone sequence:

Definition 3.1 A sequence $\{x^k\}$ is said to be *Fejér-monotone* with respect to a set C if $\|x^{k+1} - x\| \leq \|x^k - x\|$, for any $x \in C$. □

The following elementary result, that we state here without proof, will be used in the sequel.

Proposition 3.1 *Let the sequence $\{x^k\}$ be Fejér-monotone with respect to a nonempty set C. Then*

(i) $\{x^k\}$ is bounded.

(ii) If $\{x^k\}$ has an accumulation point which is in C, then the full sequence converges to a limit in C.

□

Note that a Fejér-monotone sequence *is not necessarily* convergent: a sequence $\{x^k\}$ could approach a set C *without* converging to an element of C. For our (CAS), convergence will follow from Proposition 3.1 and adequate choices of y^k and v^k in (1.4).

In (CAS) each k-step has two ending points: (0.a) and (3.b). If some k-step exits at (0.a), then the algorithm stops. If the k-step exits at (3.b), then x^{k+1} is generated and a new k-step is started with k replaced by $k+1$.

There are also two loops: (1.c)↔(1.b) on j indices; and the other (2.c)↔(2.b), iterating along l indices. We start proving that infinite loops do not occur on these indices.

Proposition 3.2 *Let x^k be the current iterate in (CAS). Then, either x^k is a solution in* (1.2) *and (CAS) stops in (0.a); or x^k is not a solution and the following holds:*

(i) Relative to the loop (1.c)↔(1.b), there exists a finite $j = j_k$ such that (1.d) is reached:

$$\|s^k\| > \tau 2^{-j_k} \quad \textit{with} \quad \|s^{k,j_k-1}\| \le \tau 2^{-j_k+1},$$

whenever $j_k > 0$.

(ii) Relative to the loop (2.c)↔(2.b), there exists a finite $l = l_k$ such that (2.d) is reached:

$$\langle v^{k,l_k}, s^k\rangle > \sigma\|s^k\|^2 \quad \textit{with} \quad \langle v^{k,l_k-1}, s^k\rangle \le \sigma\|s^k\|^2,$$

whenever $l_k > 0$.

Proof. If x^k is a solution, then $0 \in T(x^k)$ and (CAS) ends at (0.a). Otherwise, we have that $0 \notin T(x^k)$. To prove (i), suppose, for contradiction, that (CAS) loops forever in (1.c)↔(1.b). Then it generates an infinite sequence $\{s^{k,j}\}_{j\in\mathbb{N}}$ such that

$$s^{k,j} \in T^{\varepsilon 2^{-j}}(x^k) \quad \text{and} \quad \|s^{k,j}\| \le \tau 2^{-j}.$$

Letting j go to infinity and using (2.19), this implies that $0 \in T(x^k)$, a contradiction. Hence, there exists a finite j such that the loop (1.c)↔(1.b) ends. For this index j, Step (1.d) is reached, $j_k = j$ is defined and $\|s^{k,j_k}\| > \tau 2^{-j_k}$, so that (i) holds.

Now we prove (ii). If an infinite loop occurs at (2.c)↔(2.b), then an infinite sequence $\{(y^{k,l}, v^{k,l})\}_{l\in\mathbf{N}} \subseteq G(T)$ is generated. This sequence is such that

$$\lim_{l\to\infty} y^{k,l} = \lim_{l\to\infty} x^k - R2^{-l}s^k = x^k \quad \text{and} \quad \langle v^{k,l}, s^k\rangle \le \sigma\|s^k\|^2,$$

for all $l \in \mathbf{N}$. Because $\{y^{k,l}\}_{l\in\mathbf{N}}$ is bounded (it is a convergent sequence), so is $\{v^{k,l}\}_{l\in\mathbf{N}}$, by the local boundedness of T. Extracting a subsequence if needed, there exists a $\bar{v}$ such that

$$\lim_{i\to\infty} v^{k,l_i} = \bar{v} \in T(x^k) \subseteq T^{\varepsilon 2^{-j_k}}(x^k),$$

with $\langle \bar{v}, s^k\rangle \le \sigma\|s^k\|^2$. However, since s^k is the element of minimum norm in $T^{\varepsilon 2^{-j_k}}(x^k)$, we also have that $\langle \bar{v}, s^k\rangle \ge \|s^k\|^2$. Therefore, $\sigma \ge 1$, a contradiction.

Hence, there exists a finite l such that the loop (2.c)↔(2.b) ends. For this index l, Step (2.d) is reached, $l_k = l$ is defined and $\langle v^{k,l}, s^k\rangle > \sigma\|s^k\|^2$; altogether, (ii) is proved. ∎

We proved that the sequence generated by (CAS) is either finite, with last point which is a solution, or it is infinite, with no iterate solving (1.2). We are now in a position to show how (CAS) follows the scheme discussed in the introduction.

Proposition 3.3 *Let x^k be the current iterate in (CAS). Then, if x^k is not a solution in* (1.2), *after x^{k+1} is generated in (3.b), the following holds:*

1. *Consider the halfspace defined in* (1.3). *Then $x^k \notin H_{y^k,v^k}$ and $x^{k+1} = P_{H_{y^k,v^k}}(x^k)$.*
2. *For all $x^* \in S$, $\|x^{k+1} - x^*\|^2 \le \|x^k - x^*\|^2 - \|x^{k+1} - x^k\|^2$*
3. *Finally, $\|x^{k+1} - x^k\| > R\sigma\tau^2 2^{-l_k - 2j_k}/\|v^k\|$.*

Proof. To prove (i), use Proposition 3.2(ii) and (i): at step (3.b) the pair $(y^k = x^k - R2^{-l_k}s^k, v^k = v^{k,l_k})$ is such that

$$\langle x^k - y^k, v^k\rangle \ge R2^{-l_k}\sigma\|s^k\|^2 > R\sigma\tau^2 2^{-l_k-2j_k} > 0,$$

so that $x^k \notin H_{y^k,v^k}$. To see that x^{k+1} is its projection, just recall (1.4) and the definition of x^{k+1} in (CAS). Because x^{k+1} is an orthogonal projection, (ii) follows. As for (iii), it is also straightforward. ∎

We can now prove that (CAS) generates a Fejér-monotone sequence, with all the involved variables being bounded.

Lemma 3.1 *If (CAS) generates infinite x^k, then the sequence is Fejér-monotone with respect to the solution set S. Moreover, x^k, s^k, y^k, v^k, as well as the sets $\{s^{k,j}\}_{0\le j\le j_k}$ and $\{(y^{k,l}, v^{k,l})\}_{0\le l\le l_k}$, are bounded.*

Proof. It follows from Definition 3.1 and Proposition 3.3(ii) that the sequence is Fejér-monotone. Because of Proposition 3.1(i), there exists some compact set K_0 such that the (bounded) sequence $\{x^k\} \subset K_0$. Since T^ε is locally bounded, $M_0 := \sup\{\|u\| : u \in T^\varepsilon(K_0)\}$ is finite and $s^{k,j} \in T^{\varepsilon 2^{-j}}(x^k) \subseteq T^\varepsilon(x^k)$ is bounded for all $j \le j_k$:

$$\|s^{k,j}\| \le M_0 \text{ for all } 0 \le j \le j_k \text{ and } \|s^k\| \le M_0\,,$$

for all k. Now, define $K_1 := K_0 + \overline{B(0, RM_0)}$ to obtain

$$y^{k,l} \in K_1 \text{ for all } 0 \le l \le l_k \text{ and } y^k \in K_1\,,$$

for all k. Using once more the local-boundedness of T, we conclude that $\{v^{k,l}\}_{0\le l\le l_k}$ is bounded, and the proof is finished. ∎

Before proving our convergence result, we need a last technical lemma.

Lemma 3.2 *Suppose that (CAS) generates an infinite sequence $\{x^k\}$. If the sequence $\{j_k\}_{k\in N}$ is bounded, then $\{l_k\}_{k\in N}$ is also bounded.*

Proof. Let J be an upper bound for the sequence $\{j_k\}$ and let $\varepsilon_k := \varepsilon 2^{-j_k}$, then

$$\varepsilon_k \in [\underline{\varepsilon}, \bar{\varepsilon}] := [\varepsilon 2^{-J}, \varepsilon]\,. \tag{3.1}$$

Now, for those k such that $l_k > 0$, Proposition 3.2(ii) applies. In addition, the pairs $(y^{k,l_k-1}, v^{k,l_k-1})$ satisfy

$$\|y^{k,l_k-1} - x^k\| = R2^{-l_k+1}\|s^k\| \text{ and } v^{k,l_k-1} \in T(y^{k,l_k-1}) \subseteq T^{\varepsilon_k}(y^{k,l_k-1})\,. \tag{3.2}$$

Using Lemma 3.1 and (3.1), we have that $x^1 := y^{k,l_k-1}$, $x^2 := x^k$ and $\varepsilon^1 = \varepsilon^2 := \varepsilon_k$ are bounded and Theorem 2.3 applies. Then, together with (3.2), we obtain,

$$\|v^{k,l_k-1} - v^2\| \le AR2^{-l_k+1}\|s^k\|\,, \tag{3.3}$$

for a positive A and some $v^2 \in T^{\varepsilon_k}(x^k)$, for each k such that $l_k > 0$,.

Consider the projection $\bar{v}^k := P_{T^{\varepsilon_k}(x^k)}(v^{k,l_k-1})$. We have

$$\|s^k\| \le \|\bar{v}^k\|\,, \tag{3.4}$$

as well as

$$\|v^{k,l_k-1} - \bar{v}^k\| \le \|v^{k,l_k-1} - w\|\,, \tag{3.5}$$

for any $w \in T^{\varepsilon_k}(x^k)$; in particular, for $w = v^2$ from (3.3). Altogether, with (3.5) and (3.4),

$$\|v^{k,l_k-1} - \bar{v}^k\| \le AR2^{-l_k+1}\|\bar{v}^k\|\,, \tag{3.6}$$

for all k such that $l_k > 0$.

Consider now the scalar product $\langle v^{k,l_k-1}, s^k\rangle$ and apply successively Cauchy-Schwartz, (3.6) and (3.4) again to write the following chain of (in)equalities:

$$\begin{aligned}\langle v^{k,l_k-1}, s^k\rangle &= \langle v^{k,l_k-1} - \bar{v}^k, s^k\rangle + \langle \bar{v}^k, s^k\rangle \\ &\geq -AR2^{-l_k+1}\|s^k\|^2 + \|s^k\|^2 \\ &= \left(1 - AR2^{-l_k+1}\right)\|s^k\|^2.\end{aligned}$$

Because Proposition 3.2(*ii*) holds, we obtain

$$\sigma \geq \left(1 - AR2^{-l_k+1}\right),$$

which in turn yields

$$l_k \leq \max\left\{0, \log_2 \frac{2AR}{1-\sigma}\right\}$$

and the conclusion follows. ∎

Finally, we state our main convergence result.

Theorem 3.1 *Consider the sequence $\{x^k\}$ generated by (CAS). Then the sequence is either finite with last element solving* (1.2), *or it converges to a solution in* (1.2).

Proof. We already dealt with the finite case in Proposition 3.2. If (CAS) does not stop, then there is no x^k solving (1.2). Because of Proposition 3.3(*ii*), we have that $\sum \|x^{k+1} - x^k\|^2 < +\infty$ and therefore

$$\lim_{k\to\infty} \|x^{k+1} - x^k\| = 0. \tag{3.7}$$

On the other side, Proposition 3.3(*iii*) leads to

$$\|x^{k+1} - x^k\| > \frac{R\sigma\tau^2}{M} 2^{-l_k-2j_k},$$

where M is an upper bound for $\|v^k\|$ given by Lemma 3.1. Together with (3.7), we obtain

$$\lim_{k\to\infty} l_k + 2j_k = +\infty.$$

Then, by Lemma 3.2, the sequence $\{j_k\}$ must be unbounded.

Because $\{x^k\}$ is bounded (this is again Lemma 3.1), we can extract a convergent subsequence as follows

$$\begin{array}{rcl}\lim_{q\to\infty} x^{k_q} & = & \hat{x}, \\ \lim_{q\to\infty} j_{k_q} & = & +\infty, \\ j_{k_q} > 0 & \text{for all} & q.\end{array} \tag{3.8}$$

For this subsequence, using Proposition 3.2(*i*), it holds that

$$\lim_{q\to\infty} s^{k_q, j_{k_q}-1} = 0.$$

Since $s^{k_q,j_{k_q}-1} \in T^{\varepsilon 2^{-j_{k_q}+1}}(x^{k_q})$, together with (3.8) and (2.19) we get

$$(\varepsilon 2^{-j_{k_q}+1}, x^{k_q}, s^{k_q,j_{k_q}-1}) \longrightarrow (0, \hat{x}, 0) \quad \Longrightarrow \quad 0 \in T(\hat{x}),$$

so that $\hat{x} \in \mathcal{S}$.

Therefore the Fejér-monotone sequence $\{x^k\}$ has an accumulation point which is in $\mathcal{S}$. By Proposition 3.1, the proof is done. ∎

4 CONCLUDING REMARKS

We presented a conceptual algorithm for finding zeros of a maximal monotone operator based on projections onto suitable halfspaces. Our algorithmic scheme extends the pattern of ε-descent methods for nonsmooth optimization by generating directions of minimum norm in $T^{\varepsilon_k}(x^k)$, with $\varepsilon_k = \varepsilon 2^{-j_k} \downarrow 0$. To develop implementable algorithms, a constructive device to approximate the convex set $T^{\varepsilon}(x)$ is needed. This is done in [BSS97] by extending the bundle machinery of nonsmooth optimization to this more general context. An important consideration to reduce iterations in the implementable version concerns backtracking steps in *Step 1*. The following modification, suggested by one of the referees, avoids unnecessary iterations on j-indices:

Take $\varepsilon_0 = \varepsilon$.

Step 1' **(computing search direction)**

(1.a') Set $j := 0$.

(1.b') Compute $s^{k,j} := \operatorname{argmin}\{\|v\|^2 \mid v \in T^{\varepsilon_k 2^{-j}}(x^k)\}$.

(1.c') If $\|s^{k,j}\| \leq \tau 2^{-j}$ then set $j := j+1$ and LOOP to (1.b).

(1.d') Else, define $j_k := j$, $\varepsilon_{k+1} := \varepsilon_k 2^{-j_k}$ and $s^k := s^{k,j_k}$.

Finally, observe that (CAS) makes use of the extension $T^{\varepsilon}(x)$ from Definition 1.1. However, the same algorithmic pattern can be applied for any extension $E(\varepsilon, x)$ of the maximal monotone $x \hookrightarrow T(x)$, provided that it is *continuous* and a *transportation formula* like in Theorem 2.2 hold. In particular, any continuous $E(\varepsilon, x) \subset T^{\varepsilon}(x)$ for all $x \in D(T)$ and $\varepsilon \geq 0$ can be used.

Acknowledgements

We are indebted to Igor Konnov, whose interesting remarks were very helpful during our revision of this paper. We also want to thank the referees for their careful and exhaustive reading.

References

[BGLS97] F. Bonnans, J.Ch. Gilbert, C.L. Lemaréchal and C.A. Sagastizábal. *Optimisation Numérique, aspects théoriques et pratiques.* Collection "Mathématiques et applications", SMAI-Springer-Verlag, Berlin, 1997.

[BM73] D.P. Bertsekas and S.K. Mitter. A descent numerical method for optimization problems with nondifferentiable cost functionals. *SIAM Journal on Control*, 11(4):637–652, 1973.

[BR65] A. Brøndsted and R.T. Rockafellar. On the subdifferentiability of convex functions. *Proceedings of the American Mathematical Society*, 16:605–611, 1965.

[BIS97] R.S. Burachik, A.N. Iusem, and B.F. Svaiter. Enlargements of maximal monotone operators with application to variational inequalities. *Set Valued Analysis*, 5:159–180, 1997.

[BSS97] R.S. Burachik, C.A. Sagastizábal, and B. F. Svaiter. Bundle methods for maximal monotone operators. Submitted, 1997.

[HUL93] J.-B. Hiriart-Urruty and C. Lemaréchal. *Convex Analysis and Minimization Algorithms.* Number 305-306 in Grund. der math. Wiss. Springer-Verlag, 1993. (two volumes).

[Ius94] A. N. Iusem. An iterative algorithm for the variational inequality problem. *Computational and Applied Mathematics*, 13:103–114, 1994.

[IS97] A. N. Iusem. and B.F. Svaiter. A variant of Korpolevich's method for variational inequalities with a new search strategy. *Optimization*, 42:309–321, 1997.

[Kiw90] K.C. Kiwiel. Proximity control in bundle methods for convex nondifferentiable minimization. *Mathematical Programming*, 46:105–122, 1990.

[Kon97] I.V. Konnov. A combined relaxation method for variational inequalities with nonlinear constraints. *Mathematical Programming*, 1997. Accepted for publication.

[Kor76] G.M. Korpelevich. The extragradient method for finding saddle points and other problems. *Ekonomika i Matematischeskie Metody*, 12:747–756, 1976.

[Lem80] C. Lemaréchal. Extensions diverses des méthodes de gradient et applications, 1980. Thèse d'Etat, Université de Paris IX.

[LNN95] C. Lemaréchal, A. Nemirovskii, and Yu. Nesterov. New variants of bundle methods. *Mathematical Programming*, 69:111–148, 1995.

[Luc97] L.R. Lucambio Pérez. Iterative Algorithms for Nonsmooth Variational Inequalities, 1997. Ph.D. Thesis, Instituto de Matemática Pura e Aplicada, Rio de Janeiro, Brazil.

[Min62] G.L. Minty. Monotone nonlinear operators in a Hilbert space. *Duke Mathematical Journal*, 29:341–346, 1962.

[Mor65] J.J. Moreau. Proximité et dualité dans un espace hilbertien. *Bulletin de la Société Mathématique de France*, 93:273–299, 1965.

[Nur86] E.A. Nurminski. ε-subgradient mapping and the problem of convex optimization. *Cybernetics*, 21(6):796–800, 1986.

[Roc69] R.T. Rockafellar. Local boundedness of nonlinear monotone operators. *Michigan Mathematical Journal*, 16:397–407, 1969.

[SS96] M.V. Solodov and B.F. Svaiter. A new projection method for variational inequality problems. Technical Report B-109, IMPA, Brazil, 1996. *SIAM Journal on Control and Optimization*, submitted.

[SS97] M.V. Solodov and B.F. Svaiter. A hybrid projection-proximal point algorithm. Technical Report B-115, IMPA, Brazil, 1997.

[SZ92] H. Schramm and J. Zowe. A version of the bundle idea for minimizing a nonsmooth function: conceptual idea, convergence analysis, numerical results. *SIAM Journal on Optimization*, 2(1):121–152, 1992.

Reformulation: Nonsmooth, Piecewise Smooth, Semismooth and Smoothing Methods, pp. 45–63
Edited by M. Fukushima and L. Qi

A Non-Interior Predictor-Corrector Path-Following Method for LCP

James V. Burke* and Song Xu*

Abstract In a previous work the authors introduced a non-interior predictor-corrector path following algorithm for the monotone linear complementarity problem. The method uses Chen–Harker–Kanzow–Smale smoothing techniques to track the central path and employs a refined notion for the neighborhood of the central path to obtain the boundedness of the iterates under the assumption of monotonicity and the existence of a feasible interior point. With these assumptions, the method is shown to be both globally linearly convergent and locally quadratically convergent. In this paper it is shown that this basic approach is still valid without the monotonicity assumption and regardless of the choice of norm in the definition of the neighborhood of the central path. Furthermore, it is shown that the method can be modified so that only one system of linear equations needs to be solved at each iteration without sacrificing either the global or local convergence behavior of the method. The local behavior of the method is further illuminated by showing that the solution set always lies in the interior of the neighborhood of the central path relative to the affine constraint. In this regard, the method is fundamentally different from interior point strategies where the solution set necessarily lies on the boundary of the neighborhood of the central path relative to the affine constraint. Finally, we show that the algorithm is globally convergent under a relatively mild condition.

Key Words linear complementarity, smoothing methods, path-following methods

1 INTRODUCTION

Consider the *linear complementarity problem*:

*Department of Mathematics, University of Washington, Seattle WA 98195. email: burke@math.washington.edu, songxu@math.washington.edu

LCP(q, M): Find $(x, y) \in \mathbb{R}^n \times \mathbb{R}^n$ satisfying

$$Mx - y + q = 0, \tag{1.1}$$

$$x \geq 0, y \geq 0, x^T y = 0, \tag{1.2}$$

where $M \in \mathbb{R}^{n \times n}$ and $q \in \mathbb{R}^n$.

In this paper, we study extensions, refinements, and properties of a non-interior predictor-corrector path following algorithm for this problem which was recently proposed by Burke and Xu [3]. Here the path to be followed is the *central path*

$$\mathcal{C} = \{(x, y) \, : \, 0 < \mu, \; 0 < x, \; 0 < y, \; Mx - y + q = 0, \; \text{and } Xy = \mu^2 e\} \tag{1.3}$$

where, following standard usage in the interior-point literature [16], we denote by $e \in \mathbb{R}^n$ the vector each of whose components is 1 and by X the diagonal matrix whose diagonal entries are given by the vector $x \in \mathbb{R}^n$. The algorithm is based on Chen-Harker-Kanzow-Smale smoothing techniques [6, 14, 19] and as such relies on the function

$$\phi(a, b, \mu) = a + b - \sqrt{(a-b)^2 + 4\mu^2} \, . \tag{1.4}$$

This function is a member of the Chen–Mangasarian class of smoothing functions for the problem LCP(q, M) [8]. It is easily verified that for $\mu > 0$

$$\phi(a, b, \mu) = 0 \text{ if and only if } 0 < a, \; 0 < b, \; \text{and } ab = \mu^2. \tag{1.5}$$

Another function having this property is the smoothed Fischer-Burmeister function

$$\psi(a, b, \mu) = a + b - \sqrt{a^2 + b^2 + 2\mu^2} \, , \tag{1.6}$$

which is first studied by Kanzow [14]. For simplicity, in this paper, we will focus on the function ϕ. However, the same analysis can be easily carried out if the function ϕ is replaced by ψ.

Based on the functions ϕ, ψ and other smoothing functions, a number of non-interior path following algorithms have recently been proposed that are globally convergent or globally linearly convergent and possess rapid local convergence properties [2, 3, 5, 4, 6, 7, 9, 10, 12, 13, 14, 17, 18, 20, 21, 22]. Interested readers are referred to [3] for more references. In [3], Burke and Xu propose the first non-interior predictor-corrector algorithm for monotone LCP. The central idea is to apply Newton's method to equations of the form $F(x, y, \mu) = v$ for various choices of the right hand side v where the function $F : \mathbb{R}^n \times \mathbb{R}^n \times \mathbb{R}_+ \to \mathbb{R}^n \times \mathbb{R}^n \times \mathbb{R}_+$ is given by

$$F(x, y, \mu) := \begin{bmatrix} Mx - y + q \\ \Phi(x, y, \mu) \\ \mu \end{bmatrix} , \tag{1.7}$$

with

$$\Phi(x,y,\mu) = \begin{bmatrix} \phi(x_1,y_1,\mu) \\ \cdots \\ \phi(x_n,y_n,\mu) \end{bmatrix}. \tag{1.8}$$

Note that

$$F(x,y,\mu) = 0 \tag{1.9}$$

if and only if (x,y) solves LCP(q,M), and

$$F(x,y,\mu) = \begin{bmatrix} 0 \\ 0 \\ \bar{\mu} \end{bmatrix} \quad \text{with} \quad \bar{\mu} \neq 0 \tag{1.10}$$

if and only if (x,y) is on the central path $\mathcal{C}$ with corresponding smoothing parameter $\bar{\mu}$. Burke and Xu [3] show that their algorithm is both globally linearly convergent and locally quadratically convergent under standard hypotheses and requires one or two matrix factorizations in each step. In this paper, we study this algorithm further and show that it has a number of very nice properties. (1) We observe that our convergence results are valid for any norm, in particular, they are valid for the ∞-norm, which is preferred in practical implementations. (2) We show that the solution set of the LCP is contained in the interior of every slice of the neighborhood relative to the affine constraints. Thus, this neighborhood is fundamentally different from those used in the interior-point literature, where the solution set necessarily lies on the boundary of the neighborhood relative to the affine constraints. This property explains why the non-interior method can be initiated from any point in the space and why the predictor steps are so efficient. (3) We show that the algorithm can be modified so that only one matrix factorization is needed in each iteration. (4) We show that essentially the same convergence theory can be obtained if the monotonicity and strict feasibility hypotheses are replaced by the hypotheses that the matrix M is both a P_0 and an R_0 matrix. (5) Finally, we establish the global convergence for the algorithm under relatively mild conditions.

The plan of the paper is as follows. In Section 2, we study some structural properties of the neighborhood of the central path. In Section 3, we state our predictor–corrector algorithm and show that it is well–defined. Section 4 contains the convergence analysis.

A few words about our notation are in order. All vectors are column vectors with the superscript T denoting transpose. The notation $\mathbb{R}^n$ is used for real n–dimensional space and $\mathbb{R}^{n\times n}$ is used to denote the set of all $n \times n$ real matrices. We denote the non–negative orthant in $\mathbb{R}^n$ by $\mathbb{R}^n_+$ and its interior by $\mathbb{R}^n_{++}$. Given $x, y \in \mathbb{R}^n$, we write $x \leq y$ to indicate that $y - x \in \mathbb{R}^n_+$. The notation $\|\cdot\|$ is used to denote a norm. Most of the results in this paper are established for an arbitrary norm. However, certain norms do play a special role. Given $x \in \mathbb{R}^n$, we denote by $\|x\|_1$, $\|x\|_2$, and $\|x\|_\infty$ the 1–norm, 2–norm, and ∞–norm of x, respectively. A matrix $M \in \mathbb{R}^{n\times n}$ is said to be a P_0 matrix if all of its principal minors are non–negative. The matrix M is said to be an R_0

matrix if the problem LCP$(0, M)$ has the unique solution $(x, y) = (0, 0)$. If the matrix M is positive semi–definite, then the problem LCP(q, M) is said to be a monotone linear complementarity problem.

2 A NEIGHBORHOOD OF THE CENTRAL PATH

Let $\|\cdot\|$ be a given norm on $\mathbb{R}^n$. By the equivalence of norms on $\mathbb{R}^n$, there exist positive constants n_l and n_u such that

$$n_l \|x\|_\infty \leq \|x\| \leq n_u \|x\|_\infty, \quad \text{for all} \quad x \in \mathbb{R}^n. \tag{2.1}$$

For example, when $\|.\|$ is the ∞-norm, $n_l = n_u = 1$, and when $\|.\|$ is the 2-norm, $n_l = 1$ and $n_u = \sqrt{n}$. We take the set

$$\mathcal{N}(\beta) := \left\{ (x, y) \;\middle|\; \begin{array}{l} Mx - y + q = 0, \ \Phi(x, y, \mu) \leq 0, \\ \|\Phi(x, y, \mu)\| \leq \beta\mu \text{ for some } \mu > 0 \end{array} \right\}, \tag{2.2}$$

as our neighborhood of the central path, where $\beta > 0$ is given. This neighborhood can be viewed as the union of the *slices*

$$\mathcal{N}(\beta, \mu) := \{ (x, y) \, : \, Mx - y + q = 0, \ \Phi(x, y, \mu) \leq 0, \ \|\Phi(x, y, \mu)\| \leq \beta\mu \} \tag{2.3}$$

for $\mu > 0$. When the norm is chosen to be the 2-norm, the neighborhood is reduced to the one studied by Burke and Xu [3].

The neighborhood (2.2) refines the neighborhood concept introduced in [2] by requiring that all points in the neighborhood satisfy the additional inequality $\Phi(x, y, \mu) \leq 0$. It will be shown that if the algorithm is initiated in this neighborhood, then the inequality $\Phi(x, y, \mu) \leq 0$ is automatically satisfied at subsequent iterates. Hence the addition of this inequality does not complicate the structure of the algorithm. In the monotone case, this inequality is key to establishing the boundedness of the iterates. However, the boundedness of the iterates can be assured in a number of ways. For example, the assumption that the matrix M is an R_0 matrix also suffices. Thus, in order to keep the discussion at a general level, we introduce the following boundedness hypothesis.

Hypothesis (A): For any $\beta > 0$ and $\mu_0 > 0$, the set

$$\bigcup_{0<\mu\leq\mu_0} \mathcal{N}(\beta, \mu)$$

is bounded

Lemma 2.1 *[2, Proposition 2.4] If M is an R_0 matrix, then for every $\beta > 0$ and $\mu_0 > 0$ the set*

$$\bigcup_{0<\mu\leq\mu_0} \{ (x, y) \, : \, Mx - y + q = 0, \ \|\Phi(x, y, \mu)\| \leq \beta\mu \}$$

is bounded.

Remark. Clearly, the boundedness of the set given in Lemma 2.1 implies the boundedness of the set given in Hypothesis (A).

Lemma 2.2 *Assume that the problem LCP(q, M) is monotone and has a feasible interior point. Then for any $\beta > 0$ and $\mu_0 > 0$, the set*

$$\bigcup_{0<\mu\leq\mu_0} \mathcal{N}(\beta,\mu)$$

is bounded. Indeed, for any $(x,y) \in \bigcup_{0<\mu\leq\mu_0} \mathcal{N}(\beta,\mu)$, we have for $i = 1,2,\ldots,n$

$$-\frac{\beta\mu_0}{2n_l} \leq x_i \leq \frac{\bar{x}^T\bar{y} + \frac{\beta\mu_0}{2n_l}(\|\bar{x}\|_1 + \|\bar{y}\|_1) + n\max\{\mu_0^2, \frac{\beta^2\mu_0^2}{4n_l^2}\}}{\bar{y}_i}$$

$$-\frac{\beta\mu_0}{2n_l} \leq y_i \leq \frac{\bar{x}^T\bar{y} + \frac{\beta\mu_0}{2n_l}(\|\bar{x}\|_1 + \|\bar{y}\|_1) + n\max\{\mu_0^2, \frac{\beta^2\mu_0^2}{4n_l^2}\}}{\bar{x}_i},$$

where $(\bar{x},\bar{y})$ is any feasible interior point, that is, a point satisfying

$$M\bar{x} - \bar{y} + q = 0, \quad \bar{x} > 0, \; \bar{y} > 0.$$

Proof. Let $\beta > 0$, $0 < \mu \leq \mu_0$, and $(x,y) \in \mathcal{N}(\beta,\mu)$ be given, and let $(\bar{x},\bar{y})$ be a feasible interior point for LCP(q,M). First observe that if $-\delta \leq \phi(a,b,\mu)$, then $-\delta/2 < \min\{a,b\}$. To see this, note that the condition $-\delta \leq \phi(a,b,\mu)$ implies that

$$0 < \sqrt{[(a+\delta/2)-(b+\delta/2)]^2 + 4\mu^2} \leq (a+\delta/2)+(b+\delta/2). \tag{2.4}$$

Squaring both sides and cleaning up yields $0 < \mu^2 \leq (a+\delta/2)(b+\delta/2)$. Thus, since at least one of $(a+\delta/2)$ and $(b+\delta/2)$ must be positive by (2.4), both must be positive yielding $-\delta/2 < \min\{a,b\}$. It follows from (2.1) that

$$\|\Phi(x,y,\mu)\|_\infty \leq \frac{1}{n_l}\|\Phi(x,y,\mu)\| \leq \frac{\beta\mu}{n_l}.$$

This observation implies that

$$x_i > -\frac{\beta\mu}{2n_l} \geq -\frac{\beta\mu_0}{2n_l}, \text{ and } y_i > -\frac{\beta\mu}{2n_l} \geq -\frac{\beta\mu_0}{2n_l}, \text{ for } i = 1,2,\ldots,n\,. \tag{2.5}$$

Next, note that if $0 \leq a$ and $0 \leq b$, then the inequality $\phi(a,b,\mu) \leq 0$ implies that $0 \leq a+b \leq \sqrt{(a-b)^2+4\mu^2}$. Again, by squaring and cleaning up, we see that this gives $ab \leq \mu^2$. This observation implies that

$$x_iy_i \leq \mu_0^2 \text{ for each } i \in \{1,\ldots,n\} \text{ with } 0 < x_i,\; 0 < y_i\,. \tag{2.6}$$

We conclude the proof by noting that monotonicity yields $0 \leq (\bar{x}-x)^T(\bar{y}-y)$, or equivalently $\bar{x}^T y + \bar{y}^T x \leq \bar{x}^T \bar{y} + x^T y$. This inequality plus those in (2.5) and (2.6) yield the bound

$$\begin{aligned}
\sum_{y_i>0} \bar{x}_i y_i + \sum_{x_i>0} \bar{y}_i x_i &\leq \bar{x}^T \bar{y} + x^T y - \left[\sum_{y_i<0} \bar{x}_i y_i + \sum_{x_i<0} \bar{y}_i x_i\right] \\
&\leq \bar{x}^T \bar{y} + x^T y + \frac{\beta\mu_0}{2n_l}(\|\bar{x}\|_1 + \|\bar{y}\|_1) \\
&\leq \bar{x}^T \bar{y} + \sum_{\substack{x_i>0 \\ y_i>0}} x_i y_i + \sum_{\substack{x_i<0 \\ y_i<0}} x_i y_i + \frac{\beta\mu_0}{2n_l}(\|\bar{x}\|_1 + \|\bar{y}\|_1) \\
&\leq \bar{x}^T \bar{y} + n \max\{\mu_0^2, \frac{\beta^2\mu_0^2}{4n_l^2}\} + \frac{\beta\mu_0}{2n_l}(\|\bar{x}\|_1 + \|\bar{y}\|_1) .
\end{aligned}$$

It follows that if $y_i > 0$, then

$$y_i \leq \frac{\bar{x}^T \bar{y} + \frac{\beta\mu_0}{2n_l}(\|\bar{x}\|_1 + \|\bar{y}\|_1) + n \max\{\mu_0^2, \frac{\beta^2\mu_0^2}{4n_l^2}\}}{\bar{x}_i},$$

and, if $x_i > 0$, then

$$x_i \leq \frac{\bar{x}^T \bar{y} + \frac{\beta\mu_0}{2n_l}(\|\bar{x}\|_1 + \|\bar{y}\|_1) + n \max\{\mu_0^2, \frac{\beta^2\mu_0^2}{4n_l^2}\}}{\bar{y}_i}.$$

∎

An important property of the neighborhood $\mathcal{N}(\beta)$ that distinguishes it from its counter part in the interior point literature is that the solution set

$$\mathcal{S} = \{(x, y) : Mx - y + q = 0, x \geq 0, y \geq 0, x^T y = 0\}, \tag{2.7}$$

is contained in the interior of the slice $\mathcal{N}(\beta, \mu)$ relative to the affine set

$$\Lambda = \{(x, y) : Mx - y + q = 0\}, \tag{2.8}$$

for all $\mu > 0$ and $\beta > 2n_u$. This property partially explains why the non-interior path-following method can be initiated from any point in the space and why locally the Newton predictor steps are so efficient.

Theorem 2.1 *For any $\mu > 0$ and $\beta > 2n_u$, the solution set S is contained in the interior of the slice $\mathcal{N}(\beta, \mu)$ relative to the affine set Λ.*

Proof. Let

$$\mathcal{N}_\infty(\beta, \mu) := \{ (x, y) : Mx - y + q = 0, \ \Phi(x, y, \mu) \leq 0, \ \|\Phi(x, y, \mu)\|_\infty \leq \frac{\beta}{n_u}\mu\} . \tag{2.9}$$

It follows from (2.1) that

$$\mathcal{N}_\infty(\beta,\mu) \subseteq \mathcal{N}(\beta,\mu).$$

Thus, it suffices to prove that S is contained in the interior of the set $\mathcal{N}_\infty(\beta,\mu)$ relative to the affine set Λ.

First note that if $x_i y_i \le \mu^2$, then

$$\begin{aligned}\phi(x_i,y_i,\mu) &= x_i + y_i - \sqrt{(x_i-y_i)^2+4\mu^2} \\ &= x_i + y_i - \sqrt{(x_i+y_i)^2+4(\mu^2-x_iy_i)} \\ &\le 0.\end{aligned}$$

Also, if $x_i \ge 0, y_i \ge 0$ and $x_i y_i = 0$, then either $x_i = 0$ or $y_i = 0$. If $x_i = 0$ and $y_i \ge 0$, then

$$\begin{aligned}\phi(x_i,y_i,\mu) &= x_i + y_i - \sqrt{(x_i-y_i)^2+4\mu^2} \\ &= y_i - \sqrt{y_i^2+4\mu^2} \\ &\ge y_i - (y_i+2\mu) \\ &= -2\mu > -\frac{\beta}{n_u}\mu.\end{aligned}$$

Similarly, if $y_i = 0$ and $x_i \ge 0$, then again $\phi(x_i,y_i,\mu) > -\frac{\beta}{n_u}\mu$. Now if $(x^*,y^*) \in S$, then by the continuity of function ϕ, there is a $\delta > 0$ such that for all (x,y) in

$$\mathcal{O}(\delta) = \{(x,y) : \|x-x^*\|_\infty \le \delta,\ \ \|y-y^*\|_\infty \le \delta\}$$

we have $\phi(x_i,y_i,\mu) \ge -\frac{\beta}{n_u}\mu$ and $x_i y_i \le \mu^2$ for all $i = 1,\ldots,n$. Therefore for any $(x,y) \in \mathcal{O}(\delta)$, we have

$$\Phi(x,y,\mu) \le 0,\ \ \|\Phi(x,y,\mu)\|_\infty \le \frac{\beta}{n_u}\mu,$$

and so (x^*,y^*) is in the interior of the set $\mathcal{N}_\infty(\beta,\mu)$ relative to the affine set Λ. ∎

To illustrate this property, consider the problem LCP(q,M) given in [8, Example 5.1], where

$$M = \begin{bmatrix} 1 & 2 \\ 2 & 5 \end{bmatrix},\quad q = \begin{bmatrix} -1 \\ -1 \end{bmatrix}.$$

The unique solution of this problem is $(x_1,x_2) = (1,0)$ and $(y_1,y_2) = (0,1)$. For $\mu > 0$, let

$$\mathcal{N}_x(\beta,\mu) := \{\,(x_1,x_2) : \exists y \text{ such that } Mx - y + q = 0,\ \Phi(x,y,\mu) \le 0, \\ \|\Phi(x,y,\mu)\|_\infty \le \beta\mu\}$$

Figure 1.1 The nested slices of $\mathcal{N}_x(\beta, \mu)$ with $\beta = 4$ and $\mu = 15, 10, 5$.

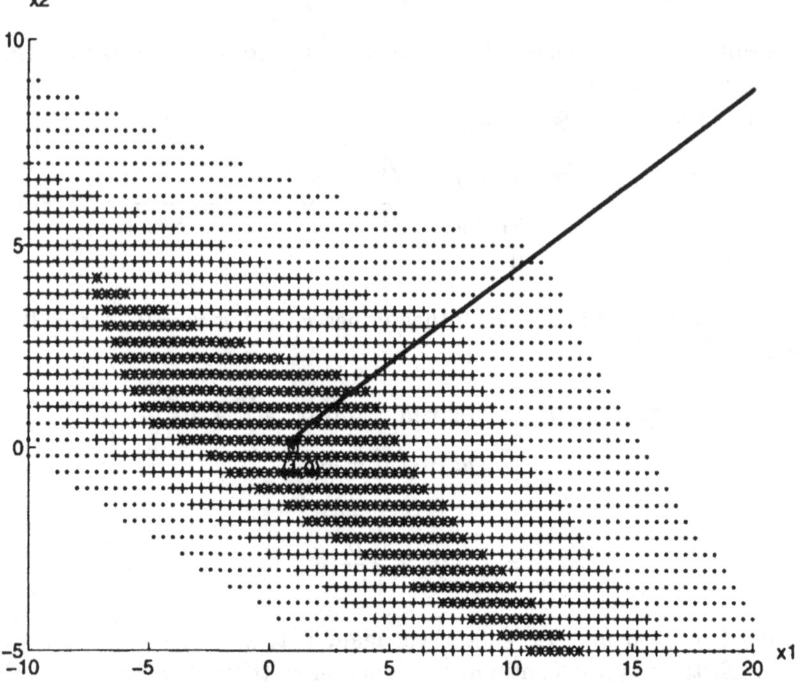

be the projection of the slice $\mathcal{N}(\beta, \mu)$ onto the x-coordinate. In figure 1, several slices of $\mathcal{N}_x(\beta, \mu)$ are drawn with $\beta = 4$ and $\mu = 15, 10, 5$ respectively. Note that the solution $(x_1, x_2) = (1, 0)$ is in the interior of these slices.

We conclude this section by cataloging a few technical properties of the function $\phi(a, b, \mu)$ for later use.

Lemma 2.3 *The function ϕ defined in (1.4) has the following properties:*

(i) *[14] The function $\phi(a, b, \mu)$ is continuously differentiable on $\mathbb{R}^2 \times \mathbb{R}_{++}$.*

(ii) *[3, Lemma 2.2] The function $\phi(a, b, \mu)$ is concave on $\mathbb{R}^2 \times \mathbb{R}_{++}$,*

(iii) *[17, Lemma 2] For any $(a, b, \mu) \in \mathbb{R}^2 \times R_{++}$, we have*

$$\left\|\nabla^2 \phi(a, b, \mu)\right\|_2 \leq \frac{4}{\sqrt{(a-b)^2 + 4\mu^2}} \leq \frac{2}{\mu}.$$

3 A PREDICTOR-CORRECTOR ALGORITHM

The Algorithm: [3]

Step 0: (Initialization)

Choose $x^0 \in I\!R^n$, set $y^0 = Mx^0 + q$, and let $\mu_0 > 0$ be such that $\Phi(x^0, y^0, \mu_0) < 0$. Choose $\beta > 2n_u$ so that $\|\Phi(x^0, y^0, \mu_0)\| \le \beta\mu_0$. We now have $(x^0, y^0) \in \mathcal{N}(\beta, \mu_0)$. Choose $\bar{\sigma}$, α_1, and α_2 from $(0,1)$.

Step 1: (The Predictor Step)

Let $(\Delta x^k, \Delta y^k, \Delta\mu_k)$ solve the equation

$$F(x^k, y^k, \mu_k) + \nabla F(x^k, y^k, \mu_k)^T \begin{bmatrix} \Delta x^k \\ \Delta y^k \\ \Delta\mu_k \end{bmatrix} = 0\,. \tag{3.1}$$

If $\|\Phi(x^k + \Delta x^k, y^k + \Delta y^k, 0)\| = 0$, **STOP**, $(x^k + \Delta x^k, y^k + \Delta y^k)$ solves LCP(q, M) ; else if

$$\|\Phi(x^k + \Delta x^k, y^k + \Delta y^k, \mu_k)\| > \beta\mu_k,$$

set

$$\hat{x}^k := x^k, \quad \hat{y}^k := y^k, \quad \hat{\mu}_k := \mu_k, \quad \text{and} \quad \eta_k = 1\,; \tag{3.2}$$

else let $\eta_k = \alpha_1^s$ where s is the positive integer such that

$$\|\Phi(x^k + \Delta x^k, y^k + \Delta y^k, \alpha_1^t\mu_k)\| \le \alpha_1^t\beta\mu_k, \tag{3.3}$$

for $t = 0, 1, \ldots, s$, and

$$\|\Phi(x^k + \Delta x^k, y^k + \Delta y^k, \alpha_1^{s+1}\mu_k)\| > \alpha_1^{s+1}\beta\mu_k. \tag{3.4}$$

Set

$$\hat{x}^k := x^k + \Delta x^k, \quad \hat{y}^k := y^k + \Delta y^k, \quad \hat{\mu}_k := \eta_k\mu_k. \tag{3.5}$$

Step 2: (The Corrector Step)

Let $(\Delta\hat{x}^k, \Delta\hat{y}^k, \Delta\hat{\mu}_k)$ solve the equation

$$F(\hat{x}^k, \hat{y}^k, \hat{\mu}_k) + \nabla F(\hat{x}^k, \hat{y}^k, \hat{\mu}_k)^T \begin{bmatrix} \Delta\hat{x}^k \\ \Delta\hat{y}^k \\ \Delta\hat{\mu}_k \end{bmatrix} = \begin{bmatrix} 0 \\ 0 \\ (1-\bar{\sigma})\hat{\mu}_k \end{bmatrix} \tag{3.6}$$

and let $\hat{\lambda}_k$ be the maximum of the value $1, \alpha_2, \alpha_2^2, \ldots$, such that

$$\left\|\Phi(\hat{x}^k + \hat{\lambda}_k\Delta\hat{x}^k, \hat{y}^k + \hat{\lambda}_k\Delta\hat{y}^k, (1-\bar{\sigma}\hat{\lambda}_k)\hat{\mu}_k)\right\| \le (1-\bar{\sigma}\hat{\lambda}_k)\beta\hat{\mu}_k. \tag{3.7}$$

Set

$$x^{k+1} = \hat{x}^k + \hat{\lambda}_k\Delta\hat{x}^k, \quad y^{k+1} = \hat{y}^k + \hat{\lambda}_k\Delta\hat{y}^k, \quad \mu_{k+1} = (1-\bar{\sigma}\hat{\lambda}_k)\hat{\mu}_k, \tag{3.8}$$

and return to Step 1.

Remarks.

- In [3], the algorithm is stated using the 2–norm. It will be shown that the algorithm and its convergence analysis remain valid regardless of the choice of norm.

- Note that if the null step (3.2) is taken in Step 1, then the Newton equations (3.1) and (3.6) have the same coefficient matrix. Therefore only one matrix factorization is needed to implement both Steps 1 and 2 in this case. Otherwise, two matrix factorizations are needed.

- The algorithm can be modified so that only one matrix factorization is needed with the modified algorithm preserving the same nice convergence properties. The modification goes as follows. If $s = 0$ in the predictor step, then use the update

$$\hat{x}^k := x^k, \hat{y}^k := y^k, \hat{\mu}_k := \mu_k$$

instead of (3.5). On the other hand, if $s \geq 1$, use the update (3.5) and skip the corrector step. In other words, use the update

$$x^{k+1} = \hat{x}^k, y^{k+1} = \hat{y}^k, \mu_{k+1} = \hat{\mu}_k,$$

in the corrector step instead of (3.8).

- In the initialization step, setting

$$\mu_0 > \sqrt{\max_{\substack{i \in \{1,\ldots,n\} \\ 0 < x_i^0,\ 0 < y_i^0}} x_i^0 y_i^0}$$

guarantees that the inequality $\Phi(x^0, y^0, \mu_0) < 0$ is satisfied. For example, one can choose $(x^0, y^0) = (0, q)$ in which case μ_0 can be taken to be any positive number.

- The condition that $\beta > 2n_u$ is only employed in the proof of local quadratic convergence. It is not required to verify the global linear convergence of the method. Theorem 2.1 motivates why this condition on β is required. Recall that when β is chosen in this way, the solution set is contained in the interior of the slices $\mathcal{N}(\beta, \mu)$ relative to the affine set Λ. Thus, eventually the Newton iterate associated with the predictor step remains in the interior of the current slice of the central path relative to the affine set Λ. Hence a full Newton step can be taken yielding the local quadratic convergence of the method.

- Observe that the function F has nonsingular Jacobian at a given point if and only if $\nabla_{(x,y)}F$ is nonsingular at that point. In [14, Theorem 3.5], it is shown that if $\mu > 0$ and the matrix M is a P_0 matrix, then $\nabla_{(x,y)}F(\bar{x}, \bar{y}, \mu)$ is nonsingular for all $(\bar{x}, \bar{y}) \in I\!R^{2n}$. Therefore, since the matrix M is assumed to be positive semi–definite and the algorithm is initiated with $\mu_0 > 0$, the Jacobians $\nabla F(x^k, y^k, \mu_k)$ and $\nabla F(\hat{x}^k, \hat{y}^k, \hat{\mu}_k)$ are always nonsingular. Hence the Newton equations (3.1) and (3.6) yield unique solutions whenever (x^k, y^k, μ_k) and $(\hat{x}^k, \hat{y}^k, \hat{\mu}_k)$ are well-defined. In addition, since $y^0 = Mx^0 + q$, we have $y^k = Mx^k + q$ and $\hat{y}^k = M\hat{x}^k + q$ for all well-defined iterates.

In analyzing the algorithm, it is helpful to take a closer look at the Newton equations (3.1) and (3.6). In (3.1) we have $\Delta\mu_k = -\mu_k$ and so (3.1) reduces to the system

$$\begin{aligned} M\Delta x^k - \Delta y^k &= 0 \\ \nabla_x \Phi(x^k, y^k, \mu_k)\Delta x^k &+ \nabla_y \Phi(x^k, y^k, \mu_k)\Delta y^k \\ &= -\Phi(x^k, y^k, \mu_k) + \mu_k \nabla_\mu \Phi(x^k, y^k, \mu_k). \end{aligned} \tag{3.9}$$

Similarly, in (3.6), $\Delta\hat{\mu}_k = -\bar{\sigma}\hat{\mu}_k$ reducing (3.6) to the system

$$\begin{aligned} M\Delta \hat{x}^k - \Delta \hat{y}^k &= 0 \\ \nabla_x \Phi(\hat{x}^k, \hat{y}^k, \hat{\mu}_k)\Delta \hat{x}^k &+ \nabla_y \Phi(\hat{x}^k, \hat{y}^k, \hat{\mu}_k)\Delta \hat{y}^k \\ &= -\Phi(\hat{x}^k, \hat{y}^k, \hat{\mu}_k) + \bar{\sigma}\hat{\mu}_k \nabla_\mu \Phi(\hat{x}^k, \hat{y}^k, \hat{\mu}_k). \end{aligned} \tag{3.10}$$

Theorem 3.1 *Consider the algorithm described above and suppose that the matrix M is a P_0 matrix. If $(x^k, y^k) \in \mathcal{N}(\beta, \mu_k)$ with $\mu_k > 0$, then either $(x^k + \Delta x^k, y^k + \Delta y^k)$ solves LCP(q, M) or both $(\hat{x}^k, \hat{y}^k, \hat{\mu}_k)$ and $(x^{k+1}, y^{k+1}, \mu_{k+1})$ are well-defined with the backtracking routines in Steps 1 and 2 finitely terminating. In the latter case, we have $(\hat{x}^k, \hat{y}^k) \in \mathcal{N}(\beta, \hat{\mu}_k)$ and $(x^{k+1}, y^{k+1}) \in \mathcal{N}(\beta, \mu_{k+1})$ with $0 < \mu_{k+1} < \hat{\mu}_k \leq \mu_k$. Since $(x^0, y^0) \in \mathcal{N}(\beta, \mu_0)$ with $\mu_0 > 0$, this shows that the algorithm is well-defined.*

Proof. Let $(x^k, y^k) \in \mathcal{N}(\beta, \mu_k)$ with $\mu_k > 0$. By the last remark given above, $(\Delta x^k, \Delta y^k, \Delta\mu_k)$ exists and is unique. Since $y^k + \Delta y^k = M(x^k + \Delta x^k) + q$, we have

$$\|\Phi(x^k + \Delta x^k, y^k + \Delta y^k, 0)\| = 0$$

if and only if $(x^k + \Delta x^k, y^k + \Delta y^k)$ solves LCP(q, M). Therefore, if $(x^k + \Delta x^k, y^k + \Delta y^k)$ does not solve LCP(q, M), then by continuity, there exist $\epsilon > 0$ and $\bar{\mu} > 0$ such that $\|\Phi(x^k + \Delta x^k, y^k + \Delta y^k, \mu)\| > \epsilon$ for all $\mu \in [0, \bar{\mu}]$. In this case, the backtracking routine described in (3.3) and (3.4) of Step 1 is finitely terminating. Hence $(\hat{x}^k, \hat{y}^k, \hat{\mu}_k)$ is well-defined, with $0 < \hat{\mu}_k \leq \mu_k$, and $(\Delta\hat{x}^k, \Delta\hat{y}^k, \Delta\hat{\mu}_k)$ is uniquely determined by (3.6). To see that the backtracking routine in Step 2 is finitely terminating, define $\theta(x, y, \mu) = \|\Phi(x, y, \mu)\|$. This is a convex composite function [1]. By (3.6)

$$\begin{aligned} &\theta'((\hat{x}^k, \hat{y}^k, \hat{\mu}_k); (\Delta\hat{x}^k, \Delta\hat{y}^k, \Delta\hat{\mu}_k)) \\ = &\inf_{\lambda > 0} \lambda^{-1} \left(\left\| \Phi(\hat{z}^k) + \lambda \nabla\Phi(\hat{z}^k)^T \begin{bmatrix} \Delta\hat{x}^k \\ \Delta\hat{y}^k \\ \Delta\hat{\mu}_k \end{bmatrix} \right\| - \|\Phi(\hat{z}^k)\| \right) \end{aligned}$$

$$
\begin{aligned}
&\leq \left\| \Phi(\hat{z}^k) + \nabla\Phi(\hat{x}^k)^T \begin{bmatrix} \Delta\hat{x}^k \\ \Delta\hat{y}^k \\ \Delta\hat{\mu}_k \end{bmatrix} \right\| - \left\|\Phi(\hat{z}^k)\right\| \\
&= -\left\|\Phi(\hat{z}^k)\right\| \\
&< 0,
\end{aligned}
$$

where $\hat{z}^k = (\hat{x}^k, \hat{y}^k, \hat{\mu}_k)$. Therefore, (3.7) can be viewed as an instance of a standard backtracking line search routine and as such is finitely terminating with $0 < \mu_{k+1} < \hat{\mu}_k \leq \mu_k$ (indeed, one can replace the value of $\bar{\sigma}$ on the right hand side of (3.7) by any number in the open interval $(0,1)$).

Since $(x^k, y^k) \in \mathcal{N}(\beta, \mu_k)$, the argument given above implies that either $(x^k + \Delta x^k, y^k + \Delta y^k)$ solves LCP(q, M) or $\hat{y}^k = M\hat{x}^k + q$ with $\left\|\Phi(\hat{x}^k, \hat{y}^k, \hat{\mu}_k)\right\| \leq \beta\hat{\mu}_k$ and $y^{k+1} = Mx^{k+1} + q$ with $\left\|\Phi(x^{k+1}, y^{k+1}, \mu_{k+1})\right\| \leq \beta\mu_{k+1}$. Thus, if $(x^k + \Delta x^k, y^k + \Delta y^k)$ does not solve LCP(q, M), we need only show that $\Phi(\hat{x}^k, \hat{y}^k, \hat{\mu}_k) \leq 0$ and $\Phi(x^{k+1}, y^{k+1}, \mu_{k+1}) \leq 0$ in order to have $(\hat{x}^k, \hat{y}^k) \in \mathcal{N}(\beta, \hat{\mu}_k)$ and $(x^{k+1}, y^{k+1}) \in \mathcal{N}(\beta, \mu_{k+1})$. First note that the componentwise concavity of Φ implies that for any $(x, y, \mu) \in \mathbb{R}^{2n+1}$ with $\mu > 0$, and $(\Delta x, \Delta y, \Delta\mu) \in \mathbb{R}^{2n+1}$ one has

$$
\Phi(x + \Delta x, y + \Delta y, \mu + \Delta\mu) \leq \Phi(x, y, \mu) + \nabla\Phi(x, y, \mu)^T \begin{bmatrix} \Delta x \\ \Delta y \\ \Delta\mu \end{bmatrix} .
$$

Hence, in the case of the predictor step, either (3.2) holds or

$$
\begin{aligned}
&\Phi(\hat{x}^k, \hat{y}^k, \hat{\mu}_k) \\
&= \Phi(x^k + \Delta x^k, y^k + \Delta y^k, \eta_k\mu_k) \\
&\leq \Phi(x^k, y^k, \mu_k) + \nabla\Phi(x^k, y^k, \mu_k)^T \begin{pmatrix} \Delta x^k \\ \Delta y^k \\ (\eta_k - 1)\mu_k \end{pmatrix} \\
&= \Phi(x^k, y^k, \mu_k) + \nabla\Phi(x^k, y^k, \mu_k)^T \begin{pmatrix} \Delta x^k \\ \Delta y^k \\ -\mu_k \end{pmatrix} + \eta_k\mu_k\nabla_\mu\Phi(x^k, y^k, \mu_k) \\
&= \eta_k\mu_k\nabla_\mu\Phi(x^k, y^k, \mu_k) \leq 0.
\end{aligned}
$$

In either case, $\Phi(\hat{x}^k, \hat{y}^k, \hat{\mu}_k) \leq 0$. For the corrector step, we have

$$
\begin{aligned}
&\Phi(x^{k+1}, y^{k+1}, \mu_{k+1}) \\
&= \Phi(\hat{x}^k + \hat{\lambda}_k\Delta\hat{x}^k, \hat{y}^k + \hat{\lambda}_k\Delta\hat{y}^k, \hat{\mu}_k + \hat{\lambda}_k\Delta\hat{\mu}_k) \\
&\leq \Phi(\hat{x}^k, \hat{y}^k, \hat{\mu}_k) + \hat{\lambda}_k\nabla\Phi(\hat{x}^k, \hat{y}^k, \hat{\mu}_k)^T \begin{bmatrix} \Delta\hat{x}^k \\ \Delta\hat{y}^k \\ -\bar{\sigma}\hat{\mu}_k \end{bmatrix} \\
&= (1 - \hat{\lambda}_k)\Phi(\hat{x}^k, \hat{y}^k, \hat{\mu}_k) \leq 0,
\end{aligned}
$$

since we have already shown that $\Phi(\hat{x}^k, \hat{y}^k, \hat{\mu}_k) \leq 0$. This completes the proof. ∎

4 CONVERGENCE

In this section, we require the following hypothesis in order to establish the global linear convergence of the method.

Hypothesis (B): Given $\beta > 0$ and $\mu_0 > 0$, there exists a $C > 0$ such that

$$\left\|\nabla_{(x,y)}F(\bar{x},\bar{y},\mu)^{-1}\right\|_\infty \leq C, \tag{4.1}$$

for all $0 < \mu \leq \mu_0$ and $(\bar{x},\bar{y}) \in \mathcal{N}(\beta,\mu)$.

In [2, Proposition 4.3], we show that a bound of this type exits under the assumption of a non–degeneracy condition due to Fukushima, Luo, and Pang [11, Assumption (A2)]. Similar results of this type have since been obtained by Chen and Xiu [7, Section 6], Tseng [20, Corollary 2], and Qi and Sun [17, Proposition 2].

In private discussions, Kanzow [15] points out that the Fukushima, Luo, and Pang non–degeneracy condition implies the uniqueness of the solution to LCP(q,M). Kanzow's proof easily extends to show that *any* condition which implies Assumption (B) also implies the uniqueness of the solution to LCP(q,M).

Proposition 4.1 *[3, Proposition 4.1] If Assumptions (A) and (B) hold, then LCP(q,M) has a unique solution.*

Theorem 4.1 *(Global Linear Convergence) Suppose that M is a P_0 matrix and hypotheses (A) and (B) hold. Let $\{(x^k,y^k,\mu_k)\}$ be the sequence generated by the algorithm. If the algorithm does not terminate finitely at the unique solution to LCP(q,M), then for $k = 0,1,\ldots,$*

$$(x^k,y^k) \in \mathcal{N}(\beta,\mu_k) \tag{4.2}$$

$$(1-\bar{\sigma}\hat{\lambda}_{k-1})\eta_{k-1}\cdots(1-\bar{\sigma}\hat{\lambda}_0)\eta_0\mu_0 \;=\; \mu_k, \tag{4.3}$$

with

$$\hat{\lambda}_k \geq \bar{\lambda} := \min\left\{1, \frac{\alpha_2(1-\bar{\sigma})\beta}{n_u(2C^2(\frac{\beta}{n_l}+2\bar{\sigma})^2+\bar{\sigma}^2)+\bar{\sigma}(1-\bar{\sigma})\beta}\right\}, \tag{4.4}$$

where C is the constant defined in (4.1). Therefore μ_k converges to 0 at a global linear rate. In addition, the sequence $\{(x^k,y^k)\}$ converges to the unique solution of LCP(q,M).

Proof. The inclusion (4.2) has already been established in Theorem 3.1 and the relation (4.3) follows by construction.

For the sake of simplicity, set $(x,y,\mu) = (\hat{x}_k,\hat{y}_k,\hat{\mu}_k)$ and $(\Delta x,\Delta y) = (\Delta\hat{x}^k,\Delta\hat{y}^k)$. Then for $i \in \{1,\ldots,n\}$ and $\lambda \in [0,1]$, Lemma 2.3 and (3.6) imply that

$$|\phi(x_i+\lambda\Delta x_i, y_i+\lambda\Delta y_i, (1-\bar{\sigma}\lambda)\mu)|$$

$$
\begin{aligned}
&= \left| \phi(x_i, y_i, \mu) + \lambda \nabla \phi(x_i, y_i, \mu)^T \begin{pmatrix} \Delta x_i \\ \Delta y_i \\ -\bar{\sigma}\mu \end{pmatrix} + \right. \\
&\quad \left. \frac{\lambda^2}{2} \begin{pmatrix} \Delta x_i \\ \Delta y_i \\ -\bar{\sigma}\mu \end{pmatrix}^T \nabla^2 \phi(x_i + \theta_i \lambda \Delta x_i, y_i + \theta_i \lambda \Delta y_i, (1 - \theta_i \bar{\sigma} \lambda)\mu) \begin{pmatrix} \Delta x_i \\ \Delta y_i \\ -\bar{\sigma}\mu \end{pmatrix} \right| \\
&\le (1 - \lambda)|\phi(x_i^k, y_i^k, \mu)| + \\
&\quad \frac{\lambda^2}{2} \left\| \nabla^2 \phi(x_i + \theta_i \lambda \Delta x_i, y_i + \theta_i \lambda \Delta y_i, (1 - \theta_i \bar{\sigma} \lambda)\mu) \right\|_2 \left\| \begin{pmatrix} \Delta x_i \\ \Delta y_i \\ -\bar{\sigma}\mu \end{pmatrix} \right\|_2^2 \\
&\le (1 - \lambda)|\phi(x_i, y_i, \mu)| + \frac{\lambda^2}{(1 - \bar{\sigma}\lambda)\mu} \|(\Delta x_i, \Delta y_i, -\bar{\sigma}\mu)\|_2^2
\end{aligned}
$$

for some $\theta_i \in [0, 1]$. Set $t_i := \|(\Delta x_i, \Delta y_i, -\bar{\sigma}\mu_k)\|_2^2$ for $i = 1, \ldots, n$, then

$$
\begin{aligned}
&\|\Phi(x + \lambda \Delta x, y + \lambda \Delta y, (1 - \bar{\sigma}\lambda)\mu)\| \\
&\le (1 - \lambda) \|\Phi(x, y, \mu)\| + \frac{\lambda^2}{(1 - \bar{\sigma}\lambda)\mu} \left\| \begin{pmatrix} t_1 \\ \cdots \\ t_n \end{pmatrix} \right\| \\
&\le (1 - \lambda) \|\Phi(x, y, \mu)\| + \frac{\lambda^2 n_u}{(1 - \bar{\sigma}\lambda)\mu} \left\| \begin{pmatrix} t_1 \\ \cdots \\ t_n \end{pmatrix} \right\|_\infty \\
&\le (1 - \lambda) \|\Phi(x, y, \mu)\| + \frac{\lambda^2 n_u}{(1 - \bar{\sigma}\lambda)\mu} \left(2 \left\| \begin{pmatrix} \Delta x \\ \Delta y \end{pmatrix} \right\|_\infty^2 + \bar{\sigma}^2 \mu^2 \right) \\
&\le (1 - \lambda)\beta\mu + \frac{\lambda^2 n_u}{1 - \bar{\sigma}\lambda} \left(2C^2 (\frac{\beta}{n_l} + 2\bar{\sigma})^2 + \bar{\sigma}^2 \right) \mu,
\end{aligned} \tag{4.5}
$$

where the last inequality follows from (3.9) and the bound

$$
\begin{aligned}
&\left\| \begin{pmatrix} \Delta x \\ \Delta y \end{pmatrix} \right\|_\infty \\
&\le \left\| \nabla_{(x,y)} F^{-1}(x, y, \mu) \right\|_\infty \left(\|\Phi(x, y, \mu)\|_\infty + \bar{\sigma}\mu \|\nabla_\mu \Phi(x, y, \mu)\|_\infty \right) \\
&\le C(\frac{\beta}{n_l} + 2\bar{\sigma})\mu.
\end{aligned} \tag{4.6}
$$

It is easily verified that

$$
(1 - \lambda)\beta\mu + \frac{\lambda^2 n_u}{1 - \bar{\sigma}\lambda} \left(2C^2 (\frac{\beta}{n_l} + 2\bar{\sigma})^2 + \bar{\sigma}^2 \right) \mu \le (1 - \bar{\sigma}\lambda)\beta\mu,
$$

whenever

$$
\lambda \le \frac{(1 - \bar{\sigma})\beta}{n_u (2C^2 (\frac{\beta}{n_l} + 2\bar{\sigma})^2 + \bar{\sigma}^2) + \bar{\sigma}(1 - \bar{\sigma})\beta}.
$$

Therefore

$$\hat{\lambda}_k \geq \min\left\{1, \frac{\alpha_2(1-\bar{\sigma})\beta}{n_u(2C^2(\frac{\beta}{n_l}+2\bar{\sigma})^2+\bar{\sigma}^2)+\bar{\sigma}(1-\bar{\sigma})\beta}\right\}.$$

To conclude, note that the sequence $\{(x^k, y^k)\}$ is bounded by hypotheses (A) and Theorem 3.1. In addition, just as in (4.6), the relations (3.9) and (3.10) yield the bounds

$$\left\|\begin{pmatrix} \Delta x^k \\ \Delta y^k \end{pmatrix}\right\|_\infty \leq C(\frac{\beta}{n_l}+2)\mu_k \quad \text{and} \quad \left\|\begin{pmatrix} \Delta \hat{x}^k \\ \Delta \hat{y}^k \end{pmatrix}\right\| \leq C(\frac{\beta}{n_l}+2)\mu_k\ ,$$

since $0 < \bar{\sigma} < 1$ and $0 < \eta_k \leq 1$ for all k. Therefore, (4.3) and (4.4) imply that

$$\begin{aligned} \|(x^{k+1}, y^{k+1}) - (x^k, y^k)\|_\infty &\leq \left\|\begin{pmatrix} \Delta x^k \\ \Delta y^k \end{pmatrix}\right\|_\infty + \hat{\lambda}_k \left\|\begin{pmatrix} \Delta \hat{x}^k \\ \Delta \hat{y}^k \end{pmatrix}\right\|_\infty \\ &\leq 2C(\frac{\beta}{n_l}+2)\mu_k \leq 2C(\frac{\beta}{n_l}+2)(1-\bar{\sigma}\bar{\lambda})^k \mu_0\ . \end{aligned}$$

Hence, $\{(x^k, y^k)\}$ is a Cauchy sequence and so must converge to the unique solution of LCP(q, M). ∎

Theorem 4.2 *(Local Quadratic Convergence) Suppose M is a P_0 matrix, assumption (B) holds and that the sequence $\{(x^k, y^k, \mu_k)\}$ generated by the algorithm converges to $\{(x^*, y^*, 0)\}$ where $\{(x^*, y^*)\}$ is the unique solution to LCP(q, M). If it is further assumed that the strict complementary slackness condition $0 < x^* + y^*$ is satisfied, then*

$$\mu_{k+1} = O(\mu_k^2), \tag{4.7}$$

that is, μ_k converges quadratically to zero.

Proof. First observe that due to the strict complementarity of $\{(x^*, y^*)\}$, Part (iii) of Lemma 2.3 indicates that there exist constants $\epsilon > 0$ and $L > 0$ such that

$$\left\|\nabla^2\phi(x, y, \mu)\right\|_2 \leq L, \quad \text{whenever} \quad \|(x, y, \mu) - (x^*, y^*, 0)\| \leq \epsilon. \tag{4.8}$$

Hence, for all k sufficient large and $\eta \in (0, 1]$, we have for each $i \in \{1, \ldots, n\}$ that

$$\begin{aligned} &|\phi(x_i^k + \Delta x_i^k, y_i^k + \Delta y_i^k, \eta\mu_k)| \\ = &\left|\phi(x_i^k, y_i^k, \mu_k) + \nabla^T\phi(x_i^k, y_i^k, \mu_k)\begin{pmatrix} \Delta x_i^k \\ \Delta y_i^k \\ (\eta - 1)\mu_k \end{pmatrix}\right| + \end{aligned}$$

$$
\begin{aligned}
& \frac{1}{2}\left|\begin{pmatrix} \Delta x_i^k \\ \Delta y_i^k \\ (\eta-1)\mu_k \end{pmatrix}^T \nabla^2\phi(z_i^k)\begin{pmatrix} \Delta x_i^k \\ \Delta y_i^k \\ (\eta-1)\mu_k \end{pmatrix}\right| \\
= & \left|\phi(x_i^k, y_i^k, \mu_k) + \nabla^T\phi(x_i^k, y_i^k, \mu_k)\begin{pmatrix} \Delta x_i^k \\ \Delta y_i^k \\ -\mu_k \end{pmatrix} + \eta\mu_k\nabla_\mu\phi(x^k, y^k, \mu_k)\right| + \\
& \frac{1}{2}\left|\begin{pmatrix} \Delta x_i^k \\ \Delta y_i^k \\ (\eta-1)\mu_k \end{pmatrix}^T \nabla^2\phi(z_i^k)\begin{pmatrix} \Delta x_i^k \\ \Delta y_i^k \\ (\eta-1)\mu_k \end{pmatrix}\right| \\
\le & \ \eta\mu_k|\nabla_\mu\phi(x^k, y^k, \mu_k)| + \frac{L}{2}\left\|(\Delta x_i^k, \Delta y_i^k, (\eta-1)\mu_k)\right\|_2^2 \\
= & \ 2\eta\mu_k + \frac{L}{2}\left(\left\|(\Delta x_i^k, \Delta y_i^k)\right\|_2^2 + (1-\eta)^2\mu_k^2\right)
\end{aligned}
$$

where $z_i^k = (x_i^k + \theta_i\Delta x_i^k, y_i + \theta_i\Delta y_i^k, (1+\theta_i(\eta-1))\mu_k)$. Now using an argument similar to that used to obtain (4.5), we have

$$
\begin{aligned}
& \left\|\Phi(x^k + \Delta x^k, y^k + \Delta y^k, \eta\mu_k)\right\| \\
\le & \ n_u\left\|\Phi(x^k + \Delta x^k, y^k + \Delta y^k, \eta\mu_k)\right\|_\infty \\
\le & \ n_u\left(2\eta\mu_k + \frac{L}{2}\left(2C^2(\frac{\beta}{n_l} + 2)^2 + 1\right)\mu_k^2\right) \\
\le & \ 2n_u\eta\mu_k + n_uL\left(C^2(\frac{\beta}{n_l} + 2)^2 + \frac{1}{2}\right)\mu_k^2. \qquad (4.9)
\end{aligned}
$$

Hence, since $\beta > 2n_u$, the inequality (3.3) in Step 1 of the algorithm holds with $t = 0$ for all k sufficiently large. It is easy to verify that

$$
2n_u\eta\mu_k + n_uL\left(C^2(\frac{\beta}{n_l} + 2)^2 + \frac{1}{2}\right)\mu_k^2 \le \eta\beta\mu_k, \qquad (4.10)
$$

whenever

$$
\eta \ge \frac{n_uL\left(C^2(\frac{\beta}{n_l} + 2)^2 + \frac{1}{2}\right)\mu_k}{\beta - 2n_u}.
$$

Hence, by (3.4), we have

$$
\alpha_1\eta_k \le \frac{n_uL\left(C^2(\frac{\beta}{n_l} + 2)^2 + \frac{1}{2}\right)\mu_k}{\beta - 2n_u},
$$

and so

$$
\eta_k \le \frac{n_uL\left(C^2(\frac{\beta}{n_l} + 2)^2 + \frac{1}{2}\right)\mu_k}{\alpha_1(\beta - 2n_u)} \qquad (4.11)
$$

for all k sufficient large. Therefore, by (3.5),

$$
\mu_{k+1} = O(\mu_k^2).
$$

■

We now show that under relatively mild condition, the algorithm is globally convergent.

Theorem 4.3 *Suppose that M is a P_0 matrix and hypothesis (A) holds. Let $\{x^k, y^k, \mu_k\}$ be a sequence generated by the Algorithm. Then*

(i) *the sequence $\{\mu_k\}$ is monotonically decreasing and convergent to 0 as $k \to \infty r$,*

(ii) *the sequence $\{(x^k, y^k)\}$ is bounded and every accumulation point of $\{(x^k, y^k)\}$ is a solution to LCP(q, M).*

Proof. Since M is a P_0 matrix, the algorithm is well defined. By the construction of the algorithm, we can see that $\mu_{k+1} < \mu_k$ for $(k = 0, 1, \ldots)$. Hence the sequence $\{\mu_k\}$ is monotonically decreasing. Since $\mu_k \geq 0$ $(k = 0, 1, \ldots)$, there is a $\bar{\mu} \geq 0$ such that $\mu_k \to \bar{\mu}$. Since $(x^k, y^k) \in \mathcal{N}(\beta, \mu_k)$, the sequence (x^k, y^k) is bounded, by taking a subsequence if necessary, we may assume that $\{(x^k, y^k)\}$ converges to some point $(\bar{x}, \bar{y})$. If $\bar{\mu} = 0$, it follows from $\{(x^k, y^k)\} \in \mathcal{N}(\beta, \mu_k)$ that $(\bar{x}, \bar{y})$ is a solution of the LCP and we obtain the desired results. Suppose that $\bar{\mu} > 0$. Since $\nabla_{(x,y)} F(\bar{x}, \bar{y}, \bar{\mu})$ is nonsingular, there exist $\epsilon > 0$, $L > 0$ and $C > 0$ such that

$$\left\|\nabla_{(x,y)} F(x, y, \mu)^{-1}\right\|_\infty \leq C, \tag{4.12}$$

$$\left\|\nabla^2 \phi(x_i, y_i, \mu)\right\|_2 \leq L, \tag{4.13}$$

for all $(x, y, \mu) \in \mathcal{O}_{(\bar{x}, \bar{y}, \bar{\mu})} = \{(x, y, \mu) : \|(x, y, \mu) - (\bar{x}, \bar{y}, \bar{\mu})\| \leq \epsilon\}$. Similar to the proof of the global linear convergence result, we can show that there exists a $\bar{\lambda}$ such that $\hat{\lambda}_k \geq \bar{\lambda}$ for sufficient large k. Therefore, for sufficient large k, $\mu_{k+1} \leq c\mu_k$ for some constant $c \in (0, 1)$, which yields a contradiction. ■

References

[1] J. Burke, *Descent methods for composite nondifferentiable optimization problems*, Mathematical Programming, 33 (1987), pp. 260—279.

[2] J. Burke and S. Xu, *The global linear convergence of a non-interior path-following algorithm for linear complementarity problem.* To appear in Mathematics of Operations Research.

[3] ——, *A non-interior predictor-corrector path following algorithm for the monotone linear complementarity problem.* Preprint, Department of Mathematics, University of Washington , Seattle, WA 98195, December, 1997.

[4] B. Chen and X. Chen, *A global and local super-linear continuation method for $P_0 + R_0$ and monotone NCP.* To appear in SIAM Journal on Optimization.

[5] ——, *A global linear and local quadratic continuation method for variational inequalities with box constraints.* Preprint, Department of Management and Systems, Washington State University, Pullman, WA 99164-4736, March, 1997.

[6] B. CHEN AND P. HARKER, *A non-interior-point continuation method for linear complementarity problems*, SIAM J. Matrix Anal. Appl., 14 (1993), pp. 1168—1190.

[7] B. CHEN AND N. XIU, *A global linear and local quadratic non-interior continuation method for nonlinear complementarity problems based on Chen-Mangasarian smoothing functions.* To appear in SIAM Journal on Optimization.

[8] C. CHEN AND O. L. MANGASARIAN, *A class of smoothing functions for nonlinear and mixed complementarity problems*, Comp. Optim. and Appl. , 5 (1996), pp. 97—138.

[9] X. CHEN, L. QI, AND D. SUN, *Global and superlinear convergence of the smoothing Newton method and its application to general box constrained variational inequalities*, Mathematics of Computation, 67 (1998), pp. 519-540.

[10] X. CHEN AND Y. YE, *On homotopy smooothing methods for variational inequalities.* To appear in SIAM J. Control Optim.

[11] M. FUKUSHIMA, Z.-Q. LUO, AND J.-S. PANG, *A globally convergent sequential quadratic programming algorithm for mathematical programs with linear complementarity constraints*, Computational Optimization and Applications, 10 (1998), 5-34.

[12] K. HOTTA AND A. YOSHISE, *Global convergence of a class of non-interior-point algorithms using Chen-Harker-Kanzow functions for nonlinear complementarity problems.* Discussion Paper Series, No. 708, University of Tsukuba, Tsukuba, Ibaraki 305, Japan, December, 1996.

[13] H. JIANG, *Smoothed Fischer-Burmeister equation methods for the complementarity problem.* Report, Department of Mathematics, University of Melbourne, Parkville, Australia, June, 1997.

[14] C. KANZOW, *Some noninterior continuation methods for linear complementarity problems*, SIAM J. Matrix Anal. Appl., 17 (1996), pp. 851—868.

[15] ——, *Private communications.* Seattle, Washington, May, 1997.

[16] M. KOJIMA, N. MEGGIDO, T. NOMA, AND A. YOSHISE, *A unified approach to interior point algorithms for linear complementarity problems*, Springer-Verlag, Berlin, 1991.

[17] L. QI AND D. SUN, *Improving the convergence of non-interior point algorithms for nonlinear complementarity problems.* To appear in Mathematics of Computation.

[18] L. QI, D. SUN, AND G. ZHOU, *A new look at smoothing Newton methods for nonlinear complementarity problems and box constrained variational*

inequalities. Applied Mathematics Report, AMR 97/13, School of Mathematics University of New South Wales, Sydney 2052, Australia, June, 1997.

[19] S. SMALE, *Algorithms for solving equations.* Proceedings of the International Congress of Mathematicians, Berkeley, California, 1986.

[20] P. TSENG, *Analysis of a non-interior continuation method based on Chen-Mangasarian smoothing functions for complementarity problems,* Reformulation - Nonsmooth, Piecewise Smooth, Semismooth and Smoothing Methods, M. Fukushima and L. Qi, eds., (Kluwer Academic Publisher, Nowell, MA. USA, 1998), 381-404.

[21] S. XU, *The global linear convergence of an infeasible non-interior path-folowing algorithm for complementarity problems with uniform P-functions.* Technique Report, Department of Mathematics, University of Washington, Seattle, WA 98195, December, 1996.

[22] ——, *The global linear convergence and complexity of a non-interior path-following algorithm for monotone LCP based on Chen-Harker-Kanzow-Smale smoothing function.* Technique Report, Department of Mathematics, University of Washington, Seattle, WA 98195, February, 1997.

Reformulation: Nonsmooth, Piecewise Smooth, Semismooth and Smoothing Methods, pp. 65–79
Edited by M. Fukushima and L. Qi

Smoothing Newton Methods for Nonsmooth Dirichlet Problems [1]

Xiaojun Chen[†], Nami Matsunaga[‡] and Tetsuro Yamamoto[§]

Abstract We apply the class of Chen-Mangasarian smooth approximation functions and the smoothing Newton method proposed in [6] to solve a class of quasi-linear nonsmooth Dirichlet problems. We show that the smoothing Newton method converges globally and superlinearly for solving the system of nonsmooth equations arising from nonsmooth Dirichlet problems. Moreover, we show that the error bound for finite difference solution of smooth Poisson-type equations in a bounded domain [15] remains the same order for a class of nonsmooth equations. It is thus expected that the smoothing Newton method works well for nonsmooth Dirichlet problems. We report encouraging numerical results for three examples arising from a model of the ideal MHD equilibria.

Key Words nonsmooth Dirichlet problem, error bound, smoothing Newton methods.

1 INTRODUCTION

We consider the following nonsmooth Dirichlet problem

$$-\Delta u + \max(0, q(u)) = \phi(s,t) \quad \text{in } \Omega, \tag{1.1}$$

$$u = \psi(s,t) \quad \text{on } \partial\Omega, \tag{1.2}$$

[1]This work was supported by the Australian Academy of Science, the Australian Research Council, the Japan Society for the Promotion of Science and the Scientific Research Grant-in-Aid from the Ministry of Education, Science, Sports and Culture of Japan.
[†]School of Mathematics, University of New South Wales, Sydney 2052, Australia. E-mail: X.Chen@unsw.edu.au
[‡]Graduate School of Science and Engineering, Doctor Course, Ehime University, Matsuyama 790, Japan. E-mail: matunaga@math.sci.ehime-u.ac.jp
[§]Department of Mathematical Sciences, Ehime University, Matsuyama 790, Japan. E-mail: yamamoto@dpc.ehime-u.ac.jp

where $\Omega \subset R^2$ is a bounded convex domain with the boundary $\partial\Omega$, ϕ and ψ are given functions, and q is a continuously differentiable function with $q'(u) \geq 0$.

We assume that the equation (1.1) – (1.2) has a solution $u \in C^2(\bar{\Omega})$.

Discretization of (1.1) – (1.2) by a finite difference scheme or a finite element scheme leads to the following system of nonsmooth equations

$$F(x) \equiv Ax + b + \max(0, g(x)) = 0, \tag{1.3}$$

where $A = (a_{ij}) \in R^{n\times n}$ is an irreducibly diagonally dominant L-matrix (in a sense of Young [16], i.e., $a_{ii} > 0$, $a_{ij} \leq 0$ $(i \neq j)$ $\forall i, j)$, $b \in R^n$ and $g : R^n \to R^n$ is continuously differentiable and monotone , i.e.,

$$(g(x) - g(y))^T(x - y) \geq 0 \quad \text{for all } x, y \in R^n.$$

In (1.3), $\max(\cdot)$ denotes the componentwise maximum operator. Furthermore, we assume that A is positive definite, i.e., $x^TAx \geq 0$ for all $x \in R^n$, which is equivalent to $x^T(A + A^T)x \geq 0$ for all $x \in R^n$. Since A is irreducibly diagonally dominant, A is nonsingular and so there is a positive constant μ such that $x^TAx \geq \mu x^Tx$ for all $x \in R^n$. It turns out that F is a strongly monotone function, i.e.,

$$(x - y)^T(F(x) - F(y)) \geq \mu\|x - y\|^2, \quad \text{for all } x, y \in R^n.$$

In the last few years, smoothing Newton methods have extensively been studied for solving nonsmooth equations arising from variational inequality problems and complementarity problems [1, 2, 4, 5, 6, 9, 11, 12]. The feature of these nonsmooth equations is their piecewise smooth property. The function of F defined by (1.3) is also piecewise smooth. This motivates us to investigate smoothing Newton methods for solving nonsmooth equations (1.3).

In this paper, we apply Chen-Mangasarian smooth functions and the smoothing Newton method proposed in [6] to solve the system of equations (1.3). We show the global and superlinear convergence of the smoothing Newton method for solving (1.3) without additional assumptions. If g is an affine function, then the convergence is finite. Moreover, we show that the error bound for the finite difference solution of smooth Poisson-type equations in a Dirichlet domain given in [15] remains the same order for the equation (1.1) – (1.2). Therefore, it is expected that the smoothing Newton method would find a good approximate solution to the original boundary problem (1.1) and (1.2). Numerical results indicate that the smoothing Newton method works very well for the large scale system of nonsmooth equations arising from a model of the ideal MHD equilibria [13].

We denote the ith row of a matrix A by A_i, and the identity matrix by I. We use $\|\cdot\|$ to denote the Euclidean norm. Furthermore, R_+ and R_{++} stand for the nonnegative and the positive orthant of R, respectively.

2 THE SMOOTHING NEWTON METHOD

Let $\rho : R \to R_+$ be a density function with a bounded absolute mean, that is

$$\kappa := \int_{-\infty}^{\infty} |s|\rho(s)ds < \infty. \tag{2.1}$$

Application of the class of Chen-Mangasarian approximation functions to the function F in (1.3) gives

$$f_i(x, \varepsilon) = (Ax + b)_i + \int_{-\infty}^{\infty} \max(0, g_i(x) - \varepsilon s)\rho(s)ds, \quad i = 1, \ldots n, \tag{2.2}$$

where $g_i(x)$ denotes the ith component of the column vector $g(x)$. By using (2.2), a smooth approximation function can be generated by an appropriate density function. The following proposition can easily be obtained from the existing results of smoothing methods for complementarity problems, for example see [4, 9].

Proposition 2.1 *The function $f : R^n \times R_{++} \to R^n$ defined by (2.2) has the following four properties.*

(i) For any $(x, \varepsilon) \in R^n \times R_{++}$,

$$|F_i(x) - f_i(x, \varepsilon)| \leq \kappa\varepsilon, \quad i = 1, 2, \ldots, n. \tag{2.3}$$

(ii) f is continuously differentiable with respect to the variable x, and for any $(x, \varepsilon) \in R^n \times R_{++}$

$$f_x(x, \varepsilon) = A + D(x)g'(x), \tag{2.4}$$

where $D(x)$ is a diagonal matrix whose diagonal elements are

$$D_{ii}(x) = \int_{-\infty}^{g_i(x)/\varepsilon} \rho(s)ds, \quad i = 1, 2, \ldots, n,$$

and $g'(x)$ stands for the Jacobian matrix of $g(x)$, which is diagonal.

(iii) For any $x \in R^n$

$$\lim_{\varepsilon \downarrow 0} f_x(x, \varepsilon) = f^0(x), \tag{2.5}$$

where for $i = 1, 2, \ldots, n$, the ith row of the matrix $f^0(x)$ is defined by

$$f_i^0(x) = \begin{cases} A_i + g_i'(x) & \text{if } g_i(x) > 0, \\ A_i & \text{if } g_i(x) < 0, \\ A_i + \left(\int_{-\infty}^{0} \rho(s)ds\right) g_i'(x) & \text{if } g_i(x) = 0. \end{cases}$$

(iv) For any $x \in R^n$, f satisfies the directional derivative consistency property at x, i.e.

$$\lim_{h \to 0} \frac{F(x + h) - F(x) - f^0(x + h)h}{\|h\|} = 0. \tag{2.6}$$

Let us denote

$$\Theta(x) = \frac{1}{2}\|F(x)\|^2$$

and

$$\theta_k(x) = \frac{1}{2}\|f(x, \varepsilon_k)\|^2.$$

The following algorithm is an application of the smoothing Newton method [6] to solve (1.3).

Algorithm 2.1 *Given* $\rho, \alpha, \eta \in (0,1)$, *and a starting point* $x^0 \in R^n$. *Choose a scalar* $\sigma \in (0, \frac{1}{2}(1-\alpha))$. *Let* $\nu = \frac{\alpha}{2\sqrt{n}\kappa}$. *Let* $\beta_0 = \|F(x^0)\|$ *and* $\varepsilon_0 = \nu\beta_0$.
For $k \geq 0$*:*

1. *Find a solution* $\hat{d}^k$ *of the system of linear equations*

$$F(x^k) + f^0(x^k)d = 0. \tag{2.7}$$

If $\|F(x^k + \hat{d}^k)\| \leq \eta\beta_k$, *let* $x^{k+1} = x^k + \hat{d}^k$. *Otherwise perform Step 2.*

2. *Find a solution* d^k *of the system of linear equations*

$$F(x^k) + f_x(x^k, \varepsilon_k)d = 0. \tag{2.8}$$

Let m_k *be the smallest nonnegative integer* m *such that*

$$\theta_k(x^k + \rho^m d^k) - \theta_k(x^k) \leq -2\sigma\rho^m\Theta(x^k). \tag{2.9}$$

Set $t_k = \rho^{m_k}$ *and* $x^{k+1} = x^k + t_k d^k$.

3. *3.1 If* $\|F(x^{k+1})\| = 0$, *terminate.*
3.2 If

$$0 < \|F(x^{k+1})\| \leq \max\{\eta\beta_k, \alpha^{-1}\|F(x^{k+1}) - f(x^{k+1}, \varepsilon_k)\|\}, \tag{2.10}$$

let

$$\beta_{k+1} = \|F(x^{k+1})\| \quad \text{and} \quad \varepsilon_{k+1} = \min\{\nu\beta_{k+1}, \frac{\varepsilon_k}{2}\}.$$

3.3 Otherwise, let $\beta_{k+1} = \beta_k$ *and* $\varepsilon_{k+1} = \varepsilon_k$.

We denote the level set of F by

$$D(\Gamma) = \{x \in R^n \ : \ \|F(x)\| \leq \Gamma\},$$

where Γ is a positive constant.

Theorem 2.1 *For any starting point* $x^0 \in R^n$, *Algorithm 1 is well defined and the generated sequence* $\{x^k\}$ *remains in* $D((1+\alpha)\|F(x^0)\|)$ *and satisfies*

$$\lim_{k\to\infty} \|F(x^k)\| = 0. \tag{2.11}$$

Proof. Since A is positive definite and g is monotone, we have that for any $\varepsilon > 0$ and $x \in R^n$, $f_x(x, \varepsilon)$ is a P-matrix, and so it is nonsingular. (A matrix is called P-matrix if all of its principal minors are positive.) Moreover, by Lemma 3.1 in [4], there exists a finite nonnegative integer m_k such that (2.9) holds. Hence the algorithm is well defined. Notice that the function F is strongly monotone. This implies that the level set is bounded for every positive number Γ. Therefore by Theorem 3.1 in [6], we obtain this theorem. ∎

Theorem 2.2 *The system (1.3) has a unique solution x^*, and $\{x^k\}$ converges to x^* superlinearly. Moreover, if g has a locally Lipschitz continuous derivative around x^*, then the convergence rate is quadratic. In addition, if g is an affine function, then the convergence is finite.*

Proof. The existence and uniqueness of the solution of (1.3) are ensured by the strong monotonicity of F. Furthermore, $\|f^0(x)^{-1}\|$ is bounded by $\|A^{-1}\|$ in R^n since $f^0(x) \in A + \partial g(x)$ and every matrix in $\partial g(x)$ is positive semidefinite. Thus the convergence rate follows from the proof of Theorem 3.2 in [6]. ∎

3 ERROR BOUNDS FOR FINITE DIFFERENCE SOLUTION

We consider a discretization of the problem (1.1) - (1.2) by the Shortley-Weller approximation [8] which includes the usual five point formula as a special case and denote by U_{ij} the numerical solution at the grid point

$$P_{ij} = (s_i, t_j) = (i\delta s, j\delta t),$$

where δs and δt denote mesh sizes for s and t directions.

We denote the solution of the nonlinear system (1.3) by

$$x^* = (\ldots, U_{ij}, \ldots)^T = (x_1^*, \ldots, x_n^*)^T.$$

Here n denotes the number of the grid points P_{ij} in Ω. We put $x = (x_1, \ldots, x_n)^T$ and

$$g(x) = (q(x_1), \ldots, q(x_n))^T.$$

Let u be the solution of (1.1) – (1.2) and u_{ij} and τ_{ij} be the exact value u and the truncation error of u at P_{ij}, respectively.

Theorem 3.1 *Let $u \in C^{2,\gamma}(\bar{\Omega})$ be a solution of (1.1) - (1.2), where γ stands for the exponent of Hölder-continuity, and* $0 < \gamma < 1$ (cf. Gilbarg and Trudinger [10]). *Then, we have*

$$|u_{ij} - U_{ij}| \le \begin{cases} O((\delta s)^\gamma + (\delta t)^\gamma) & \text{in } \Omega, \\ O((\delta s)^{1+\gamma} + (\delta t)^{1+\gamma}) & \text{near } \partial\Omega. \end{cases}$$

If $u \in C^4(\bar{\Omega})$, *then*

$$|u_{ij} - U_{ij}| \le \begin{cases} O((\delta s)^2 + (\delta t)^2) & \text{in } \Omega, \\ O((\delta s)^3 + (\delta t)^3) & \text{near } \partial\Omega. \end{cases}$$

Remark. We define that $P_{ij} \in \Omega$ is near $\partial\Omega$ if the distance between P_{ij} and $\partial\Omega$ is at most $O(\delta s + \delta t)$ (cf. [15]).

Proof. Let

$$\bar{x} = (\ldots, u_{ij}, \ldots)^T = (\bar{x}_1, \ldots, \bar{x}_n)^T.$$

By the definition of F, we have

$$F(\bar{x}) = A\bar{x} + b + \max(0, g(\bar{x})) = \tau$$

where $\tau = (\ldots, \tau_{ij}, \ldots)^T$. Hence we have

$$A(\bar{x} - x^*) + \max(0, g(\bar{x})) - \max(0, g(x^*)) = \tau. \tag{3.1}$$

Recall that

$$\max(0, g(\bar{x})) - \max(0, g(x^*)) = V(\bar{x} - x^*) \tag{3.2}$$

where

$$V \in \text{co}\{V : V \in \partial \max(0, g(w)), w \in \overline{\bar{x}x^*}\},$$

and V is a diagonal matrix whose diagonal elements are nonnegative. Here $\overline{\bar{x}x^*}$ denotes the line segment between $\bar{x}$ and x^*. See [7, Proposition 2.6.5].

Let

$$|\bar{x} - x^*| = (|\bar{x}_1 - x_1^*|, \ldots, |\bar{x}_n - x_n^*|)^T = (\ldots, |u_{ij} - U_{ij}|, \ldots)^T$$

and

$$|\tau| = (\ldots, |\tau_{ij}|, \ldots)^T.$$

It now follows from (3.1) and (3.2) that

$$(A + V)(\bar{x} - x^*) = \tau$$

and

$$|\bar{x} - x^*| = |(A + V)^{-1}\tau| \le (A + V)^{-1}|\tau| \le A^{-1}|\tau|, \tag{3.3}$$

where we have used the facts that an irreducibly diagonally dominant L-matrix is an M-matrix, so that A and $A+V$ are M-matrices and $0 \le (A+V)^{-1} \le A^{-1}$. Therefore, from (3.3), we can use the results obtained in Yamamoto [15] to establish Theorem 3, although we have $\tau_{ij} = O(\delta s + \delta t)$ at every point P_{ij} such that at least one of four neighbor points $(s_i \pm \delta s, t_j)$ and $(s_i, t_j \pm \delta t)$ does not belong to $\bar{\Omega}$. ∎

4 NUMERICAL EXPERIMENTS

Combining Theorem 3 with Theorems 1 and 2, it is now expected that a good numerical solution for (1.1) - (1.2) would be obtained by Algorithm 1. In this section, we report results of numerical experiments done for testing Algorithm 1 and the accuracy for the finite difference solution of the following problem

$$-\Delta u + a\max(0,u) = \phi(s,t) \quad \text{in } \Omega, \tag{4.1}$$

$$u = \psi(s,t) \quad \text{on } \partial\Omega, \tag{4.2}$$

where $\Omega = [0,1]\times[0,1]$, $a > 0$, ϕ and ψ are given functions. This problem arises from the ideal MHD equilibria (cf. [13]).

We consider the following three examples.

Example 1 Let

$$\phi(x,y) = \begin{cases} (2\pi^2 + a)\cos(\pi x)\cos(\pi y) & \text{if } 0 < x < 0.5, 0 < y < 0.5, \\ & \text{or } 0.5 < x < 1, 0.5 < y < 1, \\ (2\pi^2)\cos(\pi x)\cos(\pi y) & \text{otherwise} \end{cases}$$

and $\psi(x,y) = \cos(\pi x)\cos(\pi y)$. The exact solution in this problem is

$$u(x,y) = \cos(\pi x)\cos(\pi y).$$

Example 2 Let

$$\phi(x,y) = \begin{cases} \phi_1(x,y) & \text{if } 0 < x < 0.5, 0 < y < 0.1, \\ & \text{or } 0.5 < x < 1, 0.1 < y < 1 \\ \phi_2(x,y) & \text{otherwise} \end{cases},$$

where $\phi_1(x,y) = -((x-0.5)^2 + (y-0.1)^2 - a)e^{(x-0.5)(y-0.1)} - a$ and $\phi_2(x,y) = -((x-0.5)^2 + (y-0.1)^2)e^{(x-0.5)(y-0.1)}$, and $\psi(x,y) = e^{(x-0.5)(y-0.1)} - 1$. The exact solution is then

$$u(x,y) = e^{(x-0.5)(y-0.1)} - 1.$$

Example 3 Let

$$\phi(x,y) = \begin{cases} (36\pi^2(x^2+y^2) + a)\sin(6\pi xy) & \text{if } 0 < xy < 1/6, \\ & \text{or } 1/3 < xy < 1/2, \\ & \text{or } 2/3 < xy < 5/6, \\ 36\pi^2(x^2+y^2)\sin(6\pi xy) & \text{otherwise} \end{cases}$$

and $\psi(x,y) = \sin(6\pi xy)$. The exact solution is then

$$u(x,y) = \sin(6\pi xy).$$

We use the usual five point difference formula to generate a system of nonsmooth equations (1.3) from (1.1) – (1.2). We use the following three smooth approximation functions to generate (2.2) (cf. [1, 2, 4, 6, 9, 11]).

Neural networks smooth approximation function (S1)

Let the density function be

$$\rho(s) = \frac{e^{-s}}{(1+e^{-s})^2}.$$

Then, the smooth approximation function is then defined by

$$f_i(x,\varepsilon) = (Ax+b)_i + g_i(x) + \varepsilon \log(1 + e^{(-g_i(x))/\varepsilon})$$

and

$$f_x(x,\varepsilon) = A + D(x)g'(x),$$

where

$$D_{ii}(x) = 1 - e^{-g_i(x)/\varepsilon}/(1+e^{-g_i(x)/\varepsilon}).$$

Chen-Harker-Kanzow-Smale smooth approximation function (S2)

Let the density function be

$$\rho(s) = \frac{2}{(s^2+4)^{\frac{3}{2}}}.$$

The smooth approximation function is then defined by

$$f_i(x,\varepsilon) = (Ax+b)_i + \frac{1}{2}(\sqrt{g_i(x)^2 + 4\varepsilon^2} + g_i(x))$$

and

$$f_x(x,\varepsilon) = A + D(x)g'(x)$$

where

$$D_{ii}(x) = \frac{1}{2}(1 + \frac{g_i(x)}{\sqrt{g_i(x)^2 + 4\varepsilon^2}}).$$

Uniform smooth approximation function (S3)

Let the density function be

$$\rho(s) = \begin{cases} 1 & \text{if } |s| \le 0.5, \\ 0 & \text{otherwise.} \end{cases}$$

Then

$$f_i(x,\varepsilon) = \begin{cases} (Ax+b)_i + \frac{1}{2\varepsilon}(g_i(x) + \frac{\varepsilon}{2})^2 & \text{if } |g_i(x)| \le \frac{\varepsilon}{2}, \\ F_i(x) & \text{otherwise,} \end{cases}$$

and

$$f_x(x,\varepsilon) = \begin{cases} A_i + \frac{1}{\varepsilon}(g_i(x) + \frac{\varepsilon}{2})g_i'(x) & \text{if } |g_i(x)| \le \frac{\varepsilon}{2}, \\ F_i'(x) & \text{otherwise.} \end{cases}$$

The three density functions have a common property: $\rho(-s) = \rho(s)$. By (2.5), this implies that for any $x \in R^n$, the derivatives $f_x(x,\varepsilon)$ of the three smooth approximation functions have the same limit $f^0(x)$ as $\varepsilon \downarrow 0$, where

$$f_i^0(x) = \begin{cases} A_i + g_i'(x) & \text{if } g_i(x) > 0, \\ A_i & \text{if } g_i(x) < 0, \\ A_i + \frac{1}{2} g_i'(x) & \text{otherwise.} \end{cases} \tag{4.3}$$

We chose $x^0 = e$, $\rho = 0.75$, $\alpha = 0.56$, $\eta = 0.87$ and $\sigma = 0.2$ in Algorithm 1. The stopping criterion was $\|F(x^k)\| \leq 10^{-8}$. Numerical results were obtained using Fortran 90 on a Solaris workstation of Ehime University.

In Tables 1 – 3, we show the iteration number and the CPU time used to solve the problems, and the error of the final iterate x^k to the exact values $\bar{x}$ of the solution of the nonsmooth Dirichlet problem at these grid points. Figures 1 – 6 show the distribution of the error $|x_i^k - \bar{x}_i|$ in the whole domain Ω for Examples 1 – 3. Among others, Figures 1 – 4 show the behavior of the errors when the constant a varies from 1 to 1000.

We also ran the three problems with starting points

$$x = (0, 0, \ldots, 0),$$

$$x = (1, -1, 1, -1, \ldots)$$

and

$$x = (1, 0, -1, 0, 1, 0, -1, \ldots).$$

Then, Step 1 was always accepted and Step 2 was never used. The reason might be due to the strongly positive quasi-linear property of the equations (1.1) – (1.2).

There are many partial differential equations from engineering applications which are nondifferentiable [3, 12, 13]. This paper shows that smoothing Newton methods are simple and efficient methods for solving a large class of nondifferentiable partial differential equations (1.3).

References

[1] B. Chen and P.T. Harker, *Smooth approximations to nonlinear complementarity problems*, SIAM J. Optim., 7 (1997), pp. 403-420.

[2] C. Chen and O.L. Mangasarian, *A class of smoothing functions for nonlinear and mixed complementarity problems*, Comp. Optim. Appl., 5 (1996), pp. 97-138.

[3] X. Chen, *Global and superlinear convergence of inexact Uzawa methods for saddle point problems with nondifferentiable mappings*, SIAM J. Numer. Anal. (1998), to appear.

[4] X. Chen, L. Qi and D. Sun, *Global and superlinear convergence of the smoothing Newton method and its application to general box constrained variational inequalities*, Math. Comp., 67 (1998), 519-540.

Table 1.1 Example 1. Error $\|x^k - \bar{x}\|$, iterations, CPU time (sec.)

$a = 1$									
n	49×49			99×99			149×149		
S1	0.15×10^{-2}	4	5.8	0.76×10^{-3}	4	226.4	0.51×10^{-3}	4	1152.6
S2	0.15×10^{-2}	4	5.8	0.76×10^{-3}	4	225.8	0.51×10^{-3}	4	1165.8
S3	0.15×10^{-2}	4	5.8	0.76×10^{-3}	4	227.4	0.51×10^{-3}	4	1155.8
$a = 10$									
S1	0.15×10^{-2}	4	5.9	0.73×10^{-3}	4	224.9	0.49×10^{-3}	5	1440.8
S2	0.15×10^{-2}	4	5.8	0.73×10^{-3}	4	223.3	0.49×10^{-3}	5	1434.3
S3	0.15×10^{-2}	4	6.0	0.73×10^{-3}	4	223.5	0.49×10^{-3}	5	1444.8
$a = 100$									
S1	0.13×10^{-2}	5	7.3	0.66×10^{-3}	6	336.1	0.44×10^{-3}	6	1741.8
S2	0.13×10^{-2}	5	7.4	0.66×10^{-3}	6	336.2	0.44×10^{-3}	6	1740.1
S3	0.13×10^{-2}	5	7.3	0.66×10^{-3}	6	338.0	0.44×10^{-3}	6	1765.1
$a = 1000$									
S1	0.12×10^{-2}	6	8.8	0.59×10^{-3}	6	334.8	0.39×10^{-3}	7	2067.0
S2	0.12×10^{-2}	6	8.8	0.59×10^{-3}	6	334.5	0.39×10^{-3}	7	1996.0
S3	0.12×10^{-2}	6	8.8	0.59×10^{-3}	6	334.5	0.39×10^{-3}	7	2086.4

[5] X. Chen and T. Yamamoto, *Convergence domains of certain iterative methods for solving nonlinear equations*, Numer. Funct. Anal. Optim. 10 (1989), pp. 37-48.

[6] X. Chen and Y. Ye, *On homotopy-smoothing methods for box-constrained variational inequalities*, SIAM J. Control Optim., to appear.

[7] F. H. Clarke, Optimization and Nonsmooth Analysis, Reprint, SIAM, Philadelphia, 1990.

[8] G.E. Forsythe and W.R. Wasow, Finite Difference Methods for Partial Differential Equations, Wiley, 1960.

[9] S.A. Gabriel and J.J. Moré, *Smoothing of mixed complementarity problems*, in: M.C. Ferris and J.S. Pang, eds., Complementarity and Variational Problems: State of the Art, SIAM, Philadelphia, Pennsylvania, 1997, pp. 105-116.

[10] D. Gilbarg and N.S. Trudinger, Elliptic Partial Differential Equations of Second Order, Springer-Verlog, Berlin, Heidelberg, New York, 1977.

Table 1.2 Example 2 with $a = 1$. Error, iterations, CPU(sec.)

n	49×49	99×99	149×149
	$\|x^k - \bar{x}\|$		
S1	0.64×10^{-5}	0.32×10^{-5}	0.21×10^{-5}
S2	0.64×10^{-5}	0.32×10^{-5}	0.21×10^{-5}
S3	0.64×10^{-5}	0.32×10^{-5}	0.21×10^{-5}
	Iterations, CPU time (sec.)		
S1	4, 5.9	4, 224.0	3, 935.4
S2	4, 5.8	4, 223.9	3, 860.9
S3	4, 5.9	4, 223.9	3, 927.3

Table 1.3 Example 3 with $a = 1$. Error, iterations, CPU(sec.)

n	49×49	99×99	149×149
	$\|x^k - \bar{x}\|$		
S1	0.18×10^{0}	0.92×10^{-1}	0.61×10^{-1}
S2	0.18×10^{0}	0.92×10^{-1}	0.61×10^{-1}
S3	0.18×10^{0}	0.92×10^{-1}	0.61×10^{-1}
	Iterations, CPU time (sec.)		
S1	3, 4.4	3, 167.5	3, 866.3
S2	3, 4.4	3, 167.5	3, 911.0
S3	3, 4.4	3, 167.6	3, 865.1

[11] C. Kanzow, *Some noninterior continuation methods for linear complementarity problems*, SIAM J. Matrix Anal. Appl., 17 (1996), pp. 851-868.

[12] C.T. Kelley, *Identification of the support of nonsmoothness*, in: W.W. Hager, D.W. Hearn and P.M. Pardalos eds., Large Scale Optimization: State of the Art, Boston, Kluwer Academic Publishers B.V. 1993, pp. 192-205.

[13] F. Kikuchi, *An iteration scheme for a nonlinear eigenvalue problem*, Theor. Appl. Mechanics, 29 (1981) 319-333.

[14] N. Matsunaga and T. Yamamoto, Superconvergence of finite difference methods for Dirichlet problems, preprint.

[15] T. Yamamoto, *On the accuracy of finite difference solution for Dirichlet problems*, Preprint, Department of Mathematical Sciences, Ehime University, November 1997 (A summary of an invited talk in the meeting "Study of Numerical Algorithms" held at RIMS Kyoto Univ.).

[16] D.M. Young, Iterative Solution of Large Linear Systems, Academic Press, New York, London, 1971.

Figure 1.1 Example 1 with $a = 1$, S1 and $n = 99 \times 99$.

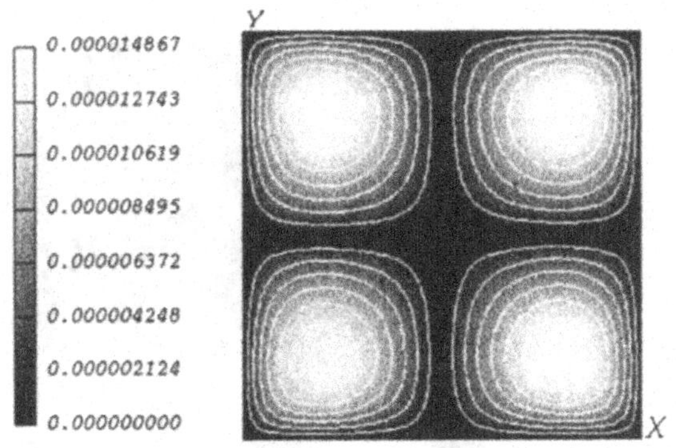

Figure 1.2 Example 1 with $a = 10$, S1 and $n = 99 \times 99$.

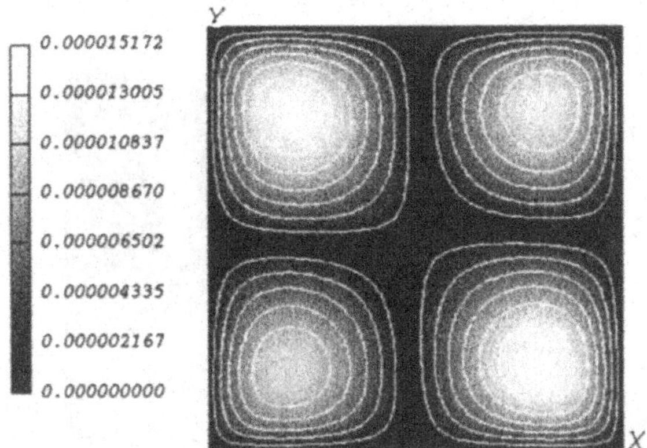

Figure 1.3 Example 1 with $a = 100$, S1 and $n = 99 \times 99$.

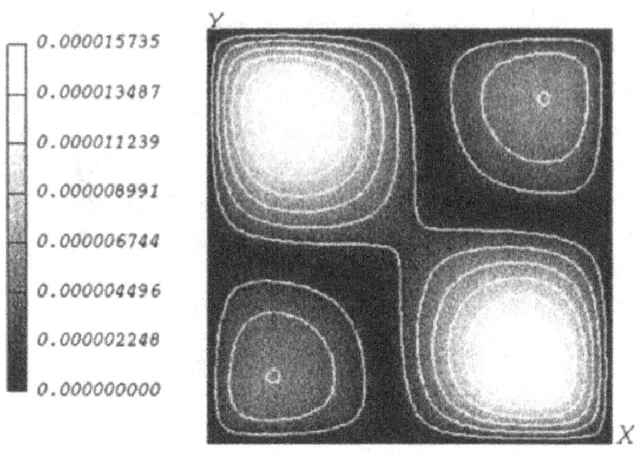

Figure 1.4 Example 1 with $a = 1000$, S1 and $n = 99 \times 99$.

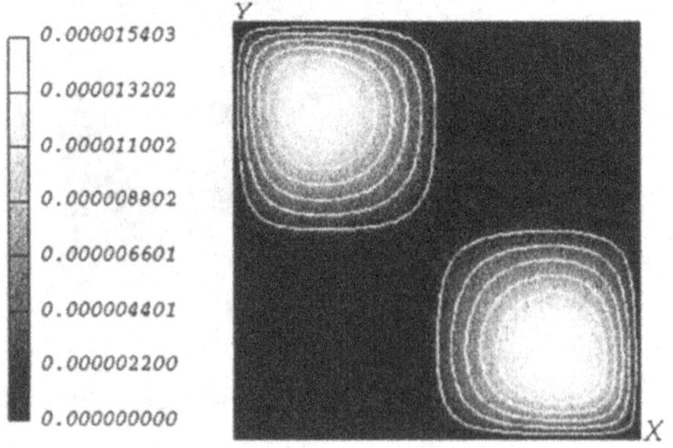

Figure 1.5 Example 2 with $a = 1$, S1 and $n = 49 \times 49$.

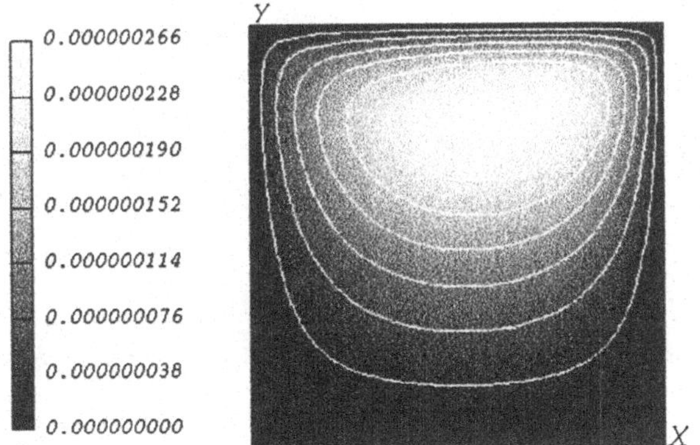

Figure 1.6 Example 3 with $a = 1$, S1 and $n = 49 \times 49$.

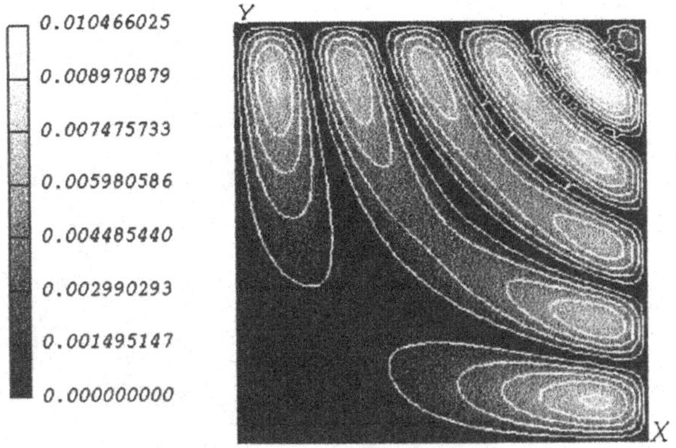

Reformulation: Nonsmooth, Piecewise Smooth, Semismooth and Smoothing Methods, pp. 81–116
Edited by M. Fukushima and L. Qi

Frictional Contact Algorithms Based on Semismooth Newton Methods

Peter W. Christensen* and Jong-Shi Pang†

Abstract In this paper, we establish that the discrete, three-dimensional, quasistatic, small-displacement, elastic-body frictional contact problem can be formulated as a system of semismooth equations. We give two such formulations, one unconstrained (i.e., no additional restriction on the variables of the equation) and the other constrained (that is, with additional nonnegativity constraints on some variables). A potential reduction Newton method for solving a constrained semismooth equation is developed and its convergence is established. This method is applied to the two formulations of the frictional contact problem and experimentally tested on several realistic contact problems with over 400 contact nodes. The numerical results demonstrate that the unconstrained formulation yields performance far superior to the constrained formulation.

Key Words frictional contact, complementarity problems, semismooth equations, Newton method, interior point method.

1 INTRODUCTION

Frictional contact problems are characterized by being highly nonlinear. This is due to the multivalued force-displacement and force-velocity relations, the nonconservative friction forces, and the intricate dependence of the contact zone on the applied loads. It is a well-known fact that for quasistatic models of the problem, non-existence and non-uniqueness of solutions may occur. For a

*Department of Mechanical Engineering, Linköping University, Linköping, S-581 83, Sweden, Email: petch@ikp.liu.se. The work of this author was based on research supported by the Center for Industrial Information Technology (CENIIT) and ABB Atom AB.
†Department of Mathematical Sciences, The Johns Hopkins University, Baltimore, Maryland 21218-2682, U.S.A., Email: jsp@vicp1.mts.jhu.edu. The work of this author was based on research supported by the United States National Science Foundation under grant CCR-9624018.

discussion of these issues, see e.g. [12, 15]. Initially, "trial-and-error" methods and penalty approaches were developed to solve these problems. For two-dimensional problems these methods often work well, but for three-dimensional problems, nonconvergence is more common.

In order to develop theoretically sound algorithms, Mathematical Programming (MP) methods were called for; see the review paper [13]. With such methods it became possible to construct algorithms with demonstrated convergence properties and obtain conditions for existence and uniqueness of solutions. As it happens, for two-dimensional problems these methods frequently coincide with some "trial-and-error" method. However, for the more complex three-dimensional problems this is not the case; see Klarbring [13] for a plausible explanation of this. In the initial attempts of using MP methods for contact problems involving friction, physically meaningless assumptions were made which ignored the load history dependence and took the (unknown) normal contact forces as given. The first time a load history dependent contact problem (although still under the assumption of constant contact surface) was solved with an MP method was by Klarbring [11]. The ideas behind this work were due to a paper on elasto-plasticity by Kaneko [10] using a formulation as a linear complementarity problem (LCP). The basic step of this approach is to approximate the nonlinear friction cone by affine functions in order to formulate the problem as a parametric LCP involving time derivatives. This problem is then solved by a principal pivot algorithm. The solution obtained will be piecewise linear in time; no discretization in time is needed.

In Klarbring and Björkman [14], this method was extended to handle arbitrary contact surfaces. Holmberg [9] later refined this method in order to reduce its need for computer storage and increase the numerical efficiency. The main disadvantage of this formulation is the fact that other nonlinearities such as plasticity, wear and large deformations cannot be dealt with.

A fundamentally different approach is to divide the loading history in time steps, or increments. Since frictional contact problems are load history dependent, small time steps may be needed. In each increment a static problem is obtained by first performing a backward Euler time discretization. A finite sequence of so-called time-incremental problems is then solved by an MP method. In an earlier paper [3], we have considered two formulations of these time-incremental problems in the case of planar contacts, one as a system of unconstrained Bouligand differentiable equations and the other as a system of constrained smooth equations. The former formulation is closely related to the augmented Lagrangian formulation of frictional contact problems studied by Alart and Curnier [1], De Saxce and Feng [4], and Simo and Laursen [24]. The latter formulation is inspired by the great success of the family of interior point methods for solving (monotone) complementarity problems. We applied a slight modification of the B-differentiable Newton (BN) method of Pang [17] to the unconstrained formulation and compared it with the the potential reduction interior point (IP) method of Wang, Monteiro and Pang [27] applied to the constrained formulation. (Previously, the IP method had also been applied for solving dynamic, multi-rigid-body, frictional contact problems in [26].) The numerical results in [3] showed that the BN method was far superior to the IP method. More seriously, when the IP method was applied to spatial frictional contact problems, extreme numerical difficulties were encountered that caused us to abandon the method for these three-dimensional problems. Various possible explanations for such difficulties exist which we refrain from making at this time. Inspired by the success of the BN method for solving the planar contact problems and a fretting problem [25], we were motivated to develop a hybrid method for solving the spatial contact problems that would improve upon the IP method of [27] for this application. In this regard, our effort is a success.

It turns out that the functions defining the two formulations of the (planar and spatial) frictional contact problems are both semismooth (formal definition to be given later). Inspired by this discovery and the recent advances in Newton methods for solving semismooth equations in general [18, 21, 22] and complementarity problems in particular (for the latter methods, see the references collected in [5] and the papers in [6]), we present a unified algorithm for solving a system of constrained semismooth equations that is applicable to both formulations of the frictional contact problem. Global convergence of this algorithm is established. For the application to the frictional contact problems, extensive computational results show that the unconstrained formulation remain superior to the constrained formulation.

2 THE MATHEMATICAL MODEL OF FRICTIONAL CONTACT

Consider two three-dimensional bodies, $\mathcal{A}$ and $\mathcal{B}$, in physical space. Body $\mathcal{A}$ is a fixed rigid foundation and body $\mathcal{B}$ is taken as isotropically linearly elastic. We want to solve the discrete small-displacement quasistatic frictional contact problem, i.e. determine displacements and contact forces as functions of time for a prescribed loading history, when inertia effects are neglected. Assuming that body $\mathcal{B}$ is either naturally discrete, e.g. a truss, or discretized by finite elements, we obtain the force equilibrium equation

$$Ku = f, \tag{2.1}$$

where $u \in \Re^{n_u}$ is the vector of displacements of the nodes of the body $\mathcal{B}$, n_u is the number of these displacement nodes, $f \in \Re^{n_u}$ is the force acting on body $\mathcal{B}$ and $K \in \Re^{n_u \times n_u}$ is the symmetric positive semidefinite stiffness matrix corresponding to body $\mathcal{B}$.

The nodes on the surface of body $\mathcal{B}$ that are allowed to come into contact with the foundation $\mathcal{A}$ are termed contact nodes. We assume that there are n_c contact nodes. At each contact node we define a normal direction and two tangential directions. Matrices C_n, C_t and C_o are introduced that transform the displacement vector u to the normal displacement vector w_n and the two tangential displacement vectors w_t and w_o:

$$w_n = C_n u \in \Re^{n_c}, \quad w_t = C_t u \in \Re^{n_c}, \quad w_o = C_o u \in \Re^{n_c}. \tag{2.2}$$

The subscripts n, t and o stand for normal, tangential and orthogonal, respectively. The rows of the matrices C_n, C_t and C_o are denoted C_{in}, C_{it} and C_{io}, respectively, for $i = 1, \ldots, n_c$.

The force in (2.1) is divided into a time dependent prescribed force, $f^{\text{ext}} \in \Re^{n_u}$, and normal and tangential contact forces $p_n \in \Re^{n_c}$, $p_t \in \Re^{n_c}$ and $p_o \in \Re^{n_c}$. These contact forces p_n, p_t and p_o are negatively work conjugate to w_n, w_t and w_o, respectively. Thus, taking into account the equation (2.1), we obtain the force equilibrium equation:

$$Ku = -C_n^T p_n - C_t^T p_t - C_o^T p_o + f^{\text{ext}}, \tag{2.3}$$

where the superscript T denotes the transpose of a vector or a matrix. The components of the vectors p_n, p_t, p_o, w_n, w_t and w_o are denoted p_{in}, p_{it}, p_{io}, w_{in}, w_{it} and w_{io}, respectively, for $i = 1, \ldots, n_c$.

We turn our attention to the laws of contact; the normal contact law and the friction law. The normal contact law used here is that of Signorini expressed as follows:

$$w_n \le g, \quad p_n \ge 0, \quad p_n^T(w_n - g) = 0, \tag{2.4}$$

where $g \in \Re^{n_c}$ are the initial gaps between the contact nodes and the rigid foundation $\mathcal{A}$. These equations state that the bodies cannot penetrate, there is no adhesion and if the bodies are not in contact then the normal contact force is zero. Similarly, if the contact force is nonzero, then the gap between the bodies must be zero.

The friction law employed is the classical Coulomb friction model in which the contact force is required to lie within or on the boundary of a quadratic cone. We assume that the static and kinetic coefficients of friction are both equal to μ_i, $i = 1, \ldots, n_c$. For our purposes it is suitable to express Coulomb's law at a contact node $i = 1, \ldots, n_c$ as a variational inequality:

$$(p_{it}, p_{io}) \in \mathcal{F}(\mu_i p_{in}) : \; \dot{w}_{it}(q_{it} - p_{it}) + \dot{w}_{io}(q_{io} - p_{io}) \le 0, \; \forall \, (q_{it}, q_{io}) \in \mathcal{F}(\mu_i p_{in}), \tag{2.5}$$

where the superposed dot denotes time derivative and

$$\mathcal{F}(\gamma) \equiv \{ (q_{it}, q_{io}) \in \Re^2 : q_{it}^2 + q_{io}^2 \le \gamma^2 \}, \quad \text{for } \gamma \in \Re_+,$$

is a set of admissible tangential contact forces. (A remark about notation: the effective domain of the (set-valued) friction map $\mathcal{F} : \Re_+ \to \Re^2$ is taken to be the nonnegative real line; thus in the Coulomb law (2.5), the normal force p_{in} is understood to be nonnegative as it must be when (2.4) holds.)

The quasistatic frictional contact problem is now defined by equations (2.1)–(2.5). Numerically, the problem is solved by applying the prescribed force $f^{\text{ext}}(\tau)$, $\tau \in [0, \mathcal{T}]$, where $[0, \mathcal{T}]$ is a given time interval, in increments. For each increment we solve a static problem which is obtained by approximating $\dot{w}_{it}$ and $\dot{w}_{io}$ by a backward Euler time discretization. Let the time interval $[0, \mathcal{T}]$ be divided into subintervals. For a specific subinterval, say $[\tau_\ell, \tau_{\ell+1}]$, the approximations of $\dot{w}_{it}$ and $\dot{w}_{io}$ are as follows:

$$\dot{w}_{it}(\tau_{\ell+1}) \approx \frac{w_{it} - \bar{w}_{it}}{\tau_{\ell+1} - \tau_\ell}, \quad \dot{w}_{io}(\tau_{\ell+1}) \approx \frac{w_{io} - \bar{w}_{io}}{\tau_{\ell+1} - \tau_\ell},$$

where $\bar{w}_{it}$ denotes $w_{it}(\tau_\ell)$ and w_{it} denotes $w_{it}(\tau_{\ell+1})$. Coulombs's law (2.5) may then be written in a time discretized form as: find $(p_{it}, p_{io}) \in \mathcal{F}(\mu_i p_{in})$ such that

$$(w_{it} - \bar{w}_{it})(q_{it} - p_{it}) + (w_{io} - \bar{w}_{io})(q_{io} - p_{io}) \leq 0, \quad \forall\, (q_{it}, q_{io}) \in \mathcal{F}(\mu_i p_{in}).$$

Using (2.2) to substitute for the w_n, w_t, and w_o variables, we obtain our time-incremental friction problem as follows. Given the inputs $f^{\text{ext}}(\tau_{\ell+1})$, $\bar{w}_{it}$ and $\bar{w}_{io}$, find the unknown vectors u, p_n, p_t and p_o so that the following conditions hold: (2.3),

$$0 \leq p_n \perp (g - C_n u) \geq 0, \tag{2.6}$$

where $a \perp b$ means that the two vectors a and b are perpendicular, and for all $i = 1, \ldots, n_c$,

$$(p_{it}, p_{io}) \in \text{ argmax } \{ q_{it}(C_{it}u - \bar{w}_{it}) + q_{io}(C_{io}u - \bar{w}_{io}) \; : \; (q_{it}, q_{io}) \in \mathcal{F}(\mu_i p_{in}) \}, \tag{2.7}$$

where "argmax" denotes the set of maximizers of an optimization problem.

3 TWO FORMULATIONS AS SEMISMOOTH EQUATIONS

Since our goal is to apply a nonsmooth Newton method for solving the frictional contact problem, we find it useful to present a quick review of some elementary concepts and basic facts in nonsmooth analysis, especially those pertaining to semismooth functions.

3.1 Semismooth function theory

A function $G : U \subseteq \Re^n \to \Re^m$ is said to be *piecewise smooth*, or PC^1, at a vector $\bar{x} \in U$ [23] if G is continuous at $\bar{x}$ and there exist an open neighborhood $\bar{U} \subseteq U$ of $\bar{x}$ and a finite family of continuously differentiable (i.e., C^1) functions $G^j : \bar{U} \to \Re^m$, $j = 1, \ldots, p$ (called the pieces of G at $\bar{x}$) such that for each $x \in \bar{U}$, $G(x) = G^j(x)$ for some j. The function $G : U \subseteq \Re^n \to \Re^m$ is said to be *B-differentiable* at a vector $\bar{x} \in U$ [17] if it is Lipschitz continuous in a neighborhood of $\bar{x}$ and directionally differentiable at $\bar{x}$. Following standard notation, the directional derivative of G at $\bar{x}$ along a direction $d \in \Re^n$ is denoted $G'(\bar{x}; d)$. The function G is said to be PC^1 (B-differentiable) in U if G is PC^1 (B-differentiable) at all points in U. A function $G : U \subseteq \Re^n \to \Re^m$ is said to be *semismooth* at a vector $\bar{x} \in U$ [21] if G is locally Lipschitz continuous in a neighborhood of $\bar{x}$ and the following limit holds:

$$\lim_{h \to 0} \frac{G'(\bar{x} + h; h) - G'(\bar{x}; h)}{\| h \|} = 0.$$

It is known that among these three properties for the function G at the point $\bar{x}$,

$$\text{piecewise smoothness} \Longrightarrow \text{semismoothness} \Longrightarrow \text{B-differentiability}.$$

We now define an important regularity concept that was first introduced in [21] where it was called BD-regularity. To define this concept, let $G : \Re^n \to \Re^n$ be a locally Lipschitz mapping. Let $\mathcal{D}_G$ denote the set of vectors where G is F-differentiable. It is well known that $\mathcal{D}_G$ is a dense subset of $\Re^n$. Let $\bar{x} \in \Re^n$ be an arbitrary vector. The B-differential of G at $\bar{x}$ is the set

$$\partial_B G(\bar{x}) \equiv \left\{ \lim_{k\to\infty} G'(x^k) \,:\, \lim_{k\to\infty} x^k = \bar{x},\, x^k \in \mathcal{D}_G \right\},$$

where $G'(\cdot)$ denote the Jacobian matrix of G at a F-differentiable point. We say that the function G is *BD-regular* at $\bar{x}$ if all matrices $Q \in \partial_B G(\bar{x})$ are nonsingular. We summarize several useful consequences of this regularity condition in the proposition below. See [21] for the proof.

Proposition 3.1 *Let $G : \Re^n \to \Re^n$ be a locally Lipschitz mapping. Suppose G is semismooth in a neighborhood of a vector $\bar{x} \in \Re^n$ and BD-regular at $\bar{x}$. Then there exist a neighborhood V of $\bar{x}$ and a constant $c > 0$ such that*

(a) for every $y \in V$ and $h \in \Re^n$, there exists a matrix $Q \in \partial_B G(y)$ such that $G'(y;h) = Qh$;

(b) for every $y \in V$ and $Q \in \partial_B G(y)$, Q is nonsingular and $\| Q^{-1} \| \le c$;

(c) for every $\varepsilon > 0$, there exists a $\delta > 0$ such that for all y with $\|y - \bar{x}\| \le \delta$ and any $Q \in \partial_B G(y)$, Q is nonsingular and

$$\| y - Q^{-1}G(y) - \bar{x} \| \le \varepsilon \| y - \bar{x} \|.$$

The above proposition has been the key to the convergence proofs of many Newton methods for solving nonsmooth equations; it remains so in the subsequent analysis.

3.2 Semismooth formulations

There are two ways to reformulate the (time-incremental) friction problem for the application of MP methods. In both formulations, the Coulomb friction constraints (2.7) are written as a system of nonsmooth equation as done in [1, 3, 4, 24]. For this purpose, we define the following planar disks for $i = 1, \ldots, n_c$

$$\mathcal{F}_i \equiv \mathcal{F}(\mu_i\,(p_{in})_+) = \{\, (q_{it}, q_{io}) \in \Re^2 \,:\, q_{it}^2 + q_{io}^2 \le (\mu_i\,(p_{in})_+)^2 \,\},$$

where τ_+ denotes the nonnegative part of the scalar τ. The reason for replacing p_{in} by $(p_{in})_+$ in $\mathcal{F}(\mu_i(p_{in})_+)$ is that in solving one of the formulations stated below, the quantity p_{in} is not required to be nonnegative during the solution process; we use its nonnegative part to define the corresponding friction cone. In particular, for a negative p_{in} (which could occur during the algorithm), the cone $\mathcal{F}_i$ reduces to the origin in the plane. Let Π_i denote the Euclidian projector onto $\mathcal{F}_i$. The discretized Coulomb law (2.7) is equivalent to: for any $r_i > 0$, $i = 1, \ldots, n_c$,

$$\begin{aligned}
& (p_{it}, p_{io}) \\
= \; & \Pi_i(p_{it}(r_i), p_{io}(r_i)) \\
= \; & \begin{cases} (p_{it}(r_i), p_{io}(r_i)) & \text{if } (p_{it}(r_i), p_{io}(r_i)) \in \mathcal{F}_i \\ \dfrac{\mu_i\,(p_{in})_+}{\sqrt{(p_{it}(r_i))^2 + (p_{io}(r_i))^2}}\,(p_{it}(r_i), p_{io}(r_i)) & \text{if } (p_{it}(r_i), p_{io}(r_i)) \notin \mathcal{F}_i, \end{cases}
\end{aligned}$$

or more compactly,

$$(p_{it}, p_{io}) = \min\left(\frac{\mu_i\,(p_{in})_+}{\sqrt{(p_{it}(r_i))^2 + (p_{io}(r_i))^2}},\, 1\right)(p_{it}(r_i), p_{io}(r_i)), \tag{3.1}$$

where 0/0 is defined to be 1 and

$$p_{it}(r_i) \equiv p_{it} + r_i\,(C_{it}u - \bar{w}_{it}) \quad \text{and} \quad p_{io}(r_i) \equiv p_{io} + r_i\,(C_{io}u - \bar{w}_{io}).$$

It should be pointed out that although it is possible to introduce standard complementarity formulations for the friction constraints (as done in [3] for planar problems and in [26, 20] for rigid-body problems), we use the nonsmooth equation formulation (3.1) for two important reasons. One, the formulation is consistent with the augmented Lagrangian formulation used in this field. Two, this formulation produces very good computational results, especially in the unconstrained approach described below.

The major difference between the two formulations of the friction problem occurs in the restatement of the normal contact condition (2.6). In one formulation (inspired by the well-known formulation of the complementarity conditions in terms of the "min" operator), this condition is written in the equivalent form:

$$\min(\,p_n,\ \rho_n \circ\ (g - C_n u)\,) = 0, \tag{3.2}$$

where $\rho_n = (\rho_{in} : i = 1, \ldots, n_c)$ is a vector of given positive scalars, the notation $a \circ b$ denotes the Hadamard product of the vectors a and b, and the minimum operator is taken componentwise. This results in the incremental friction problem being equivalent to the unconstrained system of nonsmooth equations:

$$0 = H_{\mathcal{U}}(u, p_t, p_o, p_n) \equiv \begin{pmatrix} Ku + C_n^T p_n + C_t^T p_t + C_o^T p_o - f^{\text{ext}} \\ \min(\,p_n,\ \rho_n \circ\ (g - C_n u)\,) \\ \left(\begin{pmatrix} p_{it} \\ p_{io} \end{pmatrix} - \Pi_i \begin{pmatrix} p_{it}(r_i) \\ p_{io}(r_i) \end{pmatrix}\right)_{i=1}^{i=n_c} \end{pmatrix}. \tag{3.3}$$

In the other formulation (inspired by the family of interior point methods for solving complementarity problems), the normal contact condition (2.6) is written as:

$$\begin{aligned} v_n + C_n u - g &= 0 \\ p_n \circ v_n &= 0 \\ (v_n, p_n) &\geq 0. \end{aligned}$$

This results in the incremental friction problem being equivalent to the constrained system of nonsmooth equations:

$$0 = H_{\mathcal{C}}(u, p_t, p_o, p_n, v_n) \equiv \begin{pmatrix} Ku + C_n^T p_n + C_t^T p_t + C_o^T p_o - f^{\text{ext}} \\ v_n + C_n u - g \\ p_n \circ v_n \\ \left(\begin{pmatrix} p_{it} \\ p_{io} \end{pmatrix} - \Pi_i \begin{pmatrix} p_{it}(r_i) \\ p_{io}(r_i) \end{pmatrix} \right)_{i=1}^{i=n_c} \end{pmatrix}$$

$$(v_n, p_n) \geq 0. \tag{3.4}$$

In both formulations, the projector Π_i is not Fréchet differentiable; however, it is Bouligand differentiable, as shown in [3]. For convenience, we write down the directional derivatives of the mappings $H_{\mathcal{U}}$ and $H_{\mathcal{C}}$. For the former mapping, we let z denote the generic tuple (u, p_t, p_o, p_n) and dz the direction (du, dp_t, dp_o, dp_n), and define the following sets which are motivated by the minimum operator in the expressions (3.1) and (3.2):

$$\begin{aligned} \mathcal{I}_< &\equiv \{ i : p_{in} + \rho_{in}(C_{in}u - g_i) < 0 \}, \\ \mathcal{I}_= &\equiv \{ i : p_{in} + \rho_{in}(C_{in}u - g_i) = 0 \}, \\ \mathcal{I}_> &\equiv \{ i : p_{in} + \rho_{in}(C_{in}u - g_i) > 0 \}, \\ \mathcal{J} &\equiv \{ i : p_{in} < 0 \}, \\ \mathcal{J}_< &\equiv \{ i : p_{in} > 0,\ (p_{it}(r_i))^2 + (p_{io}(r_i))^2 < (\mu_i p_{in})^2 \}, \\ \mathcal{J}_= &\equiv \{ i : p_{in} > 0,\ (p_{it}(r_i))^2 + (p_{io}(r_i))^2 = (\mu_i p_{in})^2 \}, \\ \mathcal{J}_> &\equiv \{ i : p_{in} > 0,\ (p_{it}(r_i))^2 + (p_{io}(r_i))^2 > (\mu_i p_{in})^2 \}, \\ \mathcal{K}_+ &\equiv \{ i : p_{in} = 0 < (p_{it}(r_i))^2 + (p_{io}(r_i))^2 \}, \\ \mathcal{K}_0 &\equiv \{ i : p_{in} = p_{it}(r_i) = p_{io}(r_i) = 0 \}. \end{aligned}$$

The index sets $\mathcal{I}_=$, $\mathcal{J}_=$, $\mathcal{K}_+$ and $\mathcal{K}_0$ are the source of nondifferentiability for the function $H_{\mathcal{U}}$. Indeed, $H_{\mathcal{U}}$ is strongly F-differentiable at a tuple (u, p_t, p_o, p_n) where these "equality" index sets are empty. We also let

$$dp_{it}(r_i) \equiv dp_{it} + r_i C_{it} du \quad \text{and} \quad dp_{io}(r_i) \equiv dp_{io} + r_i C_{io} du.$$

We further let

$$R_i \equiv \frac{\mu_i p_{in}}{((p_{it}(r_i))^2 + (p_{io}(r_i))^2)^{3/2}} \begin{pmatrix} (p_{io}(r_i))^2 & -p_{it}(r_i)\, p_{io}(r_i) \\ -p_{it}(r_i)\, p_{io}(r_i) & (p_{it}(r_i))^2 \end{pmatrix}, \quad \forall i \in \mathcal{J}_>,$$

and

$$\theta_i \equiv \frac{1}{\mu_i p_{in}} \min\left(\mu_i dp_{in} - \frac{p_{it}(r_i) dp_{it}(r_i) + p_{io}(r_i) dp_{io}(r_i)}{\mu_i p_{in}}, 0 \right), \ \forall i \in \mathcal{J}_=.$$

We then have

$$H'_{\mathcal{U}}(z; dz) =$$

$$\begin{pmatrix} K du + C_n^T dp_n + C_t^T dp_t + C_o^T dp_o \\ (dp_{in})_{i \in \mathcal{I}_<} \\ (\min(dp_{in}, -\rho_{in} C_{in} du))_{i \in \mathcal{I}_=} \\ (-\rho_{in} C_{in} du)_{i \in \mathcal{I}_>} \\ \begin{pmatrix} dp_{it} \\ dp_{io} \end{pmatrix}_{i \in \mathcal{J}} \\ \begin{pmatrix} -r_i C_{it} du \\ -r_i C_{io} du \end{pmatrix}_{i \in \mathcal{J}_<} \\ \left[-r_i \begin{pmatrix} C_{it} du \\ C_{io} du \end{pmatrix} - \theta_i \begin{pmatrix} p_{it}(r_i) \\ p_{io}(r_i) \end{pmatrix} \right]_{i \in \mathcal{J}_=} \\ \left[\begin{pmatrix} dp_{it} \\ dp_{io} \end{pmatrix} - D_i \begin{pmatrix} p_{it}(r_i) \\ p_{io}(r_i) \end{pmatrix} - R_i \begin{pmatrix} dp_{it}(r_i) \\ dp_{io}(r_i) \end{pmatrix} \right]_{i \in \mathcal{J}_>} \\ \left[\begin{pmatrix} dp_{it} \\ dp_{io} \end{pmatrix} - D_i^+ \begin{pmatrix} p_{it}(r_i) \\ p_{io}(r_i) \end{pmatrix} \right]_{i \in \mathcal{K}_+} \\ \left[\begin{pmatrix} dp_{it} \\ dp_{io} \end{pmatrix} - \min(D_i^+, 1) \begin{pmatrix} dp_{it}(r_i) \\ dp_{io}(r_i) \end{pmatrix} \right]_{i \in \mathcal{K}_0} \end{pmatrix},$$

where

$$D_i = \frac{\mu_i dp_{in}}{\sqrt{(p_{it}(r_i))^2 + (p_{io}(r_i))^2}}$$

and

$$D_i^+ = \frac{\mu_i (dp_{in})_+}{\sqrt{(dp_{it}(r_i))^2 + (dp_{io}(r_i))^2}}.$$

For the mapping H_C, we let w denote the generic tuple (u, p_t, p_o, p_n, v_n) and let dw represent $(du, dp_t, dp_o, dp_n, dv_n)$. We write down the directional derivative $H'_C(w; dw)$ at the tuple w with the pair (p_n, v_n) being positive. For such a tuple w, the index sets $\mathcal{J}$, $\mathcal{K}_+$ and $\mathcal{K}_0$ are all empty. We then get

$$H'_C(w; dw) =$$

$$\left(\begin{array}{c} K du + C_n^T dp_n + C_t^T dp_t + C_o^T dp_o \\ dv_n + C_n du \\ p_n \circ dv_n + v_n \circ dp_n \\ \left(\begin{array}{c} -r_i\, C_{it} du \\ -r_i\, C_{io} du \end{array} \right)_{i \in \mathcal{J}_<} \\ \left[-r_i \left(\begin{array}{c} C_{it} du \\ C_{io} du \end{array} \right) - \theta_i \left(\begin{array}{c} p_{it}(r_i) \\ p_{io}(r_i) \end{array} \right) \right]_{i \in \mathcal{J}_=} \\ \left[\left(\begin{array}{c} dp_{it} \\ dp_{io} \end{array} \right) - D_i \left(\begin{array}{c} p_{it}(r_i) \\ p_{io}(r_i) \end{array} \right) - R_i \left(\begin{array}{c} dp_{it}(r_i) \\ dp_{io}(r_i) \end{array} \right) \right]_{i \in \mathcal{J}_>} \end{array} \right),$$

where

$$D_i = \frac{\mu_i \, dp_{in}}{\sqrt{(p_{it}(r_i))^2 + (p_{io}(r_i))^2}}.$$

The lemma below establishes a key property of the mapping Π_i. (A referee suggests that the function G below is "strongly semismooth"; we have not verified this claim.)

Lemma 3.1 *For any constant $\tau > 0$, the vector function*

$$G : (a, b, c) \in \Re^3 \mapsto \left[\min \left(\frac{\tau\, a_+}{\sqrt{b^2 + c^2}},\ 1 \right) \right] (b, c) \in \Re^2$$

is everywhere semismooth, where $0/0$ is defined to be 1.

Proof. Let $(\bar{a}, \bar{b}, \bar{c})$ be arbitrary triple. We consider several cases.

Case 1. $\bar{b}^2 + \bar{c}^2 > 0$. In this case, the function G is clearly piecewise smooth in a neighborhood of of $(\bar{a}, \bar{b}, \bar{c})$; thus it is semismooth there.

Case 2. $\bar{b}^2 + \bar{c}^2 = 0$ and $\bar{a} \neq 0$. In this case, the function G in a neighborhood of $(\bar{a}, \bar{b}, \bar{c})$ is equal to either the zero function (if $\bar{a} < 0$) or the function $(a, b, c) \mapsto (b, c)$ (if $\bar{a} > 0$). In both situations, the desired semismoothness follows.

Case 3. $\bar{a} = \bar{b} = \bar{c} = 0$. It suffices to show

$$\lim_{h \to 0} \frac{G'(h; h) - G'(0; h)}{\| h \|} = 0.$$

But this follows quite easily by observing since $G(th) = tG(h)$ for $t > 0$ and all h that $G'(h; h) = G'(0; h) = G(h)$, hence the numerator is identically equal to zero. Q.E.D.

The following proposition is now an easy consequence of the above lemma. No proof is required.

Proposition 3.2 *The mapping $H_{\mathcal{U}}$ defined in (3.3) is semismooth at all tuples (u, p_t, p_o, p_n). The mapping $H_{\mathcal{C}}$ defined in (3.4) is semismooth at all tuples (u, p_t, p_o, p_n, v_n).*

4 A NEWTON METHOD FOR A SEMISMOOTH CE

In order to have a unified treatment of the two formulations (3.3) and (3.4), we propose a method that combines the BN method in [17] for unconstrained B-differentiable equations and the potential reduction interior point method in [27] for constrained smooth equations. This unified method is applicable for solving a constrained system of semismooth equations. In this section, we present the method and its convergence properties. Specialization of the methods to the two mappings $H_{\mathcal{U}}$ and $H_{\mathcal{C}}$ will be discussed in the next section.

Let $H : \Re^n \to \Re^n$ be a continuous mapping and Ω be a nonempty subset of $\Re^n$. The constrained equation (CE) defined by the pair (Ω, H) is to find a vector z such that

$$H(z) = 0, \quad z \in \Omega, \tag{4.1}$$

We adopt the framework introduced in [16] which deals with constrained smooth equations and extends the method in [27]. Generalizing to a semismooth setting, we make the following blanket assumptions on the pair (Ω, H).

(A1) Ω is a closed subset of $\Re^n$ with a nonempty interior, denoted int Ω;

(A2) there exists a closed convex set $S \subseteq \Re^n$ such that

(a) $0 \in S$;

(b) the (open) set $\Omega_{\mathcal{I}} \equiv H^{-1}(\text{int } S) \cap \text{int } \Omega$ is nonempty;

(c) the set $H^{-1}(\text{int } S) \cap \text{bd } \Omega$ is empty.

(d) H is semismooth in $\Omega_{\mathcal{I}}$, and for each $z \in \Omega_{\mathcal{I}}$, $H'(z; \cdot)$ is surjective in its second argument.

We further postulate the existence of a potential function $p : \text{int } S \to \Re$, and a vector $a \in \Re^n$ satisfying the following conditions:

(A3) for every sequence $\{u^k\} \subset \text{int } S$ such that

$$\text{either } \lim_{k\to\infty} \|u^k\| = \infty \text{ or } \lim_{k\to\infty} u^k = \bar{u} \in \text{bd } S \setminus \{0\},$$

we have

$$\lim_{k\to\infty} p(u^k) = \infty; \tag{4.2}$$

(A4) p is continuously differentiable on its domain and $u^T \nabla p(u) > 0$ for all nonzero $u \in \text{int } S$;

(A5) for all $u \in \text{int } S$, $u^T \nabla p(u) \geq \dfrac{(a^T u)\,(a^T \nabla p(u))}{\|a\|^2}$ (the right-hand fraction is taken to be zero if $a = 0$; thus if $a = 0$, this assumption is not needed as it is implied by (A4)).

Initiated at a vector $z^0 \in \Omega_{\mathcal{I}}$, the solution method for the CE (Ω, H) generates a sequence of iterates $\{z^k\}$ lying in the working set $\Omega_{\mathcal{I}}$. In each iteration, a search direction is computed by solving a certain perturbed directional Newton equation corresponding to the current iterate $z^k \in \Omega_{\mathcal{I}}$. This direction provides descent at z^k for the merit function:

$$\psi(z) \equiv p(H(z)), \quad z \in \Omega_{\mathcal{I}}.$$

The next iterate z^{k+1} is obtained by performing a line search such that sufficient decrease in ψ is obtained from its current value $\psi(z^k)$; moreover z^{k+1} is restricted to remain in $\Omega_{\mathcal{I}}$. Below is a step-by-step description of the potential reduction Newton (PRN) method for solving the CE (4.1).

Algorithm PRN: with $\Omega_{\mathcal{I}}$, p, and a satisfying (A2)–(A5).

Step 0. (Initialization) Let a vector $z^0 \in \Omega_{\mathcal{I}}$ and scalars $\rho \in (0,1)$ and $\alpha \in (0,1)$ be given. Let a sequence of scalars $\{\sigma_k\} \subset [0,1)$ be also given. Set the iteration counter $k = 0$.

Step 1. (Computing the modified Newton direction) Solve the perturbed directional Newton equation:

$$H(z^k) + H'(z^k; dz^k) = \sigma_k \frac{a^T H(z^k)}{\|a\|^2}\, a \tag{4.3}$$

to obtain a search direction dz^k (if $a = 0$, the right-hand side is taken to be zero).

Step 2. (Armijo line search) Let $\tau_k \equiv \rho^{m_k}$, where m_k is the smallest nonnegative integer m such that $z^k + \rho^m dz^k \in \Omega_{\mathcal{I}}$ and

$$\psi(z^k + \rho^m dz^k) - \psi(z^k) \leq -\alpha\, \rho^m\, (1 - \sigma_k)\, H(z^k)^T \nabla p(H(z^k)).$$

Set $z^{k+1} \equiv z^k + \tau_k dz^k$.

Step 3. (Termination test) If

$$\|\, H(z^{k+1})\,\| \leq \text{ prescribed tolerance},$$

stop; accept z^{k+1} as an approximate solution of the CE (Ω, H). Otherwise, return to Step 1 with k replaced by $k+1$.

Since $z^k \in \Omega_I$, it follows that $\psi = p(H(\cdot))$ is directionally differentiable at z^k, by assumption (A2d); this assumption further implies that the search direction dz^k exists. Moreover, by the same proof as in Lemma 2 of [16], we can deduce (provided that $H(z^k)$ is nonzero),

$$\psi'(z^k; dz^k) \leq -(1-\sigma_k)\, H(z^k)^T \nabla p(H(z^k)) < 0, \tag{4.4}$$

where the last inequality follows from assumption (A4). Thus the integer m_k is well defined and can be determined in a finite number of trials beginning with $m = 0$ and increasing by 1 at each trial.

We establish a preliminary lemma.

Lemma 4.1 *Suppose H is semismooth near a vector z^* and BD-regular at z^*. There exist a neighborhood V of z^* and positive scalars $\gamma_1 > \gamma_2$ such that for all vectors $z \in V$, scalars $\sigma \in [0,1)$ and any vector dz satisfying*

$$H(z) + H'(z; dz) = \sigma \frac{a^T H(z)}{\| a \|^2} a,$$

we have

$$\gamma_2 (1-\sigma) \| H(z) \| \leq \| dz \| \leq \gamma_1 \| H(z) \|.$$

Proof. Since H is semismooth near z^*, it is locally Lipschitz in a neighborhood of z^*. Thus if L denotes the Lipschitz modulus of H near z^*, it follows that there exists a neighborhood V_1 of z^* such that for all $z \in V_1$,

$$\| H'(z; v) \| \leq L \| v \|, \quad \forall v \in \Re^n.$$

By Proposition 3.1, there exist a neighborhood V_2 of z^* and a constant $c > 0$ such that for all $z \in V_2$,

$$\| H'(z; v) \| \geq c \| v \|, \quad \forall v \in \Re^n.$$

Consequently if $V \equiv V_1 \cap V_2$ and $z \in V$ and dz is as stated, we obtain

$$(1+\sigma) \| H(z) \| \geq \| H'(z; dz) \| \geq (1-\sigma) \| H(z) \|$$

from which the desired conclusion of the lemma follows with $\gamma_1 \equiv (1+\sigma)c^{-1}$ and $\gamma_2 \equiv L^{-1}$. Q.E.D.

In the theorem below, we summarize the properties of a sequence of vectors produced by the PRN algorithm.

Theorem 4.1 *Under assumptions (A1)-(A5), let $\{z^k\}$ be an infinite sequence generated by the above algorithm. The following statements hold.*

(A) $\{z^k\} \subset \Omega_I$ and $\{H(z^k)\}$ is bounded.

(B) If

$$\liminf_{k\to\infty} (1-\sigma_k)\, \tau_k > 0, \tag{4.5}$$

then every accumulation point of $\{z^k\}$ *is a solution of the CE* (Ω, H).

(C) Assume that $\limsup_{k\to\infty} \sigma_k < 1$. *If* z^* *is an accumulation point of* $\{z^k\}$ *and* H *has a strong Fréchet derivative at* z^*, *then* z^* *solves the CE* (Ω, H).

(D) Assume that $\lim_{k\to\infty} \sigma_k = 0$. *If* $\{z^k\}$ *has an accumulation point* z^∞ *and* H *is BD-regular near* z^∞, *then the following two statements are equivalent:*

(a) $H(z^\infty) = 0$ *and* $\{z^k\}$ *converges superlinearly to* z^∞, *that is,*

$$\lim_{k\to\infty} \frac{\| z^{k+1} - z^\infty \|}{\| z^k - z^\infty \|} = 0;$$

(b) $\lim_{k\to\infty} \tau_k = 1$.

Proof. The sequence $\{z^k\}$ is well defined and clearly contained in $\Omega_{\mathcal{I}}$. Since $\{\psi(z^k)\}$ is bounded above by $\psi(z^0)$, assumption (A3) implies that $\{H(z^k)\}$ is bounded. Thus (A) follows.

Let z^* be the limit of a convergent subsequence $\{z^k : k \in \kappa\}$. Clearly, $z^* \in \Omega$ because Ω is a closed set. Assume that $H(z^*) \neq 0$. Condition (A3) then implies that $H(z^*) \in \operatorname{int} S$. Thus $z^* \in \Omega_{\mathcal{I}}$ by (A2c). Assumption (A4) then yields

$$H(z^*)^T \nabla p(H(z^*)) > 0.$$

Since for each k

$$\psi(z^{k+1}) < \psi(z^k),$$

it follows that the sequence $\{\psi(z^k)\}$ converges because the subsequence $\{\psi(z^k) : k \in \kappa\}$ converges to $\psi(z^*)$. Consequently, the Armijo rule implies

$$\lim_{k\to\infty} (1 - \sigma_k)\, \tau_k\, H(z^k)^T \nabla p(H(z^k)) = 0.$$

If (4.5) holds, it follows that $H(z^*)^T \nabla p(z^*) = 0$ which is a contradiction. Thus (B) holds.

To prove (C), we may assume without loss of generality that $H(z^*) \neq 0$ and

$$\lim_{k(\in\kappa)\to\infty} \tau_k = 0.$$

Since H is assumed to have a strong F-derivative at z^*, condition (A2d) implies that the Jacobian matrix $H'(z^*)$ exists and is nonsingular. Thus by Lemma 4.1, it follows that the subsequence of directions $\{dz^k : k \in \kappa\}$ is bounded. Without loss of generality, we may assume that this subsequence converges to the limit dz^*; we may further assume that $\{\sigma_k : k \in \kappa\}$ converges to some $\sigma_* \in [0, 1)$. We can then show by passing to the limit $k(\in \kappa) \to \infty$ in (4.3) that

$$H(z^*) + H'(z^*; dz^*) = \sigma_* \frac{a^T H(z^*)}{\| a \|^2}\, a.$$

Lemma 2 of [16] then implies

$$\psi'(z^*; dz^*) \le -(1 - \sigma_*)H(z^*)^T \nabla p(H(z^*)) < 0.$$

Since z^* belongs to the open set $\Omega_{\mathcal{I}}$, it follows from the definition of m_k and (4.4),

$$\frac{\psi(z^k + \rho^{m_k - 1} dz^k) - \psi(z^k)}{\rho^{m_k - 1}} > \alpha\, \psi'(z^k; dz^k).$$

Taking the limit $k(\in \kappa) \to \infty$ and using the strong Fréchet differentiability of H at z^* (which implies the same property of ψ), we deduce

$$\psi'(z^*; dz^*) \ge \alpha\, \psi'(z^*; dz^*)$$

which is a contradiction because $\alpha \in (0, 1)$. This contradiction establishes (C).

Finally to prove (D), we assume that $\{\sigma_k\}$ converges to zero. We further assume that z^∞ is the limit of the convergent subsequence $\{z^k : k \in \kappa\}$. In the following proof, we do not assume that H is F-differentiable at z^∞.

(b) $\Rightarrow$ (a). Suppose that $\{\tau_k\}$ converges to unity. By (B), it follows that $H(z^\infty) = 0$. Next we show that the entire sequence $\{z^k\}$ converges to z^∞ superlinearly. Since $\{\sigma_k\}$ converges to zero, by Lemma 4.1, it follows that there exist constants $\gamma_1 > \gamma_2' > 0$ such that for all $k \in \kappa$ sufficiently large,

$$\gamma_2' \, \| H(z^k) \| \le \| dz^k \| \le \gamma_1 \, \| H(z^k) \|.$$

Let $c > 0$ and V be the scalar and neighborhood of z^∞ associated with the BD-regularity of H at z^∞. Let $\varepsilon > 0$ be a scalar to be determined. Associated with this ε, let $\delta > 0$ be as given in Proposition 3.1 for the function H at z^∞. Without loss of generality, we may take δ to be such that all vectors y with $\|y - z^\infty\| \le \delta$ belong to V.

Since $\{z^k : k \in \kappa\}$ converges to z^∞, z^k satisfies $\|z^k - z^\infty\| \le \delta$ for all $k \in \kappa$ sufficiently large. Corresponding to such an iterate z^k and the direction dz^k, there exists a matrix $Q_k \in \partial_B H(z^k)$ such that

$$H'(z^k; dz^k) = Q_k dz^k$$

which implies

$$dz^k = Q_k^{-1} \left(\sigma_k \frac{a^T H(z^k)}{\| a \|^2} a - H(z^k) \right).$$

Thus we have

$$\begin{aligned} z^{k+1} - z^\infty &= z^k + \tau_k \, dz^k - z^\infty = (z^k + dz^k - z^\infty) - (1 - \tau_k)\, dz^k \\ &= (z^k - Q_k^{-1} H(z^k) - z^\infty) + \sigma_k \frac{a^T H(z^k)}{\| a \|^2} Q_k^{-1} a - (1 - \tau_k)\, dz^k. \end{aligned}$$

By Proposition 3.1 and Lemma 4.1, it follows that

$$\begin{aligned}\| z^{k+1} - z^{\infty} \| &\leq \varepsilon \| z^k - z^{\infty} \| + \sigma_k c \| H(z^k) \| + (1 - \tau_k) \| dz^k \| \\ &\leq \varepsilon \| z^k - z^{\infty} \| + [\sigma_k c + (1 - \tau_k) \gamma_1] \| H(z^k) \| \\ &\leq [\varepsilon + L(\sigma_k c + (1 - \tau_k) \gamma_1)] \| z^k - z^{\infty} \|,\end{aligned}$$

where $L > 0$ is Lipschitz modulus of H near z^{∞}. From this last expression, we see that for any fixed but arbitrary $\varepsilon \in (0, 1)$, provided that $k \in \kappa$ is sufficiently large, it follows that $\|z^{k+1} - z^{\infty}\| \leq \delta$. Thus we can repeat the above derivation with k replaced by $k + 1$. In this way, we can establish that

$$\| z^{k+1} - z^{\infty} \| \leq [\varepsilon + L(\sigma_k c + (1 - \tau_k) \gamma_1)] \| z^k - z^{\infty} \|,$$

for all k sufficiently large. Furthermore, since ε is arbitrary, the superlinear convergence of the entire sequence $\{z^k\}$ to z^{∞} follows immediately.

(a) $\Rightarrow$ (b). We follow the proof of Corollary 2 in [18]. We first establish that

$$\lim_{k \to \infty} \frac{H(z^k) + H'(z^k; z^{k+1} - z^k)}{\| z^{k+1} - z^k \|} = 0. \tag{4.6}$$

Let $Q_k \in \partial_B H(z^k)$ be such that

$$H'(z^k; z^{k+1} - z^k) = Q_k(z^{k+1} - z^k).$$

We have

$$H(z^k) + H'(z^k; z^{k+1} - z^k) = H(z^k) - H(z^{\infty}) - Q_k(z^k - z^{\infty}) + Q_k(z^{k+1} - z^{\infty}).$$

Thus,

$$\begin{aligned}&\frac{H(z^k) + H'(z^k; z^{k+1} - z^k)}{\| z^{k+1} - z^k \|} \\ = \ &\frac{H(z^k) - H(z^{\infty}) - Q_k(z^k - z^{\infty})}{\| z^{k+1} - z^k \|} + \frac{Q_k(z^{k+1} - z^{\infty})}{\| z^{k+1} - z^k \|}.\end{aligned}$$

Furthermore we have

$$\begin{aligned}&\frac{H(z^k) - H(z^{\infty}) - Q_k(z^k - z^{\infty})}{\| z^{k+1} - z^k \|} \\ = \ &\frac{H(z^k) - H(z^{\infty}) - Q_k(z^k - z^{\infty})}{\| z^k - z^{\infty} \|} \frac{\| z^k - z^{\infty} \|}{\| z^{k+1} - z^k \|};\end{aligned}$$

in turn, since

$$z^{k+1} - z^k = (z^{k+1} - z^{\infty}) + (z^{\infty} - z^k),$$

the superlinear convergence of $\{z^k\}$ to z^{∞} implies that

$$\lim_{k \to \infty} \frac{\| z^k - z^{\infty} \|}{\| z^{k+1} - z^k \|} = 1. \tag{4.7}$$

Similarly we can show that

$$\lim_{k\to\infty} \frac{\| z^{k+1} - z^{\infty} \|}{\| z^{k+1} - z^k \|} = 0.$$

Combining these expression, the desired limit (4.6) follows.

Now since $z^{k+1} = z^k + \tau_k \, dz^k$, we deduce

$$\lim_{k\to\infty} \frac{\tau_k^{-1} \, H(z^k) + H'(z^k; dz^k)}{\| dz^k \|} = 0.$$

By the definition of dz^k, we obtain

$$\lim_{k\to\infty} \frac{(\tau_k^{-1} - 1) \, H(z^k) + \sigma_k \, \dfrac{a^T H(z^k)}{\|a\|^2} \, a}{\| dz^k \|} = 0.$$

Since $\{\sigma_k\}$ converges to zero, Lemma 4.1 implies that

$$\lim_{k\to\infty} \sigma_k \, \frac{a^T H(z^k)}{\| dz^k \|} \, \frac{a}{\| a \|^2} = 0.$$

Consequently, we have

$$\lim_{k\to\infty} \frac{(1 - \tau_k) \, H(z^k)}{\| z^{k+1} - z^k \|} = 0.$$

By Proposition 3 in [18], there exists a constant $\beta > 0$ such that for all k sufficiently large,

$$\| H(z^k) \| \geq \beta \, \| z^k - z^{\infty} \|.$$

By (4.7), it therefore follows that

$$\lim_{k\to\infty} \tau_k = 1$$

as desired. Q.E.D.

5 SPECIALIZATION TO THE FRICTION PROBLEM

The application of the PRN algorithm to the unconstrained formulation (3.3) is straightforward. For this application, the set Ω is the whole space $\Re^{n_u + 3n_c}$, so is the set S, the vector $a = 0$, and the function

$$p(y) = \tfrac{1}{2} \, y^T y, \quad y \in \Re^{n_u + 3n_c}.$$

Assumptions (A1)–(A5) (except for the surjectivity assumption in (A2d) which requires additional conditions; see the discussion below) can easily be seen to hold with these choices of Ω, S, a, and p. The method reduces to the BN method for solving the semismooth equation (3.3). The resulting directional Newton equation:

$$H_{\mathcal{U}}(z^k) + H'_{\mathcal{U}}(z^k; dz^k) = 0 \tag{5.1}$$

is a piecewise linear equation (equivalent to a certain mixed linear complementarity problem). As in the implementation for the planar problems [3], we have used a modification of the directional derivative $H'_{\mathcal{U}}(z^k; dz^k)$ in our computational experimentation. Specifically, we define

$$\tilde{H}'_{\mathcal{U}}(z; dz) \equiv$$

$$\left(\begin{array}{c} K du + C_n^T dp_n + C_t^T dp_t + C_o^T dp_o \\[1ex] (dp_{in})_{i \in \mathcal{I}_< \cup \mathcal{I}_=} \\[1ex] (-\rho_{in} C_{in} du)_{i \in \mathcal{I}_>} \\[1ex] \left(\begin{array}{c} dp_{it} \\ dp_{io} \end{array} \right)_{i \in \mathcal{J} \cup \mathcal{K}_+ \cup \mathcal{K}_0} \\[1ex] \left(\begin{array}{c} -r_i C_{it} du \\ -r_i C_{io} du \end{array} \right)_{i \in \mathcal{J}_< \cup \mathcal{J}_=} \\[1ex] \left[\left(\begin{array}{c} dp_{it} \\ dp_{io} \end{array} \right) - D_i \left(\begin{array}{c} p_{it}(r_i) \\ p_{io}(r_i) \end{array} \right) - R_i \left(\begin{array}{c} dp_{it}(r_i) \\ dp_{io}(r_i) \end{array} \right) \right]_{i \in \mathcal{J}_>} \end{array} \right), \tag{5.2}$$

where

$$D_i = \frac{\mu_i \, dp_{in}}{\sqrt{(p_{it}(r_i))^2 + (p_{io}(r_i))^2}},$$

and solve instead of (5.1) the following linear equation for the vector dz^k:

$$H_{\mathcal{U}}(z^k) + J_{\mathcal{U}} \, dz^k = 0, \tag{5.3}$$

where $J_{\mathcal{U}} \, dz^k \equiv \tilde{H}'_{\mathcal{U}}(z^k; dz^k)$ (specifically, $J_{\mathcal{U}}$ is the square matrix of order $n_u + 3n_c$ that represents the linear transformation $dz \mapsto \tilde{H}'_{\mathcal{U}}(z^k; dz)$). This modification is related to the modified BN scheme for solving the linear complementarity problem [8] whose finite convergence is recently established by Fischer and Kanzow [7].

The matrix $J_{\mathcal{U}}$ depends on the various index sets. In what follows we give a sufficient condition on the data of the friction problem that guarantees the nonsingularity of this matrix for all such index sets.

Proposition 5.1 *Suppose that the stiffness matrix K is positive definite and the transformation matrix*

$$[\, C_n^T \;\; C_t^T \;\; C_o^T \,]$$

has full column rank. Then there exists a positive scalar $\bar{\mu}$ such that the matrix $J_{\mathcal{U}}$ is nonsingular for all index sets and friction coefficients $\mu_i \in [0, \bar{\mu}]$.

Proof. Consider the matrix $J_{\mathcal{U}}$ with $\mu_i = 0$ for all i. It is easy to show that the assumptions imply that such a matrix must be nonsingular for all index sets. Since there are a finite number of different groups of index sets, a perturbation argument establishes the proposition. Q.E.D.

It should be noted that in the numerical experimentation, we have applied the PRN algorithm to problems with a singular stiffness matrix K. Although the above proposition no longer guarantees the nonsingularity of the matrix $J_{\mathcal{U}}$, the solvability of the linear equation (5.3) is never an issue in the computer runs.

Due to the above modification, the generated direction dz^k is no longer guaranteed to be a descent direction for the merit function

$$\psi_u(z) \equiv \tfrac{1}{2} H_{\mathcal{U}}(z)^T H_{\mathcal{U}}(z).$$

To prevent jamming of the line search, we impose a lower bound on the step size as done in [3] for the planar problems. The use of this modification is motivated by the fact that the set $\mathcal{I}_= \cup \mathcal{J}_= \cup \mathcal{K}_+ \cup \mathcal{K}_0$ is empty in the large majority of iterations in the practical implementation of the PRN algorithm; see Subsection 7.1. We have never experienced any convergence difficulty with this modification in the computational experiments to be reported later. In fact, this modified scheme performs very satisfactorily for the problems to which it is applied.

For the application of the PRN algorithm to the CE formulation (3.4), we take

$$\Omega \equiv \Re^{n_u+2n_c} \times \Re^{2n_c}_+, \quad S \equiv \Re^{n_u+n_c} \times \Re^{n_c}_+ \times \Re^{2n_c},$$

the vector $a \equiv (0, 0, \mathbf{1}_{n_c}, 0, 0)$, where $\mathbf{1}_{n_c}$ is the n_c-dimensional vector of all ones, and the function

$$p(b,c,d,e,f) \equiv \zeta \log(b^Tb + c^Tc + \mathbf{1}_{n_c}^T d + e^Te + f^Tf) - \sum_{i=1}^{n_c} \log d_i, \qquad (5.4)$$

$$\text{for } (b,c,d,e,f) \in \operatorname{int} S = \Re^{n_u+n_c} \times \Re^{n_c}_{++} \times \Re^{2n_c},$$

where $\zeta > n_c$ is an arbitrary scalar. The verification of assumptions (A1)–(A5) (except for the surjectivity assumption in (A2d)) with these choices of Ω, S, a is not difficult; for (A4) and (A5), see [27, 16]. We also employ a modification of the perturbed directional Newton equation (4.3) similar to that in the unconstrained formulation. Specifically, we define

$$\tilde{H}'_C(w; dw) \equiv$$

$$\begin{pmatrix} K du + C_n^T dp_n + C_t^T dp_t + C_o^T dp_o \\ dv_n + C_n du \\ p_n \circ dv_n + v_n \circ dp_n \\ \begin{pmatrix} -r_i C_{it} du \\ -r_i C_{io} du \end{pmatrix}_{i \in \mathcal{J}_< \cup \mathcal{J}_=} \\ \left[\begin{pmatrix} dp_{it} \\ dp_{io} \end{pmatrix} - D_i \begin{pmatrix} p_{it}(r_i) \\ p_{io}(r_i) \end{pmatrix} - R_i \begin{pmatrix} dp_{it}(r_i) \\ dp_{io}(r_i) \end{pmatrix} \right]_{i \in \mathcal{J}_>} \end{pmatrix}, \tag{5.5}$$

where

$$D_i = \frac{\mu_i \, dp_{in}}{\sqrt{(p_{it}(r_i))^2 + (p_{io}(r_i))^2}},$$

and solve instead of the piecewise linear equation:

$$H_C(w^k) + H'_C(w^k; dw^k) = \sigma_k \frac{a^T H_C(w^k)}{\|a\|^2} a$$

the following linear equation to obtain the search vector dw^k:

$$H_C(w^k) + J_C \, dw^k = \sigma_k \frac{(p_n^k)^T v_n^k}{n_c} (0, 0, \mathbf{1}_{n_c}, 0, 0)^T, \tag{5.6}$$

where $J_C \, dw^k \equiv \tilde{H}'_C(w^k; dw^k)$ (the interpretation of J_C is similar to that of $J_\mathcal{U}$). The merit function is

$$\psi_C(w) \equiv p(H_C(w)),$$

where p is given by (5.4). Comments made previously about the system (5.3) and the matrix $J_\mathcal{U}$ apply similarly to (5.6) the matrix J_C. These are not repeated.

6 SOME IMPLEMENTATION DETAILS

In order to determine the contact forces, it is obvious that only those displacements that are related to nodes on the contact surface have to be calculated; all the other displacements can be eliminated from the problem once and for all. For this purpose, the displacement vector u is decomposed into two subvectors, u_c and u_r, with u_c representing the displacements of nodes on the potential contact surface and u_r including all the other displacements of body $\mathcal{B}$. If there are no partially constrained nodes on the contact surface, $u_c \in \Re^{3n_c}$ and

$u_r \in \Re^{n_u - 3n_c}$. The equilibrium equation (2.1) is then written in partitioned form as

$$\begin{pmatrix} K_{c,c} & K_{c,r} \\ K_{r,c} & K_{r,r} \end{pmatrix} \begin{pmatrix} u_c \\ u_r \end{pmatrix} = \begin{pmatrix} f_c^{\rm ext} - \bar{C}_n^T p_n - \bar{C}_t^T p_t - \bar{C}_o^T p_o \\ f_r^{\rm ext} \end{pmatrix},$$

where the matrices $\bar{C}_n, \bar{C}_t, \bar{C}_o \in \Re^{n_c \times 3n_c}$ are formed from C_n, C_t and C_o by deleting the zero columns corresponding to noncontact nodes. By use of static condensation we eliminate u_r and obtain the following reduced equilibrium equation corresponding to nodes on the potential contact surface:

$$\tilde{K} u_c = f_c, \tag{6.1}$$

where

$$\begin{aligned} \tilde{K} &\equiv K_{c,c} - K_{c,r} K_{r,r}^{-1} K_{r,c}, \\ f_c &\equiv f_c^{\rm ext} - K_{c,r} K_{r,r}^{-1} f_r^{\rm ext} - \bar{C}_n^T p_n - \bar{C}_t^T p_t - \bar{C}_o^T p_o. \end{aligned}$$

In both formulations (3.3) and (3.4), this condensation is performed in the initialization step; the resulting formulations without the subvector u_r are then solved by the PRN algorithm. After the contact displacement subvector u_c is computed, the noncontact displacement subvector can be recovered from

$$u_r = K_{r,r}^{-1} (f_r^{\rm ext} - K_{r,c} u_c).$$

6.1 Unconstrained formulation

For the unconstrained formulation (UF), by examining (5.2), we note that in solving the linear equations (5.3) the components $(dp_{in})_{i \in \mathcal{I}_< \cup \mathcal{I}_=}$ and $(dp_{it}, dp_{io})_{i \in \mathcal{J} \cup \mathcal{K}_+ \cup \mathcal{K}_0}$ can easily be eliminated. In the case where $\tilde{K}$ is positive definite we can further eliminate du_c. With all these reductions, the number of unknowns in the system (5.3) to be solved in each iteration is $|\mathcal{I}_>| + 2|\mathcal{J}_< \cup \mathcal{J}_=| + 2|\mathcal{J}_>|$, where $|\mathcal{S}|$ denotes the cardinality of a set $\mathcal{S}$. This number is significantly smaller than the total number of variables in the friction problem, which is $n_u + 3n_c$. In the case where $\tilde{K}$ is positive semidefinite, we can partition $\tilde{K} \in \Re^{n_{u_c} \times n_{u_c}}$ as

$$\tilde{K} = \begin{pmatrix} \tilde{K}_{1,1} & \tilde{K}_{1,2} \\ \tilde{K}_{2,1} & \tilde{K}_{2,2} \end{pmatrix},$$

where $\tilde{K}_{1,1} \in \Re^{n_{rd} \times n_{rd}}$ and $\tilde{K}_{2,2} \in \Re^{(n_{u_c} - n_{rd}) \times (n_{u_c} - n_{rd})}$ with the latter matrix being positive definite. Here, n_{rd} is the number of infinitesimal (i.e. small displacement) rigid body displacements that body $\mathcal{B}$ is allowed to perform. The maximum value of n_{rd} is not more than six, corresponding to three translations and three rotations. In this case, the du_c variables corresponding to $\tilde{K}_{2,2}$ can be eliminated. The number of unknowns in the resulting system (5.3) actually solved is consequently $|\mathcal{I}_>| + 2|\mathcal{J}_< \cup \mathcal{J}_=| + 2|\mathcal{J}_>| + n_{rd}$.

When eliminating the du_c variables, a large number of matrix multiplications of the type

$$D \equiv A\tilde{K}^{-1}B^T \tag{6.2}$$

are performed in every iteration (for a positive semidefinite stiffness matrix, $\tilde{K}$ is replaced with $\tilde{K}_{2,2}$), where A and B are submatrices of $\bar{C}_n, \bar{C}_t$ and $\bar{C}_o$. In these calculations, the sparsity of these matrices is exploited; the maximum number of nonzero elements in each row of $\bar{C}_n, \bar{C}_t$ and $\bar{C}_o$ is three. In our implementation we calculate $\tilde{K}^{-1}$ explicitly in the initialization step. In the special case when all n, t, o systems coincide with the global x, y, z system in which the u vector is represented, D will consist of certain elements of $\tilde{K}^{-1}$. Thus, once $\tilde{K}^{-1}$ has been obtained, no additional calculations are needed to get the D matrix in (6.2). For other definitions of the n, t, o systems, the method of first determining $\tilde{K}^{-1}$ when calculating (6.2) should still be efficient.

6.2 Constrained formulation

For the constrained formulation (CF) we note that dv_n can easily be eliminated from (5.6) by multiplying the second row in (5.5) by diag(p_n) from the left and subtracting it from the third row. The du_c variables are eliminated in the same way as for the constrained formulation. Thus, the number of unknowns in the resulting system of linear equations (5.6) being solved is $3n_c + n_{rd}$. (For a positive definite stiffness matrix, n_{rd} is zero.) In most iterations, this number can be expected to be significantly larger than the corresponding number in the unconstrained formulation. Thus, the computational effort in each iteration of solving the CF is larger than that for the UF.

7 NUMERICAL EXAMPLES

Algorithm PRN has been implemented in FORTRAN 90 using the DEC OSF/1 compiler for the two different formulations. The compiler option `-tune host` is used. The examples to be described below have been executed on a DEC Alpha 200 Model 4/100 workstation with 96 megabytes of memory, using double arithmetic precision. The matrix factorizations and matrix inverses are performed by LAPACK [2] routines, which have not been fine tuned for the specific computer.

In what follows, we will present four different examples, denoted Problem 1–4, in order to test the numerical behavior of the two formulations. We study problems with both positive definite and semidefinite stiffness matrices. Various combinations of contact nodes and time increments are used. In all four examples, the continuous contact problem is solved approximately by first discretizing the elastic body by finite elements, and then solving the discrete quasistatic contact problem described in Section 2. Since the underlying problem is continuous, the contact forces are of little direct interest. Instead, we wish to determine from them the contact stresses. These stresses are obtained from the contact forces in the same way as in [14], i.e.,

$$s_n = Mp_n, \quad s_t = Mp_t, \quad s_o = Mp_o,$$

where s_n is the normal, and s_t and s_o are the two tangential contact stresses, and M is a diagonal matrix whose diagonal entries are inverses of the integration weights of the numerical integration rule used, which in our case is the trapezoidal rule; thus,

$$M_{ii} = \left(\sum_{j=1}^{f_i} \frac{A_{ij}}{4} \right)^{-1}, \quad i = 1, \ldots, n_c, \tag{7.1}$$

where f_i is the number finite elements adjacent to contact node i, and A_{ij} is the area of the jth element adjacent to contact node i. Finally, the computed contact stresses are plotted for various levels of the applied load in several figures.

7.1 Parameters for unconstrained formulation

We have used the same scalar parameters in (3.1) and (3.2); i.e. $r_i = \rho_{in} = M_{ii}^{-1}$, $i = 1, \ldots, n_c$, where M_{ii} are given in (7.1). A motivation for this choice is that the size of the first argument in (3.2), p_n, varies proportionally with the size of the finite elements on the contact surface. The second argument, $g - C_n u$, however, does not show the same strong dependence as p_n, and the same is true for the normal contact stress, $\sigma_n = Mp_n$. Thus, by choosing $\rho_{in} = M_{ii}^{-1}$, we get parameters that can be used profitably for meshes with different fine discretizations of the contact surface.

The line search in Algorithm PRN is slightly modified; we have replaced ρ^m in Step 2 with $0.99995\,\rho^m$. The reason for not taking full Newton steps is that if a contact node i belongs to the set $\mathcal{I}_<$, then in the case where no step size reduction is performed in the line search, p_{in} would turn zero, i.e. $i \in \mathcal{K}_+ \cup \mathcal{K}_0$ in the next iteration. However, from our numerical experience we have found that the algorithm performs equally well even without this modification.

The parameters in Algorithm PRN are chosen as $\sigma_k = 0 \; \forall \; k$, $\rho = 0.9$ and $\alpha = 0.1$. The line search is aborted whenever $\rho^m \le 0.1$. If this is not done, there is a major risk that the algorithm will stall. The starting point for the first increment is taken to be

$$p_{in}^0 = M_{ii}^{-1} + 10^{-9}, \quad p_{it}^0 = p_{io}^0 = 0, \quad \forall\, i, \ldots, n_c,$$
$$u_i^0 = 0, \quad \forall\, i, \ldots, 3n_c.$$

The reason for adding a small quantity in p_{in}^0 is to prevent the set $\mathcal{I}_=$ from becoming full in the first iteration of Problem 1, although it should be pointed out that the algorithm works just as well without this perturbation. For subsequent increments the solution at increment $j - 1$ is used as the starting point for increment j.

The only nondifferentiable set that occasionally is not empty in our numerical examples is $\mathcal{J}_=$. However throughout our extensive experiments, the frequency of a nonempty $\mathcal{J}_=$ is very low.

The input data and the unknowns are scaled is such a way that the contact forces are calculated in mega-newton and the displacements in 10^{-5} meters, respectively. With this simple scaling the maximum values of the unknowns will be roughly of the same magnitude. For UF, the system (5.3) will then always be well conditioned with condition numbers $\kappa_2(J_{\mathcal{U}})$ in the region $1 - 10^2$. For CF, the system (5.6) may become ill-conditioned close to the boundary of $\Omega_{\mathcal{I}}$. The condition numbers vary from slightly over 1 to, in extreme cases, 10^{13}. The highest condition numbers are obtained when at certain contact nodes, both p_{in} as well as v_{in} are small (a degenerate point); in which case $J_{\mathcal{C}}$ will be almost singular. The algorithm is terminated whenever

$$\Theta(z^{k+1}) \equiv \tfrac{1}{2} H_{\mathcal{U}}(z^{k+1})^T H_{\mathcal{U}}(z^{k+1}) \leq 10^{-8}.$$

Typical values of Θ at termination are $10^{-8} - 10^{-13}$.

7.2 Parameters for constrained formulation

The parameters of Algorithm PRN used by CF are the following: $\sigma_0 = 0.95$, $\alpha = \rho = 0.5$, and in the merit function, $\zeta = 1000\, n_c$. The centrality parameter σ is updated according to

$$\sigma_{k+1} = \begin{cases} 0.95\sigma_k & \text{if } \Theta(z^{k+1}) > 10^{-5} \\ 0.1\sigma_k & \text{otherwise.} \end{cases}$$

We have used a modified line search in Step 2 of the algorithm:

Step 2'. Determine

$$\delta_k \equiv \sup\{\tau \,:\, w^k + \tau' dw^k \in \Omega_{\mathcal{I}} = \Re^{n_u} \times \Re^{2n_c} \times \Re^{2n_c}_{++} \quad \text{for all} \quad \tau' \in (0,\tau)\};$$

set $\delta'_k \equiv \min(0.9999995\, \delta_k, 1)$. Let m_k be the smallest nonnegative integer m such that

$$\psi_c(w^k + \delta'_k \rho^m dw^k) - \psi_c(w^k) \leq -\alpha\, \delta'_k\, \rho^m\, (1 - \sigma_k)\, H_{\mathcal{C}}(w^k)^T \nabla p(H_{\mathcal{C}}(w^k)).$$

Set $w^{k+1} \equiv w^k + \delta'_k \rho^{m_k} dw^k$.

This modified step size determination rule has proven to yield faster convergence of the algorithm in our experiments.

The starting point for the first increment is the same as for UF. The additional variables, v_n, are initialized as follows: $v^0_{in} = 0.1$, $i = 1, \ldots, n_c$ for Problem 1 and Problem 2, and $v^0_{in} = 1.0$, $i = 1, \ldots, n_c$ for Problem 3 and Problem 4. For subsequent increments, the solution at increment $j-1$ is taken as starting point for increment j. However, with this choice of starting point the convergence is sometimes very slow; in extreme cases more than 100 iterations are needed. Therefore we use a strategy where we revert back to a so called *safe point*, with σ changed back to σ_0, whenever the number of iterations exceeds 15. This safe point has the property of not lying close to the boundary of $\Omega_{\mathcal{I}}$ and being of reasonable size:

$$p_{in}^{0(j)} = \begin{cases} 0.1\max(p_n^{(j-1)}) & \text{if } \max(p_n^{(j-1)}) > 10^{-1} \\ 10^{-2} & \text{otherwise,} \end{cases}$$

$$v_{in}^{0(j)} = \begin{cases} 0.1\max(v_n^{(j-1)}) & \text{if } \max(v_n^{(j-1)}) > 1.0 \\ 0.1 & \text{otherwise,} \end{cases}$$

$$p_{it}^{0(j)} = p_{it}^{(j-1)}, \quad p_{io}^{0(j)} = p_{io}^{(j-1)}, \quad u_i^{0(j)} = u_i^{(j-1)}.$$

For Problem 3 and Problem 4, $v_{in}^{0(j)}$ is instead given by

$$v_{in}^{0(j)} = \begin{cases} 0.1\max(v_n^{(j-1)}) & \text{if } \max(v_n^{(j-1)}) > 10.0 \\ 1.0 & \text{otherwise.} \end{cases}$$

The penalty parameters r_i and the termination rule are chosen in same way as for UF. Typical values of Θ at termination are $10^{-8} - 10^{-10}$.

7.3 Clamped block

In this example an elastic block of dimension $1 \times 1 \times 1$ m^3, clamped at one of its sides (hence the stiffness matrix is positive definite), and unilaterally constrained to a frictional foundation is subjected to varying loads. In Problem 1, the initial gap g between the block and the foundation is 10^{-5} m and the loads vary according to Figure 1.1(a). In Problem 2, g is zero and the loads vary as in Figure 1.1(b). The number of increments during loading and unloading, denoted "inc", is 2, 5 or 10. Thus the total number of time-incremental problems solved is $2 \times$ inc for Problem 1 and $2 \times \text{inc} + 1$ for Problem 2. The friction coefficients μ_i, $i = 1, \ldots, n_c$, are taken to be 0.5. The block is discretized by 8-noded trilinear finite elements with Young's modulus 200 GPa and Poisson's ratio 0.3. The elements are all of equal size and $5 \times 5 \times 5$, $10 \times 10 \times 10$, $15 \times 15 \times 15$ or $20 \times 20 \times 20$ elements are used. The potential contact surface is taken to be the whole bottom side of the block. The number of contact nodes, n_c, will then be 30, 110, 240 or 420. Table 1.1 contains a summary of the execution statistics for Problem 1. The following notation is used in the table:

nle: average number of iterations per increment

nls: average number of reductions of the step length (line searches) per increment

nu: average number of unknowns in (5.3) actually solved for UF in % of its maximal possible value

cpu: total computing time in seconds

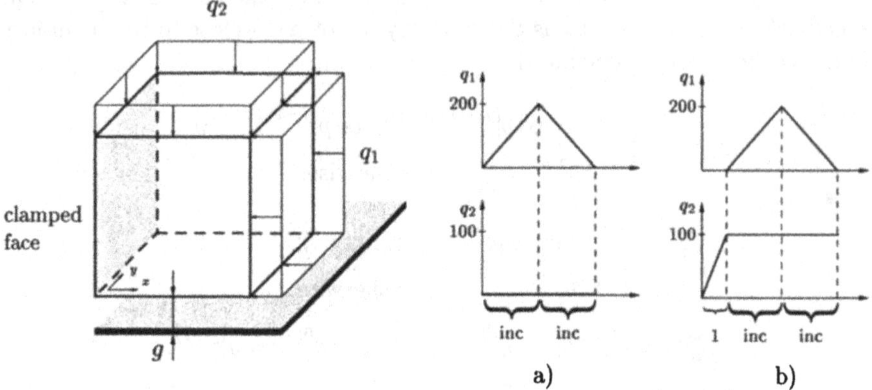

Figure 1.1 *Clamped block subjected to loads q_1 and q_2 in MPa.*

Table 1.1 Execution statistics: Problem 1.

n_c	inc	UF				CF			
		nle	nls	nu	cpu	nle	nls	nsr	cpu
	2	4.0	3.0	59.6	0.39	11.8	0.0	1	1.9
30	5	3.1	0.0	74.8	0.96	7.7	0.0	2	3.1
	10	2.7	0.0	78.3	1.7	7.1	0.0	3	5.6
	2	5.8	14.8	61.5	14.6	15.0	0.0	2	66.1
110	5	4.4	4.8	70.5	31.0	8.7	0.0	2	95.9
	10	4.1	2.0	72.7	56.6	9.7	0.2	5	211
	2	5.8	13.5	65.8	123	16.5	0.0	2	645
240	5	4.0	4.3	76.3	261	12.2	0.0	4	1185
	10	3.5	4.2	75.0	417	11.5	0.0	7	2226
	2	6.0	12.8	68.0	627	16.8	0.0	2	3165
420	5	5.3	7.1	73.1	1435	11.3	0.3	3	5236
	10	4.6	4.8	72.9	2320	12.1	0.0	6	11450

nsr: total number of safe restarts in CF.

The computing times do not include the initial static condensation discussed in Section 6. In Figure 1.2 the contact stresses for two different load levels are shown, and in Figure 1.3 the error norm for the first increment for the two methods is plotted. Note that in Figure 1.2, the zero contact stresses for nodes on the clamped face have not been plotted. In this example, the real contact area changes as the loads vary. From the table, it is evident that UF can handle these changes very well, whereas CF has problems dealing with them. For UF we note that the size of $J_{\mathcal{U}}$ is substantially reduced because many nodes belong to the set $\mathcal{I}_< \cup \mathcal{I}_=$.

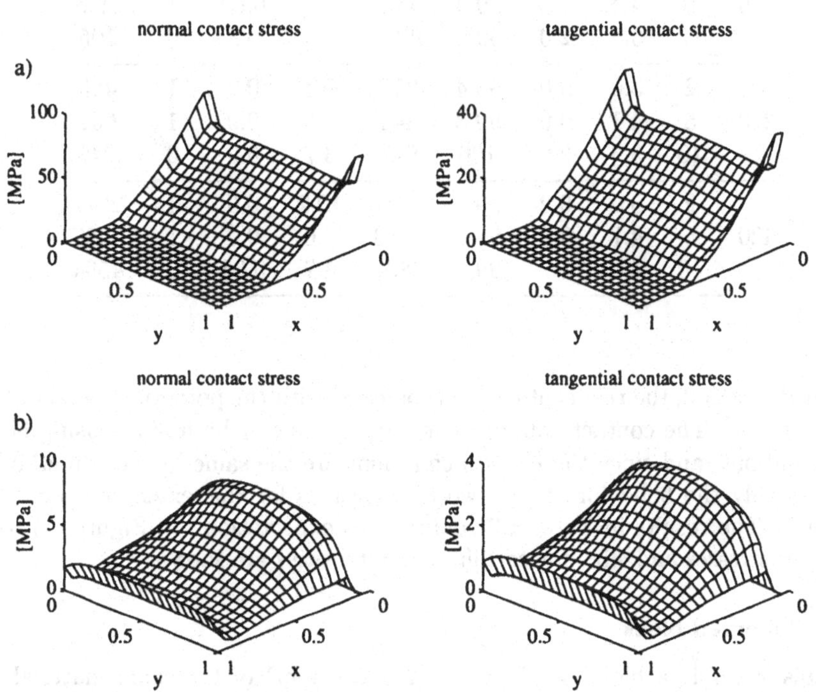

Figure 1.2 *Contact stresses when a)* $q_1 = 200$ *MPa and b)* q_1 *is unloaded to 40 MPa when* $n_c = 420$ *and inc = 10.*

Table 1.2 Execution statistics: Problem 2.

n_c	inc	UF				CF			
		nle	nls	nu	cpu	nle	nls	nsr	cpu
	2	4.2	0.0	98.9	0.97	6.2	0.2	0	1.3
30	5	3.5	0.0	100	1.7	4.1	0.0	0	1.8
	10	2.8	0.05	100	2.7	3.1	0.0	0	2.6
	2	4.8	0.0	99.1	27.8	9.4	0.2	1	51.7
110	5	4.4	2.5	99.4	55.2	6.6	0.09	1	79.8
	10	3.8	2.0	99.8	90.3	4.6	0.3	0	105
	2	5.0	0.0	99.4	252	9.4	0.0	1	455
240	5	4.1	0.0	99.8	444	6.3	0.0	1	662
	10	3.3	0.0	100	677	3.7	0.0	0	748
	2	5.0	0.2	99.0	1171	10.2	0.0	1	2377
420	5	4.7	0.6	99.7	2383	7.0	0.0	1	3597
	10	4.0	1.7	99.7	3801	4.7	0.05	0	4564

In Problem 2, the real contact area coincides with the potential contact area at all times. The contact equations should then easily be fulfilled using both formulations, and since the friction equations are the same for both methods, presumably the behavior of the two formulations for this example should be similar. The results in Table 1.2 confirm this expectation. In Figure 1.4, the contact stresses and slips for certain load intensities are shown.

7.4 Indented block

In this example a block of dimension $2 \times 2 \times 1$ m^3 of the same material as in the previous subsection is pressed onto a rigid foundation. In Problem 3, the foundation is shaped as a paraboloid, and in Problem 4 it is flat. We take advantage of the symmetry and model only a quarter of the block, of dimension $1 \times 1 \times 1$ m^3, as shown in Figure 1.5. The block is free to perform translations perpendicular to the xy plane, and hence the corresponding stiffness matrix is positive semidefinite. For Problem 3, the gap $g = 2 \cdot 10^{-4}(x^2 + y^2)$ m and the load varies as in Figure 1.5(a). The number of increments is inc $=$ 1, 2, 10 or 20. In Problem 4, $g = 0$ and the load is increased to 40 MPa in 2 increments and thereafter unloaded to 4 MPa, as indicated in Figure 1.5(b). The number of unloading steps is 2, 10 or 20. The friction coefficients for Problem 3 are 0.1 and for Problem 4, 0.2 at all nodes. The total number of time-incremental problems solved is inc for Problem 3 and inc + 2 for Problem 4. The block is discretized in the same way as in the previous subsection. The number of contact nodes, n_c, is 36, 121, 256 or 441.

Table 1.3 Execution statistics: Problem 3.

n_c	inc	UF				CF			
		nle	nls	nu	cpu	nle	nls	nsr	cpu
	1	5.0	16.0	85.1	0.36	14.0	0.0	0	0.90
36	2	5.0	6.0	78.9	0.52	18.5	0.0	1	2.2
	10	4.0	4.1	60.7	1.3	12.3	0.5	2	7.1
	20	2.9	2.0	60.2	1.8	6.7	0.3	1	7.8
	1	5.0	14.0	85.5	7.7	11.0	0.0	0	17.5
121	2	5.0	5.5	76.0	11.0	18.5	0.0	1	56.0
	10	4.7	10.4	62.6	33.9	26.7	0.2	9	389
	20	4.3	9.2	58.7	57.8	16.0	0.1	7	450
	1	6.0	13.0	85.7	66.7	10.0	0.0	0	130
256	2	5.5	5.5	77.7	90.3	18.5	0.0	1	430
	10	5.2	13.3	63.1	272	26.2	0.0	9	2948
	20	4.9	17.2	60.4	471	21.3	0.1	12	4836
	1	6.0	13.0	86.3	296	10.0	0.0	0	602
441	2	7.0	7.5	78.9	500	18.5	0.0	1	2055
	10	4.8	17.7	61.9	1057	25.8	0.0	9	13744
	20	5.2	28.4	61.2	2195	23.9	0.0	15	25422

In Table 1.3, the results for Problem 3 are summarized, and in Figure 1.6 the contact stresses and slips are shown for various load increments. Note the complex changes of the directions of the tangential contact stress and slip directions. It is obvious that CF has difficulties in solving this problem where the contact status is constantly changing; the algorithm has to revert back to a safe starting point in many increments. For inc = 2, the algorithm reverts back to a safe starting point once (for the second increment). When inc = 10 for the larger problems ($n_c \geq 121$), the contact status still changes so much between the increments that the algorithm has to revert back in nine out of the ten increments. This explains why the number of iterations per increment is considerably larger for inc = 10 than for inc = 2. The UF, on the other hand, solves this problem without any difficulties; the number of iterations is low and the size of the matrix J_u is often reduced considerably.

Table 1.4 summarizes the results of Problem 4, and Figure 1.7 shows the contact stresses at full load and after unloading. In this example, the block is always in contact with the foundation, which is why the formulations behave similarly for this problem.

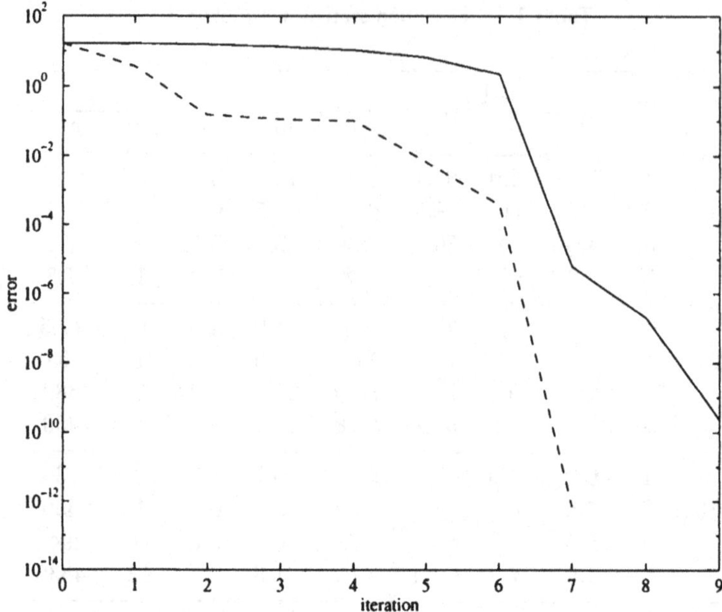

Figure 1.3 *The error norm $\Theta(z^{k+1})$ for the first increment when $n_c = 110$ and inc = 2. The solid line refers to CF and the dotted one to UF.*

Table 1.4 Execution statistics: Problem 4.

n_c	inc	UF				CF			
		nle	nls	nu	cpu	nle	nls	nsr	cpu
	2	4.3	0.0	99.2	1.1	5.0	0.3	0	1.2
36	10	3.0	0.0	100	2.5	3.2	0.0	0	2.3
	20	2.6	0.0	100	3.6	2.6	0.0	0	3.4
	2	5.0	0.0	98.6	31.2	6.3	0.3	0	37.7
121	10	3.5	0.8	100	65.2	3.5	0.0	0	61.9
	20	2.9	0.05	100	98.3	3.0	0.05	0	94.6
	2	5.0	0.0	98.8	241	10.5	0.5	1	497
256	10	3.4	0.0	100	481	3.5	0.0	0	496
	20	2.8	0.0	100	709	3.0	0.0	0	771
	2	5.3	0.0	99.0	1149	11.0	0.5	1	2402
441	10	3.5	0.0	100	2288	3.6	0.0	0	2367
	20	2.7	0.05	100	3181	2.8	0.05	0	3387

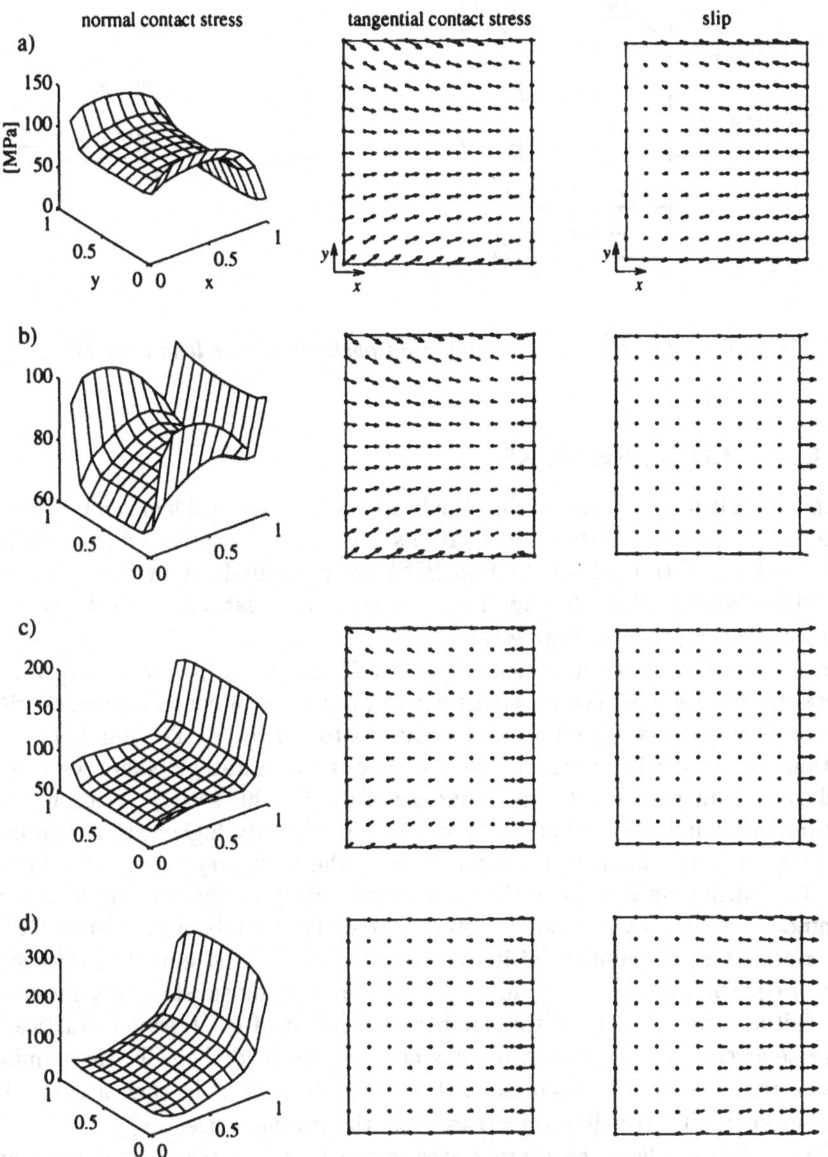

Figure 1.4 *Contact stresses and slip for various loads when $n_c = 110$ and inc = 10.*
a) $q_1 = 200$, b) $q_1 = 160$, c) $q_1 = 120$ and d) $q_1 = 0$ MPa.

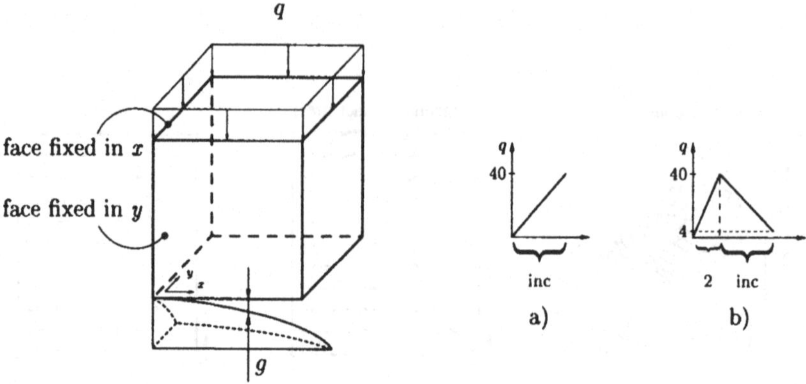

Figure 1.5 *Block pressed onto a rigid obstacle by the load q in MPa.*

8 CONCLUDING REMARKS

In this work it has been shown that the frictional contact problem can be formulated as a system of semismooth equations. Both an unconstrained formulation (UF) and a constrained formulation (CF) are presented. A unified potential reduction Newton algorithm applicable to both formulations is developed and its convergence properties are established.

From our experience it is clear that the CF has little, if any, practical advantages over the UF. For problems where the region of contact changes with time, CF needs a larger number of iterations to converge. The number of iterations are kept within reasonable size thanks to a strategy where we revert back to a point not close to the boundary of the feasible set $\Omega_{\mathcal{I}}$. Without this feature, the number of iterations would be significantly higher still. The fact that CF has difficulties when started close to the boundary is partially due to the ill-conditioning exhibited close to the boundary of the set $\Omega_{\mathcal{I}}$; a feature common to many interior point methods. For these kinds of problems, UF is superior to CF; the number of iterations is low and the size of the linear system of equations to be solved in each iteration may be substantially reduced. Our initial concern that the the number of iterations for UF would drastically increase as the problem size grows has not been verified. For CF, the number of safe restarts tends to increase with the problem size. In such a case, the number of iterations will also increase with the problem size.

For problems where the contact area is constant, the two formulations are more or less equal in performance. This is not surprising since then the gap constraints should be easily satisfied and the friction equations are the same for the two methods.

As correctly noted by two referees of this paper, there is a discrepency between Theorem 4.1 and the implemented method; the method is not exactly the PRN algorithm with which the result deals. Regrettably, we do not have a satisfactory resolution for the discrepency at this time. Nevertheless, the

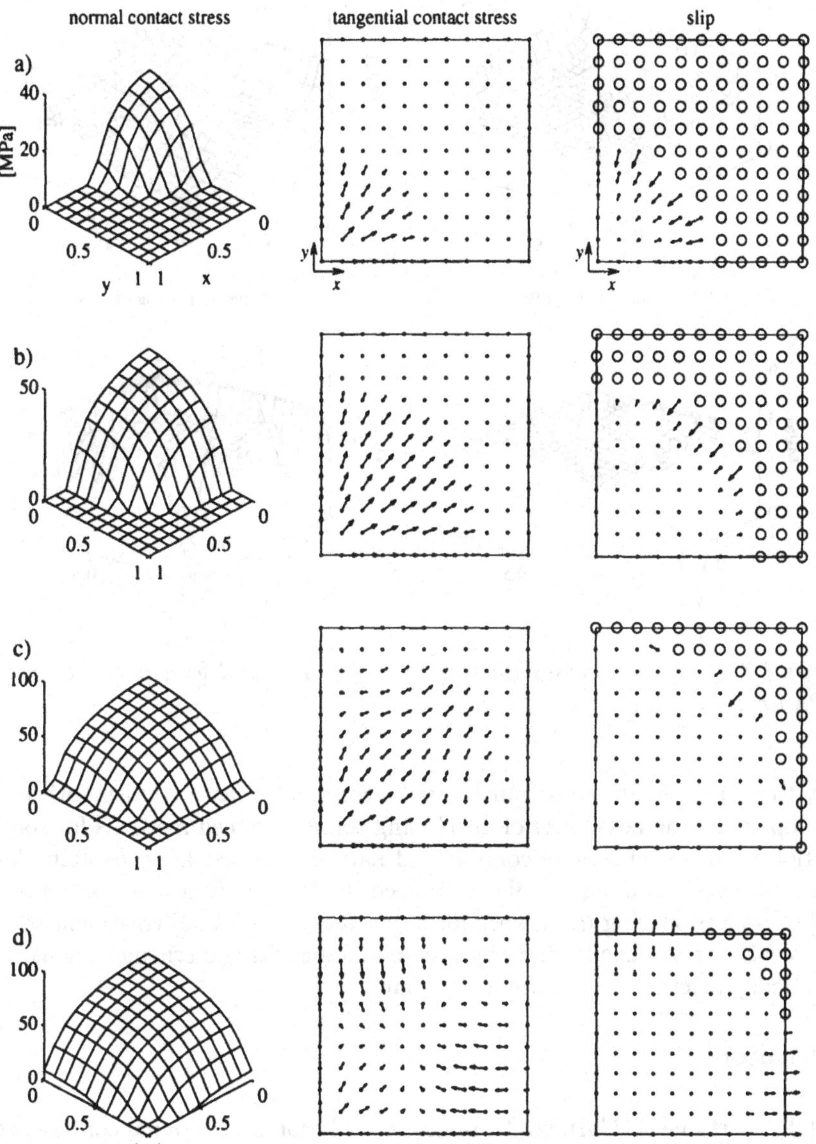

Figure 1.6 *Contact stresses and slip for various loads when $n_c = 121$ and inc $= 20$.*
a) $q = 4$, b) $q = 14$, c) $q = 25$ and d) $q = 40$ MPa. A $\circ$ represents a node with zero normal contact stress.

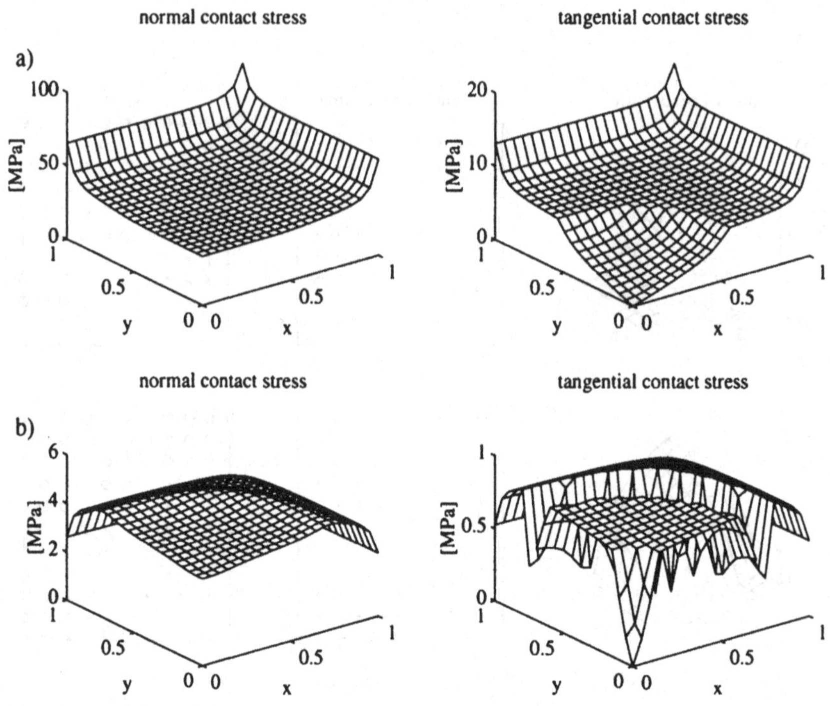

Figure 1.7 *Contact stresses when a) $q = 40$ MPa and b) q is unloaded to 4 MPa when $n_c = 441$.*

fact that the friction constraints can be formulated as a semismooth equation opens up the possibility of developing some advanced methods for solving friction problems of a more complicated nature, such as those involving large displacements and dynamic effects. Indeed, the unified treatment presented in [19] offers a promising framework for a broad computational treatment of friction problems of various kinds by a common algorithmic scheme that is both practically efficient and theoretically sound.

References

[1] P. Alart and A. Curnier, "A mixed formulation for frictional contact problems prone to Newton like solution methods", *Computational Methods in Applied Mechanics and Engineering* 92 (1991) 353–375.

[2] E. Anderson *et al.*, *LAPACK Users' Guide, 2nd edition*, SIAM Publications, Philadelphia (1995).

[3] P.W. Christensen, A. Klarbring, J.S. Pang and N. Strömberg, "Formulation and comparison of algorithms for frictional contact problems", *International Journal for Numerical Methods in Engineering*, forthcoming.

[4] G. De Saxce and Z.Q. Feng, "New inequality and functional for contact with friction: the implicit standard material approach", *Mechanical Structures & Machines* 19 (1991) 301–325.

[5] M.C. Ferris and J.S. Pang, "Engineering and economic applications of complementarity problems", *SIAM Review* 39 (1997).

[6] M.C. Ferris and J.S. Pang, editors, *Complementarity and Variational Problems: State of the Art*, SIAM Publications, Philadelphia (1997).

[7] A. Fischer and C. Kanzow, "On finite termination of an iterative method for linear complementarity problems", *Mathematical Programming* 74 (1996) 279–282.

[8] P.T. Harker and J.S. Pang, "A damped-Newton method for the linear complementarity problem, in (E. L. Allgower and K. Georg, eds.) *Computational Solution of Nonlinear Systems of Equations* [Lectures in Applied Mathematics, Volume 26], American Mathematical Society, Providence, Rhode Island (1990) pp. 265–284.

[9] G. Holmberg, "A solution scheme for three-dimensional multi-body contact problems using mathematical programming", *Computers & Structures* 37 (1990) 503–514.

[10] I. Kaneko, "Complete solutions for a class of elastic-plastic structures", *Computational Methods in Applied Mechanics and Engineering* 21 (1980) 193–209.

[11] A. Klarbring, "Contact Problems With Friction - Using a Finite Dimensional Description and the Theory of Linear Complementarity", Linköping Studies in Science and Technology, Thesis No. 20, Linköping Institute of Technology, Linköping, Sweden (1984).

[12] A. Klarbring, "Examples of non-uniqueness and non-existence to quasistatic contact problems with friction", *Ingenieur-Archiv* 60 (1990) 529–541.

[13] A. Klarbring, "Mathematical programming in contact problems", in M.H. Aliabdali and C.A. Brebbia, eds., *Computational Methods in Contact Mechanics*, Computational Mechanics Publications, Southampton (1993) pp. 233–263.

[14] A. Klarbring and G. Björkman, "A mathematical programming approach to contact problems with friction and varying contact surface", *Computers & Structures* 30 (1988) 1185–1198.

[15] A. Klarbring and J.S. Pang, "Existence of solutions to discrete semicoercive frictional contact problems", *SIAM Journal on Optimization*, forthcoming.

[16] R.D.C. Monteiro and J.S. Pang, "A potential reduction Newton method for constrained equations", manuscript, Department of Mathematical Sciences, The Johns Hopkins University, Baltimore (March 1997).

[17] J.S. Pang, "Newton's method for B-differentiable equations", *Mathematics of Operations Research* 15 (1990) 311–341.

[18] J.S. Pang and L. Qi, "Nonsmooth equations: motivation and algorithms", *SIAM Journal on Optimization* 3 (1993) 443-465.

[19] J.S. Pang and D. Stewart, "A unified approach to frictional contact problems", manuscript, Department of Mathematical Sciences, The Johns Hopkins University, Department of Mathematical Sciences, Maryland (March 1997).

[20] J.S. Pang and J.C. Trinkle, "Complementarity formulations and existence of solutions of multi-rigid-body contact problems with Coulomb friction", *Mathematical Programming* 73 (1996) 199–225.

[21] L. Qi, "Convergence analysis of some algorithms for solving nonsmooth equations", *Mathematics of Operations Research* 18 (1993) 227-244.

[22] L. Qi and J. Sun, "A nonsmooth version of Newton's method", *Mathematical Programming* 58 (1993) 353-368.

[23] S. Scholtes, "Introduction to piecewise differentiable equations," Habilitation thesis, Institut für Statistik und Mathematische Wirtschaftstheorie, Universität Karlsruhe (Karlsruhe 1994).

[24] J.C. Simo and T.A. Laursen, "An augmented Lagrangian treatment of contact problems involving friction", *Computers & Structures* 42 (1992) 97–116.

[25] N. Strömberg, "An augmented Lagrangian method for fretting problems", *European Journal of Mechanics, A/Solids*, 16 (1997) 573–593.

[26] J.C. Trinkle, J.S. Pang, S. Sudarsky and G. Lo, "On dynamic multi-rigid-body contact problems with Coulomb friction", *Zeitschrift für Angewandte Mathematik und Mechanik*, 77 (1997) 267–279.

[27] T. Wang, R.D.C. Monteiro, and J.S. Pang, "An interior point potential reduction method for constrained equations", *Mathematical Programming* 74 (1996) 159-196.

Reformulation: Nonsmooth, Piecewise Smooth, Semismooth and Smoothing Methods, pp. 117–126
Edited by M. Fukushima and L. Qi

Well-Posed Problems and Error Bounds in Optimization

Sien Deng*

Abstract For a given minimization problem with a nonempty solution set, we study the notion of well-posedness and the existence of two associated error bounds for the problem. We give sufficient conditions for the problem to be well-posed and the existence of these error bounds. We show that the well-posedness can be characterized in terms of these error bounds when the problem is a convex program. As an application, we reproduce known error bound results in mathematical programming.

Key Words well-posedness, level-coercivity, error bounds.

1 INTRODUCTION

Consider the problem

$$\text{minimize } \{f(x), \quad x \in C\}, \tag{1.1}$$

where f is a continuous function on $\mathbb{R}^n$, and C is a closed nonempty set in $\mathbb{R}^n$. Suppose that the optimal solution set S is nonempty. Without loss of generality, suppose further that $f(x) \geq 0$ for all $x \in C$ and the optimal value of (1.1) is 0. According to the well-posedness theory of Tikhonov [21], we say that (1.1) is well-posed if

$$[\,\forall\ \{x^k\}_{k=1}^{\infty} \subset C \text{ with } f(x^k) \to 0] \Rightarrow [\ dist(x^k, S) \to 0],$$

where $dist(y, S)$ denotes the Euclidean distance between the vector y and S.

Given below are examples of well-posed problems in mathematical programming.

*Department of Mathematical Sciences, Northern Illinois University, DeKalb, IL 60115, USA (deng@math.niu.edu).

Example 1: For (1.1), suppose that there is a continuous function $\phi : \mathbb{R}_+ \to \mathbb{R}_+$ with $\phi(t) = 0$ iff $t = 0$ such that

$$dist(x, S) \leq \phi(f(x)), \quad \forall x \in C. \tag{1.2}$$

Then (1.1) is well-posed. In particular, (1.1) is well-posed when (1.1) has the "weak sharp minima" considered by Burke and Ferris [4].

Example 2: Consider the linear complementarity problem (LCP): find x such that

$$Mx + q \geq 0, \quad x \geq 0, \quad \langle x, Mx + q \rangle = 0, \tag{1.3}$$

where $M \in \mathbb{R}^{n \times n}$, $q \in \mathbb{R}^n$, and $\langle \cdot, \cdot \rangle$ denotes the usual inner product on $\mathbb{R}^n$. Let S be the set of solutions to (1.3), and suppose that S is nonempty. In this case, we can choose $C = \mathbb{R}^n$, and $f(x) = \|min(x, Mx + q)\|_\infty$, where $\|\cdot\|_\infty$ denotes the ∞-norm, and the *min* operator denotes the component-wise minimum of vectors. It is well-known that $S = \text{argmin}_{x \in \mathbb{R}^n} f(x)$ [5]. Since $min(x, Mx + q)$ is a piecewise affine function of x, by Robinson's fundamental result on the upper Lipschitzian continuity of polyhedral multifunctions [18], there exist $\epsilon > 0$ and $\tau(\epsilon) > 0$ such that

$$dist(x, S) \leq \tau(\epsilon) f(x), \quad \forall x \in \mathbb{R}^n \quad \text{with } f(x) \leq \epsilon. \tag{1.4}$$

It is clear that (1.1) is well-posed.

Example 3: For (1.1), let $C = \mathbb{R}^n$, and $f(x) = \max_{1 \leq i \leq m} \{dist\ (x, C_i)\}$ where each C_i is a closed convex set. Suppose that $S = \cap_{i=1}^m C_i$ is nonempty. Then (1.1) is well-posed if and only if the collection $\{C_1, C_2, ..., C_m\}$ is regular in the sense of Bauschke and Borwein [3]. In the same reference, the authors have shown that the regularity of the collection $\{C_1, C_2, ..., C_m\}$ plays a vital role in the study of convex feasibility problems.

Example 4: For (1.1), when f and C are convex, it was proved by Lemaire [11] that (1.1) is well-posed if and only if $f + \delta_C$ is an asymptotically well-behaved proper, lsc convex function introduced by Auslender and Crouzeix [2], where δ_C denotes the indicator function of C.

Example 4 illustrates that the validity of the well-posedness of (1.1) is not a trivial issue even when (1) is a convex program since there are convex functions which are not asymptotically well-behaved (see [19], [2] for examples of such functions). For related studies on well-posed optimization problems, see Dontchev and Zolezzi [9], and Chou, Ng, and Pang [6]

It is now well understood in mathematical programming, that error bounds such as (1.2) and (1.4) are very useful for exact penalization, sensitivity analysis, the convergence and the rate of convergence of many iterative algorithms, etc. The literature on this subject is vast. We refer the reader to Pang [17] for extensive references on error bounds and their applications. An important message from this state-of-the-art survey [17] is that many fundamental notions in mathematical programming can be characterized by error bounds (either locally or globally) for given sets in terms of certain residual functions.

For (1.1), f serves as a natural residual function. The objective of this paper is to show that the well-posedness of (1.1) is closely related to the following error bounds for S in terms of f.

For any $\epsilon > 0$, there is some $\tau(\epsilon) > 0$ such that

$$dist(x, S) \leq \tau(\epsilon) f(x), \quad \forall x \in T_1(\epsilon), \tag{1.5}$$

where $T_1(\epsilon) = \{x \in C \mid f(x) \geq \epsilon\}$.

For any $\delta > 0$, there is some $\kappa(\delta) > 0$ such that

$$dist(x, S) \leq \kappa(\delta) f(x), \quad \forall x \in T_2(\delta), \tag{1.6}$$

where $T_2(\delta) = \{x \in C \mid dist(x, S) \geq \delta\}$.

The difference between (1.5) and (1.6) is the choice of test sets $T_1(\epsilon)$ and $T_2(\delta)$. Li [12] has shown that (1.5) and (1.6) always hold when f is a convex piecewise quadratic function and C is a convex polyhedral set. In this paper, we show that, for an optimization problem in the abstract form of (1.1), (1.5) and (1.6) hold when $f + \delta_C$ is level-coercive. Moreover, the existence of (1.6) is equivalent to the well-posedness of (1.1) and the existence of (1.5). When f and C are convex, the existence of (1.5), the existence of (1.6), and the well-posedness of (1.1) are equivalent. Unlike convex piecewise quadratic programming [12], in view of the discussion after Example 4, we know that (1.5) and (1.6) do not always hold even when (1.1) is a convex program. Example 4 shows that the well-posedness of (1.1) is equivalent to $f + \delta_C$ being asymptotically well-behaved. Auslender and Crouzeix [2] have shown that an asymptotically well-behaved proper lsc convex function g can be characterized in terms of error bounds for certain level-sets of g; see Lemma 2.2 for details. In contrast, we study the existence of (1.5) and the existence of (1.6) for the set S. This approach enables us to discover how (1.2), (1.5), and (1.6) are related to each other. Our analysis reveals that, among other things, when f and C are convex, the existence of (1.5) and (1.6) is necessary for the existence of a global error bound in the form of (1.2). Moreover, we illustrate that many known error bound results in mathematical programming can be obtained from this approach.

2 WELL-POSEDNESS AND ERROR BOUNDS

We begin this section by introducing the notions of recession cones and recession functions. Following [20], for any given closed nonempty set $X \subset \mathbb{R}^n$, we define the recession cone of X, denoted by X^∞, as

$$\begin{aligned} X^\infty \;=\; & \{d \mid \exists \text{ a sequence } \{x^n\}_{n=1}^\infty \text{ in } X \text{ and } \lambda_n \downarrow 0 \\ & \text{such that } \lim_{n \to +\infty} \lambda_n x^n = d\}; \end{aligned} \tag{2.1}$$

for any proper lsc function $g : \mathbb{R}^n \to \mathbb{R} \cup \{+\infty\}$, we define the recession function g^∞ as

$$\begin{aligned} g^\infty(d) \;=\; & \inf \{\beta \mid \exists \text{ some choice of } d^n \to d \text{ and } \lambda_n \downarrow 0 \\ & \text{such that } \lambda_n g(d^n/\lambda_n) \to \beta\}. \end{aligned} \tag{2.2}$$

For (1.1), we say that $f + \delta_C$ is level-coercive if $(f + \delta_C)^\infty(v) > 0$ for all $0 \neq v \in I\!R^n$; we say that $f + \delta_C$ is level-bounded if $\{x \mid f(x) \leq \alpha\} \cap C$ is bounded for all $\alpha > 0$. The following result concerning the level-coercivity of $f + \delta_C$, the level-boundedness of $f + \delta_C$, and the boundedness of S is well-known. See [20] for a complete treatment of level-coercivity, level-boundedness, and recession cones and functions.

Lemma 2.1 *For (1.1), consider the following statements.*
(a) $f + \delta_C$ is level-coercive.
(b) $f + \delta_C$ is level-bounded.
(c) S is bounded.
Then the following implications hold.
Statement (a) implies statement (b), and statement (b) implies statement (c). When f and C are convex, statements (a), (b), and (c) are equivalent.

For a proper lsc convex function g, let $m = \inf\{g(x) \mid x \in I\!R^n\}$. Following [2], we say that g is asymptotically well-behaved if

$$[\ \forall \{x^k\}_{k=1}^\infty \text{ with } dist(0, \partial g(x^k)) \to 0 \text{ as } k \to \infty] \Rightarrow [\lim_{k\to\infty} g(x^k) = m],$$

where $\partial g(x^k)$ denotes the subdifferential of g at x^k in convex analysis [19]. Denote by $\mathcal{F}$ the set of all such convex functions. We cite a characterization of $g \in \mathcal{F}$ which is due to Auslender and Crouzeix [2].

Lemma 2.2 *Suppose that g is a proper lsc convex function. Then $g \in \mathcal{F}$ if and only if for any $\lambda > m$, there is some $\tau(\lambda) > 0$ such that*

$$dist(x, S_\lambda) \leq \tau(\lambda)[g(x) - \lambda]_+, \quad \forall x \in I\!R^n,$$

where $S_\lambda = \{x \mid g(x) \leq \lambda\}$ and $[a]_+ = \max\{0, a\}$.

Quite interestingly, when f is convex differentiable and $C = I\!R^n$, Wei et al. [24] have demonstrated how to design a decent algorithm for (1.1) such that any sequence generated by the algorithm converges to S without assuming that f is asymptotically well-behaved.

When (1.1) is a convex program, Lemaire [11] has given the following characterization of the well-posedness of (1.1) in terms of asymptotic well-behavior of $f + \delta_C$ (see also Theorem 11 [1]).

Lemma 2.3 *For (1.1), suppose that f and C are convex. Then the following are equivalent.*
(a) The problem (1.1) is well-posed.
(b) The function $f + \delta_C \in \mathcal{F}$.

Now we are in the position to state the main result of this paper.

Theorem 2.1 *For (1.1), consider the following statements:*
(a) The function $f + \delta_C$ is level-coercive.

(b) The function $f + \delta_C$ is level-bounded.
(c) The error bound (1.6) holds.
(d) The problem (1.1) is well-posed.
(e) The error bound (1.5) holds.
Then the following implications hold.

Statement (a) implies statements (b) and (c). The boundedness of S and statement (c) implies statement (a). Statement (b) implies statement (d). Statement (c) is equivalent to statements (d) and (e). When f and C are convex, statements (c), (d) and (e) are equivalent.

Proof. [$(a) \Rightarrow (b)$]: This implication follows from Lemma 2.1.

[$(a) \Rightarrow (c)$]: Suppose that (1.6) does not hold for some $\delta > 0$. Then there is a sequence $\{x^k\}_{k=1}^{\infty} \subset C$ with $dist(x^k, S) \geq \delta \; \forall k$, and $f(x^k)/dist(x^k, S) \to 0$ as $k \to \infty$. It follows that $f(x^k) \to 0$ if $\{x^k\}_{k=1}^{\infty}$ is bounded. By the continuity of f and the closedness of C, we conclude that any subsequence of $\{x^k\}_{k=1}^{\infty}$ is divergent. Without loss of generality, we may assume that $\|x^k\| \to \infty$ as $k \to \infty$, and $x^k/\|x^k\| \to d \neq 0$ where $\|\cdot\|$ denotes the Euclidean norm on $\mathbb{R}^n$. Let $y^k \in S$ such that $dist(x^k, S) = \|x^k - y^k\|$. Since S is bounded, $x^k/\|x^k - y^k\| \to d \in C^{\infty}$. It follows that $1/\|x^k - y^k\|(f(x^k) + \delta_C(x^k)) \to 0$, which is a contradiction.

[[boundedness of S +(c)] $\Rightarrow$ (a)]: Since $(f + \delta_C)^{\infty}(\cdot)$ is positively homogeneous of degree 1, we only need to show that $(f + \delta_C)^{\infty}(d) > 0$ for all $d \in \mathbb{R}^n$ with $\|d\| = 1$. For $x \notin C$, $(f + \delta_C)(x) = +\infty$. Thus $(f + \delta_C)^{\infty}(d) = +\infty$ if $d \notin C^{\infty}$. For any $d \in C^{\infty}$ with $\|d\| = 1$, let $\{x^k\}_{k=1}^{\infty}$ be such that $\|x^k\| \to +\infty$ and $x^k/\|x^k\| \to d$. As we know that $f(x^k) + \delta(x^k) = +\infty$ when $x^k \notin C$, without loss of generality, we may assume that $\{x^k\}_{k=1}^{\infty} \subset C$. Since S is bounded, we may further assume that $dist(x^k, S) \geq \delta$ for some $\delta > 0$ and all k. Let $y^k \in S$ such that $dist(x^k, S) = \|x^k - y^k\|$. Since (c) holds, there is some $\kappa(\delta) > 0$ such that

$$\|x^k - y^k\| = dist(x^k, S) \leq \kappa(\delta) f(x^k).$$

It follows that

$$\liminf_{k \to +\infty} f(x^k)/\|x^k - y^k\| \geq 1/\kappa(\delta).$$

This is enough to show that $(f + \delta_C)^{\infty}(d) > 0$ for all $d \neq 0$.

[$(b) \Rightarrow (d)$]: Let $\{x^k\}_{k=1}^{\infty} \subset C$ with $f(x^k) \to 0$ as $k \to \infty$. Since (b) holds, the sequence $\{x^k\}_{k=1}^{\infty}$ is bounded. It follows from the continuity of f, and the closedness of C that $dist(x^k, S) \to 0$.

[$(c) \Rightarrow (d)$]: Suppose that (1.1) is not well-posed. Then there are some $\delta > 0$ and a sequence $\{x^k\}_{k=1}^{\infty} \subset C$ with $f(x^k) \to 0$ and $dist(x^k, S) \geq \delta$ for all $k = 0, 1, 2, \ldots$. By (c), there is some $\kappa(\delta) > 0$ such that $dist(x^k, S) \leq \kappa(\delta) f(x^k)$. Since $f(x^k) \to 0$, $dist(x^k, S) \to 0$ as $k \to \infty$, which is a contradiction.

[$(c) \Rightarrow$ (e)]: Suppose that (1.5) does not hold for some $\epsilon > 0$. Then there is a sequence $\{x^k\}_{k=1}^{\infty} \subset C$ with $f(x^k) \geq \epsilon$ and $f(x^k)/dist(x^k, S) \to 0$ as $k \to \infty$. Since $f(x^k) \geq \epsilon$, $dist(x^k, S) \to \infty$ as $k \to \infty$. Therefore, there is some $\delta > 0$ such that $dist(x^k, S) \geq \delta \; \forall k$, which shows that (1.6) does not hold on $T_2(\delta)$.

[[$(d)+(e)$] $\Rightarrow$ (c)]: Suppose that (1.6) does not hold for some $\delta > 0$. Then there is a sequence $\{x^k\}_{k=1}^{\infty} \subset C$ with $dist(x^k, S) \geq \delta \ \forall k$, and

$$f(x^k)/dist(x^k, S) \to 0 \quad \text{as } k \to \infty. \tag{2.3}$$

We may assume, without loss of generality, that

$$f(x^k) \geq \epsilon \quad \forall k; \tag{2.4}$$

otherwise, by (d), we would have a subsequence $\{x^{k_l}\}_{k_l=1}^{\infty}$ such that $dist(x^{k_l}, S) \to 0$ as $k_l \to \infty$. But (2.4) along with (2.3) contradicts (e).

Now we prove the statements (c), (d), and (e) are equivalent under the convexity assumption on f and C. We only need to show that statement (d) implies statement (c), and statement (e) implies statement (d).

[$(d) \Rightarrow (c)$]: We prove this implication by a technique used in [12]. Suppose, to the contrary, that there are some $\delta > 0$ and a sequence $\{x^k\}_{k=1}^{\infty} \subset C$ such that

$$\lim_{k\to\infty} f(x^k)/\|x^k - \Pi_S(x^k)\| \to 0 \quad \text{with } dist(x^k, S) \geq 3\delta,$$

where $\Pi_S(x^k)$ denotes the Euclidean projection of x^k onto S. Let $\lambda_k = \delta/\|x^k - \Pi_S(x^k)\| < 1/2$ and $y^k = \Pi_S(x^k) + \lambda_k(x^k - \Pi_S(x^k))$. Then $y^k \in C$ by the convexity of C. Since $y^k - \Pi_S(x^k) \in N_S(\Pi_S(x^k))$, where $N_S(\Pi_S(x^k))$ denotes the normal cone of S at $\Pi_S(x^k)$ [19], dist $(y^k, S) = \|y^k - \Pi_S(x^k)\| \geq \delta$ for all k. By the convexity of f,

$$f(y^k) \leq (1-\lambda_k)f(\Pi_S(x^k)) + \lambda_k f(x^k) \leq \delta f(x^k)/\|x^k - \Pi_S(x^k)\| \to 0 \quad \text{as } k \to \infty.$$

This contradicts that (1.1) is well-posed.

[$(e) \Rightarrow (d)$]: For any $\epsilon > 0$, let $S_\epsilon = \{x \in C \mid f(x) \leq \epsilon\}$. For $y \in C\backslash S_\epsilon$, $f(y) > \epsilon$. Let $\Pi_S(y)$ be the Euclidean projection of y onto S, and $\bar{y} = (\epsilon/f(y))y + (1 - \epsilon/f(y))\Pi_S(y)$. By the convexity of f and C, $\bar{y} \in C$ and $f(\bar{y}) \leq \epsilon$. It follows that $\bar{y} \in S_\epsilon$. Consequently,

$$\begin{aligned} dist(y, S_\epsilon) &\leq \|y - \bar{y}\| \leq (1 - \epsilon/f(y))\|y - \Pi_S(y)\| \\ &= (1 - \epsilon/f(y))dist(y, S) \leq (1 - \epsilon/f(y))\tau(\epsilon)f(y) \\ &= \tau(\epsilon)(f(y) - \epsilon), \end{aligned}$$

where the last inequality follows from (1.5). By Lemma 2.2, $f + \delta_C \in \mathcal{F}$. The desired implication follows by invoking Lemma 2.3. ∎

The boundedness of S plus (c) implying (a) was observed by Paul Tseng [22] in a private communication. This observation immediately yields the following consequence: suppose that (1) has the weak sharp minima [4] and S is bounded. Then $f + \delta_C$ must be level-coercive.

For the LCP (1.3) with f given by Example 2, by (2.2) and Proposition 1 of Gowda [10], we have $f^\infty(u) = \|min(u, Mu)\|_\infty$. With this observation, we conclude that the level-coercivity of f is equivalent to M being an R_0-matrix, that is $min(u, Mu) = 0$ iff $u = 0$. See [5] for more about R_0-matrices. In [10], Gowda studied the notion of a vector-valued recession function for a piecewise affine vector-valued function. For the LCP (1.3), it is easy to show that the ∞-norm of such a vector-valued recession function equals the recession function considered in this paper.

It is easy to construct examples where (c) holds, but (b) does not hold. We know from Example 2 that the LCP (1.3) is well-posed. Hence Example 3.1 of Luo et al. [14] shows that (d) does not imply (e) in general. Since (c) implies (e), the same example also shows that (d) does not imply (c). For the LCP (1.3), since (1.4) holds, we know that (c) is equivalent to (e). We don't know whether (c) is equivalent to (e) in general.

A useful consequence of Theorem 2.1 is that it tells us exactly how to obtain a global error bound for S in the form of (1.2). Let us consider the following local error bound for S first:

For some $\epsilon > 0$, there is some continuous function $\phi : I\!R_+ \to I\!R_+$ with $\phi(t) = 0$ iff $t = 0$ such that

$$dist(x, S) \leq \phi(f(x)), \quad \forall x \in C \backslash T_1(\epsilon). \tag{2.5}$$

Lemma 2.4 *Suppose that there is some strictly increasing and continuous function $\phi : I\!R_+ \to I\!R_+$ with $\phi(0) = 0$ such that (2.5) holds. Then there are some $\delta > 0$ and $\gamma > 0$ such that*

$$dist(x, S) \leq \gamma\phi(f(x)), \quad \forall x \in C \backslash T_2(\delta). \tag{2.6}$$

Proof. Suppose that (2.6) does not hold for any $\delta > 0$. Then there is a sequence $\{x^k\}_{k=1}^{\infty} \subset C$ such that $x^k \notin S$, $dist(x^k, S) \to 0$ and

$$\phi(f(x^k))/dist(x^k, S) \to 0 \quad \text{as } k \to \infty.$$

Hence $\phi(f(x^k)) \to 0$ as $k \to \infty$. Since ϕ is strictly increasing, $f(x^k) \to 0$ as $k \to \infty$, which is a contradiction. ∎

It is clear that the existence of (2.5) implies the well-posedness of (1.1). We now state a consequence of Theorem 2.1 when (2.5) holds. No proof is needed.

Corollary 2.1 *For (1.1), suppose that either $f + \delta_C$ is level-coercive or f and C are convex. If (2.5) holds, then there is some $\tau > 0$ such that*

$$dist\,(x, S) \leq \phi(f(x)) + \tau f(x), \quad \forall x \in C. \tag{2.7}$$

Corollary 2.1 reveals that the existence of the local error bound (2.5) is sufficient for the existence of the global error (2.7) under either the convexity or the level-coercivity assumption on f and C. Many known global error bound results in the mathematical programming literature are consequences of Corollary 2.1. For simplicity, in what follows, we only consider (1.1) with $C = \mathbb{R}^n$ and f being either the natural residual function for the LCP (1.3) considered in Example 2 or the pointwise maximum of finitely many analytic functions. For the former case, we know that (1.4) holds. For the latter case, thanks to Luo and Pang, we have the following local error bound result[15].

Lemma 2.5 *For (1.1), suppose that $C = \mathbb{R}^n$, and $f(x) = [\max_{1\le i\le m}\{f_i(x)\}]_+$, where each f_i is analytic. Then for any nonempty compact test set $X \subset \mathbb{R}^n$, there are $\tau(X) > 0$ and $\alpha > 0$ such that the Hölderian error bound holds on X*

$$dist(x, S) \le \tau(X) f(x)^\alpha, \qquad \forall x \in X. \tag{2.8}$$

In Lemma 2.5, if f is level-bounded, then the existence of (2.8) implies the existence of (2.5) with $\phi(t) = \beta t^\alpha$. In view of (1.4) and by invoking Corollary 2.1 and Lemma 2.5, we reproduce the following global error bound results for the LCP (1.3) and analytic systems. Part (i) is due to Mangasarian et al. [16], and Part (ii) is due to Deng [8].

Corollary 2.2 *For (1.1), suppose that $C = \mathbb{R}^n$, and f is level-coercive.*

(i) If f is given by Example 2 for the LCP (1.3), then there is some $\tau > 0$ such that the Lipschitzian global error bound holds for S

$$dist(x, S) \le \tau f(x), \qquad \forall x \in \mathbb{R}^n. \tag{2.9}$$

(ii) If f is given by Lemma 2.5, then there are some $\tau > 0$ and $\alpha > 0$ such that the Hölderian global error bound holds for S

$$dist(x, S) \le \tau[f(x) + f(x)^\alpha], \qquad \forall x \in \mathbb{R}^n. \tag{2.10}$$

Remark: An error bound in the form of (2.9) is more desirable than that in the form of (2.10). For the LCP (1.3), f is level-coercive if and only if M is an R_0-matrix (see the discussion after Theorem 2.1). Luo and Tseng [13] have shown that, for the LCP (1.3), M being an R_0-matrix can be characterized in terms of the existence of (2.9) for all q such that the LCP (1.3) is solvable. For the generalization of this result to the piecewise affine mapping setting, see Gowda [10]. For (ii), if f is convex, the level-coercivity assumption of f can be weaken to $0 \in ri[dom\ f^*]$ where ri denotes the relative interior of a convex set, and f^* denotes the convex conjugate function of f, see [7, 8] for the details of refinements. When each f_i in Lemma 2.5 is convex quadratic, Wang and Pang [23] have shown that (2.10) holds without any coercivity assumption.

Acknowledgements

The author thanks Professor Paul Tseng for helpful suggestions and the referees for their useful comments.

References

[1] A. Auslender, *How to deal with the unbounded in optimization: Theory and algorithms,* Mathematical Programming ,Series B, 79 (1997), pp. 3-18.

[2] A. Auslender, and J.-P. Crouzeix, *Well-behaved asymptotical convex functions,* in: Analyse Non-lineaire (Gauthier-Villars, Pairs, 1989) pp. 101-122.

[3] H. H. Bauschke and J. M. Borwein, *On projection algorithms for solving convex feasibility problems,* SIAM Review 38 (1996), pp. 367–426.

[4] J. V. Burke and M. C. Ferris, *Weak sharp minima in mathematical programming,* SIAM J. Control and Optimization 31 (1993), pp. 1340–1359.

[5] R. W. Cottle, J.-S. Pang, and R.E. Stone, *The Linear Complementarity Problem,* Academic Press, Boston, MA 1992.

[6] C.-C. Chou, K.-F. Ng, and J.-S. Pang, *Minimizing and stationary sequences of constrained optimization problems,* SIAM J. Control and Optimization, to appear.

[7] S. Deng, *Global error bounds for convex inequality systems in Banach spaces,* SIAM J. Control and Optimization, to appear.

[8] S. Deng, *Perturbation analysis of a condition number for convex inequality systems and global error bounds for analytic systems,* Mathematical Programming, to appear.

[9] A. L. Dontchev and T. Zolezzi: Well-posed Optimization Problems, Lecture Notes in Mathematics No. 1543, Springer-Verlag, Berlin, 1993.

[10] M. S. Gowda, *An analysis of zero set and global error bound properties of a piecewise affine function via its recession function,* SIAM Journal on Matrix Analysis 17 (1996), pp. 594–609.

[11] B. Lemaire, *Bonne Position, conditionnement, et bon comportement asymptotique,* Expose No. 5, Seminaire D'Analyse Convexe, Universite de Montpellier, Montpellier, 1992.

[12] W. Li, *Error bounds for piecewise convex quadratic programs and applications,* SIAM J. Control and Optimization 33 (1995), pp. 1510-1529.

[13] X. D. Luo and P. Tseng, *On a global projection-type error bound for the linear complementarity problem,* Linear Algebra and Its Applications 253 (1997), pp. 251-278.

[14] Z. Q. Luo, O. L. Mangasarian, J. Ren and M.V. Solodov, *New error bounds for the linear complementarity problem,* Mathematics of Operations Research 19 (1994), pp. 880–892.

[15] Z. Q. Luo and J.-S. Pang, *Error bounds for analytic systems and their applications,* Mathematical Programming 67 (1995), pp. 1-28.

[16] O. L. Mangasarian and J. Ren, *New improved error bounds for the linear complementarity problem,* Mathematical Programming 66 (1994), pp. 241–257.

[17] J.-S. Pang, *Error bounds in mathematical programming,* Mathematical Programming, Series B, 79 (1997), pp. 299-332.

[18] S. M. Robinson, *Some continuity properties of polyhedral multifunction,* Mathematical Programming Study 14 (1981), pp. 206–214.

[19] R. T. Rockafellar, *Convex Analysis,* Princeton University Press, Princeton, NJ, 1970.

[20] R. T. Rockafellar and R. J. B. Wets, *Variational Analysis,* Springer Verlag, Berlin, Germany, 1997.

[21] A. N. Tikhonov, *On the stability of the functional optimization problem,* Math. Phys. 6 (1966), pp. 631-634.

[22] P. Tseng, private communication, January 1998.

[23] T. Wang and J.-S. Pang, *Global error bounds for convex quadratic inequality systems,* Optimization 31 (1994), pp. 1-12.

[24] Z. Wei, L. Qi, and H. Jiang, *Some convergence properties of descent methods,* Journal of Optimization Theory and Applications 95 (1997), pp. 177-188.

Reformulation: Nonsmooth, Piecewise Smooth,
Semismooth and Smoothing Methods, pp. 127–147
Edited by M. Fukushima and L. Qi

Modeling and Solution Environments for MPEC: GAMS & MATLAB [1]

Steven P. Dirkse[†] and Michael C. Ferris[‡]

Abstract We describe several new tools for modeling MPEC problems that are built around the introduction of an MPEC model type into the GAMS language. We develop subroutines that allow such models to be communicated directly to MPEC solvers. This library of interface routines, written in the C language, provides algorithmic developers with access to relevant problem data, including for example, function and Jacobian evaluations. A MATLAB interface to the GAMS MPEC model type has been designed using the interface routines. Existing MPEC models from the literature have been written in GAMS, and computational results are given that were obtained using all the tools described.

Key Words complementarity, algorithm, MPEC, modeling.

1 INTRODUCTION

The Mathematical Program with Equilibrium Constraints (MPEC) arises when one seeks to optimize an objective function subject to equilibrium contraints. These equilibrium constraints may take the form of a variational inequality or complementarity problem, or they may be implicitly defined by a second-level optimization problem, the so-called bilevel programming problem. Problems of this form have existed for quite some time but have recently become the subject of increased interest, as evidenced by the monograph [23] describing the theoretical and algorithmic state of the art for these problems. In this paper, we briefly examine the currently available modeling and solution environments

[1]This material is based on research supported by National Science Foundation Grant CCR-9619765.
[†]GAMS Development Corporation, 1217 Potomac Street NW, Washington, D.C. 20007 (steve@gams.com).
[‡]Computer Sciences Department, University of Wisconsin - Madison, 1210 West Dayton Street, Madison, Wisconsin 53706 (ferris@cs.wisc.edu).

for MPEC problems. We then describe an extension to the GAMS [4] modeling language that aids effective modeling of the MPEC problem and software libraries that allow MPEC solvers written in C, Fortran, and MATLAB to access these MPEC models.

Currently, a person wishing to solve an MPEC problem might start by choosing a solver designed for this purpose, most likely one written in C, Fortran or MATLAB . They would then formulate their problem to interface correctly with that particular solver. This would require them to code the functions defining the problem, and perhaps first or second derivatives as well. They would need to provide this information in the (non)sparse format required by the solver, and satisfy any other requirements peculiar to their algorithm of choice. Such an approach is undesirable from a modeling standpoint simply because it ignores the organizational complexity of correctly assembling (and reassembling) data from many different sources into a single model. In addition, such an approach limits the modeler to using the target solver. If this solver fails, major changes may be required to put the problem into a form suitable for another solver.

These problems (and others) are exactly those that led to the development and use of modeling languages, first for linear and then nonlinear programming. These modeling languages help organize and automate the modeling task and allow this to be done in a machine- and solver-independent manner. In addition to GAMS, other modeling languages exist (e.g. AMPL [14], AIMMS [3]) that have a great deal in common. However, GAMS is currently the only one to support the mixed complementarity, or `MCP` [10], model type, although a similar AMPL interface [12] is currently under construction. The addition of the complementarity model to GAMS has been well received, becoming widely used in economics, structural engineering and other fields. In addition, the collection of models MCPLIB [6] and its MATLAB interface [13] has been very useful to the community of algorithm developers for the complementarity problem. It has been a convenient source of test problems, a basis for computational comparison between algorithms, and a factor in advancing the computational state of the art. We hope that the tools described in this paper can be similarly useful to those modeling and developing algorithms for MPEC problems.

The rest of the paper is organized as follows. In Section 2, we introduce terminology and provide a definition of the MPEC problem, as well as an example. Our definition is intimately related to the definition of the `MPEC` model type in GAMS, the subject of Section 3. Once a problem is formulated using the `MPEC` model type, the problem is made available to a solver via the MPECIO software library described in Section 4. It is also possible to access the problem from within MATLAB using an additional interface layer, the subject of Section 5.

As an example of how all these tools can be used, we have formulated several MPEC models from the literature in GAMS and have solved them using C, Fortran, and MATLAB solvers. Preliminary computational results are given in Section 6.

2 MPEC DEFINITION

An MPEC is a constrained nonlinear programming problem in which some or all of the constraints are formulated in terms of the solution of a second level problem. A popular form in which such constraints arise is as equilibrium constraints. A simple example is that of maximizing revenue generated from the tolling of roads on a traffic system subject to that traffic system being in equilibrium. Another example involves generating optimal taxation policies under the assumption that an economy is in equilibrium.

A crucial part of the MPEC problem definition is the system of equilibrium constraints. These can be defined in a number of ways. For example, they can arise as the solution to an optimization problem or using generalized equations [30, 29], variational inequalities [17], the min operator [17], or a complementarity problem [17]. In our case, we will assume that the equilibrium system always takes the form of a mixed complementarity problem, or MCP, defined in terms of some (possibly infinite) lower and upper bounds $\ell \in (\mathbf{R} \cup \{-\infty\})^n$ and $u \in (\mathbf{R} \cup \{+\infty\})^n$ satisfying $-\infty \leq \ell_i < u_i \leq +\infty$ and a nonlinear function $F: \mathbf{B} \to \mathbf{R}^n$. Throughout the paper, we will use the notation $\mathbf{B}$ to represent the box $\mathbf{B} := [\ell, u] = \{y \in \mathbf{R}^n: \ell_i \leq y_i \leq u_i\}$. The variable $y \in \mathbf{R}^n$ solves $MCP(F, \mathbf{B})$ if the following holds:

$$\begin{array}{llll} & F_i(y) = 0 & \text{and} & \ell_i < y_i < u_i \\ \text{or} & F_i(y) \geq 0 & \text{and} & y_i = \ell_i \\ \text{or} & F_i(y) \leq 0 & \text{and} & y_i = u_i. \end{array} \tag{2.1}$$

The MCP definition is quite compact; only F and $\mathbf{B}$ need be specified. It is entirely equivalent to the box-constrained or rectangular VI. This allows a certain simplicity in formulation and solution. However, the formulation is general enough to include as special cases the nonlinear complementarity problem (NCP) where $\mathbf{B} := [0, +\infty]$ and nonlinear systems of equations where $\mathbf{B} := [-\infty, +\infty]$. The box $\mathbf{B}$ allows both free variables and variables with one or two finite bounds, hence the mixed nature of the problem. This results in improved efficiency of modeling and problem solution, as compared with the fixed box $[0, +\infty]$ of the NCP. We will use the shorthand notation $F(y) \perp y \in \mathbf{B}$ for (2.1) as a generalization of the orthogonality that holds in the case of the NCP.

We define the MPEC over the design variables $x \in \mathbf{X} \subseteq \mathbf{R}^n$ and state variables $y \in \mathbf{Y} \subseteq \mathbf{R}^m$. There is an objective function $\theta: \mathbf{R}^{n+m} \to \mathbf{R}$ to be optimized, subject to two types of constraints. The first type of constraints require (possibly joint) feasibility of the variables (x, y) and are determined by the functions $h: \mathbf{R}^n \to \mathbf{R}^k$ and $g: \mathbf{R}^{n+m} \to \mathbf{R}^p$ and the box $\mathbf{X}$. The equilibrium constraints are defined by the box $\mathbf{Y}$ and the function $F: \mathbf{R}^{n+m} \to \mathbf{R}^m$. The latter constraints require the state variables y to solve the MCP defined by $\mathbf{Y}$ and (parametrically) by $F(x, \cdot)$. Put succinctly, we have

$$\text{minimize} \quad \theta(x, y) \tag{2.2a}$$

$$\begin{aligned} \text{subject to} \quad & h(x) \in H & \text{(2.2b)} \\ & x \in \mathbf{X} & \\ & g(x,y) \in G & \text{(2.2c)} \\ \text{and} \quad & y \text{ solves } \mathrm{MCP}(F(x,\cdot),\mathbf{Y}) & \text{(2.2d)} \end{aligned}$$

where the sets G and H are used to signify that the given constraint could either be an inequality or an equation.

Note that the definition above is quite general in allowing for feasibility constraints of the form (2.2b) and (2.2c). The models from the literature do not require constraints of the form (2.2c); many omit the constraints (2.2b) as well. This is due primarily to the scarcity of techniques applicable to problems with the joint feasibility constraint (2.2c). We include this constraint in the hope that solvers for this type of model will soon become available, and because it is a programmatically trivial task to generalize from (2.2b) to (2.2c) in the GAMS model definition and in the solver interface routines of Section 4.

As an example, the following problem exercises the full generality of our definition:

$$\begin{aligned} \text{minimize} \quad & \theta(x,y) := (x-1-y)^2 \\ \text{subject to} \quad & x^2 \leq 2 \\ & \ell \leq x \leq u \\ & (x-1)^2 + (y-1)^2 \leq 3 \\ & y - x^2 + 1 \;\perp\; y \geq 0 \end{aligned} \tag{2.3}$$

The final line above represents the fact that y solves the NCP with $F(x,y) = y - x^2 + 1$. The MPEC model as we have defined it generalizes both the nonlinear program and the MCP. Those doing modeling work in complementarity are eager to adapt existing MCP models to take advantage of this framework. For example, one can optimize revenue or congestion in traffic equilibrium problems over a set of feasible tolls [9], or convert a Nash equilibrium model to a Stackelberg (leader-follower) game [33]. Typically, the optimization problem has relatively few design variables and many more state variables. This approach has implications for algorithm design, which can exploit the fact that the number of design variables and "side constraints" h and g is very small.

One can also view the problem as an NLP generalized to include some equilibrium constraints. This approach is taken by chemical process engineers, who have equations that are valid only for certain states of the system [28]. In these models, the contraints g and h may dominate, requiring a different kind of algorithm. In either case, the problem fits well into the MPEC framework, and into the GAMS `MPEC` model, which we now describe.

3 THE MPEC MODEL TYPE

The MPEC definition (2.2) combines components of NLP and MCP definitions. In the same way, the GAMS `MPEC` model type borrows from both the `NLP` and `MCP` model types. This makes the programming task easier, but more importantly, it allows users to move from MCP or NLP to MPEC models with little difficulty. The use of the usual GAMS constructs (e.g. sets, parameters, variables, equations, control structures) is unchanged; the only changes are those made to the GAMS `model` and `solve` statements.

The `model` statement is used to associate a model name with a list of equations and equation-variable pairs that define the model. In the MPEC case, there must be one (equality constrained) equation that defines the objective variable (one specifies an objective *variable* in GAMS, not an objective row). The objective variable must appear linearly in this equation, and must not appear in any other equation of the model, so that it can be substituted out by the interface, leaving an objective function of the form (2.2a). Typically, an equation "cost" is declared of the form:

```
cost.. theta =e=  some gams expression;
```

The objective equation can appear (unpaired) anywhere in the model list. In addition, other equations defining constraints to be satisfied can be included in the model list, just as in an NLP. These define the constraints (2.2*b*) and (2.2*c*); the partition occurs automatically.

Mathematically, the equilibrium constraints (2.2d) are defined by matching the state variables y with equations. In GAMS, this is done by including an equation-variable pair in the model list. Each pair defines a complementarity relationship between the function determined by the equation and the variable in the pair. In the case of equations and variables indexed by sets, functions and variables with matching indices are paired. These define the equilibrium constraints (2.2d) in exactly the same way as for the GAMS `MCP` model type. It also specifies the partition into design and state variables. Variables appearing in pairs are state variables, while those not appearing are design variables. Note a subtle difference between the MPEC model statement and the MCP model statement: in the MCP model, there can be equations that are not explicitly paired with (free) variables. In the MPEC case, any equation which is not matched is assumed to be a side constraint, while unmatched variables are deemed to be design variables. In GAMS, variable bounds are attributes of the variables, so the boxes $\mathbf{X}$ and $\mathbf{Y}$ are given as well, and the problem (2.2) is completely specified.

From the modeler's perspective, the `solve` statement for `MPEC` models is no different than for other model types; it simply instructs GAMS to solve the model indicated. The modeler indicates that this is an `MPEC` model, the variable to optimize, and whether to maximize or minimize. GAMS will do a number of checks (bound consistency, nonlinear functions defined at initial point, functions smooth) before writing the problem to disk as a sequence of scratch files and calling an MPEC solver. This MPEC solver will use the interface library routines of Section 4 to read and interpret the scratch files, evaluate functions and gradients, and write solution data.

As an example, we include the GAMS code for the simple example (2.3) in Figure 1. The file three.gms that is depicted in Figure 1 can be obtained from

```
* simple MPEC model
variables
theta,
x               'design variables',
y               'state variables';
x.lo = -1;
x.up = 2;

equations
cost,
h,
g,
F;

cost ..  theta =e= sqr(x-1-y);

h ..  sqr(x) =l= 2;

g ..  sqr(x-1) + sqr(y-1) =l= 3;

F ..  y - sqr(x) + 1 =g= 0;


model three / cost, h, g, F.y /;

option mpec=bundle;
solve three using mpec minimizing theta;
```

Figure 1.1 GAMS Model for (2.3)

the MPEC Web site `http://www.gams.com/mpec/`. Other example files can also be found at this site, along with the codes that form the content of the next two sections.

4 MPECIO

In Section 3, we described how an MPEC model is processed by GAMS and sent to a solver. This communication is currently done via files, but could also be done in (virtual) memory, via the Internet, or using some other physical layer. This involves many details that depend on the operating system, the file system, the compilers used, etc. The GAMS I/O library insulates the solver from these details and allows access to the problem in a standard form. In this section, we describe the interface to an extension of the GAMS I/O library that will read and interpret the scratch files, evaluate functions and gradients for the MPEC problem, and write solution data. The reader is assumed to have some familiarity with C, since the interface we now describe is written in that language.

Interface Initialization

```
int mpecInit (char *controlFileName, int indexStart,
              int diagRequired, conType_t conType,
              mpecRec_t **mpec);
void sparseInit (mpecRec_t *mpec, int colPtrx[], int cpxdim,
                 int rowIdxx[], int rixdim,
                 int colPtry[], int cpydim,
                 int rowIdxy[], int riydim,
                 int colPtrh[], int cphdim,
                 int rowIdxh[], int rihdim,
                 int colPtrg[], int cpgdim,
                 int rowIdxg[], int rigdim);
```

The first task of the solver is to call the `mpecInit` routine to read in the scratch files and construct a problem of the form (2.2). If there is any inconsistency in the MPEC specification, it is detected during this initialization phase. It is here that the variables are partitioned into design and state variables, and the required maps are set up to support efficient function and gradient evaluation by other routines. Parameters to `mpecInit` exist allowing the user to specify if index ranges must begin with 0 (C style) or 1 (Fortran style), and whether or not extra space should be allocated to store all the zero elements of the diagonal of the Jacobian. Since there is some overhead in allowing for the side constraints h (2.2b) and g (2.2c) and many solvers will not be able to handle these constraints if they are present, the parameter `conType` exists to indicate whether to allow for no side constraints, those of the form (2.2b) or (2.2c) only, or both (2.2b) and (2.2c). A pointer to a record containing all information specific to this model (e.g. dimensions, nonzero estimates) is passed back to the calling routine. This pointer will be passed on all subsequent MPECIO calls.

In order to fully initialize the MPEC model, some space is required for the row indices and column pointers used to store the sparse Jacobians. Rather than allocating this space inside the library, the `mpecInit` routine returns estimates of the amount of space required for this to the calling routine. A second routine, `sparseInit`, must then be called, which passes in these arrays, as well as the number of elements actually allocated for them. This routine completes the initialization, using the space provided to it. The assumption here is that the user will not modify these arrays, as they are used by both the solver and the interface library. This assumption saves having to store two copies of the sparsity structure and copy it to the user's data structure at each derivative evaluation.

Variable Bounds and Level Values

```
void getxBounds (mpecRec_t *mpec, double lb[], double ub[]);
void getxLevels (mpecRec_t *mpec, double x[]);
void setxbar (mpecRec_t *mpec, double xbar[]);
void getyBounds (mpecRec_t *mpec, double lb[], double ub[]);
void getyLevels (mpecRec_t *mpec, double y[]);
```

The routines to obtain variable bounds and initial level values are for the most part self-explanatory. The `setxbar` routine is used to store a vector of design variables $\bar{x}$ in MPECIO for use in subsequent calls to function and gradient routines that pass only state variables y. This is useful for solvers that implement a two-level solution scheme in which the inner solver (an MCP code) has no knowledge of the variables x in the outer optimization problem. In our current implementation, the box $\mathbf{Y}$ does not depend on x, so the `getyBounds` routine would be called only once. A possible generalization is to allow $\mathbf{Y}$ to depend on x, in which case a new function with the input parameter x would be required. This function would of course be called whenever x is changed.

Function and Jacobian Evaluation

```
int getF ( mpecRec_t *mpec, double x[], double y[],
           double F[]);
int getdF (mpecRec_t *mpec, double x[], double y[],
           double F[],
           double Jx[], int colPtrx[], int rowIdxx[],
           double Jy[], int colPtry[], int rowIdxy[]);
int getFbar ( mpecRec_t *mpec, int n, double y[], double F[]);
int getdFbar (mpecRec_t *mpec, int n, int nnz, double y[],
              double F[],
              double Jy[], int colPtry[], int rowIdxy[]);
```

The routine `getF` takes the current point (x, y) as input and outputs the value of the function F at this point. The routine `getdF` computes the derivative of F as well. The derivative of F w.r.t. x and y is returned in separate matrices, both of which are stored sparsely in row index, column pointer fashion. The routines `getFbar` and `getdFbar` are similar, but in these routines, the input x is assumed to be the constant value $\bar{x}$ fixed in the previous call to `setxbar`. In this case, derivatives w.r.t. x and objective function values and derivatives are not passed back. These routines are designed for use by an algorithm solving an inner (MCP) problem.

```
int getObj (  mpecRec_t *mpec, double x[], double y[],
              double *obj);
```

```
int getdObj ( mpecRec_t *mpec, double x[], double y[],
              double *obj,
              double dObjDx[], double dObjDy[]);
```

In some cases, a solver may need information about the objective function θ only. This is supported efficiently in the current library via the `getObj` and `getdObj` routines. The derivative of θ w.r.t. x and y is returned as two dense vectors.

```
int getConEqSense (mpecRec_t *mpec, int hEqSense[],
                   int gEqSense[]);
int getCon ( mpecRec_t *mpec, double x[], double y[],
             double h[], int eqSense[]);
int getdCon (mpecRec_t *mpec, double x[], double y[],
             double h[], double g[],
             double dh[], int hColPtr[], int hRowIdx[],
             double dg[], int gColPtr[], int gRowIdx[]);
```

The sense of the upper-level constraints ($\geq, \leq, =$) is determined by a call to `getConEqSense`. In order to evaluate the side constraints h and g, one must use the `getCon` and `getdCon` routines. The level values of h and g are returned in `h[]` and `g[]`. The derivatives of h and g, if requested, are returned in row index, column pointer format. These calls return data for constraint types consistent with the model initialization (see `mpecInit` above). For example, if the model was initialized to allow for no side constraints, it is an error to call `getCon` or `getdCon`.

Solver Termination

```
void putxLevels (mpecRec_t *mpec, double x[]);
void putyLevels (mpecRec_t *mpec, double y[]);
void putObjVal (mpecRec_t *mpec, double obj);
void putStats (mpecRec_t *mpec, solStats_t *s);
int mpecClose (mpecRec_t *mpec);
```

Once a solution has been found, the solver must pass this solution on to the interface library. This is done via the `putxLevels`, `putyLevels`, and `putObjVal` routines. The `putStats` routine is used to report the model status (e.g. local optimum found, infeasible, unbounded, intermediate nonoptimal) and solver status (e.g. normal, iteration limit, out of memory, panic termination) via integer codes, as well as information about execution times, iterations used, function evaluation errors, etc. All of this information is stored in the `mpec` data structure and written to disk when the `mpecClose` routine is called. When the solver terminates, these files are read by GAMS so that the solution information is available for reporting purposes, as data to formulate other models, and to continue the execution of the GAMS model.

5 INTERFACING TO MATLAB SOLVERS

As an example of the intended use of MPECIO, we have developed an interface to the MATLAB programming environment. Our implementation is based on that developed for GAMS/MCP in [13]. We have attempted to make the interface to MPEC models more robust and portable than the MCP interface, while maintaining the ease of use of the tools within MATLAB .

The main difference between the MPEC interface and the MCP interface is in how the data is stored and accessed from within MATLAB . In the MCP interface, a large vector containing all the global data from the problem in question is copied to the user workspace. This technique is not very portable, since this large vector is stored in a machine-specific form, and requires different data files for different machine types. In addition, a special GAMS "solver" is required to create this large vector. In contrast, the MPEC interface to MATLAB accesses the data in exactly the same manner that a normal solver for GAMS does, using MPECIO as a machine-independent layer between it and the GAMS data files. Therefore, the only issues are to build or obtain MPECIO for the required platform and provide the machine dependent (or independent) files that are typically accessed by a solver.

We have provided a solver "mpecdump", callable from GAMS, that creates relevant scratch files on the user's platform in binary format. In this manner, any GAMS user can write their own MPEC problems in GAMS and generate all the relevant scratch files on their own machine. The solver mpecdump is freely available for download from the MPEC Web site `http://www.gams.com/mpec/`.

For users without access to GAMS, we have created a collection of MPEC problems that can be downloaded to any machine running MATLAB and used in conjunction with the interface routines we now describe. These problems are all stored in compressed "zip" format and can be downloaded from the MPEC Web site. For portability purposes, each zip file contains the ASCII files defining a single problem. It is intended that this collection grow into a library of test problems for MPEC. Figure 2 depicts the two ways that can be used to access the data of a particular problem.

The remainder of the MATLAB interface consists of a single MEX file and several simple accompanying m-files. The idea behind using a single MEX file, mpecfunc.c, is to allow all the m-files to access a single mpec structure, namely the mpecRec_t structure provided by MPECIO, without explicitly declaring this structure in the MATLAB workspace. Furthermore, since we envision that the structure used by MPECIO may be modified in the future to enable further enhancements to the modeling format, any changes to the MATLAB interface will be transparent to the MATLAB user. The source code of the mpecfunc routine is available via the MPEC Web site `http://www.gams.com/mpec/`. Together with the object code of MPECIO, this source allows any user to build a version of the MATLAB interface for their architecture, which they can then use to read in the problem data files they download and those they create from their own GAMS models. All of this can be done independently of the authors of this paper.

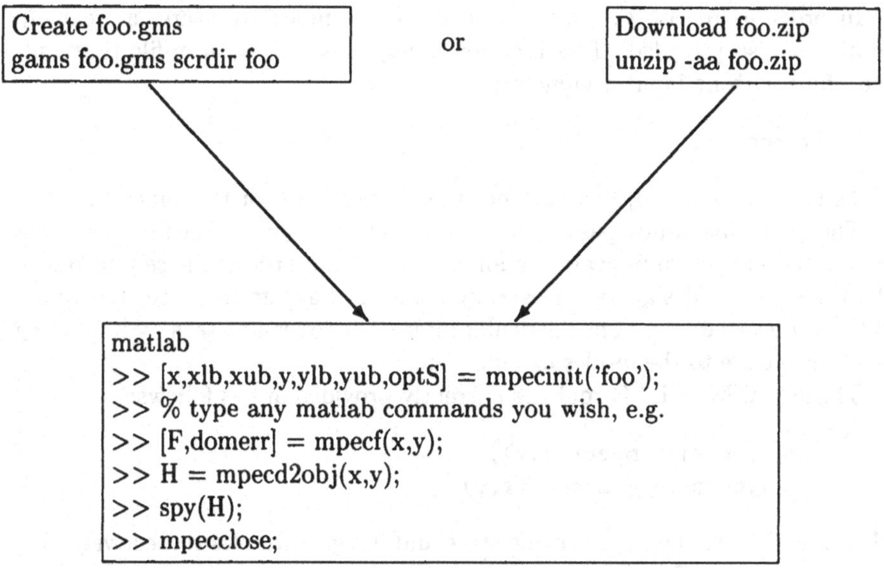

Figure 1.2 Two ways to run Matlab on a given problem.

Currently, mpecfunc can be called with 0, 2, or 3 input arguments. The call with a string argument and an integer allows a simple implementation of the initialization routine:

```
[x,xlb,xub,y,ylb,yub,optS[,gEq]] = mpecinit('foo');
```

The purpose of the routine is to set up the mpec structure and to assign values of the design and state variables into the MATLAB users workspace. First of all, the routine removes any previous mpec structure that is in the MATLAB workspace. It then checks whether a directory "foo" exists, containing the scratch files generated by GAMS, and aborts with an error message if they do not. The new mpec structure is created by calls to the MPECIO routines mpecInit and sparseInit. The bounds on the design and state variables are returned by this routine, using appropriate calls to the MPECIO routines get*Bounds and get*Levels. For simplicity, the MATLAB interface provides the constraints $h(x)$ and $g(x,y)$ in a single vector **g**. The optional return argument **gEq** informs the user if each of the constraints is a less than inequality (-1), an equality (0) or a greater than inequality (1). The mpecfunc routine allows the user to split these constraints into h and g respectively; interested users should examine the source code of mpecfunc.c.

We note here a nice feature of our m-file implementation, namely that the mpecinit m-file can be edited to explicitly call gams via a system call if the scratch files do not exist, but a model file "foo.gms" does exist. Other specializations of the MATLAB interface are just as easy to carry out.

In order to release the memory used by the mpec structure, a clean up routine is also provided. This function is implemented as an m-file that calls mpecfunc with no input arguments.

```
mpecclose;
```

As is to be expected, there are no return arguments for this function.

The remaining m-files pass three arguments to mpecfunc. The first argument is an internal code to indicate the function to be performed, the second passes the values of the design variables x, and the third argument passes the values of the state variables y. The particular internal codes can be seen by inspecting the source code to the m-files or mpecfunc.c.

The actual MATLAB routines currently provided are as follows:

```
[F,domerr] = mpecf(x,y);
[F,J,domerr] = mpecdf(x,y);
```

Here `x` and `y` are the values of the state and design variables respectively. The return arguments `F` and `J` contain the MCP function $F(x, y)$ and its Jacobian, while `domerr` is the number of domain errors encountered in the nonlinear function evaluator. Note that `J` is returned to the MATLAB user as a sparse matrix.

The following functions all return information about the objective function θ.

```
[theta,domerr] = mpecobj(x,y);
[theta,g,domerr] = mpecdobj(x,y);
H = mpecd2obj(x,y);
```

The first derivatives of θ are returned in the vector `g`; the second derivative matrix H as a sparse matrix. Note that `H` is calculated by finite differencing, since GAMS does not currently have the ability to provide second derivatives.

The side constraints are provided by the following routines, along with their derivatives.

```
[g,domerr] = mpeccon(x,y);
[g,Jg,domerr] = mpecdcon(x,y);
```

Details of the calling sequence for the MPECIO routines within the MATLAB interface can be obtained directly from the source code mpecfunc.c that can be found at `http://www.gams.com/mpec/`. Also, by editing the "m" files that implement `mpecdf`, `mpecdobj` and `mpecdcon` respectively, it is possible to return derivatives with respect to x and y to the MATLAB environment separately.

6 COMPUTATIONAL RESULTS

We have linked a number of MPEC solvers to the collection of models we have put together using the MPECIO library from Section 4 and the MATLAB interface from Section 5. This has provided a thorough test of our interface routines and has helped to modify and improve their design. In addition, it allows us to perform some computational experiments using these codes and report on the results. Before doing so, a word on interpreting our results is in order. While computational work on the MPEC has not advanced to a stage where any of the solvers can be considered well known or completely robust, the techniques used in the solvers we report on are representative of the current state of the art in this area. It is hoped that the work described in this paper will allow more algorithms to be developed that robustly solve all these problems and other problems of interest to the larger optimization community.

Two of the solvers we test use an implicit programming approach in which the MPEC is reformulated as an NLP. In order to do this, our code assumes for simplicity that there are no side constraints (2.2b) or (2.2c) and further that there is a (locally) unique solution y of the equilibrium problem $\mathrm{MCP}(F(x,\cdot),\mathbf{Y})$ for each value of x. We denote this solution by $y(x)$. Under these assumptions, the problem (2.2) is equivalent to the implicit program:

$$\begin{aligned} &\text{minimize} \quad \Theta(x) = \theta(x, y(x)) \\ &\text{subject to} \quad x \in \mathbf{X}. \end{aligned} \tag{6.1}$$

This implicit programming formulation has the advantage of simple constraints, but a rather complex, in fact nondifferentiable objective function Θ, even though the original functions θ and F may be smooth. Nonsmoothness results from the underlying equilibrium conditions.

One solution strategy for the implicit program (6.1) is to apply a "bundle method", an algorithm specifically designed to solve nonsmooth optimization problems. This idea is presented in [20, 21, 25, 26]. The implementation of the bundle method we used, `btncbc` [31], is a Fortran subroutine developed for nonconvex bound constrained problems. One of the arguments to this subroutine is a routine to evaluate the objective function and a generalized gradient at an algorithm-specified point. We coded this routine in C, using the formulas developed in [26] for the MPEC formulation (2.2) without the side constraints (2.2b) and (2.2c). This C routine uses the MPECIO library and a version of PATH [7, 8] modified to return the basis at the solution to the inner MCP.

Another solver we have implemented also uses the implicit approach to reformulate (2.2) as (6.1). In this case, the latter problem is solved by SolvOpt 1.1 [22], a C implementation of Shor's R-algorithm [32] for unconstrained, nondifferentiable optimization problems. In order to handle the constraints $x \in \mathbf{X}$, an exterior penalty amounting to the weighted maximum constraint violation was used. In version 1.0 of SolvOpt, it was necessary to add this penalty to the objective function explicitly; version 1.1 provides some support for this, including automatic adjustment of the penalty parameter.

The SolvOpt user is required to write separate routines to compute θ and elements of its subdifferential $\partial\theta$. The constraint violations are penalized through θ, but in other respects, these routines are similar to the function / generalized gradient routine passed to the bundle code `btncbc`. Both SolvOpt and the bundle code are implemented to run as GAMS subsystems, using the MPECIO library as their interface to GAMS.

In order to test the MATLAB interface of Section 5, we obtained a prototype MATLAB implementation [19] of the penalty interior point approach (PIPA) code ([23], Chapter 6) from H. Jiang and D. Ralph. This code (that we call PIPA-N) is specialized for problems whose equilibrium constraints (2.2d) take the form of an NCP. The PIPA algorithm computes a search direction via a solution to a uniquely solvable QP subproblem and performs an Armijo search using this direction. In order to use the PIPA implementation with our MATLAB interface library, it was necessary to provide m-files `f.m` (to compute the objective function θ and the equilibrium function F), `df.m` (to compute derivatives of `f.m`), and `d2f.m` (to compute second derivative information for θ only). The first two files `f.m` and `df.m` are trivial, and amount to little more than wrapper functions for our interface routines. The right choice for `d2f.m` is not so obvious.

In [23], the description of and proofs of convergence for PIPA assume only that the matrix Q in the direction-finding QP is symmetric positive semidefinite; it is not necessary to use any second-order information. Computationally, this is very appealing, since this makes `d2f.m` trivial (one can use the identity matrix, for example), but our results using the identity matrix were quite discouraging. This was also the experience of the authors of the code, hence their use of the Hessian of θ. Unfortunately, second order information is not currently available in GAMS, nor consequently to MPECIO or the MATLAB interface. To improve our results, we modified the MATLAB code to compute a BFGS approximation to the Hessian of θ. This was an improvement over using an identity matrix Q, but the results were still not what we had hoped, so we implemented a simple finite difference scheme for computing the Hessian of θ, in which our only enhancement was to recognize variables appearing linearly in θ and avoid computing the corresponding columns of the Hessian.

During our computational tests, we discovered that many of the problems we wanted to solve had equilibrium constraints that did not take the form of an NCP. In some cases, problems had variables whose lower bounds could be shifted to 0, but more often there were free variables or variables with upper and lower bounds, so that a transformation into NCP form is much less appealing. Fortunately, we were able to obtain a second implementation [19] of PIPA (that we call PIPA-M) that allowed problems of the more general form

$$\begin{array}{ll} \text{minimize} & \theta(x,y,w,z) \\ \text{subject to} & Ax \leq b \\ & \ell \leq x \leq u \\ & F(x,y,w,z) = 0 \\ & 0 \leq w \perp y \geq 0 \end{array} \tag{6.2}$$

where $x \in \mathbf{R}^n$, $y, w \in \mathbf{R}^m$, $z \in \mathbf{R}^l$, and $F: \mathbf{R}^{n+m+m+l} \to \mathbf{R}^{m+l}$. The requirements for using PIPA-M are similar to using the NCP version: three MATLAB m-files `f.m`, `df.m`, and `d2f.m`. However, the transformation from (2.2) to (6.2) is more complex. Lower-bounded variables from (2.2) need to be shifted so that their lower bound is zero, while upper bounded variables are first negated, along with their bounds, and then shifted. Free variables require no transformation. In the most difficult case, variables from (2.2) with finite upper and lower bounds are converted as follows: each scalar pair $\hat{F}(z) \perp z \in [\ell, u]$ becomes

$$\begin{aligned} 0 \le w_\ell = (z - \ell) \quad &\perp \quad y_\ell \ge 0 \\ 0 \le w_u = (u - z) \quad &\perp \quad y_u \ge 0 \\ \hat{F}(z) - y_\ell + y_u = 0 \quad &\perp \quad z \end{aligned}$$

While this transformation does increase the problem size, it does so quite sparsely, adding 2 equations and 6 nonzeroes for each double-bounded variable.

Our computational tests of these solvers and the interface tools of Section 4 and Section 5 were carried out using a preliminary version of MPECLIB, a library of MPEC models formulated in the GAMS language. We do not describe all of these models here, because most of them have been described elsewhere and due to the fact that we have made the GAMS source for these models available via the Web `http://www.gams.com/mpec/`. The file fjq1.gms contains problems 1-4 from [11], desilva.gms is described in [5], hq1.gms in [18], oz3.gms in [26], mss.gms in [24], qvi.gms in [16], and bard3.gms in [2]. For bard3 and oz3, we move the constraints (2.2b) into the complementarity problem. In the problems where the second level problem is an optimization problem or variational inequality, we add explicit multipliers to rewrite the second level problem as an MCP.

All of our computation was done on a Sun Ultra Enterprise 2 with 256 MB of RAM and two 300MHz processors running Solaris 2.5 and equipped with the standard C, F77, and MATLAB (version 5.1.0) cmex compilers. Table 1.1 contains results obtained from the GAMS subsystems BUNDLE and SolvOpt. We report the number of generalized gradient evaluations, the final objective value θ for the original problem, and the CPU time in seconds. For SolvOpt, we also report the number of evaluations of the implicit objective function in the column headed k_Θ, while in the BUNDLE code, the objective and a generalized gradient are always returned together. We report the iteration count k for BUNDLE, since some of its iterations may need to compute multiple generalized gradients. Iterations that terminated with a non-optimal solution or status are marked with a † in the grad column. In some cases, the BUNDLE and SolvOpt codes failed due to an unsolved inner problem or failed internal tolerance checks; these are indicated as "fail".

It is interesting to note that the optimal value of 0 for the oz3 problem in Table 1.1 is quite an improvement over the value of 4.999375 reported in [11].

		BUNDLE				SolvOpt			
model		k	grad	θ	secs	grad	k_Θ	θ	secs
bard3		1	1	-12.678	.01	15†	16	-11.80	
desilva		9	9	-1	.01	15	70	-1	.03
epowplan		fail				127	485	-451.5	6.0
fjq1	a	10	10	3.208	.02	2†	8	3.309	
fjq1	b	10	12	3.208	.02	1†	1	4.502	
fjq2	a	13	13	3.449	.01	14	42	3.450	.05
fjq2	b	12	12	3.449	.01	24†	28	13.60	.02
fjq3	a	8	83	4.604	.02	15†	19	5.552	.03
fjq3	b	9	86	4.604	.04	24†	27	17.70	.03
fjq4	a	13	89	6.593	.03	11	25	6.593	.03
fjq4	b	13	13	6.593	.03	12†	31	6.595	.03
gauvin		13	13	20	.01	14	66	20	.02
hq1		13	13	-3266.67	.01	16	89	-3266.67	.04
mss	1	14	15	-343.35	.03	13	62	-343.35	.09
mss	2	16	16	-3.16	.04	14	73	-3.16	.14
mss	3	11	180	-5.35	.21	11	43	-5.35	.05
oz3		2	2	0	.01	fail			
qvi		11	12	3.07e-9	.02	14	121	1e-18	.06
ralphmod		fail				fail			
tollmpec		7	7	-20.83	11.8	fail			

Table 1.1 Implicit Algorithms

In our computational work, we tested 6 distinct MATLAB codes: both the general and the NCP-constrained code were tested using a constant Q matrix for the QP subproblem, using an approximate Hessian of θ obtained via BFGS updating, and using a difference approximation to the Hessian. However, the results obtained using the constant Q were disappointing, so are not included in Table 1.2. Since many of the models were not NCP-constrained, we do not report on the BFGS version of this algorithm either. We are left with algorithm PIPA-N (the NCP-constrained version of PIPA with a difference approximation to the Hessian), PIPA-M (the MCP-constrained version of PIPA, difference approximation), and PIPA-B (the MCP-constrained version of PIPA using BFGS updates). For each of these three techniques, we report the iteration count k (maximum 1000), the final objective value θ, and the CPU time in seconds. Each iteration involves the solution of one direction-finding QP subproblem; these subproblems are solved using the standard MATLAB QP solver.

Where a model was not NCP-constrained and PIPA-N could not be applied, this is so indicated in Table 1.2. We also indicate cases where the PIPA algorithms failed to converge with a † in the seconds column. There are a number of case where this occurred. For model oz3, the PIPA-B code was getting close to the optimal θ value of 4.999375 reported in [11], but the same cannot be said of PIPA-M. Both of these algorithms had difficulty with the bilevel programs in fjq1 - fjq4, hitting the iteration limit before reaching a solution. PIPA-M seemed to get quite close to the optimal value of -12.67871 from [11], but this may or may not have been at a feasible point, while PIPA-B failed completely on this model. The electric power planning model epowplan led to immediate failure in the linesearch phase of PIPA, while the memory required to solve the tollmpec model via PIPA was too much for our computing platform. The ralphmod model is a linear complementarity constrained MPEC with a quadratic objective, so the PIPA-N and PIPA-M codes use exact Hessians.

It is clear that more algorithmic development work for MPEC problems is required. New algorithms, such as those outlined in [11, 15, 19, 27] hold great potential to take advantage of the tools we have developed here.

7 CONCLUSIONS

This paper has described some enabling technology for modeling and solving MPEC problems. The cornerstone for all of these is the addition of the MPEC model type to the GAMS language. These models are constructed and passed to the solvers by MPECIO, a new library of routines to ease problem data access. To exhibit the use of these routines we have linked various MPEC solvers directly to GAMS.

As another example of the utility of the interface library, we have provided simple MATLAB functions to access the problem data from within MATLAB, thus allowing the prototyping of algorithms within this environment.

It is our hope that the interface we have created between the MPEC modeler and MPEC algorithm developer will spur development within both of these groups by fueling a powerful synergy between them.

		PIPA-N			PIPA-M			PIPA-B	
model	k	secs	θ	k	secs	θ	k	secs	θ
bard3	MCP			1000	65†	-12.63		fail	
desilva	218	5.9	-1	135	4.9	-.999999	145	4.6	-.999999
epowplan	MCP			stepsize			stepsize		
fjq1 a	MCP			1000	62†	3.274	1000	62†	3.2320
fjq1 b	MCP			1000	60†	3.262	1000	63†	3.267
fjq2 a	MCP			1000	62†	3.450	59	1.6	3.449
fjq2 b	MCP			1000	51†	3.450	18	.38	3.449
fjq3 a	MCP			1000	54†	4.594	35	.90	4.604
fjq3 b	MCP			1000	53†	4.594	26	.61	4.604
fjq4 a	MCP			15	.37	6.593	24	.56	6.593
fjq4 b	MCP			18	.45	6.593	52	1.5	6.593
gauvin	15	.18	20	18	.37	20	18	.34	20
hq1	12	.12	-3266.67	18	.3	-3266.67	18	.26	-3266.67
mss 1	MCP			24	1.2	-343.35	29	1.3	-343.35
mss 2	MCP			52	2.8	-3.16	52	2.8	-3.16
mss 3	MCP			21	1.0	-5.35	21	1.5	-5.35
oz3	MCP			1000	65†	57.50	1000	104†	5.172
qvi	MCP			1000	18	0	1000	14	0
ralphmod	107	138	-683.03	224	1782	-683.03	1000	4488†	-563.4
tollmpec	MCP			memory use			memory use		

Table 1.2 MATLAB PIPA codes

We gratefully acknowledge those who have assisted us in this work. H. Jiang and D. Ralph provided the MATLAB codes used in our tests and were generous with comments on the MATLAB interface. A. Kuntsevich and F. Kappel have made SolvOpt publicly available. J. Outrata provided guidance on the use of bundle methods for these problems.

References

[1] A. Bachem, M. Grötchel, and B. Korte, editors. *Mathematical Programming: The State of the Art, Bonn 1982*, Berlin, 1983. Springer Verlag.

[2] J. F. Bard. Convex two-level optimization. *Mathematical Programming*, 40:15–27, 1988.

[3] J. Bisschop and R. Entriken. *AIMMS – The Modeling System.* Paragon Decision Technology, Haarlem, The Netherlands, 1993.

[4] A. Brooke, D. Kendrick, and A. Meeraus. *GAMS: A User's Guide.* The Scientific Press, South San Francisco, CA, 1988.

[5] A. H. DeSilva. *Sensitivity Formulas for Nonlinear Factorable Programming and their Application to the Solution of an Implicitly Defined Optimization Model of US Crude Oil Production.* PhD thesis, George Washington University, Washington, D.C., 1978.

[6] S. P. Dirkse and M. C. Ferris. MCPLIB: A collection of nonlinear mixed complementarity problems. *Optimization Methods and Software*, 5:319–345, 1995.

[7] S. P. Dirkse and M. C. Ferris. The PATH solver: A non-monotone stabilization scheme for mixed complementarity problems. *Optimization Methods and Software*, 5:123–156, 1995.

[8] S. P. Dirkse and M. C. Ferris. A pathsearch damped Newton method for computing general equilibria. *Annals of Operations Research*, 1996.

[9] S. P. Dirkse and M. C. Ferris. Traffic modeling and variational inequalities using GAMS. In Ph. L. Toint, M. Labbe, K. Tanczos, and G. Laporte, editors, *Operations Research and Decision Aid Methodologies in Traffic and Transportation Management*, NATO ASI Series F. Springer-Verlag, 1997.

[10] S. P. Dirkse, M. C. Ferris, P. V. Preckel, and T. Rutherford. The GAMS callable program library for variational and complementarity solvers. Mathematical Programming Technical Report 94-07, Computer Sciences Department, University of Wisconsin, Madison, Wisconsin, 1994.

[11] F. Facchinei, H. Jiang, and L. Qi. A smoothing method for mathematical programs with equilibrium constraints. Technical report, Universitá di Roma "La Sapienza", Roma, Italy, 1996.

[12] M. C. Ferris, R. Fourer, and D. M. Gay. Expressing complementarity problems and communicating them to solvers. Mathematical Programming Technical Report 98-02, Computer Sciences Department, University of Wisconsin, Madison, Wisconsin, 1998.

[13] M. C. Ferris and T. F. Rutherford. Accessing realistic complementarity problems within Matlab. In G. Di Pillo and F. Giannessi, editors, *Nonlinear Optimization and Applications*, pages 141–153. Plenum Press, New York, 1996.

[14] R. Fourer, D. Gay, and B. Kernighan. *AMPL*. The Scientific Press, South San Francisco, California, 1993.

[15] M. Fukushima, Z.-Q. Luo, and J. S. Pang. A globally convergent sequential quadratic programming algorithm for mathematical programs with linear complementarity constraints. *Computational Optimization and Applications, forthcoming*, 1997.

[16] P. T. Harker. Generalized Nash games and quasi-variational inequalities. *European Journal of Operations Research*, 54:81–94, 1987.

[17] P. T. Harker and J. S. Pang. Finite–dimensional variational inequality and nonlinear complementarity problems: A survey of theory, algorithms and applications. *Mathematical Programming*, 48:161–220, 1990.

[18] J. M. Henderson and R. E. Quandt. *Microeconomic Theory*. McGraw–Hill, New York, 3rd edition, 1980.

[19] H. Jiang and D. Ralph. QPECgen, a matlab generator for mathematical programs with quadratic objectives and affine variational inequality constraints. Technical report, The University of Melbourne, Department of Mathematics and Statistics, Parkville, Victoria, Australia, 1997.

[20] M. Kočvara and J. V. Outrata. On optimization of systems governed by implicit complementarity problems. *Numerical Fucntional Analysis and Optimization*, 15:869–887, 1994.

[21] M. Kočvara and J. V. Outrata. On the solution of optimum design problems with variational inequalities. In *Recent Advances in Nonsmooth Optimization*, pages 172–192. World Scientific Publishers, Singapore, 1995.

[22] A. Kuntsevich and F. Kappel. SolvOpt: The solver for local nonlinear optimization problems. Institute for Mathematics, Karl-Franzens University of Graz, 1997.

[23] Z.-Q. Luo, J. S. Pang, and D. Ralph. *Mathematical Programs with Equilibrium Constraints*. Cambridge University Press, 1996.

[24] Frederic H. Murphy, Hanif D. Sherali, and Allen L. Soyster. A mathematical programming approach for determining oligopolistic market equilibrium. *Mathematical Programming*, 24:92–106, 1982.

[25] J. V. Outrata. On optimization problems with variational inequality constraints. *SIAM Journal on Optimization*, 4:340–357, 1994.

[26] J. V. Outrata and J. Zowe. A numerical approach to optimization problems with variational inequality constraints. *Mathematical Programming*, 68:105–130, 1995.

[27] D. Ralph. A piecewise sequential quadratic programming method for mathematical programs with linear complementarity constraints. In *Pro-*

ceedings of the Seventh Conference on Computational Techniques and Applications (CTAC95), 1996.

[28] R. Raman. *Integration of logic and heuristic knowledge in discrete optimization techniques for process systems.* PhD thesis, Department of Chemical Engineering, Carnegie Mellon University, Pittsburgh, Pennsylvania, 1993.

[29] S. M. Robinson. Strongly regular generalized equations. *Mathematics of Operations Research*, 5:43–62, 1980.

[30] S. M. Robinson. Generalized equations. In Bachem et al. [1], pages 346–367.

[31] H. Schramm and J. Zowe. A version of the bundle idea for minimizing a nonsmooth function: Conceptual idea, convergence analysis, numerical results. *SIAM Journal on Optimization*, 2:121–152, 1992.

[32] N. Z. Shor. *Minimization Methods for Nondifferentiable Functions.* Springer-Verlag, Berlin, 1985.

[33] H. Van Stackelberg. *The Theory of Market Economy.* Oxford University Press, 1952.

Reformulation: Nonsmooth, Piecewise Smooth, Semismooth and Smoothing Methods, pp. 149–159
Edited by M. Fukushima and L. Qi

Merit Functions and Stability for Complementarity Problems

Andreas Fischer*

Abstract The parametric complementarity problem is reformulated as parametric optimization problem. Results on the quantitative stability of the latter are used to obtain such results for the former. In particular, for a new class of functions a result on the local upper Lipschitz-continuity of the solution set map belonging to the parametric complementarity problem is shown. This class of functions extends the class of uniform P-functions so that certain complementarity problems whose solution set is not a singleton can be dealt with.

Key Words merit function, growth behavior, stability, parametric complementarity problem, uniform P-function

1 INTRODUCTION

Let $(P, d(\cdot,\cdot))$ be a metric space of parameters and a function $F : \Re^n \times P \to \Re^n$ be given. Then we consider the parametric complementarity problem, i.e., the problem of finding $x \in \Re^n$ such that

$$F(x,p) \geq 0, \quad x \geq 0, \quad x^T F(x,p) = 0, \tag{1.1}$$

where $p \in P$ is arbitrary but fixed. As is well known from the usual complementarity problem we can write (1.1) as system of equations

$$H(x,p) = 0 \tag{1.2}$$

in the sense that the solution sets of (1.1) and (1.2) are equal. We will exploit this fact to present a new approach for obtaining results on quantitative stability for the parametric complementarity problem (1.1). To this end, the

*Department of Mathematics, University of Dortmund, D-44221, Germany, fischer@math.uni-dortmund.de

parametric system (1.2) will be reformulated as parametric optimization problem. Under appropriate conditions on the parametric system the stability of the parametric optimization problem can be analyzed by means of recently developed results [12, 14]. In this way we obtain stability results for the parametric system, too.

Using several conditions on the underlying complementarity problem we then show that the approach not only enables us to recover known results but also leads to new ones. In particular, we introduce a class of functions that weakens the notion of uniform P-functions. In this way certain complementarity problems whose solution sets are not a singleton can be dealt with. A result on the local upper Lipschitz-continuity of the solution set map belonging to such parametric complementarity problems is proved.

Notation. Throughout the paper let $p_0 \in P$ be a fixed parameter. We denote by $\|\cdot\|$ the Euclidean norm, by B the closed unit ball in $\Re^n$ and by $B(p_0, \gamma) \subset P$ the closed γ-neighborhood of p_0. The distance of a point $x \in \Re^n$ from a nonempty set $A \subseteq \Re^n$ is defined by $\operatorname{dist}[x, A] := \inf\{\|x - a\| \mid a \in A\}$. Moreover, we denote by C the linear space of the continuous functions from $\Re^n$ to $\Re^n$.

2 A GENERAL STABILITY RESULT

We consider problem (1.2), where $H : \Re^n \times P \to \Re^n$ is assumed to be continuous. The solution set of (1.2) depends on the parameter $p \in P$ and is denoted by

$$\mathcal{S}(p) := \{x \in \Re^n \mid H(x, p) = 0\}.$$

Let $\mathcal{S}$ denote an arbitrary, but fixed, subset of $\mathcal{S}(p_0)$. We will make use of the following assumptions:

(A1) The set $\mathcal{S}$ is nonempty and compact.

(A2) There are numbers $c > 0$, $\epsilon > 0$, and $\kappa \in [1, \infty)$, such that

$$\|H(x, p_0)\| \geq c d[x, \mathcal{S}]^\kappa \qquad \forall x \in \mathcal{S} + \epsilon B.$$

(A3) There are numbers $\epsilon_0 > 0$, $\gamma_0 > 0$, and $\mu_0 > 0$ such that

$$\|H(x, p) - H(y, p_0)\| \leq \mu_0 (d(p_0, p) + \|x - y\|)$$

$$\forall x, y \in \mathcal{S} + \epsilon_0 B,\ \forall p \in B(p_0, \gamma_0).$$

Remark. *Assumption (A2) is called* growth condition. *Together with Assumptions (A1) it implies that $\mathcal{S}$ is equal to the solution set $\mathcal{S}(p_0)$ or to an isolated part of it. Assumption (A3) in particular implies the local Lipschitz-continuity of $H(\cdot, p_0)$ on $\mathcal{S} + \epsilon_0 B$. On the other hand, the local Lipschitz-continuity of H in respect to (x, p) is sufficient for (A3).*

Associated with problem (1.2), we consider the following parametric optimization problem

$$h(x,p) := \|H(x,p)\| \to \min \quad \text{subject to} \quad x \in \text{cl}Q, \tag{2.1}$$

where $Q \subseteq \Re^n$ is any given set and cl$Q$ denotes the closure of Q. The solution set of this problem is denoted by

$$\mathcal{Z}_Q(p).$$

If $Q = \Re^n$ we simply write $\mathcal{Z}(p)$. With the notation

$$\mathcal{S}_Q(p) := \mathcal{S}(p) \cap \text{cl}Q,$$

the following proposition is evident.

Proposition 2.1 *For any $p \in P$ and any $Q \subseteq \Re^n$, it holds that $\mathcal{S}_Q(p) \neq \emptyset$ implies $\mathcal{S}_Q(p) = \mathcal{Z}_Q(p)$.*

Theorem 2.1 *Let S be a subset of $\mathcal{S}(p_0)$ satisfying Assumptions (A1) - (A3). Then, there are an open bounded set $Q \supset S$ and numbers $\gamma_1, \mu_1 > 0$ such that $S = \mathcal{Z}_Q(p_0)$ and*

$$\emptyset \neq \mathcal{Z}_Q(p) \subseteq \mathcal{Z}_Q(p_0) + \mu_1 d(p_0,p)^{\frac{1}{\kappa}} B \qquad \forall p \in B(p_0,\gamma_1).$$

Proof: The theorem immediately follows from Theorem 2.2. in Klatte [14], see also Ioffe [12]. ∎

Theorem 2.1 together with Proposition 2.1 yields the next result.

Corollary 2.1 *Let S be a subset of $\mathcal{S}(p_0)$ satisfying Assumptions (A1) - (A3). Then, with the set Q and the numbers γ_1, μ_1 from Theorem 2.1, it holds that*

$$\mathcal{S}_Q(p) \subseteq S + \mu_1 d(p_0,p)^{\frac{1}{\kappa}} B \qquad \forall p \in B(p_0,\gamma_1).$$

Remark. *It is possible to derive results analogous to Theorem 2.1 and Corollary 2.1 if problems (1.2) and (2.1) are considered subject to an additional parametric constraint $x \in \Gamma(p)$ with a point-to-set mapping $\Gamma : P \to 2^{\Re^n}$. Then, the growth condition (A2) needs to hold for $x \in (S + \epsilon B) \cap \Gamma(p_0)$ only, whereas Γ has to satisfy a so-called pseudo-Lipschitz* property, see [14] for more details. Although we do not consider this more general approach it may be useful in respect to reformulations of complementarity problems as constrained optimization problems.*

3 APPLICATION TO COMPLEMENTARITY PROBLEMS

To obtain stability results for problem (1.1) by means of the general result in the previous section we will assume that (A1) is satisfied, see Subsection 3.2.

Moreover, Assumption (A3) will also be fulfilled in the situations considered there. Thus, we have to think about the following questions:

1. Which mapping H should be used to replace the complementarity problem (1.1) by (1.2)?
2. Which conditions ensure that $\mathcal{S}_Q(p) \neq \emptyset$ for Q from Theorem 2.1?
3. Which assumptions guarantee that the growth condition (A2) is fulfilled?

Regarding Question 1 there is a number of possibilities for the definition of H (see [11, 17, 18, 19]) where, sometimes, additional constraints (like $x \in \Re^n_+$) occur.

In the remainder of this paper, we will only consider problem (1.2) with H defined by

$$H(x,p) := [\varphi(x_1, F_1(x,p)), \ldots, \varphi(x_n, F_n(x,p))]^T,$$

where $\varphi : \Re^2 \to \Re^2$ is given by

$$\varphi(a,b) := \sqrt{a^2 + b^2} - a - b,$$

see [6]. Since

$$\varphi(a,b) = 0 \quad \Leftrightarrow \quad a \geq 0,\, b \geq 0,\, ab = 0$$

it is evident that

$$h(x,p) = 0 \quad \Leftrightarrow \quad H(x,p) = 0 \quad \Leftrightarrow \quad x \text{ solves (1.1) for } p \in P.$$

Obviously, the mapping φ is globally Lipschitz-continuous. Therefore, if numbers $\epsilon_F > 0$, $\gamma_F > 0$, and $\mu_F > 0$ exist so that

$$\|F(x,p) - F(y,p_0)\| \leq \mu_F(d(p_0,p) + \|x-y\|) \quad \forall x,y \in \mathcal{S} + \epsilon_F B,\ \forall p \in B(p_0, \gamma_F), \tag{3.1}$$

Assumption (A3) holds for suitable numbers $\epsilon_0 > 0$, $\gamma_0 > 0$, and $\mu_0 > 0$.

Now, let us consider Question 2. Even if it is possible to simply assume $\mathcal{S}_Q(p) \neq \emptyset$ (with Q from Theorem 2.1) so that Corollary 2.1 can be applied it would be nice if $\mathcal{S}_Q(p) \neq \emptyset$ can be guaranteed under as weak as possible conditions. To present such a condition let the particular normed space P_C given by

$$P_C := \{p \in C \mid \sup_{x \in \Re^n} \|p(x)\| < \infty\}, \quad \|p\|_\infty := \sup_{x \in \Re^n} \|p(x)\| \quad \forall p \in P_C.$$

With the metric $d_C(\cdot,\cdot)$ induced by this norm $(P_C, d_C(\cdot,\cdot))$ is a metric space. Moreover, let $P_\Re := \{p = (M,q) \mid M \in \Re^{n \times n},\, q \in \Re^n\}$ be equipped with the metric $d_\Re(\cdot,\cdot)$ induced by any norm in $\Re^{n\times n} \times \Re^n$.

Proposition 3.1 *(i) Let $P := (P_C, d_C(\cdot,\cdot))$, $p_0 := 0$, $F_0 \in C$, and $F(\cdot, p(\cdot)) := F_0(\cdot) + p(\cdot)$. If F_0 is a P_0-function and if $\mathcal{S}(p_0)$ is nonempty and bounded then, for any open set $Q \supset \mathcal{S}(p_0)$, a number $\gamma(Q) > 0$ exists so that*

$$\mathcal{S}_Q(p) \neq \emptyset \qquad \forall p \in B(0, \gamma(Q)).$$

(ii) Let $P := (P_{\Re}, d_{\Re}(\cdot,\cdot))$, $p_0 := (M_0, q_0)$, *and* $F(x,p) := Mx + q$ *for all* $x \in \Re^n$. *If* M_0 *is a* P_0*-matrix and if* $\mathcal{S}(p_0)$ *is nonempty and bounded then, for any open set* $Q \supset \mathcal{S}(p_0)$, *a number* $\tilde{\gamma}(Q) > 0$ *exists so that*

$$\mathcal{S}_Q(p) \neq \emptyset \qquad \forall p \in B(p_0, \tilde{\gamma}(Q)).$$

This proposition can be derived from a result by Facchinei [3, Theorem 4.4 and Definition 4.2] (under the additional assumption that F_0 is continuously differentiable) or from a subsequent paper by Ravindran and Gowda [21, Corollary 6 (c) and Definition 1].

For further conditions that ensure $\mathcal{S}_Q(p) \neq \emptyset$ for p close to p_0 we refer the reader to the article by Gowda and Pang [9]. Another way to derive those conditions is to consider relations between the sets $\mathcal{Z}(p)$ and $\mathcal{S}(p)$ and to apply results by De Luca, Facchinei, and Kanzow [1] on stationary points of a merit function that is based on φ.

3.1 Ensuring the Growth Condition

Now, we will consider Question 3. The following proposition can be derived from a result by Luo and Tseng [15] or from an earlier result by Robinson [22] on polyhedral multifunctions, see Fischer [7]. A subset $\mathcal{S}$ of $\mathcal{S}(p_0)$ will be called isolated if an $\varepsilon > 0$ exists so that $\mathcal{S} = (\mathcal{S} + \varepsilon B) \cap \mathcal{S}(p_0)$.

Proposition 3.2 *Let* $F(\cdot, p_0)$ *be given by* $F(x,p_0) = M_0 x + q_0$ *for all* $x \in \Re^n$. *Then for any isolated nonempty subset* $\mathcal{S}$ *of* $\mathcal{S}(p_0)$ *numbers* $c > 0$ *and* $\epsilon > 0$ *exist so that the growth condition (A2) is satisfied with* $\kappa = 1$.

Proposition 3.3 *If* $F(\cdot, p_0)$ *is analytic then for any isolated and bounded nonempty subset* $\mathcal{S}$ *of* $\mathcal{S}(p_0)$ *numbers* $c > 0$, $\epsilon > 0$, *and* $\kappa \in [1, \infty)$ *exist such that (A2) is satisfied.*

Proof: Since $\mathcal{S}$ is bounded a number $c_0 > 0$ exists so that, for all $x \in \mathcal{S} + B$,

$$|x^T F(x,p_0)| = |\max\{x, F(x,p_0)\}^T \min\{x, F(x,p_0)\}| \leq c_0 \|\min\{x, F(x,p_0)\}\|. \tag{3.2}$$

Lemma 3.1 in Tseng [24] provides

$$|\min\{a,b\}| \leq \frac{\sqrt{2}+2}{2} |\varphi(a,b)| \qquad \forall a,b \in \Re. \tag{3.3}$$

Using this, we easily get

$$|\min\{0,a\}| + |\min\{0,b\}| \leq 2|\min\{a,b\}| \leq (\sqrt{2}+2)|\varphi(a,b)| \qquad \forall a,b \in \Re.$$

From this and (3.2), we obtain for all $x \in \mathcal{S} + B$,

$$\|\min\{0, F(x,p_0)\}\| + \|\min\{0,x\}\| + |x^T F(x,p_0)| \leq c_1 \|H(x,p_0)\|$$

with a suitably chosen $c_1 > 0$. The left hand side is the natural residual of the linear complementarity problem (1.1) for $p = p_0$. Based on a recent result by Luo and Pang [16] this residual can be bounded below such that, for all $x \in \mathcal{S} + \epsilon B$ with $\epsilon > 0$ sufficiently small,

$$\begin{aligned} c_2 \text{dist}[x, \mathcal{S}(p_0)]^\kappa &\leq \|\min\{0, F(x, p_0)\}\| + \|\min\{0, x\}\| + |x^T F(x, p_0)| \\ &\leq c_1 \|H(x, p_0)\| \end{aligned}$$

follows for a suitable $c_2 > 0$. This yields the desired result. ∎

To present a third answer to Question 3 we introduce the following class of functions.

Definition 3.1 *Let $R \subseteq \Re^n$ and $S \subseteq \Re^n$ be closed nonempty sets. A function $G : \Re^n \to \Re^n$ is called* uniform P-function w.r.t. (R, S) *if $\sigma > 0$ and a function $\pi : R \to S$ exist such that $\|x - \pi(x)\| = \text{dist}[x, S]$ for all $x \in R$ and*

$$\max_{1 \leq i \leq n} (G_i(x) - G_i(\pi(x)))(x - \pi(x))_i \geq \sigma \text{dist}[x, S]^2 \qquad \forall x \in R.$$

Obviously, for arbitrarily chosen (R, S), any uniform P-function is also a uniform P-function w.r.t. (R, S). However, as can be seen in the next example, the new class is actually larger than the class of uniform P-functions and has particular importance if G is not differentiable.

Example 3.1 *Let $n := 1$, $S := [0, 1]$, $R := [0, \infty)$ and G be defined by*

$$G(x) := \max\{0, x(x-1)\}.$$

Then, $G(x) = 0$ iff $x \in [0, 1]$ and G is uniformly P with respect to (R, S) but neither a P- nor a P_0-function. The solution set of the complementarity problem

$$G(x) \geq 0, \quad x \geq 0, \quad x^T G(x) = 0$$

is equal to S and $|\min\{x, G(x)\}| = |\min\{x, x^2 - x\}| \geq \text{dist}[x, S]$ holds.

Theorem 3.1 *Let $\mathcal{S}(p_0)$ be nonempty and bounded. If $F(\cdot, p_0)$ is a locally Lipschitz-continuous uniform P-function w.r.t. $(\Re^n_+, \mathcal{S}(p_0))$ then a number $c > 0$ exists such that (A2) is satisfied with $S = \mathcal{S}(p_0)$ and $\kappa = \epsilon = 1$.*

Proof: For any $x \in \Re^n$ let x_+ denote the projection of x onto $\Re^n_+$. Thus, $\|x - x_+\| = \text{dist}[x, \Re^n_+]$. The compactness of $\mathcal{S}(p_0)$, the local Lipschitz-continuity of $F(\cdot, p_0)$, and the global Lipschitz-continuity of φ imply

$$\|F(x, p_0) - F(\pi(x), p_0)\| \leq L \text{dist}[x, \mathcal{S}(p_0)] \quad \forall x \in \Re^n_+ \cap (\mathcal{S}(p_0) + 2B), \tag{3.4}$$

$$\|H(x_+, p_0)\| - \|H(x, p_0)\| \leq L\|x - x_+\| \qquad \forall x \in (\mathcal{S}(p_0) + B) \setminus \Re^n_+ \tag{3.5}$$

for some $L \geq 1$. Now, fix any $x \in \Re^n_+ \cap (\mathcal{S}(p_0) + 2B)$, let $r := \min\{x, F(x, p_0)\}$, and denote by i an index for which the "max" in Definition 3.1 is attained. Thus,

$$\sigma \text{dist}[x, \mathcal{S}(p_0)]^2 \leq (F_i(x, p_0) - F_i(\pi(x), p_0))(x_i - \pi_i(x)) \tag{3.6}$$

holds. We distinguish two cases:

a) $x_i \geq F_i(x, p_0)$.
Then, using $F_i(\pi(x), p_0)\pi_i(x) = 0$ and $F_i(\pi(x), p_0)x_i \geq 0$, we get from (3.6) that

$$\sigma \text{dist}[x, \mathcal{S}(p_0)]^2 \leq F_i(x, p_0)(x_i - \pi_i(x)) = r_i(x_i - \pi_i(x)) \leq \|r\| \text{dist}[x, \mathcal{S}(p_0)]$$

Thus, it follows that

$$\sigma \text{dist}[x, \mathcal{S}(p_0)] \leq \|r\|. \tag{3.7}$$

b) $x_i < F_i(x, p_0)$.
Then, using $F_i(\pi(x), p_0)\pi_i(x) = 0$ and $F_i(x, p_0)\pi_i(x) \geq 0$, we obtain from (3.6) that

$$\begin{aligned} \sigma \text{dist}[x, \mathcal{S}(p_0)]^2 &\leq (F_i(x, p_0) - F_i(\pi(x), p_0))x_i \\ &\leq \|r\| \|F(x, p_0) - F(\pi(x), p_0)\|. \end{aligned}$$

Taking into account (3.4), we therefore have

$$\frac{\sigma}{L} \text{dist}[x, \mathcal{S}(p_0)] \leq \|r\|.$$

This, (3.7), and $|\min\{a, b\}| \leq 2|\varphi(a, b)|$ for $a, b \in \Re$ yield

$$\frac{\sigma}{2L} \text{dist}[x, \mathcal{S}(p_0)] \leq \frac{1}{2}\|r\| = \frac{1}{2}\|\min\{x, F(x, p_0)\}\| \leq \|H(x, p_0)\|. \tag{3.8}$$

Whereas this holds for any $x \in \Re^n_+ \cap (\mathcal{S}(p_0) + 2B)$ let us now consider an arbitrary but fixed $x \in (\mathcal{S}(p_0) + B) \setminus \Re^n_+$. Setting $\lambda := \sigma(4L^2 + \sigma)^{-1}$ we consider two cases:

a) $\|x - x_+\| \leq \lambda \text{dist}[x, \mathcal{S}(p_0)]$.
Then, we have

$$\text{dist}[x_+, \mathcal{S}(p_0)] \leq \|x - x_+\| + \text{dist}[x, \mathcal{S}(p_0)] \leq (1 + \lambda)\text{dist}[x, \mathcal{S}(p_0)]$$

and, as $\lambda < 1$, $x_+ \in \Re^n_+ \cap (\mathcal{S}(p_0) + 2B)$ follows. Hence, (3.8) can be applied for x_+ instead of x. Together with (3.5) this yields

$$\begin{aligned} \|H(x, p_0)\| &\geq \|H(x_+, p_0)\| - L\|x - x_+\| \\ &\geq \|H(x_+, p_0)\| - \lambda L \text{dist}[x, \mathcal{S}(p_0)] \qquad (3.9) \\ &\geq \tfrac{\sigma}{2L} \text{dist}[x_+, \mathcal{S}(p_0)] - \lambda L \text{dist}[x, \mathcal{S}(p_0)]. \end{aligned}$$

Moreover, using the triangle inequality, we get

$$\begin{aligned}\mathrm{dist}[x_+, \mathcal{S}(p_0)] &= \|x_+ - \pi(x_+)\| \\ &\geq \|x - \pi(x_+)\| - \|x - x_+\| \\ &\geq \mathrm{dist}[x, \mathcal{S}(p_0)] - \|x - x_+\| \\ &\geq (1-\lambda)\mathrm{dist}[x, \mathcal{S}(p_0)].\end{aligned}$$

This, (3.9), and the definition of λ lead to

$$\|H(x,p_0)\| \geq \left(\frac{\sigma}{2L}(1-\lambda) - \lambda L\right)\mathrm{dist}[x, \mathcal{S}(p_0)] = \lambda L \mathrm{dist}[x, \mathcal{S}(p_0)]. \tag{3.10}$$

b) $\|x - x_+\| > \lambda d[x, \mathcal{S}(p_0)]$.

Taking into account relation (3.3), $|\min\{a,0\}| \leq |\min\{a,b\}|$ for $a, b \in \Re$, and $\|\min\{x,0\}\| = \|x - x_+\|$ for $x \in \Re^n$, we immediately get

$$\|H(x,p_0)\| \geq \frac{1}{2}\|\min\{x, F(x,p_0)\}\| \geq \frac{1}{2}\|\min\{x,0\}\| \geq \frac{\lambda}{2}\mathrm{dist}[x, \mathcal{S}(p_0)].$$

From this, (3.10), and (3.8) we see that the proof is complete. ■

Remark. *In a certain sense, Theorem 3.1 generalizes an error bound result given by Kanzow and Fukushima [13, Lemma 5.1] for mixed complementarity problems. In contrast to Theorem 3.1 that is based on Definition 3.1, these authors consider problems with uniform P-functions.*

It can easily be seen that different possibilities for extending Theorem 3.1 exist. So, a corresponding result can be expected for mixed complementarity problems. Moreover, assuming that $F(\cdot, p_0)$ *is a uniform P-function with respect to* $(\Re^n_+ \cap (\mathcal{S}(p_0) + \delta B), \mathcal{S}(p_0))$ *for some* $\delta > 0$ *would lead to a local version of Theorem 3.1 (with* $\epsilon > 0$ *sufficiently small). With regard to (3.4) and (3.5) the assumption on the local Lipschitz-continuity can be weakened.*

3.2 Some Consequences

Putting the previous results together we obtain the propositions below. Let us note in advance that property (3.1) can easily be verified (with appropriate numbers $\epsilon_F > 0$, $\gamma_F > 0$, and $\mu_F > 0$) for each of these propositions so that Assumption (A3) will always be satisfied.

Proposition 3.4 *Let* $P := (P_\Re, d_\Re(\cdot,\cdot))$, $p_0 := (M_0, q_0)$. *Moreover, for any* $p := (M, q) \in P_\Re$, *let* $F(\cdot, p)$ *be given by*

$$F(x,p) := Mx + q \qquad \forall x \in \Re^n.$$

If the solution set $\mathcal{S}(p_0)$ *is nonempty and bounded and if* M_0 *is a* P_0*-matrix then an open set* $Q \supset \mathcal{S}(p_0)$ *and numbers* $\gamma_4 > 0$ *and* $\mu_4 > 0$ *exist so that*

$$\emptyset \neq \mathcal{S}_Q(p) \subseteq \mathcal{S}(p_0) + \mu_4 d(p_0, p)B \qquad \forall p \in B(p_0, \gamma_4).$$

This proposition is related to a result by Robinson [23] for linear complementarity problems with a positive semidefinite matrix. To prove the analogous result for P_0-matrices one has to combine Corollary 2.1, Proposition 3.1 (ii), and Proposition 3.2.

Now, using Corollary 2.1, Proposition 3.3, and Proposition 3.1 (i) we get the next result.

Proposition 3.5 *Let $P := (P_C, d_C(\cdot,\cdot))$, $p_0 := 0$, $F_0 : \Re^n \to \Re^n$, and*

$$F(\cdot, p(\cdot)) := F_0(\cdot) + p(\cdot).$$

If the solution set $\mathcal{S}(p_0)$ is nonempty and bounded and if F_0 is an analytic P_0-function then for any isolated and bounded nonempty subset $\mathcal{S}$ of $\mathcal{S}(p_0)$ an open set $Q \supset \mathcal{S}$ and numbers $\gamma_5 > 0, \mu_5 > 0$ and $\kappa \in [1, \infty)$ exist so that

$$\emptyset \neq \mathcal{S}_Q(p) \subseteq \mathcal{S} + \mu_5 d(0,p)^{\frac{1}{\kappa}} B \qquad \forall p \in B(0, \gamma_5).$$

Proposition 3.6 *Let $P := (P_C, d_C(\cdot,\cdot))$, $p_0 := 0$, $F_0 : \Re^n \to \Re^n$, and*

$$F(\cdot, p(\cdot)) := F_0(\cdot) + p(\cdot).$$

If F_0 is locally Lipschitz-continuous and both a P_0-function and a uniform P-function w.r.t. $(\Re^n_+, \mathcal{S}(p_0))$ and if $\mathcal{S}(p_0)$ is nonempty and compact then an open set $Q \supset \mathcal{S}(p_0)$ and numbers $\gamma_6 > 0, \mu_6 > 0$ exist so that

$$\emptyset \neq \mathcal{S}_Q(p) \subseteq \mathcal{S} + \mu_6 d(0,p) B \qquad \forall p \in B(0, \gamma_6).$$

Results on the local upper Lipschitz-continuity of the point-to-set mapping $p \mapsto \mathcal{S}(p)$ can also be found in Gowda and Pang [9] and Dontchev and Rockafellar [2]. There, however, assumptions are employed which imply that $\mathcal{S}(p_0)$ or, locally, $\mathcal{S}(p)$ are single-valued.

4 FINAL REMARKS

In this paper we have shown how the local growth behavior of a suitable merit function can be exploited for obtaining results on the quantitative stability of perturbed complementarity problems. On the other hand, if the quantitative stability behavior of a perturbed problem is known one can try to find results on the local growth behavior of merit functions. For possible ways to treat this inverse question we refer the reader to [5, 10].

The approach presented in this paper raises several questions which are still open. We briefly describe some of them.

- Do reasonable assumptions exist that, for Q from Theorem 2.1, ensure $\mathcal{S}_Q(p) \neq \emptyset$ for p close to p_0 and that are significantly weaker than the P_0-properties used in Proposition 3.1?
- Can other merit functions or constrained reformulations of problem (1.1) be helpful to derive further stability results?

- Are there stronger results for simpler perturbations, for example, if $F(\cdot, p)$ is given by $F(x,p) := F(x,p_0) + p$ with $p \in \Re^n$?
- Is it possible to prove a version of Theorem 3.1 for variational inequalities over polyhedral or even more general sets?

We finally note that the principal approach in this paper can be applied to other problems. In particular, finite-dimensional variational inequalities can be reformulated as optimization problem by means of the corresponding KKT system or based on the D-gap function [13, 20], for instance.

Acknowledgements

I wish to thank Professor Diethard Klatte for many helpful discussions and comments. Moreover, I am very grateful to the referees for their suggestions that led to a number of improvements, in particular to a simplification of the proof of Theorem 3.1.

References

[1] T. De Luca, F. Facchinei, and C. Kanzow: A semismooth equation approach to the solution of nonlinear complementarity problems. *Mathematical Programming* 75 (1996), 407-439.

[2] A. L. Dontchev and R. T. Rockafellar: Characterizations of strong regularity for variational inequalities over polyhedral convex sets. *SIAM Journal on Optimization* 6 (1996), 1087-1105.

[3] F. Facchinei: Structural and stability properties of P_0 nonlinear complementarity problems. *Mathematics of Operations Research*, to appear.

[4] F. Facchinei and J. Soares: A new merit function for nonlinear complementarity problems and a related algorithm. *SIAM Journal on Optimization* 7 (1997), 225-247.

[5] F. Facchinei, A. Fischer, and C. Kanzow: On the accurate identification of active constraints. *SIAM Journal on Optimization*, to appear.

[6] A. Fischer: A special Newton-type optimization method. *Optimization* 24 (1992), 269-284.

[7] A. Fischer: On the local superlinear convergence of a Newton-type method for LCP under weak conditions. *Optimization Methods and Software* 6 (1995), 83-107.

[8] C. Geiger and C. Kanzow: On the resolution of monotone complementarity problems. *Computational Optimization and Applications* 5 (1996), 155-173.

[9] M. S. Gowda and J.-S. Pang: Stability analysis of variational inequalities and nonlinear complementarity problems, via the mixed linear complementarity problem and degree theory. *Mathematics of Operations Research* 19 (1994), 831-879.

[10] W. W. Hager and M. S. Gowda: Stability in the presence of degeneracy and error estimation. Technical Report, Department of Mathematics, University of Florida, Gainesville, USA, 1997.

[11] P. T. Harker and B. T. Xiao: Newton's method for the nonlinear complementarity problem: a B-differentiable approach. *Mathematical Programming* 48 (1990), 339-358.

[12] A. Ioffe: On sensitivity analysis of nonlinear programs in Banach spaces: the approach via composite unconstrained optimization. *SIAM Journal on Optimization* 4 (1994), 1-43.

[13] C. Kanzow and M. Fukushima: Theoretical and numerical investigation of the D-Gap Function for box constrained variational inequalities. *Mathematical Programming*, to appear.

[14] D. Klatte: On quantitative stability for non-isolated minima. *Control and Cybernetics* 23 (1994), 183-200.

[15] Z.-Q. Luo and P. Tseng: Error bound and convergence analysis of matrix splitting algorithms for the affine variational inequality problem. *SIAM Journal on Optimization* 2 (1992), 43-54.

[16] Z. Q. Luo and J.-S. Pang: Error bounds for analytic systems and their applications. *Mathematical Programming* 67 (1994), 1-28.

[17] O. L. Mangasarian and M. V. Solodov: Nonlinear complementarity as unconstrained and constrained minimization. *Mathematical Programming* 62 (1993), 277-297.

[18] R. D. C. Monteiro, J.-S. Pang, and T. Wang: A positive algorithm for the nonlinear complementarity problem. *SIAM Journal on Optimization* 5 (1995), 129-148.

[19] J. J. Moré: Global methods for nonlinear complementarity problems. *Mathematics of Operations Research* 21 (1996), 589-614.

[20] J.-M. Peng: Equivalence of variational inequality problems to unconstrained minimization. *Mathematical Programming* 78 (1997), 347-356.

[21] G. Ravindran and M. S. Gowda: Regularization of P_0-functions in box variational inequality problems. Technical Report TR 97-07, Department of Mathematics and Statistics, University of Maryland Baltimore County, Baltimore, USA, 1997.

[22] S. M. Robinson: Some continuity properties of polyhedral multifunctions. *Mathematical Programming Study* 14 (1981), 206-214.

[23] S. M. Robinson: Generalized equations and their solutions, part I: Basic theory. *Mathematical Programming Study* 10 (1979), 128-141.

[24] P. Tseng: Growth behavior of a class of merit functions for the nonlinear complementarity problem. *Journal on Optimization Theory and Applications* 89 (1996), 17-37.

[11] [illegible] and [illegible] Newton [illegible] [illegible] problems [illegible] *Mathematical Programming* 64 (1994) [illegible]

[12] [illegible] On [illegible] the [illegible] *Optimization* [illegible]

[13] [illegible] Gap Functions [illegible] *Mathematical Programming* [illegible]

[14] [illegible] On quantitative [illegible] (1990) [illegible]

[15] [illegible] and [illegible] [illegible] *SIAM Journal on Optimization* [illegible]

[16] [illegible]

[17] [illegible] (1992) [illegible]

[18] [illegible]

[19] [illegible] *Optimization* [illegible]

[20] [illegible]

[21] [illegible] Press [illegible]

[22] [illegible] *Mathematical Programming Study* 4 [illegible]

[23] [illegible] *Mathematical Programming Study* [illegible]

[24] [illegible] complementarity [illegible] *Journal of Optimization Theory and Applications* [illegible]

Reformulation: Nonsmooth, Piecewise Smooth, Semismooth and Smoothing Methods, pp. 161–179
Edited by M. Fukushima and L. Qi

Minimax and Triality Theory in Nonsmooth Variational Problems

David Yang Gao*

Abstract Dual extremum principles and minimax theory are investigated for nonsmooth variational problems. The critical points of the Lagrangian $L(u, p^*)$ for fully nonlinear (both geometrically and physically nonlinear) systems are clarified. We proved that the critical point of $L(u, p^*)$ is a saddle point if and only if the Gao-Strang's gap function is positive. In this case, the system has only one dual problem. However if this gap function is negative, the critical point of the Lagrangian is a so-called super-critical point which is equivalent to the Auchmuty's anomalous critical point in geometrically linear systems. We discover that in this case, the system may have more than one set of primal-dual problems. The critical point of the Lagrangian either minimizes or maximizes both primal and dual problems. Application in nonconvex finite deformation theory is illustrated and a pure complementary energy is proposed. It is shown that the dual Euler-Lagrange equation of nonlinear variational boundary value problem is an algebraic equation. An analytic solution is obtained for a non-convex, unilateral variational problem.

Key Words nonconvex variational problem, nonsmooth optimization, duality theory, triality, finite deformation theory, nonlinear complementarity problem.

1 INTRODUCTION

Duality theory plays an essential role in mathematics and science. Many problems arising in natural phenomena require the considerations of nondifferentiability and nonconvexity for their physical modelling and total potential (or cost) functionals (cf. Panagiotopoulos *et al*, 1985, 1993, 1996). Traditional analysis and direct variational methods for solving nonconvex, nonsmooth systems are

*Department of Mathematics, Virginia Polytechnic Institute and State University, Blacksburg, VA 24061, USA. E-mail: gao@math.vt.edu

usually very difficult, or even impossible. However, duality methods have both excellent theoretical properties and good practical performance in mathematical programming (see, Walk, 1989; Wright, 1997) and nonsmooth optimization (see, Gao, 1996, 1999). The purpose of the present paper is to study the duality and minimax theory for fully nonlinear, nonsmooth variational problem:

$$P(u) = \inf_{v} J(v, \Lambda v), \tag{1.1}$$

where $\Lambda : \mathcal{U} \to \mathcal{Y}$ is a (geometrically) nonlinear operator. For geometrically linear systems, where Λ is a linear operator, duality theory and algorithms have been well studied for both convex (cf. e.g. Tonti, 1972; Rockafellar, 1974; Ekeland & Temam, 1976; Oden & Reddy, 1983) and nonconvex variational problems (cf. e.g. Toland, 1979; Clarke, 1983; Auchmuty, 1983, 1989). The symmetry between the primal and dual variational principles is amazingly beautiful (see Strang, 1986; Sewell, 1987). However, in geometrically nonlinear systems, such a symmetry is lost. In continuum mechanics, the question whether exists a pure complementary energy principle (in terms of Kirchhoff's stress only) as the dual problem of the potential variational principle has been an open problem for more than forty years. In order to recover this broken symmetry, a so-called complementary gap function was discovered by Gao-Strang in 1989, and a general duality theory was established in geometrically nonlinear systems. They proved that if this gap function is positive on the statically admissible field, the generalized complementary energy is a saddle point functional, the total potential is convex, and its dual problem is concave. Applications of this general duality theory have been given in a series of publications on finite elastoplasticity (see Gao et al, 1989-1997).

Recently, in the study of duality theory in nonconvex finite deformation theory, it is discovered that if the gap function is negative, the total potential energy is not convex, and the system has two dual problems. A general duality theory for nonconvex variational problems has been established and an interesting triality theorem is discovered in nonconvex finite deformation systems (Gao, 1997).

The aim of this paper is to generalize the results in nonlinear mechanics to general geometrically nonlinear, nonsmooth systems. This research is directly related to those problems arising in nonsmooth optimization, gam theory, variational inequality, nonlinear complementarity problems, phase transitions and nonlinear bifurcation analysis. In the present paper, a minimax theory for fully nonlinear variational problems is established. An interesting triality theorem in nonsmooth systems is proposed, which contains a minimax complementary extremum principle and a pair of minimum and maximum complementary principles. Detailed proofs are presented to the related theorems proposed recently in large deformation mechanics. Application is illustrated by a real-life example in finite deformation theory and a pure complementary energy is proposed. It is shown that by choosing a right geometrically nonlinear transformation, the dual Euler-Lagrange equation of nonlinear variational problem is an algebraic equation. An analytic solution is obtained for a unilateral variational problem with a double-well energy.

2 FRAMEWORK AND GAP FUNCTION IN FULLY NONLINEAR SYSTEMS

Let $(\mathcal{U}, \mathcal{U}^*)$ and $(\mathcal{Y}, \mathcal{Y}^*)$ be two pairs of real vector spaces, finite- or infinite-dimensional, in duality with respect to certain bilinear forms $(*, *)$ and $\langle *, * \rangle$, respectively. Let the geometrical operator Λ be a continuous, nonlinear transformation from $\mathcal{U}$ to $\mathcal{Y}$. For any given $p = \Lambda u \in \mathcal{Y}$, the directional derivative of p at u in the direction $v \in \mathcal{U}$ is defined as

$$\delta p(u; v) := \lim_{\theta \to 0^+} \frac{p(u + \theta v) - p(u)}{\theta} = \Lambda_t(u)v, \tag{2.1}$$

where Λ_t is the Gâteaux derivative of the operator Λ at u. According to Gao-Strang (1989), the following decomposition of Λ is fundamentally important in geometrically nonlinear variational problems:

$$\Lambda = \Lambda_t + \Lambda_n, \tag{2.2}$$

where Λ_n is a complementary operator of Λ_t, which plays an essential role in nonconvex analysis.

According to nonsmooth analysis, for any given real valued function $F : \mathcal{U} \to \mathbf{R}$, the sub-differential and super-differential of F at $u \in \mathcal{U}$ can be defined by

$$\partial^- F(u) = \{u^* \in \mathcal{U}^* | F(v) - F(u) \geq (u^*, v - u) \; \forall v \in \mathcal{U}\},$$

$$\partial^+ F(u) = \{u^* \in \mathcal{U}^* | F(v) - F(u) \leq (u^*, v - u) \; \forall v \in \mathcal{U}\},$$

respectively. In convex analysis, ∂^- is usually written as ∂ and we always have $\partial^+ F = -\partial^-(-F)$. If F is smooth, Gâteaux-differentiable at $u \in \mathcal{U}$, then

$$\partial^- F(u) = \partial^+ F(u) = \{DF(u)\},$$

where $DF : \mathcal{U} \to \mathcal{U}^*$ denotes the Gâteaux derivative of F at u. Let $\mathcal{U}_a \subset \mathcal{U}$ be a so-called feasible space. For a given $\bar{u}^* \in \mathcal{U}^*$, we assume that F can be written as

$$F(u) = (\bar{u}^*, u) - \Psi_{\mathcal{U}_a}(u) = \begin{cases} (\bar{u}^*, u) & \text{if } u \in \mathcal{U}_a, \\ -\infty & \text{if } u \notin \mathcal{U}_a, \end{cases} \tag{2.3}$$

where $\Psi_{\mathcal{U}_a}(u)$ is the indicator function of the $\mathcal{U}_a$. It is obvious that if $\mathcal{U}_a$ is convex and closed, $F : \mathcal{U} \to \bar{\mathbf{R}} := \mathbf{R} \cup \{-\infty, +\infty\}$ is concave, upper semicontinuous. Then the relation between the paired spaces $\mathcal{U}$ and $\mathcal{U}^*$ can be written as

$$u^* \in \partial^+ F(u). \tag{2.4}$$

By introducing a convex, lower semicontinuous function: $W : \mathcal{Y} \to \bar{\mathbf{R}}$, the relation between the dual spaces $\mathcal{Y}$ and $\mathcal{Y}^*$ (i.e. *the constitutive equation*) can be described as

$$p^* \in \partial^- W(p), \tag{2.5}$$

The conjugate function of W is defined by the Legendre-Fenchel transformation:

$$W^*(p^*) := \sup_p\{\langle p^*, p\rangle - W(p)\}, \tag{2.6}$$

which is always convex, lower semicontinuous, and the following relations are equivalent:

$$p^* \in \partial W(p) \Leftrightarrow p \in \partial W^*(p^*) \Leftrightarrow W(p) + W^*(p^*) = \langle p^*, p\rangle. \tag{2.7}$$

Finally, relation between the bilinear forms $\langle *, *\rangle$ and $(*, *)$ can be given by the ***virtual work principle***:

$$\langle p^*, \delta p(u; v)\rangle = (\Lambda_t^*(u)p^*, v) = (\bar{u}^*, v) \;\; \forall v \in \mathcal{U}, \tag{2.8}$$

where $\Lambda_t^* : \mathcal{Y}^* \to \mathcal{U}^*$ is the adjoint operator of Λ_t, defined by the generalized Gauss-Green theorem:

$$\langle p^*, \Lambda_t(u)v\rangle = (\Lambda_t^*(u)p^*, v).$$

So the ***equilibrium equation*** can be obtained as:

$$\Lambda_t^*(u)p^* = \bar{u}^*. \tag{2.9}$$

For any given system, if we can choose a geometrical operator $\Lambda : \mathcal{U} \to \mathcal{Y}$ such that $W : \mathcal{Y} \to \bar{\mathbf{R}}$ is convex, then we have the following definition:

Definition 2.1 *The system is called the geometrically nonlinear if $\Lambda : \mathcal{U} \to \mathcal{Y}$ is nonlinear; the system is called physically nonlinear if the dual relations between $\mathcal{Y}$ and $\mathcal{Y}^*$ or between $\mathcal{U}$ and $\mathcal{U}^*$ are nonlinear; the system is called fully nonlinear if it is both geometrically nonlinear and physically nonlinear.*

The governing equations for the fully nonlinear system are listed as follows:

$$\begin{array}{ll} \text{(1) Geometrical Eqn.} & p = \Lambda u, \\ \text{(2) Physical Eqn.} & p^* \in \partial^- W(p), \quad u^* \in \partial^+ F(u), \\ \text{(3) Equilibrium Eqn.} & \Lambda_t^*(u)p^* = u^*. \end{array} \tag{2.10}$$

The framework for this fully nonlinear system is shown in Fig. 1.

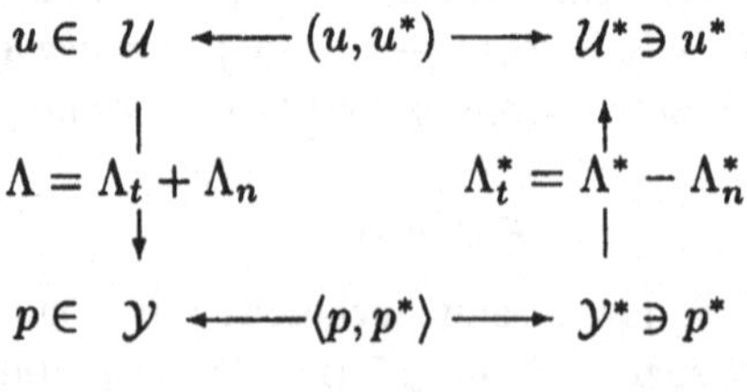

$$\begin{array}{ccccc} u \in \mathcal{U} & \longleftarrow & (u, u^*) & \longrightarrow & \mathcal{U}^* \ni u^* \\ \Big| & & & & \uparrow \\ \Lambda = \Lambda_t + \Lambda_n & & & & \Lambda_t^* = \Lambda^* - \Lambda_n^* \\ \downarrow & & & & \Big| \\ p \in \mathcal{Y} & \longleftarrow & \langle p, p^*\rangle & \longrightarrow & \mathcal{Y}^* \ni p^* \end{array}$$

Fig. 1. Framework in fully nonlinear systems.

For geometrically linear conservative systems, the values of the bilinear forms (u^*, u) and $\langle p^*, p\rangle$ are equal:

$$\langle p^*, p\rangle = \langle p^*, \Lambda u\rangle = (u^*, u). \tag{2.11}$$

If $W : \mathcal{Y} \to \mathbf{R}$ is quadratic such that $DW = Cp$, C is a linear operator, then the governing equations for linear system can be written as

$$\Lambda^* C \Lambda u = Au = u^*.$$

For conservative systems, the operator $A = \Lambda^* C \Lambda$ is usually symmetric. In the celebrated textbook by G. Strang, 1986, this symmetrical framework was discussed extensively from continuum theories to discrete systems. But for geometrically nonlinear systems, such a symmetry is broken.

Using the operator decomposition (2.2), the relation between the two bilinear forms should be

$$\langle p^*, p\rangle = \langle p^*, \Lambda_t u\rangle + \langle p^*, \Lambda_n u\rangle = (u^*, u) - G(u, p^*), \tag{2.12}$$

where $G(u, p^*)$ is the so-called complementary gap function introduced by Gao-Strang in 1989:

$$G(u, p^*) := \langle p^*, -\Lambda_n(u)u\rangle, \tag{2.13}$$

which plays a central role in nonconvex variational principles.

3 PRIMAL PROBLEM AND LAGRANGIAN

The total potential energy of the system is defined as:

$$P(u) := W(\Lambda u) - F(u). \tag{3.1}$$

The minimal potential variational problem (*inf-primal problem*) is to find $\bar{u}$ such that

$$(\mathcal{P}_{inf}) : \quad P(\bar{u}) = \inf P(u) \ \ \forall u \in \mathcal{U}. \tag{3.2}$$

For the fully nonlinear system (2.10), a point $\bar{u} \in \mathcal{U}$ is a critical point for P if

$$\Lambda_t^* \partial_p^- W(\Lambda \bar{u}) \cap \partial^+ F(\bar{u}) \neq \emptyset, \tag{3.3}$$

where ∂_p^- denotes the sub-differential of $W(\Lambda u)$ with respect to $p = \Lambda u$. On $\mathcal{U}_a$, if P is Gâteaux differentiable, the critical condition (3.3) gives the *Euler-Lagrange equation:*

$$DF(\bar{u}) = \Lambda_t^* D_p W(\Lambda \bar{u}). \tag{3.4}$$

Since $P : \mathcal{U}_a \to \mathbf{R}$ may not be convex, the critical points of P are not identical to the minimizers of P. Following theorem was proved by Gao-Strang (1989):

Theorem 3.1 *For any critical point $\bar{u}$ of P and associated $\bar{p}^* \in \partial_p^- W(\Lambda \bar{u})$, if the gap function $G(\bar{u}, \bar{p}^*) \geq 0$, then $\bar{u}$ minimizes P over $\mathcal{U}_a$. If the gap function is strictly positive, the primal problem $(\mathcal{P}_{inf})$ has at most one solution.*

This theorem shows that if the gap function has a positive sign, the potential energy P is convex. However, if the gap function has a negative sign, the total potential energy is nonconvex. Then P may have a local maximizer on a subset of $\mathcal{U}$, which could be a critical point in phase transitions, or an unstable state in post-bifurcation analysis. Let $\mathcal{U}_b$ be a subset of $\mathcal{U}_a$, we can propose the *sup-primal problem*: to find $\bar{u} \in \mathcal{U}_b$ such that

$$(\mathcal{P}_{sup}): \quad P(\bar{u}) = \sup P(u) \;\; \forall u \in \mathcal{U}_b. \tag{3.5}$$

The Lagrangian $L : \mathcal{U} \times \mathcal{Y}^* \to \bar{\mathbf{R}}$ can be given as

$$L(u, p^*) := \langle \Lambda u, p^* \rangle - W^*(p^*) - F(u). \tag{3.6}$$

A point $(\bar{u}, \bar{p}^*) \in \mathcal{U} \times \mathcal{Y}^*$ is said to be a critical point of L if L is Gâteaux-differentiable at $(\bar{u}, \bar{p}^*)$ and

$$D_u L(\bar{u}, \bar{p}^*) = 0, \quad D_{p^*} L(\bar{u}, \bar{p}^*) = 0,$$

where D_u, D_{p^*} denote partial Gateaux-derivatives on $\mathcal{U}$ and $\mathcal{Y}^*$, respectively. It is easy to find that

$$D_u L(\bar{u}, \bar{p}^*) = 0 \;\; \Rightarrow \;\; \Lambda_t^*(\bar{u})\bar{p}^* - DF(\bar{u}) = 0, \tag{3.7}$$

$$D_{p^*} L(\bar{u}, \bar{p}^*) = 0 \;\; \Rightarrow \;\; \Lambda \bar{u} - DW^*(\bar{p}^*) = 0. \tag{3.8}$$

In geometrically linear systems, the critical points of the Lagrangian was clarified by Auchmuty (1983). For geometrically nonlinear systems, we need following definitions:

Definition 3.2 *A point $(\bar{u}, \bar{p}^*)$ is said to be a saddle point of L if*

$$L(\bar{u}, p^*) \le L(\bar{u}, \bar{p}^*) \le L(u, \bar{p}^*) \quad \forall (u, p^*) \in \mathcal{U} \times \mathcal{Y}^*. \tag{3.9}$$

A point $(\bar{u}, \bar{p}^*)$ is said to be a sub-critical (or ∂^--critical) point of L if

$$L(\bar{u}, p^*) \ge L(\bar{u}, \bar{p}^*) \le L(u, \bar{p}^*) \quad \forall (u, p^*) \in \mathcal{U} \times \mathcal{Y}^*. \tag{3.10}$$

A point $(\bar{u}, \bar{p}^*)$ is said to be a super-critical (or ∂^+-critical) point of L if

$$L(\bar{u}, p^*) \le L(\bar{u}, \bar{p}^*) \ge L(u, \bar{p}^*) \quad \forall (u, p^*) \in \mathcal{U} \times \mathcal{Y}^*. \tag{3.11}$$

Proposition 3.1 *A point $(\bar{u}, \bar{p}^*)$ is a saddle point of L if and only if*

$$0 \in \partial_u^- L(\bar{u}, \bar{p}^*), \quad 0 \in \partial_{p^*}^+ L(\bar{u}, \bar{p}^*). \tag{3.12}$$

A point $(\bar{u}, \bar{p}^*)$ is a sub-critical point of L if and only if

$$0 \in \partial_u^- L(\bar{u}, \bar{p}^*), \quad 0 \in \partial_{p^*}^- L(\bar{u}, \bar{p}^*). \tag{3.13}$$

A point $(\bar{u}, \bar{p}^*)$ *is a super-critical point of* L *if and only if*

$$0 \in \partial_u^+ L(\bar{u}, \bar{p}^*), \quad 0 \in \partial_{p^*}^+ L(\bar{u}, \bar{p}^*). \tag{3.14}$$

This Proposition can be proved directly by using the definitions of sub- and super-differentials.

Remark. In geometrically linear systems ($\Lambda : \mathcal{U} \to \mathcal{Y}$ is linear), the equation (3.13) is the Auchmuty's definition of ∂-critical point (Auchmuty, 1983). By the Legendre transformation, the Hamiltonian of the system can be obtained by the Lagrangian:

$$H(u, p^*) = \langle \Lambda u, p^* \rangle - L(u, p^*) = W^*(p^*) + F(u).$$

If Λ is a linear operator, then the inequalities (3.11) are equivalent to following symmetrical canonical forms:

$$\Lambda \bar{u} \in \partial_{p^*}^- H(\bar{u}, \bar{p}^*), \quad \Lambda^* \bar{p}^* \in \partial_u^+ H(\bar{u}, \bar{p}^*),$$

which is Auchmuty's definition of *anomalous critical point.* Unfortunately this nice symmetry is lost in geometrically nonlinear systems. When Λ is nonlinear, the canonical forms should be:

$$\Lambda \bar{u} \in \partial_{p^*}^- H(\bar{u}, \bar{p}^*), \quad \Lambda_t^*(\bar{u}) \bar{p}^* \in \partial_u^+ H(\bar{u}, \bar{p}^*). \tag{3.15}$$

Theorem 3.2 *Suppose that* $\Lambda : \mathcal{U} \to \mathcal{Y}$ *is a quadratic operator,* L *is Gâteaux-differentiable on* $\mathcal{U}_a \times \mathcal{Y}^*$ *and* $(\bar{u}, \bar{p}^*)$ *is a critical point of* L. *Then* $(\bar{u}, \bar{p}^*)$ *is a saddle point of* L *if and only if* $G(u, \bar{p}^*) \geq 0 \;\; \forall u \in \mathcal{U}_a$; *while* $(\bar{u}, \bar{p}^*)$ *is a super-critical point of* L *if and only if* $G(u, \bar{p}^*) \leq 0 \;\; \forall u \in \mathcal{U}_a$.

The proof of this theorem is given in (Gao, 1997).

Let $\mathcal{Y}_a^* \subset \mathcal{Y}^*$ be a range of $\partial^- W(\mathcal{Y})$, such that $W^* : \mathcal{Y}_a^* \to \mathbf{R}$ is finite and Gâteaux differentiable. We introduce the so-called *equilibrium admissible space:*

$$\mathcal{Y}_u^* := \{(u, p^*) \in \mathcal{U}_a \times \mathcal{Y}_a^* | \;\; D_u L(u, p^*) = 0\}. \tag{3.16}$$

From Theorem 3.2, the following corollary can be easily obtained:

Corollary 3.1 *The gap function* $G(u, p^*)$ *is positive on* $\mathcal{Y}_u^*$ *if and only if all critical points of* L *are saddle points. Moreover, if the gap function is strictly positive on* $\mathcal{Y}_u^*$, *and* $W^* : \mathcal{Y}^* \to \mathbf{R}$ *is strictly convex, then* L *has at most one saddle point.*

Since $W : \mathcal{Y} \to \mathbf{R}$ is convex, the relation between the potential energy and Lagrangian is well known:

$$P(u) = \sup_{p^* \in \mathcal{Y}_a^*} L(u, p^*). \tag{3.17}$$

But the relation between the complementary energy and L will depend on the gap function.

4 DUAL EXTREMUM PRINCIPLES

According to Gao-Strang (1989), on $\mathcal{Y}_u^*$ the so-called complementary energy function is given as

$$P^c(p^*,u) = F_*(\Lambda_t^*(u)p^*) - W^*(p^*) - G(u,p^*), \tag{4.1}$$

where $F_* : \mathcal{Y}^* \to \bar{\mathbf{R}}$ is the sub-conjugate function of F defined as

$$F_*(u^*) = \inf_{u\in\mathcal{U}}\{(u^*,u) - F(u)\} = \inf_{u\in\mathcal{U}_a}(u,u^* - \bar{u}^*) = \begin{cases} 0 & \text{if } u^* = \bar{u}^*, \\ -\infty & \text{if } u^* \neq \bar{u}^*. \end{cases}$$

By the equilibrium equation (2.9), if u can be written in terms of p^*, then the pure complementary energy can be simply defined as:

$$P^*(p^*) := P^c(p^*, u(p^*)). \tag{4.2}$$

The following theorem shows how to obtain this pure complementary energy from the Lagrangian.

Theorem 4.3 *(Complementary Energy Principle) Suppose that $\Lambda : \mathcal{U} \to \mathcal{Y}$ is quadratic and $\mathcal{Y}_u^*$ is not empty. If $G(u,p^*) \geq 0 \ \forall (u,p^*) \in \mathcal{Y}_u^*$, then*

$$P^*(p^*) = \inf_u L(u,p^*). \tag{4.3}$$

If $G(u,p^*) \leq 0 \ \forall (u,p^*) \in \mathcal{Y}_u^*$, then

$$P^*(p^*) = \sup_u L(u,p^*). \tag{4.4}$$

Proof. By Theorem 3.2, if the gap function is positive on $\mathcal{Y}_u^*$, $L : \mathcal{U}_a \to \mathbf{R}$ is convex. Then any given $(u,p^*) \in \mathcal{Y}_u^*$ solves (4.3). However, if the gap function is negative on $\mathcal{Y}_u^*$, $L : \mathcal{U}_a \to \mathbf{R}$ is concave. Then any given $(u,p^*) \in \mathcal{Y}_u^*$ solves (4.4). ∎

General speaking, $P^* : \mathcal{Y}_a^* \to \bar{\mathbf{R}}$ is usually nonconvex. It may have more than one critical point. Then two extremum dual variational problems should be proposed:

(a) The *sup-dual problem*: To find $\bar{p}^*$ such that

$$(\mathcal{P}_{sup}^*): \quad P^*(\bar{p}^*) = \sup_{p^*\in\mathcal{Y}_a^*} P^*(p^*); \tag{4.5}$$

(b) The *inf-dual problem*: To find $\bar{p}^*$ such that

$$(\mathcal{P}_{inf}^*): \quad P^*(\bar{p}^*) = \inf_{p^*\in\mathcal{Y}_a^*} P^*(p^*). \tag{4.6}$$

Lemma 4.1 *If $(\bar{u},\bar{p}^*)$ is either a saddle point of L or a super-critical point of L, and L is partially Gâteaux differentiable at $(\bar{u},\bar{p}^*)$, then $(\bar{u},\bar{p}^*)$ must be* *a critical point of L. If P and P^* are Gâteaux differentiable at $\bar{u}$ and $\bar{p}^*$, respectively, then $DP(\bar{u}) = 0$, $DP^*(\bar{p}^*) = 0$, and*

$$P(\bar{u}) = L(\bar{u}, \bar{p}^*) = P^*(\bar{p}^*). \tag{4.7}$$

The proof of this lemma is similar to the proof given in (Auchmuty, 1983) for geometrically linear systems. Based on this lemma, for quadratic operator Λ, following extremal complementary principles can be proposed.

Theorem 4.4 *(Minimax Complementary Principle) Suppose that $\mathcal{Y}_u^*$ is not empty and the gap function $G(u, p^*)$ is positive on $\mathcal{Y}_u^*$. Then $(\bar{u}, \bar{p}^*) \in \mathcal{U}_a \times \mathcal{Y}_a^*$ is a critical point of L if and only if*

$$P(\bar{u}) = \inf_{u \in \mathcal{U}_a} P(u) = \sup_{p^* \in \mathcal{Y}_a^*} P^*(p^*) = P^*(\bar{p}^*). \tag{4.8}$$

Proof. By Corollary 3.1, the critical point $(\bar{u}, \bar{p}^*)$ should be a saddle point of L. For any given $u \in \mathcal{U}_a$, $L : \mathcal{Y}^* \to \mathbf{R}$ is concave, and

$$P(u) = \sup_{p^*} L(u, p^*) = L(u, \bar{p}^*) \geq L(\bar{u}, \bar{p}^*) \;\; \forall u \in \mathcal{U}_a.$$

By Lemma 4.1, $L(\bar{u}, \bar{p}^*) = P(\bar{u}) \leq P(u) \;\; \forall u \in \mathcal{U}_a$. So $\bar{u}$ is a solution of $(\mathcal{P}_{inf})$. Since the gap function is positive on $\mathcal{Y}_u^*$, $L : \mathcal{U}_a \to \mathbf{R}$ is convex, then by Theorem 4.3,

$$\begin{aligned} P^*(p^*) &= \inf_u L(u, p^*) = L(\bar{u}, p^*) \leq L(\bar{u}, \bar{p}^*) \\ &= P^*(\bar{p}^*) \;\; \forall p^* \in \mathcal{Y}_a^*. \end{aligned}$$

This shows that $\bar{p}^*$ is a solution of $(\mathcal{P}_{sup}^*)$ and Lemma 4.1 gives that $P(\bar{u}) = P^*(\bar{p}^*)$.

Conversely, since $\mathcal{Y}_u^* \neq \emptyset$, if $\bar{u}$ is a solution of $(\mathcal{P}_{inf})$ and $\bar{p}^*$ is a solution of $(\mathcal{P}_{sup}^*)$, we have, from the definition of L and the equation (4.3)

$$\begin{aligned} P(\bar{u}) &= \sup_{p^*} L(\bar{u}, p^*) = \inf_u L(u, \bar{p}^*) \leq L(\bar{u}, \bar{p}^*), \\ P^*(\bar{p}^*) &= \inf_u L(u, \bar{p}^*) = \sup_{p^*} L(\bar{u}, p^*) \geq L(\bar{u}, \bar{p}^*). \end{aligned}$$

So the equality $P(\bar{u}) = P^*(\bar{p}^*)$ shows that $(\bar{u}, \bar{p}^*)$ is a saddle point of L. By Lemma 4.1, it must be a critical point of L. ∎

This theorem shows that for a given Λ, if the gap function is positive on $\mathcal{Y}_u^*$, the system has only one potential extremum principle $(\mathcal{P}_{inf})$ and only one complementary extremum principle[1] $(\mathcal{P}_{sup}^*)$. However, if the gap function is negative on $\mathcal{Y}_u^*$, the system may have more than one primal-dual problems.

[1]The multi-duality for different Λ was discussed by Gao-Yang, 1995.

Let $\mathcal{U}_b$, $\mathcal{Y}_b^*$ be the subspaces of $\mathcal{U}_a$ and $\mathcal{Y}_a^*$, respectively. Then for nonconvex systems, we have the following results.

Theorem 4.5 *(Maximum Complementary Principle) Suppose that $\mathcal{Y}_u^*$ is not empty and the gap function $G(u,p^*) \leq 0\ \forall(u,p^*) \in \mathcal{Y}_u^*$. Then on $\mathcal{U}_a \times \mathcal{Y}_a^*$, $\sup P(u) = \sup P^*(p^*)$. Moreover, if $(\bar{u},\bar{p}^*)$ is a critical point of L and $\mathcal{U}_b \times \mathcal{Y}_b^*$ is its neighborhood such that P and P^* have only one critical point on $\mathcal{U}_b$ and $\mathcal{Y}_b^*$, respectively, then $(\bar{u},\bar{p}^*)$ maximizes L on $\mathcal{U}_b \times \mathcal{Y}_b^*$ if and only if*

$$P(\bar{u}) = \sup_{u\in\mathcal{U}_b} P(u) = \sup_{p^*\in\mathcal{Y}_b^*} P^*(p^*) = P^*(\bar{p}^*). \tag{4.9}$$

Proof. By the definition of L and Lemma 4.1, for negative gap function on $\mathcal{Y}_u^*$,

$$\sup_{u\in\mathcal{U}_a} P(u) = \sup_u \sup_{p^*} L(u,p^*),$$
$$\sup_{p^*\in\mathcal{Y}_a^*} P^*(p^*) = \sup_{p^*} \sup_u L(u,p^*).$$

So $\sup P = \sup P^*$ as one can take supremum in either order on $\mathcal{U}_a \times \mathcal{Y}_a^*$.

Suppose that $(\bar{u},\bar{p}^*)$ maximizes L on $\mathcal{U}_b\times\mathcal{Y}_b^*$. Since $\mathcal{U}_b\times\mathcal{Y}_b^*$ is a neighborhood of the critical point $(\bar{u},\bar{p}^*)$, there exists at least one point $(u,p^*) \in \mathcal{U}_b \times \mathcal{Y}_b^*$ such that $W(\Lambda u) = \sup_{p^*}\{\langle\Lambda u, p^*\rangle - W^*(p^*)\}$ is finite. So on $\mathcal{U}_b \times \mathcal{Y}_b^*$, we have

$$L(\bar{u},\bar{p}^*) = \sup_u \sup_{p^*} L(u,p^*) = \sup_{u\in\mathcal{U}_b} P(u), \tag{4.10}$$
$$L(\bar{u},\bar{p}^*) = \sup_{p^*} \sup_u L(u,p^*) = \sup_{p^*\in\mathcal{Y}_b^*} P^*(p^*). \tag{4.11}$$

Then Lemma 4.1 gives $\sup P(u) = P(\bar{u}) = L(\bar{u},\bar{p}^*) = P^*(\bar{p}^*) = \sup P^*(p^*)$ on $\mathcal{U}_b \times \mathcal{Y}_b^*$.

Conversely, if $P(\bar{u}) = \sup P(u)$ and $P^*(\bar{p}^*) = \sup P^*(p^*)$, then for any given $(u,p^*) \in \mathcal{U}_b \times \mathcal{Y}_b^*$, such that the gap function is negative, we have

$$P(\bar{u}) \geq P(u) = \sup_{p^*\in\mathcal{Y}_b^*} L(u,p^*) \geq L(u,\bar{p}^*),$$
$$P^*(\bar{p}^*) \geq P^*(p^*) = \sup_{u\in\mathcal{U}_b} L(u,p^*) \geq L(\bar{u},p^*).$$

Thus, by Lemma 4.1, on $\mathcal{U}_b \times \mathcal{Y}_b^*$,

$$L(\bar{u},p^*) \leq P^*(\bar{p}^*) = L(\bar{u},\bar{p}^*) = P(\bar{u}) \geq L(u,\bar{p}^*).$$

So $(\bar{u},\bar{p}^*)$ maximizes L on $\mathcal{U}_b \times \mathcal{Y}_b^*$ as required. ∎

Theorem 4.6 *(Minimum Complementary Principle) Suppose that $(\bar{u},\bar{p}^*)$ is a critical point of L, $G(\bar{u},\bar{p}^*) \leq 0$ and $\mathcal{U}_b \times \mathcal{Y}_b^*$ is a neighborhood of $(\bar{u},\bar{p}^*)$ such that P and P^* have only one critical point on $\mathcal{U}_b$ and $\mathcal{Y}_b^*$, respectively. Then*

$(\bar{u}, \bar{p}^*)$ mini-maximizes L on $\mathcal{U}_b \times \mathcal{Y}_b^*$ in either order ($\inf_u \sup_{p^*}$ or $\inf_{p^*} \sup_u$) if and only if

$$P(\bar{u}) = \inf_{u \in \mathcal{U}_b} P(u) = \inf_{p^* \in \mathcal{Y}_b^*} P^*(p^*) = P^*(\bar{p}^*). \tag{4.12}$$

Proof. From Theorem 3.2, $(\bar{u}, \bar{p}^*)$ must be a ∂^+-critical point of L. If $(\bar{u}, \bar{p}^*)$ minimaximizes L on $\mathcal{U}_b \times \mathcal{Y}_b^*$ in the order of $\inf_u \sup_{p^*}$, then

$$L(\bar{u}, \bar{p}^*) = \inf_u \sup_{p^*} L(u, p^*) = \inf P(u).$$

By Lemma 4.1, we have $P(\bar{u}) = L(\bar{u}, \bar{p}^*) = \inf P(u)$. Since $\bar{p}^*$ is a critical point of P^* on the open domain $\mathcal{Y}_b^*$, it should be either a local extremum point or a local saddle point of P^*. If $\bar{p}^*$ maximizes P^* on $\mathcal{Y}_b^*$, then

$$\begin{aligned} P^*(\bar{p}^*) &= \sup P^*(p^*) = \sup_{p^*} \sup_u L(u, p^*) \\ &= \sup_u \sup_{p^*} L(u, p^*) = \sup_u P(u). \end{aligned} \tag{4.13}$$

But $P^*(\bar{p}^*) = P(\bar{u})$, and $\bar{u}$ minimizes P on $\mathcal{U}_b$. So this contradiction shows that $\bar{p}^*$ can not be a local maximizer of P^*. If $\bar{p}^*$ is a saddle point of P^* and it maximizes P^* in the direction p_o^* such that $P^*(\bar{p}^*) = \sup_{\theta \in \mathbf{R}} P^*(\bar{p}^* + \theta p_o^*)$. Substituting $p^* = \bar{p}^* + \theta p_o^*$ into (4.13) we get the contradiction too. So the critical point $\bar{p}^*$ of P^* must be a local minimizer on $\mathcal{Y}_b^*$.

Now if $(\bar{u}, \bar{p}^*)$ minimaximizes L on $\mathcal{U}_b \times \mathcal{Y}_b^*$ in the order of $\inf_{p^*} \sup_u$, then

$$L(\bar{u}, \bar{p}^*) = \inf_{p^*} \sup_u L(u, p^*) = \inf P^*(p^*).$$

Then Lemma 4.1 gives that $P^*(\bar{p}^*) = L(\bar{u}, \bar{p}^*) = \inf P^*(p^*)$. Similarly, we can prove that the critical point $\bar{u}$ of P must be a local minimizer on $\mathcal{U}_b$.

Conversely, if $P(\bar{u}) = \inf P(u)$ and $P^*(\bar{p}^*) = \inf P^*(p^*)$, then on $\mathcal{U}_b \times \mathcal{Y}_b^*$,

$$\begin{aligned} P(\bar{u}) &= \inf_{u \in \mathcal{U}_b} P(u) = \inf_u \sup_{p^*} L(u, p^*), \\ P^*(\bar{p}^*) &= \inf_{p^* \in \mathcal{Y}_b^*} P^*(p^*) = \inf_{p^*} \sup_u L(u, p^*). \end{aligned}$$

By Lemma 4.1, the condition $P(\bar{u}) = P^*(\bar{p}^*)$ shows that the critical point $(\bar{u}, \bar{p}^*)$ of L minimaximizes L in either order. ∎

In nonconvex systems, the Lagrangian L could have several critical points. The gap function could be positive on one critical point and be negative on other. Combining the above results, we have the following result:

Theorem 4.7 *(Triality Theorem) Suppose that $(\bar{u}, \bar{p}^*)$ is a critical point of L, and $\mathcal{U}_b \times \mathcal{Y}_b^*$ is a neighborhood of $(\bar{u}, \bar{p}^*)$ such that P and P^* have only one critical point on $\mathcal{U}_b$ and $\mathcal{Y}_b^*$, respectively. If $G(\bar{u}, \bar{p}^*) \geq 0$, then*

$$P(\bar{u}) = \inf_{u \in \mathcal{U}_b} P(u) \;\Leftrightarrow\; P^*(\bar{p}^*) = \sup_{p^* \in \mathcal{Y}_b^*} P^*(p^*); \tag{4.14}$$

If $G(\bar{u},\bar{p}^*) \le 0$, *then*

$$P(\bar{u}) = \inf_{u\in\mathcal{U}_b} P(u) \quad \Leftrightarrow \quad P^*(\bar{p}^*) = \inf_{p^*\in\mathcal{Y}_b^*} P^*(p^*), \tag{4.15}$$

$$P(\bar{u}) = \sup_{u\in\mathcal{U}_b} P(u) \quad \Leftrightarrow \quad P^*(\bar{p}^*) = \sup_{p^*\in\mathcal{Y}_b^*} P^*(p^*). \tag{4.16}$$

Proof. From Theorem 3.2, if the gap function is positive on $\mathcal{U}_b \times \mathcal{Y}_b^*$, $(\bar{u},\bar{p}^*)$ must be a saddle point of L on $\mathcal{U}_b \times \mathcal{Y}_b^*$. So Theorem 4.4 shows that $\bar{u}$ is a solution of $(\mathcal{P}_{inf})$ on $\mathcal{U}_b \subset \mathcal{U}_a$ if and only if $\bar{p}^*$ is a solution of $(\mathcal{P}^*_{sup})$ on $\mathcal{Y}_b^* \subset \mathcal{Y}_a^*$.

If the gap function $G(u,p^*)$ is negative on $\mathcal{U}_b \times \mathcal{Y}_b^*$, then $(\bar{u},\bar{p}^*)$ is a super-critical point of L on $\mathcal{U}_b \times \mathcal{Y}_b^*$:

$$L(\bar{u},p^*) \le L(\bar{u},\bar{p}^*) \ge L(u,\bar{p}^*) \;\; \forall (u,p^*) \in \mathcal{U}_b \times \mathcal{Y}_b^*.$$

If $\bar{u}$ maximizes P on $\mathcal{U}_b$, one has

$$\begin{aligned} P(\bar{u}) &= \sup P(u) = \sup_u \sup_{p^*} L(u,p^*) \\ &= \sup_{p^*} \sup_u L(u,p^*) = \sup_{p^*} P^*(p^*) = P^*(\bar{p}^*). \end{aligned}$$

This shows that $P(\bar{u}) = \sup P(u)$ is equivalent to $P^*(\bar{p}^*) = \sup P^*(p^*)$.

If $\bar{u}$ minimizes P on $\mathcal{U}_b$,

$$P(\bar{u}) = \inf_{u\in\mathcal{U}_b} P(u) = \inf_{u\in\mathcal{U}_b} \sup_{p^*\in\mathcal{Y}_b^*} L(u,p^*) = L(\bar{u},\bar{p}^*),$$

i.e. $(\bar{u},\bar{p}^*)$ mini-maximizes L on $\mathcal{U}_b \times \mathcal{Y}_b^*$. By Theorem 4.6, one knows that $\bar{p}^*$ must minimize P^* on $\mathcal{Y}_b^*$. Conversely, if $\bar{p}^*$ minimizes P^* on $\mathcal{Y}_b^*$, then $(\bar{u},\bar{p}^*)$ mini-maximizes L on $\mathcal{U}_b \times \mathcal{Y}_b^*$ in the order of $\inf_{p^*}\sup_u$. Theorem 4.6 assures that $\bar{u}$ minimizes P on $\mathcal{U}_b$. ∎

Remark. If Λ is a linear operator, then the Problem $(\mathcal{P}^*_{inf})$ is the Auchmuty's anomalous dual problem. Since $F : \mathcal{U}_a \to \mathbf{R}$ given here is linear, the critical point $(\bar{u},\bar{p}^*)$ of L is both the saddle point and the anomalous point of L.

From the triality theorem, it is easy to get the following result:

Corollary 4.2 *Suppose that* $\mathcal{U}_b \times \mathcal{Y}_b^*$ *is a neighborhood of a critical point of* L. *If* $G(u,p^*) \ge 0 \;\; \forall (u,p^*) \in \mathcal{U}_b \times \mathcal{Y}_b^*$, *then*

$$\inf_u \sup_{p^*} L(u,p^*) = \sup_{p^*} \inf_u L(u,p^*) \;\;\; \forall u \in \mathcal{U}_b, \;\; p^* \in \mathcal{Y}_b^*. \tag{4.17}$$

If $G(u,p^*) \le 0 \;\; \forall (u,p^*) \in \mathcal{U}_b \times \mathcal{Y}_b^*$ *then*

$$\inf_u \sup_{p^*} L(u,p^*) = \inf_{p^*} \sup_u L(u,p^*) \;\;\; \forall u \in \mathcal{U}_b, \;\; p^* \in \mathcal{Y}_b^*, \tag{4.18}$$

$$\sup_u \sup_{p^*} L(u,p^*) = \sup_{p^*} \sup_u L(u,p^*) \;\;\; \forall u \in \mathcal{U}_b, \;\; p^* \in \mathcal{Y}_b^*. \tag{4.19}$$

5 APPLICATIONS IN FINITE DEFORMATION THEORY

Let $\Omega \subset \mathbf{R}^n$ be an open, simply connected, bounded domain with boundary $\partial\Omega = \Gamma_t \cup \Gamma_u$, $\Gamma_t \cap \Gamma_u = \emptyset$. On Γ_t, the surface traction $\bar{\mathbf{t}}$ is given; while on the remaining part Γ_u, the deformation $\bar{\mathbf{u}}$ is prescribed. The feasible space $\mathcal{U}_a$ is a set of all kinetically admissible deformations $\mathbf{u}(\mathbf{x})$ from $\Omega \subset \mathbf{R}^n$ to a deformed configuration in $\mathbf{R}^m$. For example,

$$\mathcal{U}_a = \{\mathbf{u} \in \mathcal{W}^{1,\alpha}(\Omega; \mathbf{R}^m) | \ \mathbf{u}(\mathbf{x}) = \bar{\mathbf{u}}(\mathbf{x}) \ \forall \mathbf{x} \in \Gamma_u\}, \tag{5.1}$$

where $\mathcal{W}^{1,\alpha}(\Omega; \mathbf{R}^m)$ is a standard Sobolev space with domain Ω and range $\mathbf{R}^m$, $1 < \alpha < +\infty$. The dual space $\mathcal{U}^*$ of $\mathcal{U}$ should be the force space. If $\mathbf{f} = \mathbf{u}^* \in \mathcal{U}^*$ is specified as the body force $\bar{\mathbf{b}}$ in Ω, and boundary force $\mathbf{t}$ on $\partial\Omega$, then the bilinear form $(\mathbf{u}, \mathbf{f})$ represents the *external work*:

$$(\mathbf{u}, \mathbf{f}) = \int_{\Omega} \mathbf{u} \cdot \bar{\mathbf{b}} \, d\Omega + \int_{\partial\Omega} \mathbf{u} \cdot \mathbf{t} \, d\Gamma.$$

For a given finite deformation operator Λ, we let $w : \mathcal{Y} \to \mathbf{R}$ be the so-called stored energy such that the internal energy $W(\Lambda\mathbf{u}) = \int_{\Omega} w(\Lambda\mathbf{u}) d\Omega$. Then on $\mathcal{U}_a$, the total potential energy can be written as

$$P(\mathbf{u}) = W(\Lambda\mathbf{u}) - F(\mathbf{u}) = \int_{\Omega} w(\Lambda\mathbf{u}) d\Omega - \int_{\Omega} \mathbf{u} \cdot \bar{\mathbf{b}} \, d\Omega - \int_{\Gamma_t} \mathbf{u} \cdot \bar{\mathbf{t}} \, d\Gamma. \tag{5.2}$$

Its critical point $\bar{\mathbf{u}}$ solves the following mixed boundary value problem: To find $\bar{\mathbf{u}} \in \mathcal{U}_a$ such that

$$(MBVP): \quad \Lambda_t^* \partial_p^- w(\Lambda\bar{\mathbf{u}}) = \bar{\mathbf{f}} = \begin{cases} \bar{\mathbf{b}} & \text{in } \Omega, \\ \bar{\mathbf{t}} & \text{on } \Gamma_t. \end{cases} \tag{5.3}$$

If we let $p = \Lambda\mathbf{u} = \nabla\mathbf{u}$, then $\Lambda : \mathcal{U} \to \mathcal{Y}$ is a geometrically linear operator. In continuum mechanics, $p = \nabla\mathbf{u} \in \mathbf{R}^{m\times n}$ is called the deformation gradient, denoted by $\mathbf{F}$. Its conjugate variable $\boldsymbol{\tau} \in \partial^- w(\mathbf{F})$ is the so-called Piola stress tensor (cf. e.g. Ogden, 1984). In this case, $\Lambda = \Lambda_t = \text{grad}$, the equilibrium equation (2.9) is linear:

$$\Lambda^* \boldsymbol{\tau} = \bar{\mathbf{f}} = \begin{cases} -\nabla \cdot \boldsymbol{\tau}^t = \bar{\mathbf{b}} & \text{in } \Omega, \\ \mathbf{n} \cdot \boldsymbol{\tau}^t = \bar{\mathbf{t}} & \text{on } \Gamma_t. \end{cases} \tag{5.4}$$

By the Fenchel-Rockafellar duality, if $\mathcal{U}_a$ is defined by (5.1), we have

$$W^*(\boldsymbol{\tau}) = \sup_{\mathbf{F}} \{\int_{\Omega} \text{tr}(\boldsymbol{\tau} \cdot \mathbf{F}) d\Omega - \int_{\Omega} w(\mathbf{F}) d\Omega\} = \int_{\Omega} w^*(\boldsymbol{\tau}) d\Omega, \tag{5.5}$$

$$\begin{aligned} F_*(\Lambda^*\boldsymbol{\tau}) &= \inf_{\mathbf{u}\in\mathcal{U}_a} \{(\mathbf{u}, \Lambda^*\boldsymbol{\tau}) - \int_{\Omega} \mathbf{u} \cdot \bar{\mathbf{b}} \, d\Omega - \int_{\Gamma_t} \mathbf{u} \cdot \bar{\mathbf{t}} \, d\Gamma\} \\ &= \int_{\Gamma_u} \bar{\mathbf{u}} \cdot \boldsymbol{\tau} \cdot \mathbf{n} d\Gamma - \Psi_{\mathcal{T}_a}(\Lambda^*\boldsymbol{\tau}), \end{aligned} \tag{5.6}$$

where $\Psi_{\mathcal{T}_a}$ is the indicator function of the statically admissible set $\mathcal{T}_a$ defined by

$$\mathcal{T}_a = \{\boldsymbol{\tau} \in \mathcal{L}^{\beta}(\Omega; \mathbf{R}^{m\times n}) | \ -\nabla \cdot \boldsymbol{\tau}^t = \bar{\mathbf{b}} \ \text{ in } \Omega; \ \mathbf{n}\cdot\boldsymbol{\tau}^t = \bar{\mathbf{t}} \ \text{ on } \Gamma_t\}, \quad (5.7)$$

$\mathcal{L}^{\beta}$ is a Lebesgue integrable space, $\beta = \alpha/(\alpha-1)$. So on $\mathcal{T}_a$, the Fenchel-Rockafellar conjugate functional for geometrically linear operator $\Lambda\mathbf{u} = \nabla\mathbf{u}$ is

$$P^c_{L-Z}(\boldsymbol{\tau}) = \int_{\Gamma_u} \bar{\mathbf{u}}\cdot\boldsymbol{\tau}\cdot\mathbf{n}d\Gamma - \int_{\Omega} w^*(\boldsymbol{\tau})d\Omega, \quad (5.8)$$

which is the well-known *Levinson-Zubov's* energy in finite deformation theory (see, Gao, 1992). It is obvious that $P^c_{L-Z} : \mathcal{T}_a \to \mathbf{R}$ is always concave, upper semicontinuous. But the following problem

$$(\mathcal{P}^c_{L-Z}): \quad P^c_{L-Z}(\bar{\boldsymbol{\tau}}) = \sup P^c_{L-Z}(\boldsymbol{\tau}) \ \ \forall \boldsymbol{\tau} \in \mathcal{T}_a \quad (5.9)$$

is not equivalent to the primal problem since $\mathbf{F} = \nabla\mathbf{u}$ is not a strain measure and the stored energy $w(\mathbf{F})$ is usually nonconvex (see Ogden, 1984). There exists a dual gap between the minimal potential problem and the Levinson-Zubov's complementary energy problem (5.9) (see, Gao, 1992), i.e.

$$\inf_{\mathbf{u}\in\mathcal{U}_a} P(\mathbf{u}) \geq \sup_{\boldsymbol{\tau}\in\mathcal{T}_a} P^c_{L-Z}(\boldsymbol{\tau}). \quad (5.10)$$

If we let Λ be a quadratic operator such that $\Lambda\mathbf{u}$ is the so-called *right Cauchy-Green strain tensor*

$$\mathbf{C} = \Lambda\mathbf{u} = \frac{1}{2}(\nabla\mathbf{u})^t \cdot (\nabla\mathbf{u}) \in \mathbf{R}^{n\times n}, \quad (5.11)$$

then Λ_t and Λ_n can be given as

$$\Lambda_t(\mathbf{u})\mathbf{v} = \frac{1}{2}[(\nabla\mathbf{v})^t\cdot(\nabla\mathbf{u}) + (\nabla\mathbf{u})^t\cdot(\nabla\mathbf{v})],$$

$$\Lambda_n(\mathbf{u})\mathbf{v} = -\frac{1}{4}[(\nabla\mathbf{v})^t\cdot(\nabla\mathbf{u}) + (\nabla\mathbf{u})^t\cdot(\nabla\mathbf{v})].$$

For hyperelasticity, the stored strain energy w is usually a convex, Gâteaux differentiable function of $\mathbf{C}$. The conjugate variable of $\mathbf{C}$ should be the *Kirchhoff stress* $p^* = \mathbf{S} = Dw(\mathbf{C})$. Then we have the following relations:

$$\mathbf{S} = Dw(\mathbf{C}) \ \Leftrightarrow \ \mathbf{C} = Dw^*(\mathbf{S}) \ \Leftrightarrow \ w(\mathbf{C}) + w^*(\mathbf{S}) = \text{tr}(\mathbf{C}\cdot\mathbf{S}). \quad (5.12)$$

The Lagrangian $L(u, p^*)$ in this case can be written as

$$L(\mathbf{u}, \mathbf{S}) = \int_{\Omega} [\text{tr}(\mathbf{C}(\mathbf{u})\cdot\mathbf{S}) - w^*(\mathbf{S})]d\Omega - \int_{\Omega} \mathbf{u}\cdot\bar{\mathbf{b}}\, d\Omega - \int_{\Gamma_t} \mathbf{u}\cdot\bar{\mathbf{t}}\, d\Gamma. \quad (5.13)$$

The critical conditions (3.7) and (3.8) give the equilibrium equation

$$\Lambda_t^*(\bar{\mathbf{u}})\bar{\mathbf{S}} = \begin{cases} -\nabla\cdot(\nabla\bar{\mathbf{u}}\cdot\bar{\mathbf{S}})^t = \bar{\mathbf{b}} & \text{in } \Omega, \\ \mathbf{n}\cdot(\nabla\bar{\mathbf{u}}\cdot\bar{\mathbf{S}})^t = \bar{\mathbf{t}} & \text{on } \Gamma_t. \end{cases} \quad (5.14)$$

and the inverse geometric-constitutive equation:

$$\mathbf{C}(\bar{\mathbf{u}}) = Dw^*(\bar{\mathbf{S}}) \quad \text{in } \Omega.$$

Since $w(\mathbf{C})$ is convex, so the critical point $(\bar{\mathbf{u}}, \bar{\mathbf{S}})$ of L solves the mixed boundary value problem (5.3).

In continuum mechanics, $L(\mathbf{u}, \mathbf{S})$ is the well-known Hellinger-Reissner generalized complementary energy. The extremum property of this energy functional has been an open problem for more than 40 years, which has many consequences in mechanics. This open problem now is solved by the triality theorem, i.e. if the gap function

$$G(\mathbf{u}, \mathbf{S}) = \langle \mathbf{S}, \Lambda_t(\mathbf{u})\mathbf{u} \rangle = \frac{1}{2} \int_\Omega \mathrm{tr}[(\nabla \mathbf{u}) \cdot \mathbf{S} \cdot (\nabla \mathbf{u})^t] d\Omega \tag{5.15}$$

is positive, $L(\mathbf{u}, \mathbf{S})$ is a saddle point energy and the pure complementary energy can be given by

$$P^*(\mathbf{S}) = \inf_{\mathbf{u} \in \mathcal{U}_a} L(\mathbf{u}, \mathbf{S}). \tag{5.16}$$

However, if $G(\mathbf{u}, \mathbf{S})$ is negative, then $L(\mathbf{u}, \mathbf{S})$ is a super-critical point functional. In this case, the total potential $P(\mathbf{u})$ is nonconvex and the pure complementary energy is given by

$$P^*(\mathbf{S}) = \sup_{\mathbf{u} \in \mathcal{U}_a} L(\mathbf{u}, \mathbf{S}). \tag{5.17}$$

Let $\mathcal{Y}_a^*$ be a subspace of the range $\partial w(\mathbf{C})$:

$$\mathcal{Y}_a^* = \{\mathbf{S} = Dw(\mathbf{C}) \ \forall \mathbf{C} \in \mathcal{L}^\alpha(\Omega; \mathbf{R}^{n \times n}) | \ \det \mathbf{S}(x) \neq 0 \ \text{ a.e. in } \Omega\}. \tag{5.18}$$

Then on $\mathcal{Y}_a^*$, this pure complementary energy can be obtained as:

$$P^*(\mathbf{S}) = \int_{\Gamma_u} \bar{\mathbf{u}} \cdot \boldsymbol{\tau} \cdot \mathbf{n} d\Gamma - \int_\Omega w^*(\mathbf{S}) d\Omega - \int_\Omega \frac{1}{2} \mathrm{tr}[\mathbf{S}^{-1} \cdot \boldsymbol{\tau}^t \cdot \boldsymbol{\tau}] d\Omega. \tag{5.19}$$

The critical condition of P^* gives the *dual Euler-Lagrange* equation:

$$\mathbf{S} \cdot [Dw^*(\mathbf{S})] \cdot \mathbf{S} = \frac{1}{2} \boldsymbol{\tau}^t \cdot \boldsymbol{\tau} \quad \text{in } \Omega. \tag{5.20}$$

This is an algebraic equation! If $\boldsymbol{\tau} \in \mathcal{T}_a$ is a critical point of P^c_{L-Z}, then the solution $\bar{\mathbf{S}}$ of this algebraic equation solves the mixed boundary value problem (5.3). Since $\nabla \bar{\mathbf{u}} = \bar{\boldsymbol{\tau}} \cdot \bar{\mathbf{S}}^{-1}$ (cf. Gao, 1992), if $\bar{\boldsymbol{\tau}} \cdot \bar{\mathbf{S}}^{-1}$ is a conservative field, then for any given $\bar{\mathbf{u}}(\mathbf{x}_0)$ $\forall \mathbf{x}_0 \in \Gamma_u$, the following path-independent integral

$$\bar{\mathbf{u}}(\mathbf{x}) = \int_{\mathbf{x}_0}^{\mathbf{x}} \bar{\boldsymbol{\tau}} \cdot \bar{\mathbf{S}}^{-1} \cdot d\mathbf{x} + \bar{\mathbf{u}}(\mathbf{x}_0), \tag{5.21}$$

gives an analytic solution for the mixed boundary value problem (5.3). Its properties are characterized by the triality theorem. By Lemma 4.1, we have

$$P(\bar{\mathbf{u}}) = L(\bar{\mathbf{u}}, \bar{\mathbf{S}}) = P^*(\bar{\mathbf{S}}),$$

i.e. there is no dual gap between the potential energy $P(\mathbf{u})$ and the complementary energy $P^*(\mathbf{S})$. Detailed study and applications of this analytic solution and the pure complementary energy in finite deformation theory are given in (Gao, 1997, 1998).

Now let us apply the results presented in this paper to solve a real-life one-dimensional nonconvex, unilateral variational problem:

$$(\mathcal{P}_u): \quad P(u) = \int_0^1 w(\Lambda u)dx - \int_0^1 f(x)u dx \ \to \min \ \forall u \in \mathcal{U}_a, \qquad (5.22)$$

where Λ is a pure quadratic operator: $\Lambda u = \frac{1}{2}(u_{,x})^2$; $w(p)$ is a convex, lower semicontinuous function of $p = \Lambda u$. For unilateral variational problem, the feasible space $\mathcal{U}_a$ can be defined as

$$\mathcal{U}_a^+ = \{u \in \mathcal{W}^{1,\alpha}[0,1] | \ u(x) \geq 0 \ \forall x \in (0,1), \quad u(0) = 0\},$$

which is a closed, convex subset of the Sobolev space $\mathcal{W}^{1,\alpha}$. As an example, here we simply let $w(p) = \frac{1}{2}E(p-\lambda)^2$, with a given parameter $\lambda \in \mathbf{R}$ and a material constant $E > 0$. Then the stored potential energy

$$W(p) = \int_0^1 \frac{1}{2}E(p-\lambda)^2 dx$$

is convex, Gâteaux differentiable. However, in terms of the linear deformation $\epsilon = u_{,x}$, $w(p(\epsilon)) = \frac{1}{2}(\frac{1}{2}\epsilon^2 - \lambda)^2$ is a double-well function for $\lambda > 0$. The nonconvex variational problem with double-well energy appears in many physical systems such as phase transitions, hysteresis, nonconvex optimal design and post-buckling analysis (cf. e.g. Ericksen, 1975; Gao, 1997). By the Kuhn-Tucker theory, the critical condition of the variational problem (5.22) is equivalent to the following nonlinear complementarity problem (NCP):

$$u(x) \geq 0 \qquad \forall x \in [0,1], \qquad (5.23)$$

$$[E(\tfrac{1}{2}u_{,x}^2 - \lambda)u_{,x}]_{,x} + f \leq 0 \qquad \forall x \in [0,1], \qquad (5.24)$$

$$u(x)\{[E(\tfrac{1}{2}u_{,x}^2 - \lambda)u_{,x}]_{,x} + f\} = 0 \quad \forall x \in [0,1]. \qquad (5.25)$$

The direct methods for solving this geometrically nonlinear complementarity problem are very difficult. But by using the duality theory, we are able to find the analytic solution. For the pure quadratic operator $p = \Lambda u = \frac{1}{2}(u_{,x})^2$, we have

$$\Lambda_t u = u_{,x}^2, \quad \Lambda_n u = -\frac{1}{2}u_{,x}^2. \qquad (5.26)$$

The dual variable of p is given by

$$p^* = DW(p) = E(p - \lambda).$$

The complementary energy is then

$$W^*(p^*) = \sup_p\{\int_0^1 p^* p dx - W(p)\} = \int_0^1 (\frac{1}{2E}p^{*2} + \lambda p^*)dx.$$

Since $W(p(\epsilon))$ is nonconvex in the linear deformation $\epsilon = u_{,x}$, the Piola stress τ is given by

$$\tau = D_\epsilon W(p(\epsilon)) = E\epsilon(\frac{1}{2}\epsilon^2 - \lambda) = \epsilon p^*. \quad (5.27)$$

For any given source function $f(x)$, the sub-conjugate function of $F(u) = \int_0^1 fudx$ should be

$$\begin{aligned} F_*(u^*) &= \inf_{u \in \mathcal{U}_a^+} \{\int_0^1 uu^* dx - \int_0^1 fudx\} \\ &= \begin{cases} 0 & \text{if } u^*(x) - f(x) \geq 0 \ \forall x \in (0,1), \\ -\infty & \text{otherwise.} \end{cases} \end{aligned}$$

So the statically admissible space $\mathcal{T}_a$ in this unilateral variational problem should be

$$\mathcal{T}_a = \{\tau \in \mathcal{W}^{1,\beta}(0,1) | \ \tau_{,x} + f \leq 0 \ \forall x \in (0,1), \ \tau(1) = 0\}. \quad (5.28)$$

It is easy to find out that

$$\tau(x) \leq \int_0^x -f(t)dt + \int_0^1 f(t)dt \ \ \forall x \in [0,1] \quad (5.29)$$

is statically admissible. The gap function in this problem should be

$$G(u, p^*) = \langle p^*, -\Lambda_n u \rangle = \frac{1}{2}\int_0^1 p^* u_{,x}^2 dx.$$

Since $p = \frac{1}{2}u_{,x}^2 \geq 0 \ \forall u \in \mathcal{U}_a^+$ the range of $p^* = E(p - \lambda)$ should be

$$\mathcal{Y}_a^* = \{p^* \in \mathcal{L}^\beta[0,1] | -\lambda E \leq p^* < +\infty, \ p^*(x) \neq 0 \ \forall x \in (0,1)\}. \quad (5.30)$$

($p^* = 0$ implies that $p = \frac{1}{2}u_{,x}^2 = \lambda$.) If $\lambda < 0$, then the gap function is positive on $\mathcal{Y}_a^*$. In this case, $P(u)$ is convex, and problem has a unique solution. If $\lambda > 0$, the gap function could be negative on $\mathcal{Y}_a^*$. In this case, $P(u)$ is nonconvex and the primal problem may have more than one solutions.

The Lagrangian $L : \mathcal{U}_a^+ \times \mathcal{Y}_a^* \to \mathbf{R}$ for this problem should be

$$L(u, p^*) = \int_0^1 [(\frac{1}{2}u_{,x}^2 - \lambda)p^* - \frac{1}{2E}p^{*2} - fu]dx. \quad (5.31)$$

The pure complementary energy P^* for this one-dimensional problem is

$$P^*(p^*) = -\int_0^1 [\frac{1}{2E}p^{*2} + \lambda p^* + \frac{1}{2}p^{*-1}\tau^2]dx, \quad (5.32)$$

which is well-defined on $\mathcal{Y}_a^*$. The dual Euler-Lagrange equation in this example is a cubic algebraic equation:

$$p^{*2}(\frac{1}{E}p^* + \lambda) = \frac{1}{2}\tau^2 \ \ \forall x \in (0,1). \quad (5.33)$$

For a given $f(x)$ such that $\tau(x)$ is obtained by (5.29), this equation has at most three solutions p_i^*, $i = 1, 2, 3$. Since $\tau = u_{,x} p^*$, $u(0) = 0$, the analytic solution (5.21) in this example is then

$$u_i(x) = \int_0^x \frac{\tau(t)}{p_i^*(t)} \mathrm{d}t. \tag{5.34}$$

By Lemma 4.1, those $u_i(x)$ in $\mathcal{U}_a^+$ should be the analytical solution of the nonconvex unilateral variational problem $(\mathcal{P}_u)$. The property of this solution is characterized by the triality theorem.

For fully nonlinear, convex finite deformation systems, a so-called complementary finite element method and algorithm have been developed for solving nonsmooth mechanics problems (Gao, 1996). The triality theory proved in this paper can be used to develop algorithms for robust numerical solutions in nonconvex, nonsmooth systems.

Acknowledgements

The author is indebted to Professor G. Auchmuty for several important discussions, and to two anonymous referees, for their careful reading and many helpful suggestions. Constant support and valuable comments from Professor M. Day and Professor G. Strang are also gratefully acknowledged.

References

[1] Auchmuty, G (1983), Duality for non-convex variational principles, *J. Diff. Eqns*, Vol. 50, pp. 80-145.

[2] Auchmuty, G (1989), Duality algorithms for nonconvex variational principles, *Numerical Funct. Anal. & Optimization*, **10**, pp. 211-264.

[3] Clarke, FH (1985), The dual action, optimal control, and generalized gradients, *Mathematical Control Theory*, Banach Center Publ., 14, PWN, Warsaw, pp. 109-119.

[4] Dem'yanov, VF, Stavroulakis, GE, Polyakova, LN and Panagiotopoulos, PD (1996), *Quasidifferentiability and Nonsmooth Modelling in Mechanics, Engineering and Economics*, Kluwer Academic Publishers, Dordrecht / Boston / London, 348pp.

[5] Ekeland, E and Temam, R (1976), *Convex Analysis and Variational Problems*, Dunod, Paris.

[6] Ericksen, JL (1975), Equilibrium of bars, *J. Elasticity*, **5**, pp. 191-202.

[7] Gao, DY (1992), Global extremum criteria for finite elasticity, *ZAMP*, **43**, pp. 924-937.

[8] Gao, DY (1994), Stability and extremum principles for post yield analysis of finite plasticity, *Acta Mech Sinica*, **10**, pp. 311-325.

[9] Gao, DY (1995), Duality theory in nonlinear buckling analysis for von Karman equations,*Studies in Applied Mathematic*, **94** (1995), pp. 423-444.

[10] Gao, DY (1996), Complementary finite-element method for finite deformation nonsmooth mechanics, *J. Eng. Math.*, **30**, pp. 339-353.

[11] Gao, DY (1997) Dual extremum principles in finite deformation theory with applications in post-buckling analysis of nonlinear beam model, *Appl. Mech. Reviews, ASME*, Vol. 50, part 2, November, S64-S71.

[12] Gao, DY (1998a), Duality in nonconvex finite deformation theory: A survey and unified approach, to appear in *From Convexity to Nonconvexity, A Volume dedicated to the memory of Professor Gaetano Fichera*, ed. by R. Gilbert, P.D. Panagiotopoulos and P. Pardalos, Kluwer Academic Publishers, 1998.

[13] Gao, DY (1998b), Complementary extremum principles in nonconvex parametric variational problems with applications, to appear in *IMA J. Appl. Math.*

[14] Gao, DY (1999), *Duality Principles in Nonconvex Systems: Theory, Methods and Applications*, to be published by Kluwer Academic Publishers.

[15] Gao, DY and Strang, G (1989), Geometric nonlinearity: Potential energy, complementary energy, and the gap function, *Quart. Appl. Math.*, **XLVII**(3), pp. 487-504.

[16] Gao, DY and Yang, WH (1995), Multi-duality in minimal surface type problems, *Studies in Appl Math*, **95**, pp. 127-146.

[17] Oden, JT and Reddy, JN (1983), *Variational Methods in Theoretical Mechanics*, Springer-Verlag.

[18] Ogden, RW (1984), *Non-linear elastic deformations*, Ellis Horwood Ltd, Chichester, 417pp.

[19] Panagiotopoulos, PD (1985), *Inequality Problems in Mechanics and Applications*, Birkhäuser, Boston.

[20] Panagiotopoulos, PD (1993), *Hemivariational Inequalities: Applications in Mechanics and Engineering*, Springer-Verlag, 451pp.

[21] Rockafellar, RT (1974), Conjugate Duality and Optimization, SIAM, J.W. Arrowsmith Ltd., Bristol 3, England.

[22] Sewell, MJ (1987), *Maximum and Minimum Principles*, Cambridge Univ. Press, 468pp.

[23] Strang, G (1986), *Introduction to Applied Mathematics*, Wellesley-Cambridge Press, 758pp.

[24] Toland, JF (1979), A duality principle for non-convex optimization and the calculus of variations, *Arch. Rational Mech. Anal.* **71**, 41-61

[25] Tonti, E (1972), A mathematical model for physical theories, *Fisica Matematica, Serie VIII, Vol. LII.*

[26] Walk, M (1989), *Theory of duality in mathematical programming*, Springer-Verlag, Wien / New York.

[27] Wright, SJ (1997), *Primal-Dual Interior-Point Methods*, SIAM, Philadelphia, 289pp.

Reformulation: Nonsmooth, Piecewise Smooth, Semismooth and Smoothing Methods, pp. 181–209
Edited by M. Fukushima and L. Qi

Global and Local Superlinear Convergence Analysis of Newton-Type Methods for Semismooth Equations with Smooth Least Squares [1]

Houyuan Jiang† and Daniel Ralph†

Abstract The local superlinear convergence of the generalized Newton method for solving systems of nonsmooth equations has been proved by Qi and Sun under the semismooth condition and nonsingularity of the generalized Jacobian at the solution. Unlike the Newton method for systems of smooth equations, globalization of the generalized Newton method seems difficult to achieve in general. However, we show that global convergence analysis of various traditional Newton-type methods for systems of smooth equations can be extended to systems of nonsmooth equations with semismooth operators whose least squares objective is smooth. The value of these methods is demonstrated from their applications to various semismooth equation reformulations of nonlinear complementarity and related problems.

Key Words nonsmooth equation, semismooth operator, Newton method, Gauss-Newton method, global convergence, superlinear convergence, complementarity problem.

1 INTRODUCTION

Let $H : \Re^n \to \Re^m$ be locally Lipschitz continuous. We are concerned with the solution of the following system of nonsmooth equations

$$H(x) = 0. \tag{1.1}$$

[1]This work is supported by the Australian Research Council.
†Department of Mathematics and Statistics, The University of Melbourne, Parkville, Victoria 3052, Australia. E-mail: jiang@mundoe.maths.mu.OZ.AU, danny@mundoe.maths.mu.OZ.AU.

If $m = n$ and H happens to be smooth on $\Re^n$, it is well-known that local and superlinear convergence of Newton's method as well as global convergence of the damped Newton and the damped Gauss-Newton methods can be established under some standard assumptions. If $m \neq n$, (1.1) is either an underdetermined or an overdetermined system. Local and global convergence of Newton-type methods has also been well studied. See [6, 12, 28].

When H is not smooth, one may not expect local or global convergence of generalized Newton methods in general. Nontheless, in the case of $m = n$, Qi and Sun [37] and later Qi [33] proved local and superlinear convergence of generalized Newton methods, where a linear system is solved at each iteration, under the semismooth condition and nonsingularity of the generalized Jacobian at the solution. In order to enlarge the convergence domain, some global strategies have to be imposed on the generalized Newton method. However, the generalized Newton method of Qi and Sun has not been proved to enjoy global convergence under the semismooth and nonsingularity conditions though some encouraging results have been obtained.

In the last decade, nonsmooth equation reformulations have proved to be one important class of tools for solving problems arising from nonlinear programming, complementarity and variational inequality problems; see [32] and [7, 13, 29, 30, 35, 31, 38, 39, 40]. In particular, a number of authors have used the so-called Fischer-Burmeister functional to reformulate the nonlinear complementarity problem as a system of nonsmooth equation. It turns out that this system is semismooth. Therefore, the generalized Newton method of Qi and Sun [37] can be applied to this system for the solution of the nonlinear complementarity problem.

The use of Fischer-Burmeister functional has been extended to other problems related to the nonlinear complementarity problem. Some efforts have been made towards establishing global convergence of the damped generalized Newton method on this special semismooth equation; see for example [5, 10, 18, 43]. Jiang [15] has proved the global convergence of the damped generalized Newton method and the damped modified Gauss-Newton method for this special semismooth equation. This realization heavily depends on two important properties that this semismooth equation possesses. Firstly, the equation is semismooth which makes the local convergence possible. Secondly, the least square form of the equation is smooth which facilitates the global convergence analysis.

In the mean time, due to the successful applications of Fischer-Burmeister functional, other nonsmooth equations have been proposed. Some of them do have two properties mentioned above. Therefore, we feel it is necessary to give a unified treatment for global convergence analysis of different Newton-type methods to semismooth equations with smooth least squares. We illustrate the applications of this general approach using a number of examples based on nonlinear complementarity and related problems.

The rest of the paper is organized as follows. In the next section, the generalized Newton method of Qi [33] for solving semismooth systems is introduced. Some preliminary results are collected. We study the damped generalized Newton and the damped modified Gauss-Newton methods and their global as well local superlinear convergence for semismooth systems with smooth least squares in Sections 3 and 4. In Section 5, a generalized trust region method is presented and its global and local superlinear convergence is explained. The damped modified Gauss-Newton proposed in Section 4 is extended to the overdetermined semismooth systems in Section 6. In Section 7, we generalize the damped modified Gauss-Newton method to the semismooth systems with polyhedral constraints. Global and local superlinear convergence is also established. Section 8 is devoted to several applications of our theory to the variational inequality problem and its special cases such as the nonlinear complementarity and box constrained variational inequality problem. These applications justify the usefulness of the theory established in the previous sections. We conclude the paper by offering some remarks in Section 9.

2 GENERALIZED NEWTON METHOD AND PRELIMINARIES

Let $H : \Re^n \to \Re^m$ be locally Lipschitz continuous at $x \in \Re^n$. Then the Clarke generalized Jacobian $\partial H(x)$ of H at x is well-defined and can be characterized by the convex hull of the following set

$$\partial_B H(x) = \{ \lim_{x^k \to x} H'(x^k) | \quad H \text{ is differentiable at } x^k \in \Re^n\},$$

where $H'(x)$ denotes the Jacobian of H at x. $\partial H(x)$ is a nonempty, convex and compact set for any fixed x [4]. The nonemptyness of $\partial H(x)$ clearly implies that $\partial_B H(x)$ is nonempty too. H is said to be semismooth at $x \in \Re^n$ if it is directionally differentiable at x, i.e., $H'(x; d)$ exists for any $d \in \Re^n$, and if

$$Vd - H'(x, d) = o(\|d\|)$$

for any $d \to 0$ and $V \in \partial H(x + d)$. H is said to be strongly semismooth at x if it is semismooth at x and

$$Vd - H'(x, d) = O(\|d\|^2).$$

The above semismooth concept for the vector-valued function was introduced by Qi and Sun [37]. But semismoothness for functionals (or scalar-valued functions) was originally introduced by Mifflin [27]. For some fine properties of semismooth functions, see [27, 37, 33, 11]. See also a survey paper [19] on applications of of semismooth properties in superlinear convergence of generalized Newton methods for different optimization problems. Here we recall from [37] two key properties which will be used in the sequel.

Proposition 2.1 *Suppose $H : \Re^n \to \Re^m$ is locally Lipschitz at $x \in \Re^n$. If H is semismooth at x, then*

$$H(x + d) - H(x) - Vd = o(\|d\|),$$

for any d $(\in \Re^n) \to 0$ and $V \in \partial H(x + d)$; and if H is strongly semismooth at x, then

$$H(x + d) - H(x) - Vd = O(\|d\|^2).$$

For ease of presentation, we shall assume from now on that $m = n$ unless otherwise it is stated. H is called strongly BD-regular at x if each element of $\partial_B H(x)$ is nonsingular. The following is a perturbation result on strong BD-regularity established in [33].

Proposition 2.2 *If H is strongly BD-regular at x, then there is a neighborhood $\mathcal{N}$ of x and a constant c such that for any $y \in \mathcal{N}$ and $V \in \partial_B H(y)$, V is nonsingular and*

$$\|V^{-1}\| \leq c.$$

We now restate a generalized version of the Newton method for the solution of the system (1.1) from Qi [33].

Algorithm 1 (`Generalized Newton Method`)

Step 1 (**Initialization**) Choose a starting point $x^0 \in \Re^n$. Let $k := 0$.

Step 2 (**Search direction**) Choose $V_k \in \partial_B H(x^k)$ and solve the generalized Newton equation

$$V_k d + H(x^k) = 0. \tag{2.1}$$

Let d^k be a solution of the above equation (we call it the generalized Newton direction). If $d = 0$ is a solution of the generalized Newton equation, the algorithm terminates. Otherwise, go to Step 3.

Step 3 (**Update**) Let $x^{k+1} := x^k + d^k$ and $k := k + 1$. Go to Step 2.

When H is smooth, Algorithm 1 reduces to the Newton method for systems of smooth equations; see [6]. It is known that there is no general local convergence of Algorithm 1; see Example 2.4 of [19]. Nontheless, the semismooth and strong BD-regularity assumptions introduced above can in fact guarantee local superlinear convergence of Algorithm 1. The following is such a result established in [33].

Theorem 2.1 *Suppose x^* is a solution of (1.1). Suppose $H : \Re^n \to \Re^n$ is semismooth and strongly BD-regular at x^*. Then Algorithm 1 is well-defined and the sequence $\{x^k\}$ converges to x^* when x^0 is chosen sufficiently close to x^*. Moreover, the convergence rate is Q-superlinear, i.e.,*

$$\lim_{k\to\infty} \frac{\|x^{k+1} - x^*\|}{\|x^k - x^*\|} = 0,$$

and the convergence rate is Q-quadratic, i.e.,

$$\lim_{k\to\infty} \frac{\|x^{k+1} - x^*\|}{\|x^k - x^*\|^2} < \infty$$

if H is strongly semismooth at x^*.

Remark. Prior to [33], Qi and Sun [37] proposed another generalized Newton method in which the generalized Jacobian V_k used in (2.1) is chosen from $\partial H(x^k)$. The local superlinear convergence of the corresponding generalized Newton method is established under the semismoothness and nonsingularity of the Clarke generalized Jacobian at the solution. Independently, Kummer [25] also presented a general analysis of superlinear convergence for the latter generalized Newton methods studied in [24]. Algorithm 1 is favourable over the one proposed in [25, 37] because V is chosen in a smaller set in Algorithm 1 which increases solvability of the generalized Newton equation and the convergence theory requires the less restrictive assumption of strong BD-regularity at the solution.

3 DAMPED GENERALIZED NEWTON METHOD

As is seen in the last section, under certain conditions, the generalized Newton method converges provided the starting point is sufficiently close to the solution of (1.1). However, no one knows where the solution is before it is found. Therefore, it is necessary to enlarge the domain of convergence of the generalized Newton method. Traditionally, certain line search imposed on the generalized Newton direction can fulfill this purpose. We demonstrate in this section that this technique is extendable to the nonsmooth system (1.1) when H has some further properties. To this end, let us define the least square merit function of (1.1), which we also call the "least square of H":

$$\theta(x) \equiv \frac{1}{2}\|H(x)\|^2.$$

Clearly, any solution of (1.1) is a global minimizer of θ over $\Re^n$. Conversely, any local solution x of θ such that $\theta(x) = 0$ is a solution of (1.1). This indicates that the value of $\theta(x)$ can be used to measure the quality of how close between x and a solution of (1.1).

Before presenting our damped generalized Newton method, we assume that

(A1) H is semismooth on $\Re^n$.
(A2) θ is continuously differentiable on $\Re^n$.

The reader may argue that the assumption (A2) is somewhat strong when H is only assumed to be semismooth on $\Re^n$. We shall demonstrate in Section 8 that this assumption is satisfied for some special systems of nonsmooth equations arising from nonlinear complementarity and related problems. As a consequence of the assumption (A2), we have

$$\nabla\theta(x) = V^T H(x),$$

for any $V \in \partial H(x)$.

Another question is: Why don't we just apply smooth optimization techniques to the following minimization problem?

$$\begin{array}{ll} \min & \theta(x) \\ \text{s.t.} & x \in \Re^n \end{array}$$

There is no doubt that optimization methods can find stationary points of the above minimization problem. But one may not expect fast local convergence results since θ is usually not twice differentiable when H is only semismooth. Furthermore, solving the above optimization problem for the solution of (1.1) may not be preferred even if H is smooth; see [6].

Algorithm 2: (Damped Generalized Newton Method)

Step 1 (Initialization) Choose an initial starting point $x^0 \in \Re^n$, two scalars $\sigma, \rho \in (0,1)$, and let $k := 0$.

Step 2 (Search direction) If $H(x^k) = 0$, then the algorithm terminates. Otherwise, choose $V_k \in \partial_B H(x^k)$, solve the generalized Newton equation (2.1) at x^k. Let d^k be a solution of (2.1) at x^k and go to Step 3.

Step 3 (Line search) Let $\lambda_k = \rho^{i_k}$ where i_k is the smallest nonnegative integer i such that

$$\theta(x^k + (\rho)^i d^k) - \theta(x^k) \leq \sigma(\rho)^i \nabla\theta(x^k)^T d^k.$$

Step 4 (Update) Let $x^{k+1} := x^k + \lambda_k d^k$ and $k := k+1$. Go to Step 2.

When H is continuously differentiable on $\Re^n$, Algorithm 2 reduces to the damped Newton method for systems of smooth equations. If there is no line search step, i.e., the full Newton step is always accepted, then Algorithm 2 reduces to Algorithm 1. We remark that the generalized Newton equation is not always solvable when V_k is singular. This question will lead to the damped modified Gauss-Newton method which will be proposed in the next section. In the sequel, the algorithm is assumed not to terminate in finitely many steps since it is easy to see that x^k is a solution of (1.1) if the algorithm terminates in Step 2.

We are now in a position to present global convergence of Algorithm 2.

Theorem 3.1 *Suppose the assumptions (A1) and (A2) hold. Suppose the generalized Newton equation in Step 2 is solvable for each k. Assume that x^* is an accumulation point of $\{x^k\}$ generated by the damped generalized Newton method. Then the following statements hold:*

(i) x^* is a solution of (1.1) if $\{d^k\}$ is bounded.

(ii) x^* is a solution of (1.1) and $\{x^k\}$ converges to x^* Q-superlinearly if H is strongly BD-regular and $\sigma \in (0, \frac{1}{2})$; furthermore the convergence rate is Q-quadratic if H is strongly semismooth on $\Re^n$.

Proof. The following proof is similar to that of Theorem 4.1 in Jiang [15] although H has a special form and the generalized Jacobian is chosen from $\partial H(x)$ at x in [15]. For sake of completeness, we still include it here.

(i) The generalized Newton direction in Step 2 is well-defined by the solvability assumption of the generalized Newton equation. By the generalized Newton equation and (A2), we have

$$\nabla\theta(x^k)^T d^k = H(x^k)^T V_k d^k = -\|H(x^k)\|^2 = -2\theta(x^k) < 0.$$

In view that $d^k \neq 0$ and that $d = 0$ is not a solution of the generalized Newton equation, it follows that d^k is a descent direction of the merit function at x^k. Therefore, the well-definedness of the line search step (Step 3) and the algorithm follows from differentiability of the merit function θ.

Without loss of generality, we may assume that x^* is the limit of the subsequence $\{x^k\}_{k\in K}$ where K is a subsequence of $\{1, 2, \ldots\}$. If $\{\lambda_k\}_{k\in K}$ is bounded away from zero, using a standard argument from the decreasing property of the merit function after each iteration and nonnegativeness of the merit function over $\Re^n$, then $\sum_{k\in K} -\lambda_k \nabla\theta(x^k)^T d^k < +\infty$, which implies that $\sum_{k\in K} \theta(x^k) < +\infty$. Hence, $\lim_{k\to+\infty, k\in K} \theta(x^k) = \theta(x^*) = 0$ and x^* is a solution of (1.1). On the other hand, if $\{\lambda_k\}_{k\in K}$ has a subsequence converging to zero, we may pass to the subsequence and assume that $\lim_{k\to\infty, k\in K} \lambda_k = 0$. From the line search step, we may show that for all sufficiently large $k \in K$

$$\theta(x^k + \lambda_k d^k) - \theta(x^k) \leq \sigma\lambda_k \nabla\theta(x^k)^T d^k,$$

$$\theta(x^k + \rho^{-1}\lambda_k d^k) - \theta(x^k) > \sigma\rho^{-1}\lambda_k \nabla\theta(x^k)^T d^k.$$

Since $\{d^k\}$ is bounded, by passing to the subsequence, we may assume that $\lim_{k\to+\infty, k\in K} d^k = d^*$. By some algebraic manipulations and passing to the subsequence, we obtain

$$\nabla\theta(x^*)^T d^* = \sigma\nabla\theta(x^*)^T d^*,$$

which means that $\nabla\theta(x^*)^T d^* = 0$. By the generalized Newton equation, it follows that

$$H(x^k)^T H(x^k) + H(x^k)^T V_k d^k = H(x^k)^T H(x^k) + \nabla\theta(x^k)^T d^k = 0.$$

This shows that $\lim_{k\to\infty, k\in K} H(x^k)^T H(x^k) = H(x^*)^T H(x^*) = 0$, namely, x^* is a solution of (1.1).

(ii) Since H is strongly BD-regular, it follows from Proposition 2.2 that

$$\|(V_k)^{-1}\| \leq c,$$

for some positive constant c and all sufficiently large $k \in K$. The generalized Newton equation implies that $\{d^k\}_{k\in K}$ is bounded. Therefore, the proof of (i) implies that $H(x^*) = 0$.

We next turn to the convergence rate. From Proposition 2.1, for any sufficiently large $k \in K$,

$$\begin{aligned} H(x^k + d^k) &= H(x^* + x^k + d^k - x^*) - H(x^*) \\ &= U(x^k + d^k - x^*) + o(\|x^k + d^k - x^*\|), \end{aligned}$$

where $U \in \partial H(x^k + d^k)$ and

$$\begin{aligned} H(x^k) &= H(x^* + x^k - x^*) - H(x^*) \\ &= V(x^k - x^*) + o(\|x^k - x^*\|), \end{aligned}$$

where $V \in \partial H(x^k)$. Let $V = V_k$ in the last equality. Then the generalized Newton equation and uniform nonsingularity of V_k $(k \in K)$ imply that

$$\|x^k + d^k - x^*\| = o(\|x^k - x^*\|), \tag{3.1}$$

and $\|d^k\| = \|x^k - x^*\| + o(\|x^k - x^*\|)$ which implies that $\lim_{k\to\infty, k\in K} d^k = 0$. Consequently, it follows from Proposition 2.2, for any sufficiently large $k \in K$

$$\lim_{k\to\infty, k\in K} \frac{\|H(x^k)\|}{\|x^k - x^*\|} > 0,$$

$$\lim_{k\to\infty, k\in K} \frac{\|H(x^k + d^k)\|}{\|x^k + d^k - x^*\|} > 0.$$

Hence, (3.1) shows that

$$\|H(x^k + d^k)\| = o(\|H(x^k)\|).$$

By the generalized Newton equation and $\sigma \in (0, \frac{1}{2})$, we obtain that $\lambda_k = 1$ for all sufficiently large $k \in K$, i.e., the full generalized Newton step is taken. In other words, when k is sufficiently large, both x^k and $x^k + d^k$ are in a small neighborhood of x^* by (3.1), and the damped Newton method becomes the generalized Newton method. Then convergence and the convergence rate follow from Theorem 2.1. ∎

Remark. We have presented a simple scheme whose performance would no doubt be improved by using a nonmonotone line search with backtracking [7] and some heuristic strategies. However even such a simple scheme is practical as demonstrated for instance by the solution of contact problems in mechanics [2, 3]. These successful applications use the standard Armijo line search as in Step 3 of Algorithm 2 but also enforce a minimum stepsize (upper bound on the index i) as a heuristic to improve algorithm performance.

4 DAMPED MODIFIED GAUSS-NEWTON METHOD

The damped generalized Newton method proposed in the last section requires the solvability of the generalized Newton equation which might be too strong in general. Once again, we borrow a technique used in solving systems of smooth equations. It turns out that the modified Gauss-Newton method is a remedy to avoid this strict requirement [6, 28]. Our third algorithm can be stated as follows.

Algorithm 3: (Damped Modified Gauss-Newton Method)

Step 1 (Initialization) Choose a starting point $x^0 \in \Re^n$, the scalars $\sigma, \rho \in (0,1)$, $\nu_0 > 0$, and let $k := 0$.

Step 2 (Search direction) Choose $V_k \in \partial_B H(x^k)$ and solve the following modified Gauss-Newton equation

$$V_k^T H(x^k) + (V_k^T V_k + \nu_k I)d = 0,$$

with I the identity matrix in $\Re^{n\times n}$. Let d^k be the solution of the above equation. If $d^k = 0$, the algorithm terminates. Otherwise, go to Step 3.

Step 3 (Line search) Let $\lambda_k = \rho^{i_k}$ where i_k is the smallest nonnegative integer i such that

$$\theta(x^k + (\rho)^i d^k) - \theta(x^k) \leq \sigma(\rho)^i \nabla\theta(x^k)^T d^k.$$

Step 4 (Update) Choose $\nu_{k+1} > 0$. Let $x^{k+1} := x^k + \lambda_k d^k$ and $k := k+1$. Go to Step 1.

When H is continuously differentiable on $\Re^n$, the above algorithm reduces to the classical damped Gauss-Newton method [6, 28]. The modified Gauss-Newton equation is now always solvable and has a unique solution at any point x in view that $V_k^T V_k + \nu_k I$ is positive definite. In the sequel, the algorithm is assumed not to terminate in finitely many steps since it is easy to see that x^k is a stationary point of the merit function θ, i.e., $\nabla\theta(x^k) = V_k^T H(x^k) = 0$, if the algorithm terminates in Step 2.

Theorem 4.1 *Suppose the assumptions (A1) and (A2) hold. Assume that x^* is an accumulation point of $\{x^k\}$ generated by the damped modified Gauss-Newton method. Then the following conclusions hold:*

(i) x^* is a stationary point of the merit function θ if both $\{\nu_k\}$ and $\{d^k\}$ are bounded. Moreover, x^* is a solution of (1.1) if there exists a nonsingular element in the generalized Jacobian of H at x^*.

(ii) x^* is a stationary point of the merit function θ if $\bar{\nu} > \nu_k > \underline{\nu}$ for some $\bar{\nu} > \underline{\nu} > 0$.

(iii) Let $\nu_k = \min\{\theta(x^k), \|\nabla\theta(x^k)\|\}$. Then x^* is a stationary point of the merit function θ; furthermore, x^* is a solution of (1.1) and $\{x^k\}$ converges to x^* Q-superlinearly if $\sigma \in (0, \frac{1}{2})$ and H is strongly BD-regular at x^*. The convergence rate is Q-quadratic if H is strongly semismooth at x^*.

Proof. The following proof is basically a restatement of that given by Jiang [15] though a special function H is considered and the generalized Jacobian is chosen differently here. We again include it for completeness.

(i) Clearly, the modified Gauss-Newton direction d^k is well-defined by the positiveness of ν_k and d^k is a descent direction of the merit function by the positive definiteness of the matrix $V_k^T V_k + \nu_k I$. By the same argument in Theorem 3.1 and the fact that $\nabla\theta^T(x^k)^T d^k = -(d^k)^T(V_k^T V_k + \nu_k I)d^k < 0$, the line search step, and hence the algorithm is well-defined.

Once again, we assume that $\lim_{k\to\infty,k\in K} x^k = x^*$. Analogous to that in the proof of Theorem 3.1, by the boundedness of $\{d^k\}$, we may prove that either $\lim_{k\to\infty,k\in K} \theta(x^k) = \theta(x^*) = 0$ or

$$\lim_{k\to\infty,k\in K} \nabla\theta(x^k)^T d^k = \nabla\theta(x^*)^T d^* = 0.$$

If $\theta(x^*) = 0$, x^* is apparently a stationary point of θ. Otherwise, by passing to the subsequence again, we may assume that $\lim_{k\to\infty,k\in K} \nu_k = \nu^*$, $\lim_{k\to\infty,k\in K} V_k = V^*$. Then it follows from the generalized Gauss-Newton equation that

$$\nabla\theta(x^*)^T d^* + (d^*)^T((V^*)^T V^* + \nu^* I)d^* = 0,$$

which implies that

$$(d^*)^T((V^*)^T V^* + \nu^* I)d^* = 0.$$

The symmetric property of the matrix $(V^*)^T V^* + \nu^* I$ shows that there exists a matrix A such that

$$A^T A = (V^*)^T V^* + \nu^* I,$$

which implies that $Ad^* = 0$. By passing to the limit, the generalized Gauss-Newton equation gives us $\nabla\theta(x^*) = \lim_{k\to\infty,k\in K} V_k^T H(x^k) = -A^T A d^* = 0$. This proves that x^* is a stationary point of the merit function θ.

(ii) Since $\bar{\nu} > \nu_k > \underline{\nu}$, the matrices $\{(V_k)^T V_k + \nu_k I\}_{k\in K}$ are uniformly positive definite for all k. It follows that $\{d^k\}_{k\in K}$ is bounded. The desired results follows from (i).

(iii) Suppose x^* is not a stationary point of the merit function θ. Then $\|\nabla\theta(x^*)\| > 0$, $\theta(x^*) > 0$ and $\nu^* > 0$. Similar to (ii), it can be shown that $\{d^k\}_{k\in K}$ is bounded. Obviously, $\{\nu_k\}$ is bounded since the sequence $\{\theta(x^k)\}$ is nonincreasing. By (i), x^* must be a stationary point, a contradiction. Therefore, x^* is a stationary point of the merit function θ, and a solution of (1.1) if H is strongly BD-regular at x^*. Consequently,

$$\lim_{k\to\infty,k\in K} \nu_k = \lim_{k\to\infty,k\in K} \min\{\theta(x^k), \|\nabla\theta(x^k)\|\} = 0.$$

By Proposition 2.2 $\{d^k\}_{k\in K}$ is bounded. The modified Gauss-Newton equation implies that

$$\lim_{k\to\infty,k\in K} \frac{\|V_k^T(H(x^k) + V_k d^k)\|}{\|d^k\|} = 0,$$

$$\lim_{k\to\infty,k\in K} \frac{\|H(x^k) + V_k d^k\|}{\|d^k\|} = 0,$$

and

$$\lim_{k\to\infty,k\in K} d^k = 0,$$

since $\lim_{k\to\infty,k\in K} H(x^k) = 0$. Next, we prove that for any sufficiently large $k \in K$,

$$\|x^k + d^k - x^*\| = o(\|x^k - x^*\|). \tag{4.1}$$

Since V_k is nonsingular for all large $k \in K$, we may rewrite the modified Gauss-Newton equation as follows

$$H(x^k) + V_k d^k + \nu_k (V_k^T)^{-1} d^k = 0,$$

which is equivalent to

$$H(x^k) - H(x^*) - V_k(x^k - x^*) + \nu_k (V_k^T)^{-1} d^k + V_k(x^k + d^k - x^*) = 0.$$

By the strong BD-regularity at x^*, $\nu_k = o(1)$, and Proposition 2.1, we may show that

$$\|x^k + d^k - x^*\| = o(\|d^k\|) + o(\|x^k - x^*\|).$$

Clearly, the modified Gauss-Newton equation implies that

$$\|d^k\| = O(\|H(x^k)\|) = O(\|H(x^k) - H(x^*)\|) = O(\|x^k - x^*\|).$$

Therefore, (4.1) follows.

The rest is similar to the counterpart of Theorem 3.1. ∎

Notice that a stationary point of the merit function θ is not necessarily a solution of (1.1). However, this drawback cannot be avoided even when H is continuously differentiable. Luckily, stationary points and solutions of (1.1) do coincide for some special forms of (1.1); see Section 8. For instance, in Example 1 of Section 8, if the nonlinear complementarity problem (NCP) considered has P_0 property, then any stationary point of the least square merit function of the nonsmooth equation reformulated from the NCP in this example is a solution of this nonsmooth equation, hence a solution of the NCP.

5 GENERALIZED TRUST REGION METHOD

Trust region methods are another important class of global methods for solving systems of smooth equations. In this section, we show that trust region method is also extendable to the nonsmooth system (1.1) under the assumptions (A1) and (A2).

The original version of the following algorithm is presented in [17] for solving the generalized complementarity problem. We note that the proposed algorithm is actually suitable for the general system (1.1) under the assumption (A1) and (A2). Furthermore, the convergence theory established in [17] also applies here. Therefore, we repropose this algorithm with some minor modifications (mostly on notation) and restate convergence results without offering any proof. See [17] for more details. Let $|||\cdot|||$ denote any norm operator.

Algorithm 4: (`Generalized Trust Region Method`)

Step 0 (Initialization) Let $\alpha_1, \alpha_2, \alpha_3, \alpha_4, \rho_1, \rho_2, \Delta_{\min}$ and Δ_1 be such that $0 < \alpha_1 < \alpha_2 < 1 < \alpha_3 < \alpha_4$, $0 < \rho_1 < \rho_2 < 1$, $\Delta_{\min} > 0$ and $\Delta_1 > 0$. Let $x^0 \in \Re^n$ be a starting point. Set $k := 0$.

Step 1 (Termination check) If $\nabla\theta(x^k) = 0$, stop. Otherwise, let $\hat{\Delta} := \max\{\Delta_{\min}, \Delta_k\}$ and choose $V_k \in \partial_B H(x^k)$.

Step 2 (Subproblem) Let $\hat{s}$ be a solution of the minimization problem

$$\begin{array}{ll} \min & \nabla\theta(x^k)^T s + \frac{1}{2} s^T V_k^T V_k s \\ \text{s.t.} & |||s||| \le \hat{\Delta}. \end{array}$$

Step 3 (Update) Let

$$\hat{r} := \frac{\theta(x^k + \hat{s}) - \theta(x^k)}{\frac{1}{2}\|H(x^k) + V_k \hat{s}\|^2 - \theta(x^k)}.$$

If $\hat{r} \ge \rho_1$, then let $s^k := \hat{s}$, $x^{k+1} := x^k + s^k$, $\delta_k := \hat{\Delta}$. For any $\alpha \in [\alpha_3, \alpha_4]$, let

$$\Delta_{k+1} := \begin{cases} \hat{\Delta} & \text{if} \quad \rho_1 \le \hat{r} < \rho_2, \\ \alpha\hat{\Delta} & \text{if} \quad \hat{r} \ge \rho_2. \end{cases}$$

Let k be replaced by $k+1$ and return to Step 1. Otherwise, choose $\Delta \in [\alpha_1\hat{\Delta}, \alpha_2\hat{\Delta})$. Let $\hat{\Delta} := \Delta$ and repeat Step 2.

Remarks. (i) Unlike the damped Gauss-Newton method discussed in the last section, we did not introduce a parameter in order to ensure solvability of the generalized Gauss-Newton equation. However, the existence of a solution to the subproblem in Step 2 can be always guaranteed from the boundedness of the constraints though this minimization problem may have several solutions. (ii) The main difference between the proposed trust region method and the classical trust region methods lies in the updating rule for the trust region radius at the beginning of each iteration. More precisely, at the beginning of each iteration k, the trust region radius is always set greater than the fixed positive constant $\Delta_{\min}$, rather than solely updated from the final trust region radius of iteration $k-1$ as in the classical trust region methods. This special strategy enables us not only to establish global convergence but also to recover local superlinear convergence of the algorithm under some conditions, even though (1.1) is a system of nonsmooth equations. At the present, we are not able to establish the same convergence results if the classical updating rule for the trust region radius is used. We point out that these remarks follow closely from [17].

Theorem 5.1 *Let $\{x^k\}$ be generated by the algorithm. Assume that x^* is an accumulation point of $\{x^k\}$ generated by the generalized trust region method. Then the following statements hold:*

(i) x^* is a stationary point of θ.

(ii) If $\{V_k\}$ is bounded, then $\{\nabla\theta(x^k)\}$ is not bounded away from zero, that is,

$$\liminf_{k\to\infty} \|\nabla\theta(x^k)\| = 0.$$

(iii) If H is BD-regular at x^*, then the entire sequence $\{x^k\}$ converges to x^* Q-superlinearly. Moreover, the convergence rate is Q-quadratic if H is strongly semismooth at x^*.

6 DAMPED MODIFIED GAUSS-NEWTON METHOD FOR OVERDETERMINED SYSTEMS

When $m > n$, (1.1) is called an overdetermined system. The Gauss-Newton method and its globalization have been studied if H is smooth; see [6, 12]. As will be seen later, nonsmooth overdetermined systems also arise from nonlinear complementarity and related problems. Thus in this section, we offer a general convergence analysis of damped modified Gauss-Newton methods for overdetermined semismooth systems with smooth least squares.

Though the system (1.1) is not square when $m > n$, the least square merit function $\theta(x) = \frac{1}{2}\|H(x)\|^2$ is still well-defined. Similar to Section 4, the counterpart of the damped modified Gauss-Newton method can be proposed for an overdetermined system. Since they are the same conceptually, we omitted details. One point worthy to be mentioned is that the modified Gauss-Newton equation is a square system of dimension n though $m > n$.

Algorithm 5: `Damped Modified Gauss-Newton Method for Overdetermined Systems`

It is exactly as the same as Algorithm 3 but noting that $m > n$.

If $m = n$, Algorithm 5 is exactly Algorithm 3. If H is smooth on $\Re^n$, then Algorithm 5 is exactly a damped version of modified Gauss-Newton methods for overdetermined systems of equations, see [6, 12]. The global convergence is also a straightforward extension of Theorem 4.1 due to the fact that the semismooth properties are also valid for nonsquare systems. We summarize these results in the following theorem with a short proof.

Theorem 6.1 *Suppose the assumptions (A1) and (A2) hold. Assume that x^* is an accumulation point of $\{x^k\}$ generated by Algorithm 5. Then the following conclusions hold:*

(i) x^* is a stationary point of the merit function θ if both $\{\nu_k\}$ and $\{d^k\}$ are bounded.

(ii) x^* is a stationary point of the merit function θ if $\bar{\nu} > \nu_k > \underline{\nu}$ for some $\bar{\nu} > \underline{\nu} > 0$.

(iii) Let $\nu_k = \min\{\theta(x^k), \|\nabla\theta(x^k)\|\}$. Then x^* is a stationary point of the merit function θ; furthermore, $\{x^k\}$ converges to x^* superlinearly if x^* is

a solution of (1.1), $\sigma \in (0, \frac{1}{2})$ and if every element of $\partial H_B(x^*)$ has full column rank.

Proof. (i), (ii) and first part of (iii) can be proved in the exactly same way as in Theorem 4.1. Since the system considered here is not square, the technique used in the proof of the second part of (iii) in Theorem 4.1 cannot be used here. Therefore, we next prove the second part of (iii).

By the modified Gauss-Newton equation, boundedness of matrices $\{V_k\}$, full column rank of the generalized Jacobian $\partial_B H(x^*)$, Proposition 2.1, $\nu_k = o(1)$ and the fact that x^* is a solution of (1.1) (i.e., $H(x^*) = 0$), we obtain

$$\begin{aligned}
& \|x^k + d^k - x^*\| \\
= \; & \|x^k - x^* - (V_k^T V_k + \nu_k I)^{-1} V_k^T H(x^k)\| \\
\leq \; & \|(V_k^T V_k + \nu_k I)^{-1}\| \|V_k^T H(x^k) - (V_k^T V_k + \nu_k I)(x^k - x^*)\| \\
= \; & \|(V_k^T V_k + \nu_k I)^{-1}\| \|V_k^T [H(x^k) - V_k(x^k - x^*)] + \nu_k (x^k - x^*)\| \\
= \; & \|(V_k^T V_k + \nu_k I)^{-1}\| \|V_k^T [H(x^k) - H(x^*) - V_k(x^k - x^*)] + \nu_k (x^k - x^*)\| \\
= \; & o(\|x^k - x^*\|).
\end{aligned}$$

The rest is similar to the counterpart of Theorem 4.1 but bearing in mind that the system considered is not square and the Jacobian $\partial_B H(x^*)$ is of full column rank. ∎

Remark. We should point out that the results in the above theorem are weaker than that in Section 4. For example, it cannot be deduced that a stationary point x^* of θ is a solution of (1.1) even if each element of $\partial_B H(x^*)$ has full column rank. This is a big difference between square systems and overdetermined systems when the damped modified Gauss-Newton method is applied.

7 EXTENSION TO SYSTEMS WITH POLYHEDRAL CONSTRAINTS

Sometimes additional constraints are required to be met besides the system (1.1). In this case, none of the previous methods applies. In this section, we show that some suitable extensions of Gauss-Newton methods can find solutions for new systems.

Consider the square system of (1.1) with additional polyhedral constraints defined by

$$\begin{aligned} H(x) &= 0 \\ Ax + a &\leq 0, \end{aligned} \tag{7.1}$$

where $A \in \Re^{l \times n}$ and $a \in \Re^l$. (7.1) can be viewed as a kind of overdetermined system containing both equalities and inequalities. Our strategy to solve (7.1) is to solve the following linearly constrained least square problem:

$$\begin{aligned} \min \quad & \theta(x) = \tfrac{1}{2}\|H(x)\|^2 \\ \text{s.t.} \quad & Ax + a \leq 0, \end{aligned} \tag{7.2}$$

Obviously, x is a global solution of the minimization (7.2) with $\theta(x) = 0$ if x is a solution of (7.1); and conversely, x is a solution of (7.1) if x is a solution of the minimization (7.2) with $\theta(x) = 0$. It is known that many optimization methods can only locate at stationary points of (7.2). x is said to be a stationary point of (7.2) if there exists a vector $\xi \in \Re^l$ such that the following so-called Karush-Kuhn-Tucker or KKT system (called a KT system in [41]) holds:

$$\begin{aligned} &\nabla\theta(x) + A^T\xi = 0 \\ &Ax + a \le 0,\ \xi \ge 0,\ (Ax + a)^T\xi = 0. \end{aligned}$$

Nevertheless, a stationary point (7.2) is sometimes a solution of (7.1) under certain conditions for some special forms of H; see Example 7 of Section 8 and its associated reference [8].

Our damped modified Gauss-Newton method for the solution of (7.1) is stated as follows.

Algorithm 6: `Damped Modified Gauss-Newton for Constrained Systems`

Step 1 (Initialization) Choose a feasible starting point $x^0 \in \Re^n$ so that $Ax^0 + a \le 0$, choose the scalars $\sigma, \rho \in (0,1)$, $\nu_0 > 0$, and let $k := 0$.

Step 2 (Search direction) Choose $V_k \in \partial_B H(x^k)$ and solve the following quadratic program

$$\begin{array}{ll} \min & \frac{1}{2}\|H(x^k) + V_k d\|^2 + \frac{1}{2}\nu_k d^T d \\ \text{s.t.} & A(x^k + d^k) + a \le 0. \end{array} \tag{7.3}$$

Let d^k be the unique solution of the above QP, and ξ^k be one of its multipliers. If $d^k = 0$, the algorithm terminates. Otherwise, go to Step 3.

Step 3 (Line search) Let $\lambda_k = \rho^{i_k}$ where i_k is the smallest nonnegative integer i such that

$$\theta(x^k + (\rho)^i d^k) - \theta(x^k) \le \sigma(\rho)^i \nabla\theta(x^k)^T d^k.$$

Step 4 (Update) Choose $\nu_{k+1} > 0$. Let $x^{k+1} := x^k + \lambda_k d^k$ and $k := k + 1$. Go to Step 1.

If there is no constraints set $Ax + a \le 0$, the above algorithm reduces to Algorithm 3 by noting that, in this case, solving the QP program (7.3) is equivalent to solving the modified Gauss-Newton equation when $\nu_k > 0$. This indicates that the results presented in Section 4 can be treated as corollaries of the main result in this section. The reason we consider Algorithm 3 in Section 4 separately is two-fold. Firstly, unconstrained systems of nonsmooth equations are so important that they deserve to be considered separately. Secondly, the proof techniques in this section use optimization methodology while the proof techniques in Section 4 are borrowed from systems of equations.

The introduction of the regularized parameter $\nu_k > 0$ ensures the existence of a unique solution to the QP (7.3). If $d^k = 0$, x^k must be a KKT point of

(7.2). Therefore, we assume that Algorithm 6 does not terminate in finitely many steps.

Theorem 7.1 *Suppose the assumptions (A1) and (A2) hold. Suppose the linear system $Ax + a \leq 0$ has a nonempty solution set. Assume that x^* is an accumulation point of $\{x^k\}$ generated by Algorithm 6. Then the following conclusions hold:*

(i) x^* is a KKT point of the minimization problem (7.2) if both $\{\nu_k\}$ and $\{d^k\}$ are bounded.

(ii) x^* is a KKT point of (7.2) if $\bar{\nu} > \nu_k > \underline{\nu}$ for some $\bar{\nu} > \underline{\nu} > 0$.

(iii) Let $\nu_k = \theta(x^k)$. Then x^* is a KKT point of (7.2). If x^* is a solution of (7.1), i.e., $\theta(x^*) = 0$, H is strongly BD-regular at x^* and $\sigma \in (0, \frac{1}{2})$, then $\{x^k\}$ converges to x^* Q-superlinearly. The convergence rate is Q-quadratic if H is strongly semismooth at x^*.

Proof. Since $Ax + a \leq 0$ is consistent, Algorithm 6 can choose a feasible starting point. The positiveness of $\nu_k > 0$ implies that the search direction is well-defined for each k. The fact that d^k is a descent direction for θ at x^k is standard in the context of nonlinear programming. We give a brief explanation: Note that if $d^k \neq 0$, then it is a descent direction of the objective function of the convex quadratic program (7.3); as a consequence, d^k is also a descent direction for θ at x^k, i.e., $\nabla\theta(x^k)^T d^k < 0$ (cf. the proof of Theorem 3.1(i)).

Using the KKT condition and the previous standard analysis presented in the proof of Theorem 3.1(i), the line search step and hence Algorithm 6 are well-defined.

(i) Suppose x^* is an accumulation point of $\{x^k\}$, i.e., there exists a subset K such that $\lim_{k\to\infty, k\in K} x^k = x^*$. By the boundedness of $\{\nu_k\}$ and $\{d^k\}$, we may assume, by passing to the subsequence, that $\lim_{k\to\infty, k\in K} V_k = V^*$, $\lim_{k\to\infty, k\in K} \nu_k = \nu^*$ and $\lim_{k\to\infty, k\in K} d^k = d^*$.

Let $\mathcal{F} = \{x | Ax + a \leq 0\}$. Obviously, $x^*, x^* + d^* \in \mathcal{F}$. Note that d^k is the (unique) solution of the QP problem (7.3). Therefore, the optimality condition of this convex QP implies that the gradient of the objective function in QP at $d = d^k$ is normal to the feasible set (see Theorem 27.4 of [41]), i.e.,

$$(\nabla\theta(x^k) + (V_k^T V_k + \nu_k I)d^k)^T (d - d^k) \geq 0,$$

for any d such that $A(x^k + d) + a \leq 0$. In other words,

$$(\nabla\theta(x^k) + (V_k^T V_k + \nu_k I)d^k)^T (x - (x^k + d^k)) \geq 0,$$

for any $x \in \mathcal{F}$. Since $\mathcal{F}$ is polyhedral, by passing to the limit, we obtain that

$$(\nabla\theta(x^*) + ((V^*)^T V^* + \nu^* I)d^*)^T (x - (x^* + d^*)) \geq 0,$$

for any $x \in \mathcal{F}$. Then Theorem 27.4 of [41] implies that $x^* + d^* \in \mathcal{F}$ is an optimal solution of the following linear program:

$$\begin{array}{ll} \min & (\nabla\theta(x^*) + ((V^*)^T V^* + \nu^* I)d^*)^T x \\ \text{s.t.} & Ax + a \leq 0. \end{array}$$

By the optimality condition of the above linear program, there exists a Lagrange multiplier ξ^* such that

$$\begin{array}{l}\nabla\theta(x^*) + ((V^*)^T V^* + \nu^* I)d^* + A^T\xi^* = 0\\ A(x^* + d^*) + a \leq 0,\ \xi^* \geq 0,\ (A(x^* + d^*) + a)^T\xi^* = 0.\end{array} \tag{7.4}$$

Following standard analysis, it is easy to deduce from the line search rule that $\nabla\theta(x^*)^T d^* = 0$. Multiplying the first equation of (7.4) by $(d^*)^T$, it follows from (7.4) and the fact that $x^* \in \mathcal{F}$ that

$$(d^*)^T((V^*)^T V^* + \nu^* I)d^* = 0,$$

and

$$(d^*)^T A^T \xi^* = 0.$$

It is easy to show that $((V^*)^T V^* + \nu^* I)d^* = 0$, which implies from the first equation of (7.4) that

$$\nabla\theta(x^*) + A^T\xi^* = 0. \tag{7.5}$$

In view that $\{x^*\} \in \mathcal{F}$, the second equation of (7.4) yields that

$$Ax^* + a \leq 0,\ \xi^* \geq 0, (Ax^* + a)^T\xi^* = 0,$$

where the last equality is implied by $(d^*)^T A^T\xi^* = 0$. Combining with (7.5), it becomes clear that x^* is a KKT point of (7.2).

(ii) Algorithm 6 can be viewed as an SQP method for solving the linearly constrained minimization (7.2). Then (ii) may follow from some standard convergence results of SQP methods for linearly constrained nonlinear programming problems. Instead, we prove the desired results from (i) by verifying that $\{d^k\}_{k\in K}$ is a bounded sequence, where K is the index set such that $\lim_{k\to\infty, k\in K} x^k = x^*$. By the KKT condition of the QP (7.3), there exists a KKT multiplier ξ^k for each k such that

$$\begin{array}{l}\nabla\theta(x^k) + (V_k^T V_k + \nu_k I)d^k + A^T\xi^k = 0\\ A(x^k + d^k) + a \leq 0,\ \xi^k \geq 0,\ (A(x^k + d^k) + a)^T\xi^k = 0.\end{array}$$

Multiplying the first equation by d^k, we obtain that

$$\nabla\theta(x^k)^T d^k + (d^k)^T(V_k^T V_k + \nu_k I)d^k + (\xi^k)^T A d^k = 0,$$

which, together with the complementarity condition, implies that

$$\nabla\theta(x^k)^T d^k + (d^k)^T(V_k^T V_k + \nu_k I)d^k = (Ax^k + a)^T\xi^k \leq 0,$$

where the last inequality follows from the feasibility of x^k and nonnegativeness of ξ^k. Then from the continuity of θ and the assumption $\bar{\nu} > \nu_k > \underline{\nu} > 0$, it is easy to prove that $\{d^k\}_{k\in K}$ is bounded.

(iii) If x^* is a not a KKT point of (7.2), then $\theta(x^*) > 0$. Note that $\theta(x^k)$ is a monotonically decreasing sequence. We obtain that the condition on ν_k in (ii) is satisfied. It follows that x^* is a KKT point, a contradiction.

Note that

$$(x^* - x^k - d^k)^T A^T \xi^k = (Ax^* + a)^T \xi^k \leq 0.$$

Recalling the optimality condition for (7.3) at the k-th iteration, we obtain

$$(x^k + d^k - x^*)^T(\nabla\theta(x^k) + (V_k^T V_k + \nu_k I)d^k) \leq 0,$$

which implies that

$$\begin{aligned}
&(x^k + d^k - x^*)^T(V_k^T V_k + \nu_k I)(x^k + d^k - x^*) \\
&\quad \leq -(x^k + d^k - x^*)^T(\nabla\theta(x^k) + (V_k^T V_k + \nu_k I)(x^* - x^k)) \\
&\quad = -(x^k + d^k - x^*)^T[V_k^T(H(x^k) - V_k(x^k - x^*)) + \nu_k(\|x^k - x^*\|)] \\
&\quad = -(x^k + d^k - x^*)^T[V_k^T(H(x^*) + o(\|x^k - x^*\|)) + \nu_k(\|x^k - x^*\|)] \\
&\quad = -(x^k + d^k - x^*)^T[V_k^T H(x^*) + o(\|x^k - x^*\|)],
\end{aligned} \tag{7.6}$$

where the last equality follows from the uniform boundedness of V_k and $\nu_k = o(1)$. Since $H(x^*) = 0$ is assumed, the above inequality shows that

$$(x^k + d^k - x^*)^T(V_k^T V_k + \nu_k I)(x^k + d^k - x^*) \leq \|x^k + d^k - x^*\| o(\|x^k - x^*\|).$$

Then by strong BD-regularity of H at x^* and the fact that

$$\lim_{k\to\infty, k\in K} \nu_k = \lim_{k\to\infty, k\in K} \theta(x^k) = 0,$$

we immediately deduce that for all sufficiently large $k \in K$

$$\|x^k + d^k - x^*\| \leq o(\|x^k - x^*\|),$$

and

$$\|d^k\| = \|x^k - x^*\| + o(\|x^k - x^*\|).$$

The rest can be proved similarly to that of Theorem 3.1 by noting that the line search does not affect the feasibility of the iteration sequence. ∎

Remark. One may propose trust region methods for the solution of (7.1). This should be a straightforward extension of Section 5. Furthermore, similar line search and trust region methods can be developed for overdetermined systems with polyhedral constraints. We leave these for the interested reader to explore.

8 APPLICATIONS TO NCP AND RELATED PROBLEMS

In this section, we present several examples arising from nonlinear complementarity, variational inequality and related problems [14, 31]. These examples will demonstrate that our convergence theory developed in the previous sections are significant.

Let $F : \Re^n \to \Re^n$, $g : \Re^n \to \Re^p$. A variational inequality problem (VI) is to find a vector $x \in C$ such that

$$F(x)^T(y - x) \geq 0, \quad \forall y \in C,$$

where $C = \{z : g(z) \geq 0\}$. A special case of the VI is so-called box-constrained VI (Box-VI for short) in which the constraints set $C = [l, u]$, where $l, u \in \Re^n \cup \{-\infty\} \cup \{+\infty\}$ and $l < u$. When $l_i = 0$ and $u_i = +\infty$ for $i = 1, \ldots, n$, the Box-VI reduces to the famous nonlinear complementarity problem (NCP) which of course in turn includes the linear complementarity problem as special problems. More precisely, the Box-VI is to find a vector $x \in [l, u]$ such that

$$\begin{array}{lll} x_i = l_i & \Longrightarrow & F_i(x) \geq 0 \\ l_i < x_i < u_i & \Longrightarrow & F_i(x) = 0 \\ x_i = u_i & \Longrightarrow & F_i(x) \leq 0, \end{array}$$

and the NCP is to find a vector $x \in \Re^n$ so that

$$0 \leq x \perp F(x) \geq 0,$$

where $\perp$ denotes the orthogonal operator.

Example 1. Let $\phi : \Re^2 \to \Re$ be defined by

$$\phi(b, c) = \sqrt{b^2 + c^2} - (b + c).$$

This function is now known as the Fischer-Burmeister functional [9]. Solving the NCP is equivalent to solving the following system of semismooth equations:

$$H(x) = \begin{pmatrix} \phi(x_1, F_1(x)) \\ \vdots \\ \phi(x_n, F_n(x)) \end{pmatrix} = 0.$$

H is semismooth on $\Re^n$ if F is continuously differentiable on $\Re^n$, and H is strongly semismooth at x if ∇F is Lipschitz at x. The least square objective associated with H is smooth on $\Re^n$ if F is smooth on $\Re^n$. Therefore, the assumptions (A1) and (A2) are satisfied. In fact, this reformulation has been used extensively in the past few years after Fischer [9] used it for nonlinear programming problems. See [5, 10, 15, 18, 43] for some references.

Let us introduce the variable $y \in \Re^n$ and define

$$G(x, y) = \begin{pmatrix} F(x) - y \\ \phi(x_1, y_1) \\ \vdots \\ \phi(x_n, y_n) \end{pmatrix} = 0.$$

Obviously, x is a solution of the NCP if and only if (x, y) is a solution of the equation $G(x, y) = 0$. Furthermore, the assumptions (A1) and (A2) hold for this latter system. One possible disadvantage of the latter is that the dimension is doubled over the former, and one possible advantage of the latter is a complete characterization of the generalized Jacobian of G at any (x, y), which is not available for the former.

Example 2. A more general class of functions than the Fischer-Burmeister functional has been introduced by Kanzow and Kleinmichel [22]. Let $\lambda \in (0,4)$ be a parameter. $\phi_\lambda : \Re^2 \to \Re$ is defined by

$$\phi_\lambda(b,c) = \sqrt{(b-c)^2 + \lambda bc} - (b+c).$$

When $\lambda = 2$, ϕ_λ reduces to the Fischer-Burmeister functional. Kanzow and Kleinmichel [22] have proved that the solution set of the NCP coincides with the solution set of the following system of nonsmooth equations:

$$H(x) = \begin{pmatrix} \phi_\lambda(x_1, F_1(x)) \\ \vdots \\ \phi_\lambda(x_n, F_n(x)) \end{pmatrix} = 0.$$

Moreover, H is semismooth on $\Re^n$ for any fixed $\lambda \in (0,4)$ if F is smooth on $\Re^n$, and the least square of H is smooth on $\Re^n$ for any fixed $\lambda \in (0,4)$. Hence, the assumptions (A1) and (A2) are satisfied for this nonsmooth system.

Example 3. Let $\psi : \Re^3 \to \Re$ be a function defined by

$$\psi(a,b,c) = \sqrt{a^2 + b^2 + c^2} - (b+c).$$

Note that this function is introduced by Kanzow [20] to propose continuation methods for solving linear complementarity problems. It is proved in Jiang [16] that x solves the NCP if and only x is a solution of the following system of nonsmooth equations:

$$H(\mu, x) = \begin{pmatrix} e^\mu - 1 \\ \psi(\mu, x_1, F_1(x)) \\ \vdots \\ \psi(\mu, x_n, F_n(x)) \end{pmatrix} = 0,$$

where $\mu \in \Re$ and e is the Euler constant. The assumptions (A1) and (A2) are satisfied when F is smooth. It should be pointed out that the function $e^\mu - 1$ can be replaced by other suitable functions such as $\mu + \mu^2$; see [16].

Example 4. Combining Examples 2 and 3, define $\psi_\lambda : \Re^3 \to \Re$ by

$$\psi_\lambda(a,b,c) = \sqrt{a^2 + (b-c)^2 + \lambda bc} - (b+c),$$

where $\lambda \in (0,4)$ is a parameter. Similarly, solving the NCP is equivalent to solving the following system of nonsmooth equation

$$H(\mu, x) = \begin{pmatrix} e^\mu - 1 \\ \psi_\lambda(\mu, x_1, F_1(x)) \\ \vdots \\ \psi_\lambda(\mu, x_n, F_n(x)) \end{pmatrix} = 0,$$

for any fixed $\lambda \in (0,4)$. One may prove that the assumptions (A1) and (A2) hold if F is smooth on $\Re^n$.

Example 5. Sun and Womersley [42] proposed the following function $\theta : \Re^2 \to \Re$:

$$\eta(b,c) = \min\{\max\{-\phi(b,c),0\},b\},$$

where ϕ is the Fischer-Burmeister-functional defined in Example 1. It is proved [42] that x is a solution of the Box-VI if and only if x is a solution of the following overdetermined system of nonsmooth equations:

$$H(x) = \begin{pmatrix} \eta(x_1 - l_1, F_1(x)) \\ \vdots \\ \eta(x_n - l_n, F_n(x)) \\ -\eta(u_1 - x_1, -F_1(x)) \\ \vdots \\ -\eta(u_n - x_n, -F_n(x)) \end{pmatrix} = 0.$$

Furthermore, H is semismooth and its norm squared is smooth on $\Re^n$ if F is smooth on $\Re^n$. This implies that the assumptions (A1) and (A2) hold.

Example 6. From Example 1, it is easy to see that solving the NCP is equivalent to finding a solution for the following simply constrained system

$$\begin{aligned} &H(x) = 0 \\ &x \geq 0, \end{aligned}$$

where H is defined in Example 1. The assumptions (A1) and (A2) are apparently satisfied. The introduction of the constraints $\{x : x \geq 0\}$ has two motivations: (i) F is sometimes not well-defined outside of $\{x : x \geq 0\}$; (ii) Enforcing $x \geq 0$ restricts possible limit points of the sequence produced by a Newton-type method to be nonnegative and stationary, as are solutions points of the NCP. See Kanzow [21]. Analogously, other constrained nonsmooth systems can be developed from Examples 2, 3 and 4.

Example 7. Under some suitable constraint qualification, the KKT condition for the VI at a solution x can be written as follows:

$$\begin{aligned} &F(x) - (g'(x))^T\xi = 0 \\ &g(x) \geq 0,\ \xi \geq 0,\ g(x)^T\xi = 0 \end{aligned}$$

where $\xi \in \Re^p$ is the Lagrange or KKT multiplier. By the property of the Fischer-Burmeister functional, solving the above KKT system is equivalent to solving any of the following two nonsmooth systems:

$$H(x,\xi) = 0,$$

or

$$H(x, \xi) = 0,$$
$$\xi \geq 0.$$

Here $H : \Re^{n+p} \to \Re^{n+p}$ is the following function:

$$H(x, \xi) = \begin{pmatrix} F(x) - (g'(x))^T \xi \\ \phi(\xi_1, g_1(x)) \\ \vdots \\ \phi(\xi_p, g_p(x)) \end{pmatrix}.$$

By the semismoothness of ϕ and smoothness of ϕ^2, it follows that the assumptions (A1) and (A2) hold if F is continuously differentiable on $\Re^n$ and g is twice continuously differentiable on $\Re^p$. See for example [8] for some details. When $F(x) = \nabla f(x)$ for a function $f : \Re^n \to \Re$, then the above KKT system is the KKT condition of the following nonlinear programming problem:

$$\begin{array}{ll} \min & f(x) \\ \text{s.t.} & g(x) \geq 0. \end{array}$$

Qi and Jiang [36] have used the above semismooth equation to propose quasi-Newton methods.

One may write many more equivalent systems of nonsmooth equations for the KKT condition for the VI using techniques from the previous examples. We omit the details here.

Example 8. For simplicity of notation, we assume there exist integers n_1, n_2 and n_3 satisfying $0 \leq n_1 \leq n_2 \leq n_3 \leq n$ such that the Box-VI has the following partition:

$$\begin{array}{l} l_i = -\infty,\ u_i = +\infty,\ 1 \leq i \leq n_1 \\ l_i > -\infty,\ u_i = +\infty,\ n_1 + 1 \leq i \leq n_2 \\ l_i = -\infty,\ u_i < +\infty,\ n_2 + 1 \leq i \leq n_3 \\ l_i > -\infty,\ u_i < +\infty,\ n_3 + 1 \leq i \leq n. \end{array}$$

Let

$$H(x) = \begin{pmatrix} H_{\text{I}} \\ H_{\text{II}} \\ H_{\text{III}} \\ H_{\text{IV}} \end{pmatrix},$$

where

$$H_{\text{I}} = \begin{pmatrix} F_1(x) \\ \vdots \\ F_{n_1}(x) \end{pmatrix}$$

$$H_{\text{II}} = \begin{pmatrix} \phi(x_{n_1+1} - l_{n_1+1}, F_{n_1+1}(x) \\ \vdots \\ \phi(x_{n_2} - l_{n_2}, F_{n_2}(x) \end{pmatrix}$$

$$H_{\mathrm{III}} = \begin{pmatrix} \phi(u_{n_2+1} - x_{n_2+1}, -F_{n_2+1}(x)) \\ \vdots \\ \phi(u_{n_3} - x_{n_3}, -F_{n_3}(x)) \end{pmatrix}$$

$$H_{\mathrm{IV}} = \begin{pmatrix} \phi(x_{n_3+1} - l_{n_3+1}, (u_{n_3+1} - x_{n_3+1})F_{n_3+1}(x)) \\ \phi(u_{n_3+1} - x_{n_3+1}, -(x_{n_3+1} - l_{n_3+1})F_{n_3+1}(x)) \\ \vdots \\ \phi(x_n - l_n, (u_n - x_n)F_n(x)) \\ \phi(u_n - x_n, -(x_n - l_n)F_n(x)). \end{pmatrix}$$

It can be verified that x is a solution of the Box-VI if and only if $H(x) = 0$, which is an overdetermined system if $n - n_3 > 0$ and a square system if $n = n_3$. By the property of the Fischer-Burmeister functional, the assumptions (A1) and (A2) are also satisfied. But further study is needed before deciding whether this reformulation is useful or not for the solution of the Box-VI.

Example 9. Qi [34] defined a function $\kappa : \Re^2 \to \Re$ as follows:

$$\kappa(a,b) = \begin{cases} a - \frac{a^2}{3b} & \text{if } |a| \le |b| \text{ and } b > 0 \\ b - \frac{b^2}{3a} & \text{if } |b| \le |a| \text{ and } a > 0 \\ 0 & \text{if } a = b = 0 \\ 2a + b + \frac{b^2}{3a} & \text{if } |b| \le |a| \text{ and } a < 0 \\ a + 2b + \frac{a^2}{3b} & \text{if } |a| \le |b| \text{ and } b < 0. \end{cases}$$

It is proved in [34] that x is a solution of the NCP if and only if x solves the following system of nonsmooth equations:

$$H(x) = \begin{pmatrix} \kappa(x_1, F_1(x)) \\ \vdots \\ \kappa(x_n, F_n(x)) \end{pmatrix} = 0.$$

Furthermore, the assumptions (A1) and (A2) are satisfied when F is smooth. Note that the function κ is rational unlike the Fischer-Burmeister functional. Qi [34] proposed and reviewed several more nonsmooth equation reformulations of the NCP and Box-VI satisfying the assumptions (A1) and (A2).

Example 10. Let $H : \Re^n \to \Re^{n+1}$ be defined by

$$H(x) = \sqrt{2} \begin{pmatrix} \sqrt{\psi_0(x^T F(x))} \\ \sqrt{\psi_1(-x_1, -F_1(x))} \\ \vdots \\ \sqrt{\psi_n(-x_n, -F_n(x))} \end{pmatrix},$$

where $\psi_0 : \Re \to [0,\infty)$, $\psi_1, \ldots, \psi_n : \Re^2 \to [0,\infty)$ are continuous functions that are zero on the nonpositive orthant only. The least square of H is

$$\theta(x) = \psi_0(x^T F(x)) + \sum_{i=1}^{n} \psi_i(-x_i, -F_i(x)).$$

This merit function was introduced by Luo and Tseng [26], who proved that it is smooth under certain conditions. One may prove that solving the NCP is equivalent to solving the nonsquare system $H(x) = 0$. See [26, 23] for more details. If furthermore, $\sqrt{\psi_0}$ and $\sqrt{\psi_1}$, ..., $\sqrt{\psi_n}$ are assumed to be semismooth, then the assumptions (A1) and (A2) are satisfied.

Example 11. Kanzow, Yamashita and Fukushima [23] investigate the nonsmooth equation $H : \Re^n \to \Re^{2n}$:

$$H(x) = \sqrt{2} \begin{pmatrix} \sqrt{\psi_0(x_1 F_1(x))} \\ \vdots \\ \sqrt{\psi_0(x_n F_n(x))} \\ \\ \sqrt{\psi_1(-x_1, -F_1(x))} \\ \vdots \\ \sqrt{\psi_n(-x_n, -F_n(x))} \end{pmatrix},$$

where $\psi_0 : \Re \to [0,\infty)$, and $\psi_1, \ldots, \psi_n : \Re^2 \to [0,\infty)$ are suitably defined continuous functions (we use the notation in Example 10, which differs from that in [23]). It is easy to see that the function H is somewhat different from that defined in Example 10. Under further conditions, it is shown in [23] that the least square of H is smooth, and x is a solution of the NCP if and only if x is a solution of the nonsquare system $H(x) = 0$. If furthermore, $\sqrt{\psi_0}$ and $\sqrt{\psi_1}$, ..., $\sqrt{\psi_n}$ are assumed to be semismooth, then the assumptions (A1) and (A2) are satisfied.

Example 12. Define the function $H : \Re^n \to \Re^{n+1}$ by

$$H(x) = \begin{pmatrix} \psi_0(xF(x)) \\ \phi(x_1, F_1(x)) \\ \vdots \\ \phi(x_n, F_n(x)) \end{pmatrix},$$

where $\psi_0 : \Re \to \Re$ is defined by $\psi_0(t) = \frac{1}{\sqrt{2}} \max\{0, t\}^2$, and ϕ is the Fischer-Burmeister functional. Yamashita and Fukushima [44] proved that the least square of H is smooth, and x is a solution of the NCP if and only if x is a solution of the nonsquare system $H(x) = 0$. Clearly, the assumptions (A1) and (A2) are satisfied.

Note that the merit function $\frac{1}{2}\|H(x)\|^2$ is designed for solving semidefinite complementarity problems [44]. It would be interesting to see how the Newton-type methods proposed in the previous sections are able to be extended for solving semidefinite complementarity problems. In particular, the question of the whether the functions ψ_i $(i = 0, \ldots, n)$ are semismooth when acting on symmetric matrices appears to need investigation.

Example 13. Consider the function $\phi_\lambda : \Re^2 \to \Re$ defined by

$$\phi_\lambda(b, c) = \lambda\phi(b, c) + (1 - \lambda)\max\{0, b\}\max\{0, c\},$$

where $\lambda \in (0, 1)$ is a parameter and ϕ is the Fischer-Burmeister functional. Chen, Chen and Kanzow [1] have proved that the solution set of the NCP coincides with the solution set of the following system of nonsmooth equations:

$$H_\lambda(x) = \begin{pmatrix} \phi_\lambda(x_1, F_1(x)) \\ \vdots \\ \phi_\lambda(x_n, F_n(x)) \end{pmatrix} = 0,$$

for any fixed $\lambda \in (0, 1)$. Moreover, H_λ is semismooth on $\Re^n$ for any fixed $\lambda \in (0, 1)$ if F is smooth on $\Re^n$, and the least square of H_λ is smooth on $\Re^n$ for any fixed $\lambda \in (0, 1)$. Hence, the assumptions (A1) and (A2) are satisfied.

9 FINAL REMARKS

We have developed global and local fast convergence theory of various generalized Newton-type methods for systems of semismooth equations with smooth least squares. These are extensions of the classical theory of Newton-type methods for smooth systems. We mention that our trust region methods are not direct extensions of the classical trust region methods. Therefore, it would be interesting to see whether the same theory of the classical trust region methods could be developed for our nonsmooth systems.

The theory presented is useful from a pure mathematical point of view. But more significantly, it has many applications from optimization as demonstrated in Section 8. Hopefully, our efforts will stimulate investigations and proposals of more nonsmooth systems satisfying the assumptions (A1) and (A2), and also facilitate convergence analysis when generalized Newton-type methods are considered.

As has been observed from computation, nonmonotone line search or nonmonotone trust region strategies can accelerate computation speed, but also enhance the possibility of avoiding some local solutions. We conjecture that all theory presented are valid if nonmonotone strategies are incorporated.

Acknowledgement

The authors are thankful to Liqun Qi for his helpful comments.

References

[1] B. Chen, X. Chen and C. Kanzow, A penalized Fischer-Burmeister NCP-function: Theoretical investigation and numerical results, Preprint 126, Institute of Applied Mathematics, University of Hamburg, Hamburg, Germany, 1997.

[2] P.W. Christensen, A. Klarbring, J.S. Pang and N. Strömberg, Formulation and comparison of algorithms for frictional contact problems, manuscript, Department of Mechanical Engineering, Linköping University, S-581 83, Linköping (November 1996).

[3] P.W. Christensen and J.S. Pang, Frictional contact algorithms based on semismooth Newton methods, *Reformulation - Nonsmooth, Piecewise Smooth, Semismooth and Smoothing Methods*, M. Fukushima and L. Qi, eds., (Kluwer Academic Publisher, Nowell, MA. USA, 1998), 81-116.

[4] F.H. Clarke, *Optimization and Nonsmooth Analysis*, Wiley, New York, 1983.

[5] T. De Luca, F. Facchinei and C. Kanzow, A semismooth equation approach to the solution of nonlinear complementarity problems, *Mathematical Programming* 75 (1996) 407-439.

[6] J.E. Dennis and R.B. Schnabel, *Numerical Methods for Unconstrained Optimization and Nonlinear Equation*, Prentice Hall, Englewood Cliffs, New Jersey, 1983.

[7] S.P. Dirkse and M.C. Ferris, The PATH Solver: A non-monotone stabilization scheme for mixed complementarity problems, *Optimization Methods and Software* 5 (1995) 123-156.

[8] F. Facchinei, A. Fischer, C. Kanzow and J. Peng, A simply constrained optimization reformulation of KKT systems arising from variational inequalities, *Applied Mathematics and Optimization*, to appear.

[9] A. Fischer, A special Newton-type optimization method, *Optimization* 24 (1992) 269-284.

[10] A. Fischer, A Newton-type method for positive semidefinite linear complementarity problems, *Journal of Optimization Theory and Applications* 86 (1995) 585-608.

[11] A. Fischer, Solution of monotone complementarity problems with locally Lipschitzian functions, *Mathematical Programming* 76 (1997) 513-532.

[12] R. Fletcher, *Practical Methods of Optimization* John Wiley, 2nd Edition, 1987.

[13] S.P. Han, J.-S. Pang and N. Rangaraj, Globally convergent Newton methods for nonsmooth equations, *Mathematics of Operations Research* 17 (1992) 586-607.

[14] P.T. Harker and J.-S. Pang, Finite-dimensional variational inequality and nonlinear complementarity problem: A survey of theory, algorithms and applications, *Mathematical Programming* 48 (1990) 161-220.

[15] H. Jiang, Global convergence analysis of the generalized Newton and Gauss-Newton methods for the Fischer-Burmeister equation for the complementarity problem, *Mathematics of Operations Research*, to appear.

[16] H. Jiang, Smoothed Fischer-Burmeister equation methods for the nonlinear complementarity problem, Manuscript, Department of Mathematics, The University of Melbourne, June 1997.

[17] H. Jiang, M. Fukushima, L. Qi and D. Sun, A trust region method for solving generalized complementarity problems, *SIAM Journal on Optimization* 8 (1998) 140-157.

[18] H. Jiang and L. Qi, A new nonsmooth equations approach to nonlinear complementarity problems, *SIAM Journal on Control and Optimization* 35 (1997) 178-193.

[19] H. Jiang, L. Qi, X. Chen and D. Sun, Semismoothness and superlinear convergence in nonsmooth optimization and nonsmooth equations, in: G. Di Pillo and F. Giannessi editors, *Nonlinear Optimization and Applications*, Plenum Publishing Corporation, New York, 1996, pp. 197-212.

[20] C. Kanzow, Some noninterior continuation methods for linear complementarity problems, *SIAM Journal on Matrix Analysis and Applications* 17 (1996) 851-868.

[21] C. Kanzow, An inexact QP-based method for nonlinear complementarity problems, Preprint 120, Institute of Applied Mathematics, University of Hamburg, Hamburg, Germany, February, 1997.

[22] C. Kanzow and H. Kleinmichel, A new class of semismooth Newton-type methods for nonlinear complementarity problems, *Computational Optimization and Applications*, to appear.

[23] C. Kanzow, N. Yamashita and M. Fukushima, New NCP-functions and their properties, *Journal of Optimization Theory and Applications* 94 (1997) 115-135.

[24] B. Kummer, Newton's method for nondifferentiable functions, in: Guddat, J. et al., eds., *Mathematical Research: Advances in Mathematical Optimization*, Akademie' Verlag, Berlin, 1988, 114-125.

[25] B. Kummer, Newton's Method Based on Generalized Derivatives for Nonsmooth Functions: Convergence Analysis, In: W. Oettli and D. Pallaschke, eds., *Advances in Optimization*, Springer-Verlag, Berlin, 1992, pp. 171-194.

[26] Z.-Q. Luo and P. Tseng, A new class of merit functions for the nonlinear complementarity problem, in: M.C. Ferris and J.-S. Pang, eds., *Complementarity and Variational Problems: State of the Art*, SIAM Publications, 1996.

[27] R. Mifflin, Semismooth and semiconvex functions in constrained optimization, *SIAM Journal on Control and Optimization* 15 (1977) 959-972.

[28] J.M. Ortega and W.C. Rheinboldt, *Iterative Solution of Nonlinear Equations in Several Variables*, Academic Press, New York, 1970.

[29] J.-S. Pang, Newton's methods for B-differentiable equations, *Mathematics of Operations Research* 15 (1990) 311-341.

[30] J.-S. Pang, A B-differentiable equation based, globally and locally quadratically convergent algorithm for nonlinear programs, complementarity, and variational inequality problems, *Mathematical Programming* 51 (1991) 101-131.

[31] J.-S. Pang, Complementarity problems, in: R. Horst and P. Pardalos, eds., *Handbook of Global Optimization*, Kluwer Academic Publishers, Boston, 1994, pp. 271-338.

[32] J.-S. Pang and L. Qi, Nonsmooth equations: Motivation and algorithms, *SIAM Journal on Optimization* 3 (1993) 443-465.

[33] L. Qi, Convergence analysis of some algorithms for solving nonsmooth equations, *Mathematics of Operations Research* 18 (1993) 227-244.

[34] L. Qi, Regular almost smooth NCP and BVIP functions and globally and quadratical convergent generalized Newton methods for complementarity and variational inequality problems, Technical Report AMR 97/14, University of New South Wales, June 1997.

[35] L. Qi and X. Chen, A globally convergent successive approximation methods for nonsmooth equations, *SIAM Journal on Control and Optimization* 33 (1995) 402-418.

[36] L. Qi and H. Jiang, Semismooth Karush–Kuhn–Tucker equations and convergence analysis of Newton methods and Quasi–Newton methods for solving these equations, *Mathematics of Operations Research* 22 (1997) 301-325.

[37] L. Qi and J. Sun, A nonsmooth version of Newton's method, *Mathematical Programming* 58 (1993) 353-368.

[38] D. Ralph, Global convergence of damped Newton's method for nonsmooth equations, via the path search, *Mathematics of Operations Research* 19 (1994) 352-389.

[39] S.M. Robinson, Generalized equations, in: A. Bachem, M. Grötschel and B. Korte, eds., *Mathematical Programming: The State of the Art*, Springer-Verlag, Berlin, 1983, pp. 346-367.

[40] S.M. Robinson, Newton's method for a class of nonsmooth equations, *Set-valued Analysis* 2 (1994) 291-305.

[41] R.T. Rockafellar, *Convex Analysis*, Princeton, New Jersey, 1970.

[42] D. Sun and R.S. Womersley, A new unconstrained differentiable merit function for box constrained variational inequality problems and a damped Gauss-Newton method, *SIAM Journal on Optimization*, to appear.

[43] N. Yamashita and M. Fukushima, Modified Newton methods for solving a semismooth reformulation of monotone complementarity problems, *Mathematical Programming* 76 (1997) 469-491.

[44] N. Yamashita and M. Fukushima, A new merit function and a descent method for semidefinite complementarity problems, *Reformulation – Nonsmooth, Piecewise Smooth, Semismooth and Smoothing Methods*, M. Fukushima and L. Qi, eds., (Kluwer Academic Publisher, Nowell, MA. USA, 1998), 405-420.

Reformulation: Nonsmooth, Piecewise Smooth,
Semismooth and Smoothing Methods, pp. 211–233
Edited by M. Fukushima and L. Qi

Inexact Trust-Region Methods for Nonlinear Complementarity Problems

Christian Kanzow* and Martin Zupke*

Abstract In order to solve the nonlinear complementarity problem, we first reformulate it as a nonsmooth system of equations by using a recently introduced NCP-function. We then apply a trust-region-type method to this system of equations. Our trust-region method allows an inexact solution of the trust-region-subproblem. We show that the algorithm is well-defined for a general nonlinear complementarity problem and that it has some nice global and local convergence properties. Numerical results indicate that the new method is quite promising.

Key Words nonlinear complementarity problem, trust-region method, nonsmooth Newton method, global convergence, quadratic convergence.

1 INTRODUCTION

Let $F : \mathbb{R}^n \to \mathbb{R}^n$ be continuously differentiable. We consider the problem of finding a solution $x^* \in \mathbb{R}^n$ of the following system of equations and inequalities:

$$x_i \geq 0, \;\; F_i(x) \geq 0, \;\; x_i F_i(x) = 0 \quad \forall i \in I := \{1, \ldots, n\}.$$

This is a *nonlinear complementarity problem* and we will denote it by NCP(F) throughout this paper. It has many important applications, and we refer the reader to the recent survey paper [11] by Ferris and Pang for a brief description of several economic and engineering applications.

Due to the importance of the complementarity problem, there exists a huge number of methods for solving problem NCP(F). Here we concentrate on a particular approach which is based on a reformulation of NCP(F) as a nonlinear system of equations. This reformulation is done by means of a so-called *NCP-function* $\varphi : \mathbb{R}^2 \to \mathbb{R}$ which is defined by the following characterization of its

*University of Hamburg, Institute of Applied Mathematics, Bundesstrasse 55, D-20146 Hamburg, Germany. e-mail: kanzow@math.uni-hamburg.de, zupke@math.uni-hamburg.de

zeros:

$$\varphi(a,b) = 0 \Longleftrightarrow a \geq 0, \ \ b \geq 0, \ \ ab = 0.$$

In this paper, we will use

$$\varphi_\lambda(a,b) := \sqrt{(a-b)^2 + \lambda ab} - a - b$$

as an NCP-function, where $\lambda \in (0,4)$ is a given parameter. This function was introduced by Kanzow and Kleinmichel [18] and covers the well-known Fischer-Burmeister function [12] by taking $\lambda = 2$ as well as the minimum function [26] in the limiting case $\lambda \to 0$.

Since φ_λ is an NCP-function, it is easy to see that a vector $x^* \in \mathbb{R}^n$ solves the nonlinear complementarity problem if and only if it is a solution of the nonsmooth system of equations

$$\Phi_\lambda(x) = 0, \tag{1.1}$$

where $\Phi_\lambda : \mathbb{R}^n \to \mathbb{R}^n$ is defined by

$$\Phi_\lambda(x) := \begin{pmatrix} \varphi_\lambda(x_1, F_1(x)) \\ \vdots \\ \varphi_\lambda(x_n, F_n(x)) \end{pmatrix}.$$

Alternatively, we may view

$$\min \ \ \Psi_\lambda(x), \quad x \in \mathbb{R}^n,$$

with $\Psi_\lambda : \mathbb{R}^n \to \mathbb{R}$ given by

$$\Psi_\lambda(x) := \frac{1}{2}\Phi_\lambda(x)^T \Phi_\lambda(x)$$

as an unconstrained minimization reformulation of NCP(F). We stress that, despite the nonsmoothness of the equation-operator Φ_λ, it is not difficult to see that its merit function Ψ_λ is continuously differentiable [18, Theorem 3.1].

Many algorithms have been proposed for solving problem NCP(F) by exploiting one of these reformulations, see, e.g., [10, 5, 18]. Most of these algorithms are globalized by using an Armijo-type line search. Another well-known globalization strategy is based on the trust-region idea. Due to the nonsmoothness of the operator Φ_λ and related reformulations of NCP(F), however, it is not possible to use a standard trust-region method from smooth optimization in order to solve the complementarity problem. In particular, the nonsmoothness of Φ_λ would cause some problems in proving local fast convergence.

So one has to modify these smooth trust-region methods in order to take into account the nonsmoothness of Φ_λ or similar operators. Modifications of this kind were described by Gabriel and Pang [15], Friedlander, Martínez and Santos [14] and Martínez and Santos [23]. On the other hand, our merit function Ψ_λ is smooth, and this additional property was fully exploited in the recent paper [17] by Jiang, Fukushima, Qi and Sun.

The latter paper will therefore be the basis of our current research. In fact, we borrow many ideas from [17]. In contrast to [17], however, our approach is based on the reformulation (1.1) (Jiang et al. [17] use the Fischer-Burmeister function) and, more important, we follow an idea by Shultz, Schnabel and Byrd [32, 2] and allow an inexact solution of the trust-region subproblem. By doing this, we improve the overall efficiency of our algorithm.

Another motivation for this research derives from the fact that trust-region methods are often said to be more reliable than corresponding line-search methods for smooth problems [13]. Hence there seems to be some hope to improve at least the robustness of existing line-search methods by using a trust-region-type globalization.

The paper is organized as follows: In Section 2, we mainly restate some properties of the functions Φ_λ and Ψ_λ from [18]. The inexact trust-region method as well as its global and fast local convergence are considered in Section 3. In Section 4, we give some details about our implementation of the inexact trust-region method and report extensive numerical results. We conclude this paper with some final remarks in Section 5.

Notation: All vector norms used in this paper are Euclidian norms. Matrix norms are assumed to be consistent with the Euclidian vector norm. A mapping $G : \mathbb{R}^n \to \mathbb{R}^m$ is called a C^1 function if G is continuously differentiable, and LC^1 function if G is differentiable with a locally Lipschitzian Jacobian. If G is a C^1 mapping, we denote its Jacobian at a point $x \in \mathbb{R}^n$ by $G'(x)$, whereas $\nabla G(x)$ denotes the transposed Jacobian. In particular, if G is a real-valued mapping, we view its gradient $\nabla G(x)$ as a column vector. If G is only locally Lipschitzian, its Jacobian may not exist on a set of measure zero. In this case, let us denote by D_G the set of differentiable points of G. If $x \notin D_G$, we may view either the B-subdifferential

$$\partial_B G(x) := \{H \in \mathbb{R}^{m\times n} | \exists \{x^k\} \subseteq D_G : \{x^k\} \to x, G'(x^k) \to H\}$$

or Clarke's [3] generalized Jacobian

$$\partial G(x) := \text{conv}\partial_B G(x)$$

of G at a point $x \in \mathbb{R}^n$ as a suitable substitute of the Jacobian. In the case $m = n$, we call a solution x^* of the system $G(x) = 0$ BD-regular if all elements in the B-subdifferential $\partial_B G(x^*)$ are nonsingular.

2 PRELIMINARIES

In this section, we recall some of the basic properties of the functions Φ_λ and Ψ_λ and refer to [18] for the corresponding proofs. In addition, we slightly improve the stationary point result from [18] by providing a full characterization for a stationary point of Ψ_λ to be a solution of NCP(F).

We first restate the following overestimation of the generalized Jacobian of Φ_λ at an arbitrary point $x \in \mathbb{R}^n$, see [18, Proposition 2.5].

Proposition 2.1 *For an arbitrary $x \in \mathbb{R}^n$, we have*

$$\partial\Phi_\lambda(x) \subseteq D_a(x) + D_b(x)F'(x),$$

where $D_a(x) = diag(a_1(x), \dots, a_n(x)), D_b(x) = diag(b_1(x), \dots, b_n(x)) \in \mathbb{R}^{n \times n}$ are diagonal matrices whose ith diagonal element is given by

$$a_i(x) = \frac{2(x_i - F_i(x)) + \lambda F_i(x)}{2\sqrt{(x_i - F_i(x))^2 + \lambda x_i F_i(x)}} - 1,$$

$$b_i(x) = \frac{-2(x_i - F_i(x)) + \lambda x_i}{2\sqrt{(x_i - F_i(x))^2 + \lambda x_i F_i(x)}} - 1$$

if $(x_i, F_i(x)) \neq (0,0)$, and by

$$a_i(x) = \xi_i - 1, \;\; b_i(x) = \chi_i - 1 \text{ for any } (\xi_i, \chi_i) \in \mathbb{R}^2 \text{ such that } \|(\xi_i, \chi_i)\| \leq \sqrt{c_\lambda}$$

if $(x_i, F_i(x)) = (0,0)$, where c_λ is a certain constant (depending on λ) with $c_\lambda \in (0,2)$.

We next want to give a characterization for stationary points of our merit function Ψ_λ to be a solution of the complementarity problem. To this end, we follow the approach by De Luca et al. [5] and define the three index sets

$$\begin{array}{ll} \mathcal{C}(x) := \{i \in I \,|\, x_i \geq 0, F_i(x) \geq 0, x_i F_i(x) = 0\} & (\text{"\underline{c}omplementary indices"}), \\ \mathcal{P}(x) := \{i \in I \,|\, x_i > 0, F_i(x) > 0\} & (\text{"\underline{p}ositive indices"}), \\ \mathcal{N}(x) := I \setminus (\mathcal{C}(x) \cup \mathcal{P}(x)) & (\text{"\underline{n}egative indices"}), \end{array}$$

where $x \in \mathbb{R}^n$ is any given vector. It is easy to see that the following relationships hold which also motivate the names for these index sets:

$$\begin{array}{lllll} [D_a(x)\Phi_\lambda(x)]_i > 0 & \Longleftrightarrow & [D_b(x)\Phi_\lambda(x)]_i > 0 & \Longleftrightarrow & i \in \mathcal{P}(x), \\ [D_a(x)\Phi_\lambda(x)]_i = 0 & \Longleftrightarrow & [D_b(x)\Phi_\lambda(x)]_i = 0 & \Longleftrightarrow & i \in \mathcal{C}(x), \\ [D_a(x)\Phi_\lambda(x)]_i < 0 & \Longleftrightarrow & [D_b(x)\Phi_\lambda(x)]_i < 0 & \Longleftrightarrow & i \in \mathcal{N}(x). \end{array} \tag{2.1}$$

We will sometimes write $\mathcal{C}, \mathcal{P}$ and $\mathcal{N}$ instead of $\mathcal{C}(x), \mathcal{P}(x)$ and $\mathcal{N}(x)$, respectively, if the vector x is clear from the context.

Based on these index sets, we follow De Luca et al. [5] and call a vector $x \in \mathbb{R}^n$ *regular* for NCP(F) if, for any nonzero vector $z \in \mathbb{R}^n$ with

$$z_\mathcal{C} = 0, \quad z_\mathcal{P} > 0, \quad z_\mathcal{N} < 0,$$

there exists a vector $y \in \mathbb{R}^n$ such that

$$y_\mathcal{P} \geq 0, \quad y_\mathcal{N} \leq 0, \quad y_{\mathcal{P}\cup\mathcal{N}} \neq 0$$

and

$$y^T \nabla F(x) z \geq 0.$$

The regularity of a vector plays a central role in the following characterization result.

Theorem 2.1 *A vector $x \in \mathbb{R}^n$ is a solution of NCP(F) if and only if x is a regular stationary point of Ψ_λ.*

Proof. Based on the relations (2.1), the proof is essentially the same as the one given in [5, Theorem 14].

We mention two simple consequences of this result. These consequences are based on certain classes of matrices for which we refer the reader to the excellent book [4] by Cottle, Pang and Stone.

Corollary 2.1 *Assume that $x^* \in \mathbb{R}^n$ is a stationary point of Ψ_λ.*

(a) If $F'(x^*)$ is a P_0-matrix, then x^* solves NCP(F).

(b) If $F'(x^*)$ is semimonotone and x^* is feasible for NCP(F), then x^* solves NCP(F).

Proof. Part (a) can be verified similarly to [20, Lemma 5.2]. In order to prove part (b), we first note that $\mathcal{N}(x^*) = \emptyset$ due to the feasibility of x^*. Let $z \in \mathbb{R}^n$ be any nonzero vector such that

$$z_{\mathcal{C}} = 0 \quad \text{and} \quad z_{\mathcal{P}} > 0.$$

Since $F'(x^*)$ is semimonotone by assumption, its transpose $\nabla F(x^*)$ is also semimonotone by [4, Corollary 3.9.7]. Hence there exists an index $i_0 \in \mathcal{P}$ such that

$$z_{i_0}[\nabla F(x^*)z]_{i_0} \geq 0.$$

Let $y \in \mathbb{R}^n$ be defined by $y_{i_0} := z_{i_0}$ and $y_i := 0$ for all $i \neq i_0$. Then

$$y^T \nabla F(x^*)z = \sum_{i=1}^{n} y_i[\nabla F(x^*)z]_i = z_{i_0}[\nabla F(x^*)z]_{i_0} \geq 0.$$

Hence x^* is a regular stationary point of Ψ_λ and therefore a solution of NCP(F) by Theorem 2.1.

Before restating the next result from [18], let us define the index sets

$$\begin{aligned}
\alpha &:= \{i \in I \,|\, x_i^* > 0 = F_i(x^*)\}, \\
\beta &:= \{i \in I \,|\, x_i^* = 0 = F_i(x^*)\}, \\
\gamma &:= \{i \in I \,|\, x_i^* = 0 < F_i(x^*)\},
\end{aligned}$$

where $x^* \in \mathbb{R}^n$ is a fixed solution of the complementarity problem. Then x^* is called an *R-regular* solution of NCP(F) if the submatrix $F'(x^*)_{\alpha\alpha}$ is nonsingular and the Schur-complement (see [4])

$$F'(x^*)_{\beta\beta} - F'(x^*)_{\beta\alpha}F'(x^*)_{\alpha\alpha}^{-1}F'(x^*)_{\alpha\beta} \in \mathbb{R}^{|\beta|\times|\beta|}$$

is a P-matrix, cf. Robinson [31]. Using the concept of R-regularity, the following result was shown in [18, Theorem 2.7].

Proposition 2.2 *Assume that $x^* \in \mathbb{R}^n$ is an R-regular solution of NCP(F). Then all elements in the generalized Jacobian $\partial\Phi_\lambda(x^*)$ are nonsingular.*

We finally mention a result which follows immediately from the (strong) semismoothness of the equation operator Φ_λ and well-known results for (strongly) semismooth functions, see [18, 30].

Proposition 2.3 *Let $x^* \in \mathbb{R}^n$ be the limit point of a sequence $\{x^k\} \subseteq \mathbb{R}^n$. Then*

$$\|\Phi_\lambda(x^k) - \Phi_\lambda(x^*) - H_k(x^k - x^*)\| = o(\|x^k - x^*\|)$$

for any $H_k \in \partial\Phi_\lambda(x^k)$. If, in addition, F is an LC^1 mapping, then

$$\|\Phi_\lambda(x^k) - \Phi_\lambda(x^*) - H_k(x^k - x^*)\| = O(\|x^k - x^*\|^2).$$

3 INEXACT TRUST-REGION METHODS

3.1 Algorithm

In this subsection, we first state our algorithm and then show that it is well-defined for an arbitrary complementarity problem.

To this end, we recall that NCP(F) can be written as the unconstrained minimization problem

$$\min \ \Psi_\lambda(x), \quad x \in \mathbb{R}^n.$$

Using a Gauss-Newton-type linearization around a given iterate $x^k \in \mathbb{R}^n$, we obtain the following quadratic approximation of the function $\Psi_\lambda(x^k + \cdot)$:

$$\Psi_\lambda(x^k + d) \approx q_k(d) := \frac{1}{2}\|\Phi_\lambda(x^k) + H_k d\|^2,$$

where $H_k \in \partial_B\Phi_\lambda(x^k)$. Since $\nabla\Psi_\lambda(x^k) = H_k^T\Phi_\lambda(x^k)$ for any $H_k \in \partial\Phi_\lambda(x^k)$ by Theorem 3.1 in [18], it follows immediately that q_k can be written as

$$q_k(d) = \Psi_\lambda(x^k) + \nabla\Psi_\lambda(x^k)^T d + \frac{1}{2}d^T H_k^T H_k d;$$

this representation of q_k will be more convenient in our subsequent analysis.

We are now able to state our trust-region-type method.

Algorithm 3.1 *(Inexact Trust-Region Method)*

(S.0) (Initialization)
Choose $\lambda \in (0,4)$, $x^0 \in \mathbb{R}^n$, $\Delta_0 > 0$, $0 < \rho_1 < \rho_2 < 1$, $0 < \sigma_1 < 1 < \sigma_2$, $\Delta_{\min} > 0$, $\varepsilon \geq 0$, and set $k := 0$.

(S.1) (Termination Criterion)
If $\|\nabla\Psi_\lambda(x^k)\| \leq \varepsilon$: STOP.

(S.2) (Subproblem Solution)
Select an element $H_k \in \partial_B \Phi_\lambda(x^k)$. Compute a solution $d^k \in \mathbb{R}^n$ of the subspace trust-region problem

$$\min \; q_k(d) \quad s.t. \quad \|d\| \le \Delta_k, \; d \in V_k, \tag{3.1}$$

where $V_k \subseteq \mathbb{R}^n$ denotes a suitable subspace which will be specified later.

(S.3) (Updates)
Compute

$$r_k := \frac{\Psi_\lambda(x^k) - \Psi_\lambda(x^k + d^k)}{\Psi_\lambda(x^k) - q_k(d^k)}.$$

If $r_k \ge \rho_1$, we call the iteration k successful *and set $x^{k+1} := x^k + d^k$; otherwise we set $x^{k+1} := x^k$.*

(a) If $r_k < \rho_1$, set $\Delta_{k+1} := \sigma_1 \Delta_k$.

(b) If $r_k \in [\rho_1, \rho_2)$, set $\Delta_{k+1} := \max\{\Delta_{\min}, \Delta_k\}$.

(c) If $r_k \ge \rho_2$, set $\Delta_{k+1} := \max\{\Delta_{\min}, \sigma_2 \Delta_k\}$.

Set $k \leftarrow k+1$ and go to (S.1).

Basically, Algorithm 3.1 is a standard trust-region method for the solution of nonlinear systems of equations. The only difference lies in the updating rule for the trust-region radius after a successful iteration; Algorithm 3.1 uses a minimal radius $\Delta_{\min} > 0$ as a lower bound for the new radius. This idea was also used by Jiang et al. [17] and some other works on nonsmooth trust-region methods [15, 14, 23]. Since Φ_λ is not differentiable everywhere, this modified updating rule is necessary in order to ensure global and local superlinear/quadratic convergence of our nonsmooth trust-region method.

Throughout this section, we assume that the termination parameter ε is equal to 0 and that the algorithm does not terminate after a finite number of iterations. We further assume that the subspace V_k from Step (S.2) of Algorithm 3.1 contains at least the gradient direction $\nabla \Psi_\lambda(x^k)$. Then the following result is well-known, see Powell [28, Theorem 4].

Lemma 3.1 *Assume that $d^k \in \mathbb{R}^n$ is a solution of the subproblem (3.1). Then*

$$\Psi_\lambda(x^k) - q_k(d^k) \ge \frac{1}{2}\|\nabla \Psi_\lambda(x^k)\| \min\left\{\Delta_k, \frac{\|\nabla \Psi_\lambda(x^k)\|}{\|H_k^T H_k\|}\right\}.$$

Note that the denominator on the right-hand side of the displayed formula in Lemma 3.1 is always nonzero; otherwise, we would have $H_k = 0$ which would imply $\nabla \Psi_\lambda(x^k) = H_k^T \Phi_\lambda(x^k) = 0$, so that the algorithm would have stopped in Step (S.1).

As an immediate consequence of Lemma 3.1, we obtain that Algorithm 3.1 can be applied to an arbitrary complementarity problem.

Proposition 3.1 *Algorithm 3.1 is well-defined.*

Proof. We only have to show that the denominator

$$\Psi_\lambda(x^k) - q_k(d^k)$$

in the definition of r_k in Step (S.3) is nonzero for all $k \in \mathbb{N}$. Assume this is not true. Then $\nabla\Psi_\lambda(x^k) = 0$ because of Lemma 3.1. Hence the algorithm would have stopped at this iteration in Step (S.1).
From Lemma 3.1 and the proof of Proposition 3.1, we actually obtain that

$$\Psi_\lambda(x^k) - q_k(d^k) > 0$$

for all $k \in \mathbb{N}$. This elementary inequality will be used several times in our convergence theory without referring to it explicitly.

3.2 Global Convergence

The aim of this subsection is to present two global convergence results for Algorithm 3.1. We start our analysis with the following lemma.

Lemma 3.2 *Let $\{x^k\}$ be any sequence generated by Algorithm 3.1, and let $\{x^k\}_K$ be a subsequence converging to a point $x^* \in \mathbb{R}^n$. If x^* is not a stationary point of Ψ_λ, then*

$$\liminf_{k\to\infty, k\in K} \Delta_k > 0.$$

Proof. Define the index set

$$\bar{K} := \{k-1 \mid k \in K\}.$$

Then the subsequence $\{x^{k+1}\}_{k\in\bar{K}}$ converges to x^*. We have to show that

$$\liminf_{k\to\infty, k\in\bar{K}} \Delta_{k+1} > 0. \tag{3.2}$$

Suppose (3.2) does not hold. Subsequencing if necessary, we can therefore assume that

$$\lim_{k\to\infty, k\in\bar{K}} \Delta_{k+1} = 0. \tag{3.3}$$

In view of the updating rule for the trust-region radius (note that the lower bound $\Delta_{\min} > 0$ plays a central role here), this implies that all iterations $k \in \bar{K}$ with k sufficiently large are not successful. Hence we have

$$r_k < \rho_1 \tag{3.4}$$

and $x^k = x^{k+1}$ for all $k \in \bar{K}$ large enough. Since $\{x^{k+1}\}_{k\in\bar{K}}$ converges to x^* by assumption, this implies that also the subsequence $\{x^k\}_{k\in\bar{K}}$ converges to x^*. Since $\Delta_{k+1} = \sigma_1\Delta_k$ for nonsuccessful iterations, we obtain

$$\lim_{k\to\infty, k\in\bar{K}} \Delta_k = 0 \tag{3.5}$$

from (3.3).

In view of our assumptions, the limit point x^* is not a stationary point of Ψ_λ. Hence there exists a constant $\beta_1 > 0$ such that

$$\|\nabla\Psi_\lambda(x^k)\| \geq \beta_1 \tag{3.6}$$

for all $k \in \bar{K}$. Due to the upper semicontinuity of the generalized Jacobian (see [3, Proposition 2.6.2 (c)]), there is also a constant $\beta_2 > 0$ with

$$\|H_k^T H_k\| \leq \beta_2 \tag{3.7}$$

for all $k \in \bar{K}$. Using Lemma 3.1, we obtain from (3.5), (3.6) and (3.7) for all $k \in \bar{K}$ sufficiently large:

$$\begin{array}{rcl} \Psi_\lambda(x^k) - q_k(d^k) & \geq & \frac{1}{2}\|\nabla\Psi_\lambda(x^k)\| \min\left\{\Delta_k, \frac{\|\nabla\Psi_\lambda(x^k)\|}{\|H_k^T H_k\|}\right\} \\ & \geq & \frac{1}{2}\beta_1 \min\left\{\Delta_k, \frac{\beta_1}{\beta_2}\right\} \\ & = & \frac{1}{2}\beta_1\Delta_k \\ & \geq & \frac{1}{2}\beta_1\|d^k\|. \end{array} \tag{3.8}$$

Since Ψ_λ is continuously differentiable, there exists for each $k \in \mathbb{N}$ a vector $\xi^k = x^k + \theta_k d^k, \theta_k \in (0,1)$, such that

$$\Psi_\lambda(x^k + d^k) = \Psi_\lambda(x^k) + \nabla\Psi_\lambda(\xi^k)^T d^k. \tag{3.9}$$

Obviously, we have $\{\xi^k\}_{k\in\bar{K}} \to x^*$. It then follows from (3.6)—(3.9) and the Cauchy-Schwarz inequality that

$$\begin{array}{rl} & |r_k - 1| \\ = & \left|\frac{\Psi_\lambda(x^k) - \Psi_\lambda(x^k+d^k)}{\Psi_\lambda(x^k) - q_k(d^k)} - 1\right| \\ = & \left|\frac{q_k(d^k) - \Psi_\lambda(x^k+d^k)}{\Psi_\lambda(x^k) - q_k(d^k)}\right| \\ = & \frac{|\Psi_\lambda(x^k) + \nabla\Psi_\lambda(x^k)^T d^k + \frac{1}{2}(d^k)^T H_k^T H_k d^k - \Psi_\lambda(x^k) - \nabla\Psi_\lambda(\xi^k)^T d^k|}{\Psi_\lambda(x^k) - q_k(d^k)} \\ \leq & \frac{2}{\beta_1\|d^k\|}\left|\nabla\Psi_\lambda(x^k)^T d^k - \nabla\Psi_\lambda(\xi^k)^T d^k + \frac{1}{2}(d^k)^T H_k^T H_k d^k\right| \\ \leq & \frac{2}{\beta_1\|d^k\|}\left(\|\nabla\Psi_\lambda(x^k) - \nabla\Psi_\lambda(\xi^k)\|\,\|d^k\| + \frac{1}{2}\|H_k^T H_k\|\,\|d^k\|^2\right) \\ \leq & \frac{1}{\beta_1}\left(2\|\nabla\Psi_\lambda(x^k) - \nabla\Psi_\lambda(\xi^k)\| + \beta_2\|d^k\|\right) \\ \to_{\bar{K}} & 0 \end{array}$$

for all $k \in \bar{K}$ sufficiently large. Hence the subsequence $\{r_k\}_{k\in\bar{K}}$ converges to 1, a contradiction to (3.4).

The following result is an important consequence of Lemma 3.2.

Lemma 3.3 *Let $\{x^k\}$ be any sequence generated by Algorithm 3.1. Then there are infinitely many successful iterations.*

Proof. Assume the number of successful iterations is finite. Then there is an index $k_0 \in \mathbb{N}$ such that $r_k < \rho_1$ and $x^k = x^{k_0}$ for all $k \in \mathbb{N}$ with $k \geq k_0$. Hence $\{\Delta_k\} \to 0$ and $\{x^k\}$ converges to x^{k_0}. Since $\nabla\Psi_\lambda(x^{k_0}) \neq 0$ (otherwise the algorithm would have stopped), this contradicts Lemma 3.2.
We are now able to prove our first global convergence result.

Theorem 3.1 *Let $\{x^k\}$ be any sequence generated by Algorithm 3.1. Then every accumulation point of $\{x^k\}$ is a stationary point of Ψ_λ.*

Proof. Let $x^* \in \mathbb{R}^n$ denote an accumulation point of $\{x^k\}$, and let $\{x^k\}_K$ be a subsequence converging to x^*. Since $x^{k+1} = x^k$ for all nonsuccessful iterations k and since there are infinitely many successful iterations by Lemma 3.3, we can assume without loss of generality that all iterates x^k with $k \in K$ are successful.

Suppose that $\nabla\Psi_\lambda(x^*) \neq 0$. Using this and the upper semicontinuity of the generalized Jacobian [3, Proposition 2.6.2 (c)], it follows that

$$\|\nabla\Psi_\lambda(x^k)\| \geq \beta_1 \quad \text{and} \quad \|H_k^T H_k\| \leq \beta_2$$

for all $k \in K$, where $\beta_1 > 0$ and $\beta_2 > 0$ are suitable constants. Since the iterations $k \in K$ are successful, we have $r_k \geq \rho_1$ for all $k \in K$. Lemma 3.1 therefore yields

$$\begin{aligned} \Psi_\lambda(x^k) - \Psi_\lambda(x^{k+1}) &\geq \rho_1\left(\Psi_\lambda(x^k) - q_k(d^k)\right) \\ &\geq \tfrac{1}{2}\rho_1\|\nabla\Psi_\lambda(x^k)\|\min\left\{\Delta_k, \tfrac{\|\nabla\Psi_\lambda(x^k)\|}{\|H_k^T H_k\|}\right\} \\ &\geq \tfrac{1}{2}\rho_1\beta_1\min\left\{\Delta_k, \tfrac{\beta_1}{\beta_2}\right\} \end{aligned} \tag{3.10}$$

for all $k \in K$. Since the entire sequence $\{\Psi_\lambda(x^k)\}$ is obviously decreasing and bounded from below, it is convergent. Using (3.10), we thus obtain

$$\begin{aligned} \frac{1}{2}\rho_1\beta_1\sum_{k\in K}\min\left\{\Delta_k, \frac{\beta_1}{\beta_2}\right\} &\leq \sum_{k\in K}\left(\Psi_\lambda(x^k) - \Psi_\lambda(x^{k+1})\right) \\ &\leq \sum_{k=0}^{\infty}\left(\Psi_\lambda(x^k) - \Psi_\lambda(x^{k+1})\right) \\ &< \infty. \end{aligned}$$

This implies $\{\Delta_k\}_{k\in K} \to 0$, a contradiction to Lemma 3.2.
Although Theorem 3.1 guarantees only convergence to stationary points of Ψ_λ, we stress that Theorem 2.1 and its Corollary 2.1 provide relatively mild conditions for such a stationary point to be a solution of the nonlinear complementarity problem itself. We also note that, for example, F being a uniform

P-function ensures the existence of an accumulation point due to Theorem 3.7 in [18].

A second global convergence result is given by the following theorem. Note that this result also applies to possibly unbounded sequences $\{x^k\}$.

Theorem 3.2 *Let $\{x^k\}$ be any sequence generated by Algorithm 3.1, and suppose that the sequence $\{H_k\}$ is bounded. Then*

$$\liminf_{k\to\infty} \|\nabla\Psi_\lambda(x^k)\| = 0.$$

Proof. The proof is similar to the one of Theorem 5.2 in [17], so we omit it here.

3.3 Local Convergence

This subsection considers the local rate of convergence of Algorithm 3.1. In order to prove our main result, we need the following two lemmas. The first one is due to Moré and Sorensen [25], whereas the second one was shown by Facchinei and Soares [10] (see also [19]); however, it is restated here under a slightly weaker condition, cf. [8].

Lemma 3.4 *Assume that $x^* \in \mathbb{R}^n$ is an isolated accumulation point of a sequence $\{x^k\}$ (not necessarily generated by Algorithm 3.1), and suppose that $\{\|x^{k+1} - x^k\|\}_K \to 0$ for any subsequence $\{x^k\}_K$ converging to x^*. Then the entire sequence $\{x^k\}$ converges to x^*.*

Lemma 3.5 *Assume that $G : \mathbb{R}^n \to \mathbb{R}^n$ is semismooth and suppose that $x^* \in \mathbb{R}^n$ is a BD-regular solution of $G(x) = 0$. Let $\{x^k\}$ and $\{d^k\}$ be any two sequences (not necessarily generated by Algorithm 3.1) such that*

$$\{x^k\} \to x^* \quad \textit{and} \quad \|x^k + d^k - x^*\| = o(\|x^k - x^*\|).$$

Then

$$\|G(x^k + d^k)\| = o(\|G(x^k)\|).$$

These two lemmas enable us to prove the following main local convergence result. In this result, we assume that the subspace V_k contains at least the gradient direction $\nabla\Psi_\lambda(x^k)$ and, if existing, the generalized Newton direction $d_N^k := -H_k^{-1}\Phi_\lambda(x^k)$.

Theorem 3.3 *Let $\{x^k\}$ be any sequence generated by Algorithm 3.1. If $\{x^k\}$ has an accumulation point x^* which is a BD-regular solution of $\Phi_\lambda(x) = 0$, then the following statements hold:*

(a) The entire sequence $\{x^k\}$ converges to x^*.

(b) The generalized Newton direction $d_N^k := -H_k^{-1}\Phi_\lambda(x^k)$ exists and is the unique solution of subproblem (3.1) for all $k \in \mathbb{N}$ sufficiently large.

(c) Eventually all iterations are successful, and we have $x^{k+1} = x^k + d_N^k$ for all $k \in \mathbb{N}$ large enough.

(d) The sequence $\{x^k\}$ converges Q-superlinearly to x^*; if F is an LC^1 mapping, the rate of convergence is Q-quadratic.

Proof. Let x^* be an accumulation point of the sequence $\{x^k\}$ which is a BD-regular solution of the system $\Phi_\lambda(x) = 0$.

(a) We prove statement (a) by verifying the assumptions of Lemma 3.4. To this end, we first note that the BD-regularity implies that x^* is an isolated solution of NCP(F) by [29, Proposition 2.5]. Since $\{\Psi_\lambda(x^k)\}$ is monotonically decreasing, it is easy to see that x^* is therefore also an isolated accumulation point of the sequence $\{x^k\}$.

Now let $\{x^k\}_K$ be any subsequence converging to x^*. For the solution $d^k \in \mathbb{R}^n$ of the subspace trust-region subproblem (3.1), we have

$$\Psi_\lambda(x^k) + \nabla\Psi_\lambda(x^k)^T d^k + \frac{1}{2}(d^k)^T H_k^T H_k d^k = q_k(d^k) \le q_k(0) = \Psi_\lambda(x^k)$$

and therefore

$$\frac{1}{2}(d^k)^T H_k^T H_k d^k \le -\nabla\Psi_\lambda(x^k)^T d^k \le \|\nabla\Psi_\lambda(x^k)\|\,\|d^k\|. \tag{3.11}$$

The assumed BD-regularity of x^* further implies that there is a constant $\alpha > 0$ such that

$$\alpha\|d^k\|^2 \le \frac{1}{2}(d^k)^T H_k^T H_k d^k$$

for all $k \in K$ sufficiently large. In view of (3.11), we thus obtain

$$\|d^k\| \le \frac{1}{\alpha}\|\nabla\Psi_\lambda(x^k)\| \tag{3.12}$$

for all $k \in K$ large enough.

Due to our global convergence result from Theorem 3.1, we have

$$\{\nabla\Psi_\lambda(x^k)\}_K \to \nabla\Psi_\lambda(x^*) = 0.$$

Hence $\{d^k\}_K \to 0$ by (3.12) so that

$$\{\|x^{k+1} - x^k\|\}_K \to 0$$

since $\|x^{k+1} - x^k\| \le \|d^k\|$. Part (a) therefore follows from Lemma 3.4.

(b), (c): We verify statements (b) and (c) simultaneously. The proof is devided into three steps, where we denote by $\bar{K}$ the set of indices $k \in \mathbb{N}$ such that the $(k-1)$st iteration is successful. In step (i), we show that the generalized Newton direction $d_N^k := -H_k^{-1}\Phi_\lambda(x^k)$ exists and is the unique solution of the subproblem (3.1) for all $k \in \bar{K}$ sufficiently large. Part (ii) shows that, if the $(k-1)$st iteration is successful, then also the kth iteration is successful. Statements (b) and (c) are then shown by using an induction argument in Step (iii).

(i) From the BD-regularity of the solution x^* and statement (a), it follows that there is a constant $c > 0$ such that, for all $k \in \mathbb{N}$ sufficiently large, the matrix H_k is nonsingular with

$$\|H_k^{-1}\| \leq c. \tag{3.13}$$

In particular, the generalized Newton direction $d_N^k := -H_k^{-1}\Phi_\lambda(x^k)$ exists for $k \in \mathbb{N}$ large enough. Since $\{x^k\} \to x^*$ by part (a), the continuity of Φ_λ and (3.13) imply that

$$\|d_N^k\| = \|H_k^{-1}\Phi_\lambda(x^k)\| \leq c\|\Phi_\lambda(x^k)\| \to 0$$

so that

$$\|d_N^k\| \leq \Delta_{\min}$$

for all $k \in \mathbb{N}$ sufficiently large. The definition of the set $\bar{K}$ together with the updating rule for the trust-region radius therefore gives

$$\|d_N^k\| \leq \Delta_{\min} \leq \Delta_k$$

for all $k \in \bar{K}$ large enough. Since $d_N^k \in V_k$ by assumption, this implies that the direction d_N^k is at least (strictly) feasible for the subproblem (3.1). Since the matrix $H_k^T H_k$ is positive definite due to the BD-regularity of the solution x^*, it follows for all $k \in \bar{K}$ sufficiently large that d_N^k is indeed the unique solution of the subspace trust-region subproblem (3.1).

(ii) Assume that $k \in \bar{K}$ so that the $(k-1)$st iteration is successful. We want to show that the kth iteration is also successful, i.e., we want to show that $r_k \geq \rho_1$.

Since x^* is a solution of the nonlinear system of equations $\Phi_\lambda(x) = 0$, we obtain from (3.13), $\{x^k\}_{\bar{K}} \to x^*$, Proposition 2.3 and part (i) of this proof:

$$\begin{aligned}
\|x^k + d^k - x^*\| &= \|x^k + d_N^k - x^*\| \\
&= \left\|-(H_k^{-1})\left(\Phi_\lambda(x^k) - \Phi_\lambda(x^*) - H_k(x^k - x^*)\right)\right\| \\
&\leq c\|\Phi_\lambda(x^k) - \Phi_\lambda(x^*) - H_k(x^k - x^*)\| \\
&= o(\|x^k - x^*\|).
\end{aligned}$$

Since Φ_λ is semismooth by [18, Theorem 2.3], Lemma 3.5 therefore implies

$$\|\Phi_\lambda(x^k + d^k)\| = o(\|\Phi_\lambda(x^k)\|).$$

Hence

$$\Psi_\lambda(x^k + d^k) = o(\Psi_\lambda(x^k)).$$

In particular, we have

$$\Psi_\lambda(x^k + d^k) \le (1 - \rho_1)\Psi_\lambda(x^k)$$

for $k \in \bar{K}$ sufficiently large since $\rho_1 < 1$. From $q_k(d^k) = q_k(d_N^k) = 0$ we therefore get

$$\begin{aligned} r_k &= \frac{\Psi_\lambda(x^k) - \Psi_\lambda(x^k + d^k)}{\Psi_\lambda(x^k) - q_k(d^k)} \\ &= 1 - \frac{\Psi_\lambda(x^k + d^k)}{\Psi_\lambda(x^k)} \\ &\ge \rho_1 \end{aligned}$$

for all $k \in \bar{K}$ large enough, i.e., the kth iteration is also successful.

(iii) It follows from parts (i), (ii) and an induction argument that, for all $k \in \mathbb{N}$ sufficiently large, the iteration k is successful and the trust-region subproblem (3.1) is solved by the generalized Newton direction d_N^k.

(d) In view of part (c), we have $x^{k+1} = x^k + d_N^k$ for all $k \in \mathbb{N}$ sufficiently large. Hence our inexact trust region method 3.1 reduces to the standard nonsmooth Newton method [30, 29] for the solution of a semismooth system of equations. The Q-superlinear/Q-quadratic rate of convergence follows therefore directly from known results by Qi [29].
We finally note that the assumed BD-regularity condition in Theorem 3.3 is satisfied, in particular, if x^* is an R-regular solution of NCP(F), see Proposition 2.2.

4 NUMERICAL RESULTS

4.1 Implementation Issues

This subsection provides a short description of our implementation of Algorithm 3.1. In particular, we give some details on the solution of the subspace trust-region subproblem (3.1).

To this end, let us define the two search directions

$$d_G^k := -\nabla\Psi_\lambda(x^k)$$

and

$$d_N^k := -H_k^{-1}\Phi_\lambda(x^k),$$

where $H_k \in \partial_B\Phi_\lambda(x^k)$. Note that H_k can be computed in essentially the same way as described in [5] for the Fischer-Burmeister function, see [34] for further details.

Assuming that the generalized Newton direction d_N^k exists, we define the subspace $V_k \subseteq \mathbb{R}^n$ from Step (S.2) of Algorithm 3.1 as

$$V_k := \text{span}\{d_G^k, d_N^k\}.$$

If d_N^k cannot be computed, we just set

$$V_k := \text{span}\{d_G^k\}.$$

Due to the theoretical results from the previous section, this choice of V_k guarantees the global and local fast convergence of our inexact trust region method.

Using this set V_k, it is well-known (see, e.g., [32, 24, 21]) that our subspace trust-region subproblem

$$\min \ q_k(d) := \Psi_\lambda(x^k) + \nabla\Psi_\lambda(x^k)^T d + \frac{1}{2} d^T H_k^T H_k d \quad \text{s.t.} \quad \|d\| \leq \Delta_k, \ d \in V_k \tag{4.1}$$

can be reformulated as a standard trust-region subproblem of dimension $r_k := \dim V_k \in \{1, 2\}$. More precisely, if $r_k = 2$ (the case $r_k = 1$ is obvious) and $\{v^{k,1}, v^{k,2}\}$ denotes an orthonormal basis of V_k, a vector $d^k \in \mathbb{R}^n$ solves problem (4.1) if and only if

$$d^k = \alpha_1^k v^{k,1} + \alpha_2^k v^{k,2},$$

where $\alpha^k := (\alpha_1^k, \alpha_2^k)^T$ is a solution of the 2-dimensional problem

$$\min \ \tilde{q}_k(\alpha) := f^k + (g^k)^T \alpha + \frac{1}{2} \alpha^T M_k \alpha \quad \text{s.t.} \quad \|\alpha\| \leq \Delta_k; \tag{4.2}$$

here, we used the notation

$$\begin{aligned} f^k &:= \Psi_\lambda(x^k) \in \mathbb{R}, \\ B_k &:= (v^{k,1}, v^{k,2}) \in \mathbb{R}^{n\times 2}, \\ g^k &:= B_k^T \nabla\Psi_\lambda(x^k) \in \mathbb{R}^2, \\ M_k &:= B_k^T H_k^T H_k B_k \in \mathbb{R}^{2\times 2}. \end{aligned}$$

It is therefore sufficient to solve the small-dimensional problem (4.2) in order to get a vector d^k as required in Step (S.2) of Algorithm 3.1.

In our MATLAB implementation of Algorithm 3.1, we modified the termination criterion in Step (S.1) and stop the iteration if

$$\Psi_2(x^k) < \varepsilon \quad \text{or} \quad \Delta_k < \Delta_{\text{stop}} \quad \text{or} \quad k > k_{\max},$$

where

$$\varepsilon := 10^{-12}, \quad \Delta_{\text{stop}} := 10^{-15}, \quad k_{\max} := 300,$$

i.e., we terminate the iteration if either an approximate solution was found or if the trust-region radius Δ_k becomes too small or if we need too many iterations. The other parameters from Algorithm 3.1 were chosen as follows:

$$\Delta_0 := 10, \ \rho_1 = 10^{-4}, \ \rho_2 = 0.75, \ \sigma_1 = 0.5, \ \sigma_2 = 2, \ \Delta_{\min} = 10^{-2}.$$

Apart from the stopping criterion, the implementation of our inexact trust-region method differs from its description in Subsection 3.1 in the following points:

(a) We used a nonmonotone trust-region strategy as described by Toint [33] (which, in turn, is based on an idea by Grippo, Lampariello and Lucidi [16]). We do not give a precise description of this strategy here, instead we only note that the basic idea of a nonmonotone trust-region method is to accept more iterations as being successful. Compared to the standard (monotone) trust-region method, this nonmonotone variant needs less many iterations, and the improvement is sometimes considerable, see [34] for a numerical comparison.

(b) We incorporated a simple heuristic backtracking strategy in order to avoid possible domain violations, see [9] for a more detailed description.

(c) Similar to [18], we used a dynamic choice of the parameter λ. The main idea is to start the iteration with $\lambda = 2$ (which seems to give the best global behaviour) and to reduce $\lambda \to 0$ if we get close to a solution (this seems to improve the local behaviour to some extent). We refer to [34] for further details. Note that the strategy used here is a heuristic, but that some justifications for taking $\lambda \approx 0$ locally were also given in the recent paper [22] by Lopes, Martínez and Pérez.

4.2 Computational Results

We tested the previously described implementation of Algorithm 3.1 on a SUN SPARC 20 station. We used all complementarity problems and all available starting points from the MCPLIB test problem collection by Dirkse and Ferris [7]. Our results are summarized in Tables 1.1–1.2, where we present the following data:

problem:	name of test example in MCPLIB
n:	dimension of test example
SP:	number of starting point in the M-file `cpstart.m`
k:	number of iterations
k_s:	number of successful iterations
N:	number of Newton steps
NF:	number of function evaluations of F
NF':	number of Jacobian evaluations of F
Ψ_2:	value of $\Psi_2(x)$ at the final iterate $x = x^f$
$\|\nabla\Psi_2\|$:	value of $\|\nabla\Psi_2(x)\|$ at the final iterate $x = x^f$
B:	number of iterations using a backtracking step.

The results in Tables 1.1–1.2 are quite promising. The number of iterations seems comparable to the full-dimensional trust-region method by Jiang et al. [17] (note that Jiang et al. [17] only count the number of successful iterations in their paper). Also the robustness of the two algorithms seems to be very similar, whereas our method is more efficient for obvious reasons.

On the other hand, if we compare our algorithm with the corresponding line search method from [18], we usually need a few more iterations. However, the

Table 1.1 Numerical results for Algorithm 3.1

problem	n	SP	k	k_s	N	NF	NF'	Ψ_2	$\|\nabla\Psi_2\|$	B
bertsekas	15	1	34	20	7	35	21	5.7e-25	9.6e-11	0
bertsekas	15	2	32	20	6	33	21	1.3e-16	9.2e-7	0
bertsekas	15	3	20	18	12	21	19	1.1e-28	9.1e-13	0
billups	1	1	—	—	—	—	—	—	—	—
colvdual	20	1	—	—	—	—	—	—	—	—
colvdual	20	2	105	76	40	106	77	4.8e-14	7.9e-5	0
colvnlp	15	1	30	19	6	31	20	8.5e-18	4.3e-7	0
colvnlp	15	2	14	11	4	15	12	2.8e-18	2.4e-7	0
cycle	1	1	7	5	6	8	6	6.2e-22	3.5e-11	0
explcp	16	1	25	19	21	26	20	2.4e-18	2.2e-9	0
hanskoop	14	1	10	10	9	12	11	3.1e-20	2.9e-9	1
hanskoop	14	2	13	13	12	14	14	1.3e-21	6.1e-10	0
hanskoop	14	3	15	11	10	16	12	3.1e-20	2.8e-9	0
hanskoop	14	4	11	8	8	13	9	7.2e-13	8.2e-6	1
hanskoop	14	5	11	11	10	13	12	7.4e-23	9.1e-11	1
josephy	4	1	14	8	11	15	9	7.6e-27	1.1e-12	0
josephy	4	2	10	6	7	11	7	6.5e-17	1.0e-7	0
josephy	4	3	54	36	48	55	37	1.6e-15	5.1e-7	0
josephy	4	4	5	5	5	6	6	8.2e-17	1.2e-7	0
josephy	4	5	4	4	4	5	5	1.5e-16	1.6e-7	0
josephy	4	6	14	7	8	15	8	6.2e-23	1.0e-10	0
kojshin	4	1	16	9	11	17	10	1.7e-19	7.2e-9	0
kojshin	4	2	11	7	7	12	8	1.1e-15	5.8e-7	0
kojshin	4	3	16	16	12	17	17	1.5e-20	1.6e-9	0
kojshin	4	4	4	4	4	5	5	2.1e-13	6.8e-6	0
kojshin	4	5	11	5	10	12	6	7.7e-27	1.1e-12	0
kojshin	4	6	13	8	10	14	9	2.0e-24	2.5e-11	0

number of function evaluations is often considerably lower than for the method in [18]. Moreover, our trust-region method seems to be more reliable than the line search method from [18].

In fact, our method has only two failures: One for the example by Billups [1], which is impossible to solve for almost all state-of-the-art solvers since the starting point is very close to a local-nonglobal minimum of basically any reasonable merit function, and one for problem `colvdual` (first starting point) which is also known to be a very difficult test problem.

The difficulty with the Billups example, given by

$$F(x) := (x-1)^2 - 1.01,$$

Table 1.2 Numerical results for Algorithm 3.1 (continued)

problem	n	SP	k	k_s	N	NF	NF'	Ψ_2	$\|\nabla\Psi_2\|$	B
mathinum	3	1	4	4	4	5	5	5.1e-16	6.4e-8	0
mathinum	3	2	5	5	5	6	6	2.2e-14	5.8e-7	0
mathinum	3	3	12	7	6	13	8	2.9e-13	1.5e-6	0
mathinum	3	4	7	7	7	8	8	6.3e-19	3.1e-9	0
mathisum	4	1	9	5	7	10	6	6.9e-13	2.8e-6	0
mathisum	4	2	6	6	6	7	7	1.6e-13	1.3e-6	0
mathisum	4	3	13	8	12	14	9	1.3e-23	8.8e-12	0
mathisum	4	4	6	6	6	7	7	2.6e-23	1.7e-11	0
nash	10	1	8	8	8	9	9	6.7e-24	2.7e-10	0
nash	10	2	21	11	10	22	12	5.9e-19	8.2e-8	0
pgvon105	105	1	53	45	47	103	46	5.1e-13	4.3e-2	49
pgvon106	106	1	30	30	27	58	31	8.3e-13	2.3e+05	27
powell	16	1	10	7	9	11	8	6.0e-20	2.5e-9	0
powell	16	2	9	9	8	10	10	3.5e-15	7.3e-7	0
powell	16	3	13	13	12	14	14	1.1e-22	1.2e-10	0
powell	16	4	10	10	9	11	11	8.1e-15	1.0e-6	0
scarfanum	13	1	11	8	8	12	9	1.6e-19	9.2e-9	0
scarfanum	13	2	19	12	7	20	13	2.6e-18	3.7e-8	0
scarfanum	13	3	13	9	11	15	10	3.3e-20	4.1e-9	1
scarfasum	14	1	12	5	11	13	6	1.5e-13	3.2e-5	0
scarfasum	14	2	15	9	8	17	10	7.5e-19	7.2e-8	1
scarfasum	14	3	21	14	6	22	15	4.9e-18	1.8e-7	0
scarfbnum	39	1	29	25	21	30	26	7.2e-28	3.9e-12	0
scarfbnum	39	2	28	26	25	29	27	1.5e-21	4.8e-9	0
scarfbsum	40	1	27	20	16	28	21	2.7e-27	1.4e-11	0
scarfbsum	40	2	30	22	21	31	23	5.7e-27	1.6e-11	0
sppe	27	1	9	9	6	10	10	3.4e-16	3.8e-8	0
sppe	27	2	8	8	6	9	9	4.7e-29	1.7e-14	0
tobin	42	1	10	10	7	11	11	3.7e-13	1.3e-6	0
tobin	42	2	11	11	8	12	12	1.0e-13	6.1e-7	0

is illustrated in Figures 1.1 and 1.2. Figure 1.1 shows the graph of our merit function Ψ_λ for $\lambda = 2$; obviously, the function Ψ_2 has two minimizers which both seem to be solutions of the complementarity problem. However, in Figure 1.2 we take a closer look at the minimum x^* which is close to the origin, and it turns out that the function value $\Psi_2(x^*)$ is not equal to 0, i.e., x^* is not a solution of NCP(F) (note the different scale of the axes in Figure 1.2). On the other hand, the standard starting point for this example is the origin, so our method (as well as almost all other methods) converges to the local-nonglobal minimizer x^*. This clearly shows why we are not able to solve this example even if it is only of dimension $n = 1$.

Figure 1.1 Graph of Ψ_2 for the example by Billups

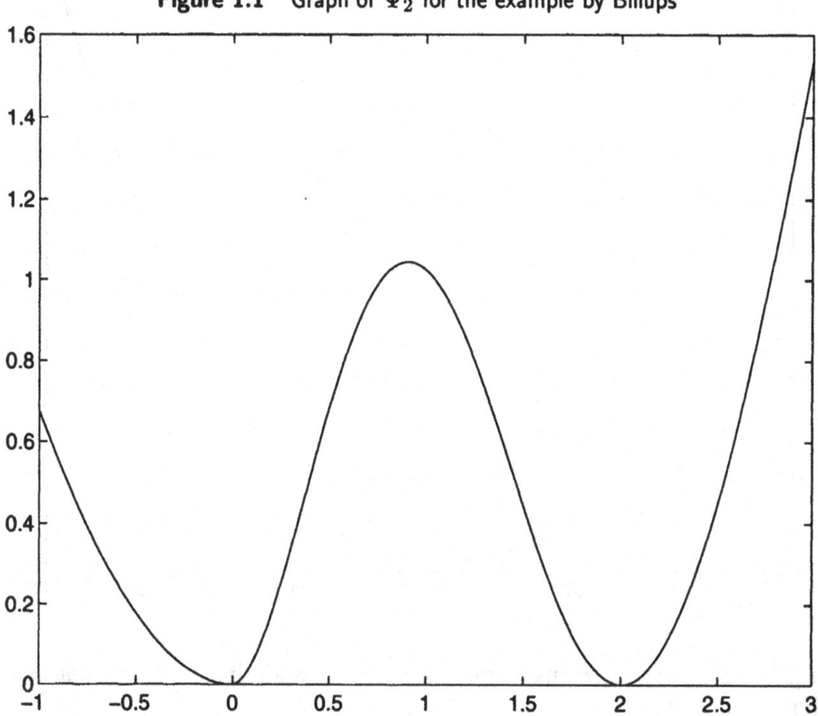

5 FINAL REMARKS

In this paper, we investigated the theoretical and numerical properties of a nonsmooth Newton-type method for the solution of nonlinear complementarity problems. In contrast to many earlier works [10, 5, 18], we used a trust-region-type globalization instead of a line search globalization. Our nonsmooth trust-region method is very similar to the one of Jiang et al. [17]. However, in order to reduce the computational overhead, we did not solve the trust-region subproblem exactly like in [17]. Instead, we followed an idea from Shultz et al. [32, 2] and solved the trust-region problem only on a subspace. Alternatively, we could have used the dogleg step by Powell [27] or the double dogleg step by Dennis and Mei [6]; however, the subspace trust-region method is more general.

From a theoretical point of view, our inexact trust-region method has the same convergence properties as the (exact) trust-region method by Jiang et al. [17]. The convergence properties are also similar to those of the corresponding line search method by Kanzow and Kleinmichel [18]. From a numerical point of view, the new method should be considerably more efficient than the algorithm

Figure 1.2 A closer look around the origin

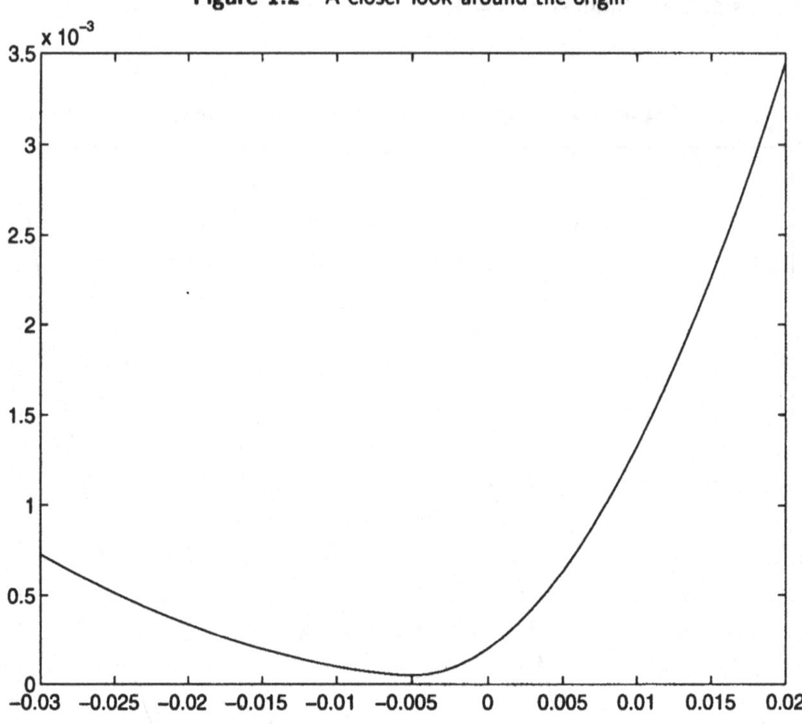

from [17]. Moreover, it seems to be more reliable and almost as efficient as the method from [18].

We therefore believe that a sophisticated implementation of a suitable trust-region method could actually outperform corresponding line search methods. At least we think that future studies in the numerical solution of complementarity problems should put more attention on trust-region-type methods than this was done before.

References

[1] S.C. Billups: *Algorithms for Complementarity Problems and Generalized Equations.* Ph.D. Thesis, Computer Sciences Department, University of Wisconsin, Madison, WI, August 1995.

[2] R.H. Byrd, R.B. Schnabel and G.A. Shultz: *Approximate solution of the trust region problem by minimization over two-dimensional subspaces.* Mathematical Programming 40, 1988, pp. 247–263.

[3] F.H. Clarke: *Optimization and Nonsmooth Analysis.* John Wiley & Sons, New York, NY, 1983 (reprinted by SIAM, Philadelphia, PA, 1990).

[4] R.W. Cottle, J.-S. Pang and R.E. Stone: *The Linear Complementarity Problem.* Academic Press, Boston, 1992.

[5] T. De Luca, F. Facchinei and C. Kanzow: *A semismooth equation approach to the solution of nonlinear complementarity problems.* Mathematical Programming 75, 1996, pp. 407–439.

[6] J.E. Dennis and H.H.W. Mei: *Two new unconstrained optimization algorithms which use function and gradient values.* Journal of Optimization Theory and Applications 28, 1979, pp. 453–482.

[7] S.P. Dirkse and M.C. Ferris: *MCPLIB: A collection of nonlinear mixed complementarity problems.* Optimization Methods and Software 5, 1995, pp. 123–156.

[8] F. Facchinei, A. Fischer and C. Kanzow: *Inexact Newton methods for semismooth equations with applications to variational inequality problems.* In: G. Di Pillo and F. Giannessi (eds.): *Nonlinear Optimization and Applications.* Plenum Press, New York, NY, 1996, pp. 125–139.

[9] F. Facchinei, A. Fischer and C. Kanzow: *A semismooth Newton method for variational inequalities: The case of box constraints.* In: M.C. Ferris and J.-S. Pang (eds.): *Complementarity and Variational Problems. State of the Art.* SIAM, Philadelphia, PA, 1997, pp. 76–90.

[10] F. Facchinei and J. Soares: *A new merit function for nonlinear complementarity problems and a related algorithm.* SIAM Journal on Optimization 7, 1997, pp. 225–247.

[11] M.C. Ferris and J.-S. Pang: *Engineering and economic applications of complementarity problems.* SIAM Review, to appear.

[12] A. Fischer: *A special Newton-type optimization method.* Optimization 24, 1992, pp. 269–284.

[13] R. Fletcher: *Practical Methods of Optimization.* John Wiley & Sons, Chichester, England, 1987.

[14] A. Friedlander, J.M. Martínez and A. Santos: *A new algorithm for bound constrained minimization.* Journal of Applied Mathematics and Optimization 30, 1994, pp. 235–266.

[15] S.A. Gabriel and J.-S. Pang: *A trust region method for constrained nonsmooth equations.* In: W.W. Hager, D.W. Hearn and P.M. Pardalos (eds.): *Large Scale Optimization. State of the Art.* Kluwer Academic Publishers, Dordrecht, 1994, pp. 155–181.

[16] L. Grippo, F. Lampariello and S. Lucidi: *A nonmonotone linesearch technique for Newton's method.* SIAM Journal on Numerical Analysis 23, 1986, pp. 707–716.

[17] H. Jiang, M. Fukushima, L. Qi and D. Sun: *A trust region method for solving generalized complementarity problems.* SIAM Journal on Optimization, to appear.

[18] C. Kanzow and H. Kleinmichel: *A new class of semismooth Newton-type methods for nonlinear complementarity problems.* Computational Optimization and Applications, to appear.

[19] C. KANZOW AND H.-D. QI: *A QP-free constrained Newton-type method for variational inequality problems.* Preprint 121, Institute of Applied Mathematics, University of Hamburg, Hamburg, Germany, March 1997 (revised November 1997).

[20] C. KANZOW, N. YAMASHITA AND M. FUKUSHIMA: *New NCP-functions and their properties.* Journal of Optimization Theory and Applications 94, 1997, pp. 115–135.

[21] O. KNOTH: *Marquardt-ähnliche Verfahren zur Minimierung nichtlinearer Funktionen.* Ph.D. Thesis, Martin-Luther-University, Halle-Wittenberg, Germany, 1983 (in German).

[22] V.L. LOPES, J.M. MARTÍNEZ AND R. PÉREZ: *On the local convergence of quasi-Newton methods for nonlinear complementarity problems.* Technical Report, Institute of Applied Mathematics, University of Campinas, Campinas, Brazil, 1997.

[23] J.M. MARTÍNEZ AND A. SANTOS: *A trust region strategy for minimization on arbitrary domains.* Mathematical Programming 68, 1995, pp. 267–302.

[24] J.J. MORÉ: *Recent developments in algorithms and software for trust region methods.* In: A. BACHEM, M. GRÖTSCHEL AND B. KORTE (eds.): *Mathematical Programming. The State of the Art.* Springer-Verlag, Berlin, 1983, pp. 258–287.

[25] J.J. MORÉ AND D.C. SORENSEN: *Computing a trust-region step.* SIAM Journal on Scientific and Statistical Computing 4, 1983, pp. 553–572.

[26] J.-S. PANG: *Newton's method for B-differentiable equations.* Mathematics of Operations Research 15, 1990, pp. 311–341.

[27] M.J.D. POWELL: *A new algorithm for unconstrained optimization.* In: J.B. ROSEN, O.L. MANGASARIAN AND K. RITTER (eds.): *Nonlinear Programming.* Academic Press, New York, NY, 1970, pp. 31–65.

[28] M.J.D. POWELL: *Convergence properties of a class of minimization algorithms.* In: O.L. MANGASARIAN, R.R. MEYER AND S.M. ROBINSON (eds.): *Nonlinear Programming 2.* Academic Press, New York, NY, 1975, pp. 1–27.

[29] L. QI: *Convergence analysis of some algorithms for solving nonsmooth equations.* Mathematics of Operations Research 18, 1993, pp. 227–244.

[30] L. QI AND J. SUN: *A nonsmooth version of Newton's method.* Mathematical Programming 58, 1993, pp. 353–367.

[31] S.M. ROBINSON: *Strongly regular generalized equations.* Mathematics of Operations Research 5, 1980, pp. 43–62.

[32] G.A. SHULTZ, R.B. SCHNABEL AND R.H. BYRD: *A family of trust-region-based algorithms for unconstrained minimization with strong global convergence properties.* SIAM Journal on Numerical Analysis 22, 1985, pp. 47–67.

[33] PH.L. TOINT: *Non-monotone trust-region algorithms for nonlinear optimization subject to convex constraints.* Mathematical Programming 77, 1997, pp. 69–94.

[34] M. ZUPKE: *Trust-Region-Verfahren zur Lösung nichtlinearer Komplementaritätsprobleme.* Diploma Thesis, Institute of Applied Mathematics, University of Hamburg, Hamburg, Germany, July 1997 (in German).

Reformulation: Nonsmooth, Piecewise Smooth,
Semismooth and Smoothing Methods, pp. 235-257
Edited by M. Fukushima and L. Qi

Regularized Newton Methods for Minimization of Convex Quadratic Splines with Singular Hessians

Wu Li† and John Swetits*

Abstract A quadratic spline function on $\mathbb{R}^n$ is a differentiable piecewise quadratic function. Many convex quadratic programming problems can be reformulated as the problem of unconstrained minimization of convex quadratic splines. Therefore, it is important to investigate efficient algorithms for the unconstrained minimization of a convex quadratic spline. In this paper, for a convex quadratic spline $f(x)$ that has a matrix representation and is bounded below, we show that one can find a global minimizer of $f(x)$ in finitely many iterations even though the global minimizers of $f(x)$ might form an unbounded set. The new idea is to use a regularized Newton direction when a Hessian matrix of $f(x)$ is singular. Applications to quadratic programming are also included.

Key Words regularized Newton method, regularized Cholesky factorization, quadratic spline, quadratic programming, unconstrained reformulation, piecewise affine equation

1 INTRODUCTION

Three different types of convex quadratic programming problems can be reformulated as unconstrained minimization problems with convex quadratic splines (*i.e.*, convex differentiable piecewise quadratic functions) as objective functions: least distance problems [27, 28, 5, 9, 18, 29, 14, 6], convex quadratic programming problems with simple bound constraints [19, 20, 14, 25], and strictly convex quadratic programming problems [19, 20]. See also [17] for more details. Therefore, there is a need to study numerical algorithms for finding a (global)

*Department of Mathematics and Statistics, Old Dominion University, Norfolk, VA 23529, USA, wuli@math.odu.edu, swetits@math.odu.edu

minimizer of a convex quadratic spline. Such algorithms can immediately be used for solving the unconstrained reformulations of quadratic programming problems mentioned above. This provides a new paradigm for finding innovative ways of solving convex quadratic programming problems. Note that finding a minimizer of a strictly convex quadratic spline $f(x)$ is an easy problem ([19, Theorem 3.2] and [33, Theorem 3]), since it only requires a standard Newton method with exact line search. However, if $f(x)$ is only convex, then Hessian matrices of $f(x)$ might be singular and the set of minimizers of $f(x)$ might be unbounded. In this case, there is no general theory about finding a minimizer of $f(x)$ in finitely many iterations. Madsen and Nielsen [22] proposed a hybrid Newton method for solving Huber M-estimator problem [13] – finding a minimizer of a special convex quadratic spline generated by Huber's cost function. The main idea is to use different descent directions depending on whether the Newton equation is consistent or not: the steepest descent direction if the Newton equation is inconsistent or the least norm solution of the Newton equation otherwise. This leads to a finite algorithm for Huber M-estimator problem [22]. Madsen and Nielsen's hybrid Newton method can be used to solve linear ℓ_1 estimation problems [23, 21], linear programming problems [24], and strictly convex quadratic programming problems with simple bound constraints [25]. Clark and Osborne also had a finite method based on simplex approach for solving Huber M-estimator problem [7]. In [18], Li, Pardalos, and Han introduced a linearly convergent Gauss-Seidel method for finding a minimizer of the convex quadratic spline derived from the unconstrained reformulation of the least distance problem. Other linearly convergent algorithms for finding minimizers of convex quadratic splines derived from unconstrained reformulations of various convex quadratic programming problems or Huber M-estimator problem can be found in [32, 12, 14, 15, 9]. From the theoretical point of view, finding a minimizer of a convex quadratic spline is an easy problem. Berman, Kovoor, and Pardalos [1] proposed a combinatorial approach for finding a zero of the gradient of a convex quadratic spline based on a multi-dimensional search method introduced by Dyer [10, 11] and Megiddo [26]. Their main result is that a minimizer of a convex quadratic spline can be found in linear time with respect to the number of variables (cf. also [16]). In particular, they proved that the least distance problem with one equality constraint and simple bound constraints can be solved in linear time (see also [2, 3, 31, 16]).

In this paper, we propose a regularized Newton method for finding a minimizer of a convex quadratic spline $f(z)$ that has the following structure:

$$f(z) := \frac{1}{2}z^T Pz + z^T p + \frac{1}{2}\left\|(Qz+q)_+\right\|^2 \quad \text{for } z \in \mathbf{R}^r, \tag{1.1}$$

where P is an $r \times r$ symmetric positive semidefinite matrix, Q is an $s \times r$ matrix, $p \in \mathbf{R}^r$, $q \in \mathbf{R}^s$, $(x)_+$ denotes a vector whose i-th component is $\max\{x_i, 0\}$, and $\|\cdot\|$ is the 2-norm for vectors. By [20, Lemma 2.1], $f(z)$ is a convex function. Therefore, $\bar{z}$ is a minimizer of $f(z)$ if and only if $\bar{z}$ is a zero of its gradient $\nabla f(z)$. That is,

$$f(\bar{z}) = \min_{z \in \mathbf{R}^r} f(z) \tag{1.2}$$

if and only if $\bar{z}$ is a solution of the following system of piecewise affine equations:

$$Pz + p + Q^T(Qz + q)_+ = 0. \tag{1.3}$$

Throughout this paper, we always assume that $f(z)$ is bounded below on $\mathbb{R}^r$. If P is positive definite, then $f(z)$ is strictly convex [20, Lemma 2.1] and it is easy to use a Newton method to find the unique minimizer of $f(z)$ in finitely many steps (cf. [19, Theorem 3.2] and [33, Theorem 3]). Note that $\nabla f(z)$ is a piecewise affine function and is not differentiable at some points in $\mathbb{R}^r$. However, $\nabla f(z)$ is a semismooth function and its generalized Jacobian $\partial^2 f(z)$ is a set of $r \times r$ matrices (cf. Section 2.6.1 of [7] or Proposition 2.1 of [30]):

$$\partial^2 f(z) \subset \mathrm{co}\left\{P + Q^T D_s(J) Q : J \subset J(z)\right\}, \tag{1.4}$$

where "co" denotes the convex hull,

$$\begin{array}{c} D_s(J) \text{ denotes an } s \times s \text{ diagonal matrix whose} \\ i\text{-th diagonal element is 1 if } i \in J \text{ and 0 if } i \notin J, \end{array} \tag{1.5}$$

and $J(z)$ is the "active" index set at z:

$$J(z) = \{i : (Qz + q)_i \geq 0\}. \tag{1.6}$$

Note that each index set J is associated with a polyhedral region:

$$W_J = \{z : (Qz + q)_i \geq 0 \text{ for } i \in J, (Qz + q)_i \leq 0 \text{ for } i \notin J\}. \tag{1.7}$$

On each W_J, f is a convex quadratic function. For convenience, we call W_J a *solution region* of f if W_J contains a minimizer of f. The first key step in finding a minimizer of f is to identify a solution region. This is similar to identification of active constraints when solving constrained minimization problems. For example, if z^k is an iterate such that $W_{J(z^k)}$ is a solution region of f (*i.e.*, the iterate z^k identifies a solution region of f), then the following system of linear equations is consistent:

$$\left[P + Q^T D_s\left(J(z^k)\right) Q\right] d = -\nabla f(z^k). \tag{1.8}$$

Moreover, there exists a solution d^k of (1.8) such that $(z^k + d^k)$ is a minimizer of f. In fact, let z^* be a minimizer of f in $W_{J(z^k)}$. Then

$$\left[P + Q^T D_s\left(J(z^k)\right) Q\right] (z^* - z^k) = -\nabla f(z^k). \tag{1.9}$$

When $[P + Q^T D_s(J(z^k))Q]$ is nonsingular, the unique solution d^k of (1.8) yields a minimizer $(z^k + d^k)$ of f. That is the main reason why one can easily find a minimizer of a strictly convex and differentiable piecewise quadratic function in finitely many iterations by using a Newton method with exact line search (cf. [19, Theorem 3.2] and [33, Theorem 3]).

In essence, there are two stages in a numerical method for finding a minimizer of f in finitely many iterations: (i) finding an iterate z^k that identifies a solution region $W_{J(z^k)}$ of f and (ii) using the Newton method to find a minimizer z^* of f in $W_{J(z^k)}$.

For the convex quadratic spline f, it is very easy to find an iterate z^k that identifies a solution region $W_{J(z^k)}$ of f. The only requirement is (cf. [20, Lemma 3.7]) that

$$\lim_{k\to\infty} f(z^k) = f_{\min}, \tag{1.10}$$

where

$$f_{\min} := \min_{z\in\mathbf{R}^r} f(z). \tag{1.11}$$

However, when $\left[P + Q^T D_s(J(z^k))Q\right]$ is singular, finding a minimizer of f in the solution region $W_{J(z^k)}$ is a nontrivial task. In fact, we have to solve the following feasibility problem after z^k identifies a solution region of f:

$$\begin{aligned}
&\left[P + Q^T D_s(J(z^k))Q\right] d = -\nabla f(z^k),\\
&\left[Q(z^k + d) + q\right]_i \geq 0 \text{ for } i \in J(z^k),\\
&\left[Q(z^k + d) + q\right]_i \leq 0 \text{ for } i \notin J(z^k).
\end{aligned} \tag{1.12}$$

Note that $W_{J(z^k)}$ contains a minimizer of f if and only if (1.12) has a solution. Moreover, if d^k is a solution of (1.12), then $(z^k + d^k)$ is a minimizer of f. However, any attempt to deal directly with (1.12) is not practical. Even though we know that (1.12) is feasible for some large k, we have no easy way of checking the feasibility of (1.12) except solving it. Fortunately, for the convex quadratic spline f, the feasibility problem will be automatically solved by using so-called regularized solutions of (1.8) (cf. Section 4).

The paper is organized as follows. In Section 2, we formulate a general hybrid Newton method (called **QSPLINE** method) for finding a minimizer of f given in (1.1). The moral of **QSPLINE** method is the following: one can find a minimizer of $f(z)$ in finite iterations by any descent method with exact line search that uses a solution of (1.8) as a descent direction when (1.8) is consistent or any descent direction otherwise, provided that the norm of the descent direction is proportional to $\|\nabla f(z^k)\|$ (cf. also Lemma 3.1). The proof of the finite termination of **QSPLINE** method is given in Section 3. In Section 4, we introduce regularized solutions of a singular system of linear equations. This leads to a regularized Cholesky factorization method for solving (1.8). The regularized Cholesky factorization method produces a regularized Newton direction that is always a descent direction for $f(z)$ at z^k, no matter whether (1.8) is consistent or not. As a consequence, we have two special versions of **QSPLINE** method that take advantage of the structure of $f(z)$. Finally, in Section 5, we show how the proposed algorithms can be used to solve three types of convex quadratic programming problems.

2 QSPLINE METHOD

In the next algorithm, we formulate a general hybrid Newton method (called **QSPLINE** method) for finding a minimizer of f and then explain the underlying philosophy for such an algorithm.

Algorithm 2.1 (**QSPLINE** method) *Suppose that P is an $r \times r$ symmetric positive semidefinite matrix, ϵ and γ are positive constants, and $f(z)$ is defined in (1.1). Let $\mathcal{H}$ be a collection of $r \times r$ positive definite matrices such that*

$$z^T H z \geq \epsilon \|z\|^2 \text{ and } \|Hz\| \leq \frac{1}{\epsilon}\|z\| \quad \text{for } z \in \mathbf{R}^r, H \in \mathcal{H}. \tag{2.1}$$

For any given $z^0 \in \mathbf{R}^r$, generate a sequence of iterates z^{k+1} for $k = 0, 1, 2, \ldots$ as follows:

Step 1: *if (1.8) is consistent, choose a solution d^k of (1.8) such that*

$$\|d^k\| \leq \gamma \|\nabla f(z^k)\|; \tag{2.2}$$

Step 2: *if (1.8) is inconsistent, choose a positive definite matrix H_k in $\mathcal{H}$ and let $d^k := -H_k^{-1}\nabla f(z^k)$;*

Step 3: *set $t_k := \min\{1, \bar{t}_k\}$ where $\bar{t}_k > 0$ satisfies $(d^k)^T \nabla f(z^k + \bar{t}_k d^k) = 0$ (i.e., $\bar{t}_k$ is a stepsize by exact line search);*

Step 4: *compute the next iterate $z^{k+1} := z^k + t_k d^k$.*

Remark. (i) We formulate **QSPLINE** method in such a way that it allows very flexible choices of H_k (when (1.8) is inconsistent) and d^k (when (1.8) is consistent). In general, H_k can be viewed as a regularization of the generalized Hessian $\partial^2 f(z^k)$. If (1.8) is inconsistent, then $W_{J(z^k)}$ is not a solution region. In this case, our objective is to get a descent direction d^k and to reduce the objective function value. In the case that (1.8) is consistent but has multiple solutions, we want to use a solution d^k of (1.8) as the descent direction. However, all solutions of (1.8) form an unbounded set and a large norm of d^k is not desirable. The purpose of (2.2) is to make sure that the norm of d^k is *proportional to* the norm of the gradient $\nabla f(z^k)$. In practice, we do not need to specify the value of γ. For example, let d^k be the least norm solution of (1.8), i.e., $d^k := -\left[P + Q^T D_s(J(z^k))Q\right]^+ \nabla f(z^k)$, where B^+ denotes the pseudoinverse of B. Then (2.2) is automatically satisfied with γ being the maximum of $\left\|\left[P + Q^T D_s(J)Q\right]^+\right\|$ for all possible index set J.

(ii) It is very easy to find a minimizer $\bar{t}_k$ of the convex quadratic spline $f(z^k + td^k)$. In particular, one could find $\bar{t}_k$ in $\mathcal{O}(r^2 + rs)$ operations (cf. [2, 3, 31, 16]).

In general, it is possible that the minimizers of $f(z^k+td^k)$ form an unbounded set. Therefore, we take precaution to use $t_k = \min\{1, \bar{t}_k\}$ as the stepsize. Also it is important for finite termination of **QSPLINE** method. (See the next section for details.) Theoretically, the algorithm might still work if we choose t_k to be the smallest minimizer of $f(z^k + td^k)$. Chen and Pinar used this strategy for a Newton method for solving Huber's M-estimator problem that terminates in finitely many iterations [4].

Note $f(z^k) \geq f(z^k+t_kd^k) \geq f(z^k+\bar{t}_kd^k)$. It is possible that $f(z^k+t_kd^k) > f(z^k+\bar{t}_kd^k)$. However, we prefer to use the Newton step $t_k = 1$ when possible.

(iii) For convex quadratic splines constructed as penalty/merit functions for convex quadratic programs, we can always choose $H_k \in \mathcal{H}$ such that $d^k := -H_k^{-1}\nabla f(z^k)$ is *automatically* a solution of (1.8) whenever (1.8) is consistent (cf. Subsections 4.2 and 4.3). In this case, Steps 1 and 2 can be naturally combined into one step in implementation.

(iv) In practice, we do not need to know the exact value of ϵ. In fact, in our applications (cf. Section 5), $\mathcal{H}$ is a collection of finitely many positive definite matrices (which are closely related to $[P + Q^T D_s(J)Q]$'s), say, $\mathcal{H} = \{M_1, \ldots, M_n\}$. Since M_i is positive definite, there exists a positive constant ϵ_i such that

$$z^T M_i z \geq \epsilon_i \|z\|^2 \quad \text{and} \quad \|M_i z\| \leq \frac{1}{\epsilon_i}\|z\| \quad \text{for } z \in \mathbf{R}^r. \tag{2.3}$$

Thus, (2.1) is *automatically* satisfied with $\epsilon = \min\{\epsilon_1, \ldots, \epsilon_n\}$.

Moreover, if we use the strategy mentioned in **(iii)**, then we have

$$\|d^k\| \cdot \|\nabla f(z^k)\| \geq -(d^k)^T \nabla f(z^k) = (d^k)^T H_k d^k \geq \epsilon \|d^k\|^2,$$

where the first inequality is Cauchy-Schwarz inequality, the equality follows from $d^k = -H_k^{-1}\nabla f(z^k)$ and the second inequality is a consequence of (2.1). Thus, in this case, we *automatically* have (2.2) with $\gamma = \frac{1}{\epsilon}$.

(v) Convex spline functions can be viewed as special cases of so-called SC^1 functions. An SC^1 function is continuously differentiable and its gradient function is semismooth. A globally convergent Newton method for convex SC^1 minimization problems with linear constraints is proposed by Pang and Qi [30]. When applied to our special case, Pang and Qi's algorithm can be described as follows.

Step 0. Initialization. Let $\rho, \sigma \in (0,1)$ be given scalars; let I be the identity matrix; and $z^0 \in \mathbf{R}^r$ be given. Set $k = 0$.

Step 1. Direction Generation. Pick an arbitrary matrix $V_k \in \nabla^2 f(z^k)$ and a scalar $\epsilon_k > 0$. Let $d^k := -(V_k + \epsilon_k I)^{-1}\nabla f(z^k)$.

Step 2. Armijo Line Search. Let m_k be the smallest nonnegative integer m such that $f(z^k + \rho^m d^k) - f(z^k) \leq \sigma\rho^m (d^k)^T \nabla f(z^k)$. Set $z^{k+1} := x^k + \rho^{m_k} d^k$.

Step 3. Termination Check. If z^{k+1} satisfies a prescribed stopping rule, terminate; otherwise, return to **Step 1** with k replaced by $(k+1)$.

Note that the line search strategy and the Newton direction used in **QSPLINE** method are different from the ones in Pang and Qi's algorithm. Due to the simpler structure of $f(z)$, we can require that d^k be a solution of (1.8) when (1.8) is consistent. Moreover, exact line search is very cheap for the spline function $f(z^k + td^k)$ as pointed out in Remark (ii) given above.

It is worth mentioning that if $(Qx+q)_i > 0$ for $i \in J(z^k)$ has no solution, then $[P + Q^T D_s(J(z^k))Q] \notin \nabla^2 f(z^k)$ (cf. Section 2.6.1 of [7] or Proposition 2.1 of [30]).

3 FINITE TERMINATION OF QSPLINE METHOD

3.1 Preliminary Lemmas

The purpose for imposing (2.1) and (2.2) in **QSPLINE** method is to insure that $\|d^k\|^2$, $\|\nabla f(z^k)\|^2$, and $-(d^k)^T\nabla f(z^k)$ are proportional to each other as shown in the following lemma.

Lemma 3.1 *Suppose that z^k and d^k are generated by* **QSPLINE** *method. Then there exists a positive constant η (independent of k) such that*

$$\frac{1}{\eta}\|\nabla f(z^k)\|^2 \le -(d^k)^T\nabla f(z^k) \le \eta\|\nabla f(z^k)\|^2, \tag{3.1}$$

$$\frac{1}{\eta}\|d^k\|^2 \le -(d^k)^T\nabla f(z^k) \le \eta\|d^k\|^2. \tag{3.2}$$

Proof. If (1.8) is not consistent, we have $\nabla f(z^k) = -H_k d^k$, which, along with (2.1), implies

$$\|\nabla f(z^k)\| \le \frac{1}{\epsilon}\|d^k\|. \tag{3.3}$$

Thus,

$$\epsilon\|d^k\|^2 \le (d^k)^T H_k d^k = -(d^k)^T\nabla f(z^k) \le \|d^k\|\cdot\|\nabla f(z^k)\| \le \frac{1}{\epsilon}\|d^k\|^2, \tag{3.4}$$

where the first inequality follows from (2.1), the second inequality is Cauchy-Schwartz inequality, and the last inequality is (3.3). Also we obtain from (3.4) that

$$\|d^k\| \le \frac{1}{\epsilon}\|\nabla f(z^k)\|. \tag{3.5}$$

Therefore, it follows from (3.3)–(3.5) that (3.1) and (3.2) hold with $\eta = \frac{1}{\epsilon^3}$.

If (1.8) is consistent, then $-\nabla f(z^k) = (P + Q^T D_s(J(z^k))Q)d^k$. Since $(P + Q^T D_s(J(z^k))Q)$ is a symmetric positive semidefinite matrix, there exists a positive constant δ_k such that for all $z \in \mathbf{R}^r$,

$$z^T[P + Q^T D_s(J(z^k))Q]z \ge \delta_k\|[P + Q^T D_s(J(z^k))Q]z\|^2$$

and

$$\|[P + Q^T D_s(J(z^k)]Q)z\| \le \frac{1}{\delta_k}\|z\|.$$

Since there are only finitely many different $(P + Q^T D_s(J(z^k))Q)$'s, we can choose a positive constant δ small enough such that

$$\begin{aligned} z^T(P + Q^T D_s(J(z^k))Q)z &\ge \delta\|(P + Q^T D_s(J(z^k))Q)z\|^2, \\ \|(P + Q^T D_s(J(z^k))Q)z\| &\le \tfrac{1}{\delta}\|z\| \end{aligned} \tag{3.6}$$

for $k = 1, 2, \ldots,\ z \in \mathbf{R}^r$. Replacing z in (3.6) by d^k we get

$$-(d^k)^T \nabla f(z^k) \ge \delta\|\nabla f(z^k)\|^2 \quad \text{and} \quad \|\nabla f(z^k)\| \le \frac{1}{\delta}\|d^k\| \quad \text{for } k = 1, 2, \ldots \tag{3.7}$$

By (3.7) we obtain

$$\begin{aligned} \frac{\delta}{\gamma^2}\|d^k\|^2 &\le \delta\|\nabla f(z^k)\|^2 \le -(d^k)^T \nabla f(z^k) \\ &\le \|d^k\| \cdot \|\nabla f(z^k)\| \le \frac{1}{\delta}\|d^k\|^2 \le \frac{\gamma^2}{\delta}\|\nabla f(z^k)\|^2, \end{aligned} \tag{3.8}$$

where the the first and last inequalities are by (2.2). It follows from (3.8) that (3.1) and (3.2) hold with $\eta = \max\left\{\frac{1}{\delta}, \frac{\gamma^2}{\delta}\right\}$.

As a consequence, (3.1) and (3.2) hold with $\eta = \max\left\{\frac{\gamma^2}{\delta}, \frac{1}{\epsilon^3}, \frac{1}{\delta}\right\}$. ■

The next step in proving finite termination of **QSPLINE** method is to show that (1.10) holds. But this requires the following two lemmas about iterates generated by descent methods for minimizing a convex quadratic spline.

Lemma 3.2 [20, Lemma 3.1] *Suppose that $g(z)$ is a convex function, its gradient $\nabla g(z)$ is Lipschitz continuous, and $0 < \beta < 1$. Then there exists a positive constant λ (depending only on g and β) such that*

$$\left(\frac{d^T \nabla g(z)}{\|d\|}\right)^2 \le \lambda(g(z) - g(z + td)), \tag{3.9}$$

whenever $t > 0$ and $0 \ge d^T \nabla g(z + td) \ge \beta \cdot d^T \nabla g(z)$.

Lemma 3.3 [20, Lemma 3.3] *Suppose that $g(z)$ is a convex quadratic spline bounded below on $\mathbf{R}^r$ and $g(z^{k+1}) \le g(z^k)$ for $k = 0, 1, \ldots$ If there exists a subsequence $\{k_j\}$ such that $\lim_{j\to\infty} \|\nabla g(z^{k_j})\| = 0$, then $\lim_{k\to\infty} g(z^k) = g_{\min}$, where $g_{\min} = \min\{g(z) : z \in \mathbf{R}^r\}$.*

Lemma 3.4 *If $\{z^k\}$ is the sequence defined by* **QSPLINE** *method and $f(z)$ is bounded below on $\mathbf{R}^r$, then*

$$\lim_{k\to\infty} f(z^k) = f_{\min}, \tag{3.10}$$

where $f_{\min} := \min\{f(z) : z \in \mathbf{R}^r\}$.

Proof. We consider two cases. If $0 \geq (d^k)^T \nabla f(z^k + t_k d^k) \geq \frac{1}{2}(d^k)^T \nabla f(z^k)$, then by Lemma 3.2 there is $\lambda > 0$ (independent of k) such that

$$\frac{\left((d^k)^T \nabla f(z^k)\right)^2}{\|d^k\|^2} \leq \lambda\left(f(z^k) - f(z^k + t_k d^k)\right) = \lambda\left(f(z^k) - f(z^{k+1})\right).$$

This, along with Lemma 3.1, yields

$$\frac{1}{\eta^2}\|\nabla f(z^k)\|^2 \leq \lambda\left(f(z^k) - f(z^{k+1})\right). \tag{3.11}$$

Otherwise, we have $t_k = 1$ and

$$(d^k)^T \nabla f(z^k + d^k) \leq \frac{1}{2}(d^k)^T \nabla f(z^k).$$

Thus,

$$\begin{aligned} 0 &\geq \frac{1}{2}(d^k)^T \nabla f(z^k) \geq (d^k)^T \nabla f(z^k + d^k) \geq f(z^k + d^k) - f(z^k) \\ &= f(z^{k+1}) - f(z^k), \end{aligned} \tag{3.12}$$

where the third inequality is by convexity of $f(z)$.

Note that $\{f(z^k)\}$ converges because $f(z^{k+1}) \leq f(z^k)$ and f is bounded below. Thus,

$$\lim_{k\to\infty}\left(f(z^k) - f(z^{k+1})\right) = 0.$$

It follows from either (3.11) or (3.12) that $\nabla f(z^k) \to 0$. Thus, (3.10) holds (cf. Lemma 3.3). ■

3.2 Finite Termination

The importance of (3.10) is the identification of solution regions through iterates z^k as shown in the following lemma.

Lemma 3.5 [20, Lemma 3.7] *Suppose that $g(z)$ is a convex quadratic spline bounded below on $\mathbf{R}^r$ and $W_1, \ldots, W_m$ are polyhedral sets such that $\mathbf{R}^r = \bigcup_{i=1}^m W_i$ and g is a quadratic function on each W_i. If $\lim_{k\to\infty} g(z^k) = g_{\min}$, where $g_{\min} = \min\{g(z) : z \in \mathbf{R}^r\}$, then there is $k_0 > 0$ such that W_i contains a minimizer of $g(z)$ whenever $z^k \in W_i$ with $k \geq k_0$.*

Once we know that $W_{J(z^k)}$ contains a minimizer of $f(z)$, we have to solve the feasibility problem (1.12). If (1.8) has a unique solution d^k, then d^k is also a solution of (1.12) and $(z^k + d^k)$ is a minimizer of $f(z)$. However, in general, (1.8) has many solutions. Thus, even if d^k is a solution of (1.8), $(z^k + d^k)$ might not be a minimizer of $f(z)$. However, with $t_k \le 1$, **QSPLINE** method will generate an iterate z^m such that $J(z^k) \subset J(z^m)$ for any solution region $W_{J(z^k)}$. In this case, for any solution d^m of (1.8), $(z^m + d^m)$ is a minimizer of $f(z)$. This is the key idea in proving the finite termination of **QSPLINE** method.

Theorem 3.1 *Suppose that P is an $r \times r$ symmetric positive semidefinite matrix and $f(z)$ defined in (1.1) is bounded below on $\mathbf{R}^r$. If $\{z^k\}$ is the sequence generated by* **QSPLINE** *method, then there exists a positive integer k_0 and a minimizer z^* of $f(z)$ such that $z^k = z^*$ for $k \ge k_0$. That is,* **QSPLINE** *method finds a minimizer of $f(z)$ in finitely many iterations.*

Proof. By Lemmas 3.4 and 3.5, there exists k_0 such that $W_{J(z^k)}$ contains a minimizer of f for all $k \ge k_0$. In particular, (1.8) is consistent for all $k \ge k_0$. Let z^* be a minimizer of f that lies in $W_{J(z^k)}$. Then, with d^k as defined in **QSPLINE** method, we have

$$\left(P + Q^T D_s \left(J\left(z^k\right)\right) Q\right) d^k = -\nabla f\left(z^k\right)$$

and

$$\left(P + Q^T D_s \left(J\left(z^k\right)\right) Q\right) \left(z^* - z^k\right) = -\nabla f\left(z^k\right).$$

Thus,

$$\left(d^k - z^* + z^k\right)^T \left(P + Q^T D_s \left(J\left(z^k\right)\right) Q\right) \left(d^k - z^* + z^k\right) = 0.$$

Since P and $Q^T D_s \left(J\left(z^k\right)\right) Q$ both are symmetric positive semidefinite, we have

$$(d^k - z^* + z^k)^T Q^T D_s \left(J\left(z^k\right)\right) Q(d^k - z^* + z^k) = 0,$$

which implies

$$D_s \left(J\left(z^k\right)\right) Q(d^k - z^* + z^k) = 0.$$

From **Step 4** of **QSPLINE** method and the above equation, we derive

$$\begin{aligned} D_s \left(J\left(z^k\right)\right) Q z^{k+1} &= D_s \left(J\left(z^k\right)\right) Q(z^k + t_k d^k) \\ &= D_s \left(J\left(z^k\right)\right) \left(Q z^k + t_k Q \left(z^* - z^k\right)\right) \\ &= D_s \left(J\left(z^k\right)\right) \left((1 - t_k) Q z^k + t_k Q z^*\right). \end{aligned} \tag{3.13}$$

If $i \in J(z^k)$, then $(Qz^k + q)_i \ge 0$ and $(Qz^* + q)_i \ge 0$ (since $z^* \in W_{J(z^k)}$). Since $0 \le t_k \le 1$, it follows from (3.13) that $(Qz^{k+1} + q)_i \ge 0$. Therefore, $J(z^k) \subset J(z^{k+1})$. Thus there exists an integer m such that $D_s\left(J\left(z^{m+1}\right)\right) = D_s\left(J\left(z^m\right)\right)$. Now we claim that z^{m+1} is a minimizer of f. Since $z^{m+1} \in W_{J(z^m)}$ and $\nabla f(z)$ is affine on $W_{J(z^m)}$, we have

$$\nabla f(z^{m+1}) = \nabla f(z^m + t_m d^m) = \nabla f(z^m) + t_m \left(P + Q^T D_s \left(J\left(z^k\right)\right) Q\right) d^m. \tag{3.14}$$

Since d^m is a solution of (1.8) with $k = m$, by (3.14), we get

$$\nabla f(z^{m+1}) = (1 - t_m) \nabla f(z^m). \tag{3.15}$$

If $\nabla f(z^m) \neq 0$ and $t_m < 1$, then

$$0 = \left|(d^m)^T \nabla f(z^{m+1})\right| = (1 - t_m)\left|(d^m)^T \nabla f(z^m)\right| \geq \frac{1 - t_m}{\eta} \|\nabla f(z^m)\|^2 > 0, \tag{3.16}$$

where the first equality is from **Step 3** in **QSPLINE** method and $t_m = \hat{t}_m$, the second equality is by (3.15), and the first inequality is (3.1) in Lemma 3.1. The contradiction in (3.16) proves that either $t_m = 1$ or $\nabla f(z^m) = 0$. Therefore, by (3.15), $\nabla f(z^{m+1}) = 0$ and z^{m+1} is a minimizer of $f(z)$. ■

4 CHOICES OF NEWTON DIRECTIONS

Note that **QSPLINE** method generates a descent direction d^k according to whether (1.8) is inconsistent or not. For the convex quadratic spline $f(z)$ given in (1.1), we can use a regularized solution of (1.8) (cf. Subsection 4.1) and combine the first two steps in **QSPLINE** method into one. The key idea is to solve a nonsingular system of linear equations derived from the original singular system in such a way that the solution of the nonsingular system is actually a solution of the original system whenever the original system is consistent.

In this section, we shall use the following vector and matrix notations. Let x be a vector and let A be a matrix. For any index set J, x_J (or A_J) denotes the vector (or the matrix) consisting of the components of x (or the rows of A) corresponding to the index set J. We use $A_{J\hat{J}}$ to denote a submatrix of A obtained by deleting rows of A whose indices are not in J as well as columns of A whose indices are not in $\hat{J}$. In particular, A_{JJ} is the principal submatrix of A corresponding to the index set J. An identity matrix of appropriate dimension is denoted by I.

4.1 Regularized Solutions of Singular Systems

Consider the following system of linear equations:

$$Ax = b, \tag{4.1}$$

where A is an $n \times n$ matrix and $b \in \mathbf{R}^n$. If A is a singular matrix, then (4.1) has either no solution or infinitely many solutions. In this subsection, we define regularized solutions of (4.1) and give some basic properties of these regularized solutions.

Lemma 4.1 *Let $J \subset \{1, \ldots, n\}$. Then A_{JJ} is nonsingular if and only if the matrix $(I - D_n(J) + D_n(J)A)$ is nonsingular. Moreover, when A_{JJ} or $(I - D_n(J) + D_n(J)A)$ is nonsingular,*

$$\bar{x} = (I - D_n(J) + D_n(J)A)^{-1} b$$

is the unique solution of the following system of linear equations:

$$x_i = b_i \text{ for } i \notin J \quad \text{and} \quad A_J x = b_J. \tag{4.2}$$

Proof. The linear system (4.2) has the following matrix representation:

$$(I - D_n(J) + D_n(J)A)x = b. \tag{4.3}$$

Thus, if $(I - D_n(J) + D_n(J)A)$ is nonsingular, then

$$\bar{x} = (I - D_n(J) + D_n(J)A)^{-1} b$$

is the unique solution of (4.3) or (4.2).

Note that $(I - D_n(J) + D_n(J)A)x = 0$ if and only if $x_i = 0$ for $i \notin J$ and $A_{JJ}x_J = 0$. Thus, $(I - D_n(J) + D_n(J)A)x = 0$ has a nonzero solution x if and only if $A_{JJ}x_J = 0$ has a nonzero solution x_J. This proves that A_{JJ} is nonsingular if and only if $(I - D_n(J) + D_n(J)A)$ is nonsingular. ■

Definition 4.1 *Suppose that there is a subset J of $\{1, \ldots, n\}$ such that A_{JJ} is nonsingular and has the same rank as A. Then $(I - D_n(J) + D_n(J)A)^{-1} b$ will be called the regularized solution of (4.1) corresponding to the index set J.*

Remark. In the case that (4.1) is consistent, it can be reduced to the following form:

$$A_J x = b_J. \tag{4.4}$$

Since A_{JJ} is nonsingular, for any given values x_i with $i \notin J$, we can solve (4.4) for x_J and get a solution of (4.1). Among all these solutions, we choose the regularized solution of (4.1) corresponding to J to be the one that has the same components as b outside the index set J. A singular system might have more than one regularized solution as shown in the following example.

Consider the linear system (4.1) with

$$A := \begin{pmatrix} 1 & 0 & 1 \\ -1 & 1 & 0 \\ 0 & 1 & 1 \end{pmatrix} \quad \text{and} \quad b := \begin{pmatrix} 1 \\ 1 \\ 2 \end{pmatrix}. \tag{4.5}$$

Then it is easy to verify that (4.1) has 3 regularized solutions:

$$\begin{pmatrix} 1 \\ 2 \\ 0 \end{pmatrix}, \quad \begin{pmatrix} 0 \\ 1 \\ 1 \end{pmatrix}, \quad \begin{pmatrix} -1 \\ 0 \\ 2 \end{pmatrix},$$

corresponding to index sets $\{2,3\}$, $\{3,1\}$, and $\{1,2\}$, respectively.

However, if (4.1) is not consistent, it might still have regularized solutions. For example, let

$$A := \begin{pmatrix} 1 & 0 & 1 \\ -1 & 1 & 0 \\ 0 & 1 & 1 \end{pmatrix} \quad \text{and} \quad b := \begin{pmatrix} 1 \\ 1 \\ 1 \end{pmatrix}. \tag{4.6}$$

It is easy to verify that (4.1) still has 3 regularized solutions:

$$\begin{pmatrix} 1 \\ 2 \\ -1 \end{pmatrix}, \quad \begin{pmatrix} 1 \\ 1 \\ 0 \end{pmatrix}, \quad \begin{pmatrix} 0 \\ 1 \\ 1 \end{pmatrix},$$

corresponding to index sets $\{2,3\}$, $\{3,1\}$, and $\{1,2\}$, respectively. But none of these regularized solutions is a solution of (4.1).

Next we show that a regularized solution of (4.1) is also a solution of (4.1) when (4.1) is solvable.

Theorem 4.1 *Suppose that there is a subset J of $\{1,\ldots,n\}$ such that A_{JJ} is nonsingular and has the same rank as A. Then $(I - D_n(J) + D_n(J)A)^{-1}b$ is a solution of (4.1) whenever (4.1) is consistent.*

Proof. By Lemma 4.1, $\bar{x}$ is the unique solution of the following linear system:

$$x_i = b_i \text{ for } i \notin J \quad \text{and} \quad A_J x = b_J. \tag{4.7}$$

Since (4.1) is consistent and A_J has the same rank as A, we know that any solution of $A_J x = b_J$ is a solution of (4.1). In particular,

$$(I - D_n(J) + D_n(J)A)^{-1}\, b$$

is a solution of (4.1). ■

One most important property of $[I - D_n(J) + D_n(J)A]$ is that it is positive definite under some assumptions on A and J, which will be crucial in deriving various special versions of **QSPLINE** method.

Theorem 4.2 *Suppose that A is symmetric positive semidefinite with eigenvalues in the interval $[0,1]$ and J is a subset of $\{1,\ldots,n\}$ such that A_{JJ} is nonsingular. Then $(I - D_n(J) + D_n(J)A)$ is positive definite.*

Proof. Since A is symmetric positive semidefinite with eigenvalues between 0 and 1, $(I - A)$ is positive semidefinite. Let $z \in \mathbb{R}^n$ with $z_i = 0$ for $i \in J$. Let A_{ij} be the element of A in the i-th row and j-th column. Then

$$\sum_{i \notin J} z_i^2 - \sum_{i \notin J} \sum_{j \notin J} A_{ij} z_i z_j = z^T (I - A) z \geq 0. \tag{4.8}$$

Thus,

$$\begin{aligned} & x^T (I - D_n(J) + D_n(J)A)\, x \\ = \;& \sum_{i \notin J} x_i^2 + \sum_{i \in J} \sum_{j=1}^{n} A_{ij} x_i x_j \\ \geq \;& \frac{1}{2} \sum_{i \notin J} x_i^2 + \frac{1}{2} \sum_{i \notin J} \sum_{j \notin J} A_{ij} x_i x_j + \sum_{i \in J} \sum_{j=1}^{n} A_{ij} x_i x_j. \end{aligned}$$

$$= \frac{1}{2}\left(\sum_{i\notin J} x_i^2 + x_J^T A_{JJ} x_J + x^T A x\right)$$

$$\geq \frac{1}{2}\left(\sum_{i\notin J} x_i^2 + x_J^T A_{JJ} x_J\right),$$

where the first inequality follows from (4.8), the second equality is by the symmetry of A and the second inequality is by the positive semidefiniteness of A. Since A is symmetric positive semidefinite and A_{JJ} is nonsingular, A_{JJ} is positive definite. For $x \neq 0$, $x_J^T A_{JJ} x_J > 0$ if $x_J \neq 0$, while $\sum_{i\notin J} x_i^2 > 0$ if $x_J = 0$. Therefore, for $x \neq 0$, we have $x^T (I - D_n(J) + D_n(J)A)\, x > 0$. ■

In general, a singular linear system might not have any regularized solution. For example, let

$$A := \begin{pmatrix} 1 & 1 & 0 \\ 1 & 1 & 0 \\ 1 & -1 & 0 \end{pmatrix} \quad \text{and} \quad b := \begin{pmatrix} 1 \\ 1 \\ 1 \end{pmatrix}. \tag{4.9}$$

Note that (4.1) is consistent for A and b given above. It is easy to verify that all 2×2 principal submatrices of A in (4.9) are singular, but A has rank 2. Thus, (4.1) does not have any regularized solution.

However, if A is symmetric and has rank r, then it is well-known that A has an $r \times r$ principal submatrix with rank r. Thus, *if A is symmetric, (4.1) always has a regularized solution.* When A is actually symmetric positive semidefinite, we can find a regularized solution of (4.1) by using the following regularized Cholesky factorization.

Algorithm 4.1 (Regularized Cholesky Factorization) *Suppose that A is an $n \times n$ symmetric positive semidefinite matrix and $b \in \mathbf{R}^n$.*

Step 1. *Let $\mathcal{J}$ be an empty index set. If $A_{11} \leq 0$, then*

add the index 1 to $\mathcal{J}$; $y_{11} = 1$ and $y_{i1} = 0$ for $i = 2, \ldots, n$;

else

$y_{11} = \sqrt{A_{11}}$ and $y_{i1} = A_{i1}/y_{11}$ for $i = 2, \ldots, n$.

Step 2. *For $k = 2, \ldots, n-1$, if $A_{kk} - \sum_{j=1}^{k-1} y_{kj}^2 \leq 0$, then*

add the index k to $\mathcal{J}$; $y_{kk} = 1$, $y_{ki} = 0$ for $i = 1, \ldots, k-1$, and $y_{ik} = 0$ for $i = k+1, \ldots, n$;

else

$y_{kk} = \left(A_{kk} - \sum_{j=1}^{k-1} y_{kj}^2\right)^{\frac{1}{2}}$ and $y_{ik} = (A_{ik} - \sum_{j=1}^{k-1} y_{kj} y_{ij})/y_{kk}$ for $i = k+1, \ldots, n$.

Step 3. *If* $A_{nn} - \sum_{j=1}^{n-1} y_{nj}^2 \leq 0$, *then*

add the index n *to* $\mathcal{J}$; $y_{nn} = 1$ *and* $y_{ni} = 0$ *for* $i = 1, \ldots, n-1$;

else

$$y_{nn} = \left(A_{nn} - \sum_{j=1}^{n-1} y_{nj}^2\right)^{\frac{1}{2}}.$$

Step 4. *For* $k \in \mathcal{J}$, *let* $\bar{b}_k = b_k$; *and for* $k \notin \mathcal{J}$, *let*

$$\bar{b}_k := b_k - \sum_{i \in \mathcal{J}} A_{ki} b_i.$$

Step 5. *Let* $z_1 = \bar{b}_1 / y_{11}$ *and* $z_k := (\bar{b}_k - \sum_{j=1}^{k-1} y_{kj} z_j)/y_{kk}$ *for* $k = 2, \ldots, n$.

Step 6. *Let* $x_n = z_n / y_{nn}$ *and* $x_k := (z_k - \sum_{j=k+1}^{n} y_{jk} z_j)/y_{kk}$ *for* $k = n - 1, \ldots, 1$.

Remark. Let J be the set of indices not in $\mathcal{J}$ after **Step 3**. Then **Steps 1-3** give a Cholesky factorization of the matrix $[I - D_n(J) + D_n(J)AD_n(J)]$. Since A is symmetric positive semidefinite, $D_n(J)AD_n(J)$ has the same rank as A. Thus, the solution of (4.2) is a regularized solution of (4.1) (cf. Lemma 4.1). However, we can rewrite (4.2) as follows:

$$x_{J^c} = b_{J^c} \quad \text{and} \quad A_{JJ} x_J = b_J - A_{JJ^c} b_{J^c}, \tag{4.10}$$

where J^c denotes the set of indices not in J. Using the vector $\bar{b}$ constructed in **Step 4**, (4.10) can be reformulated as $[I - D_n(J) + D_n(J)AD_n(J)]x = \bar{b}$, which is solved based on the Cholesky factorization of $[I - D_n(J) + D_n(J)AD_n(J)]$ (cf. **Steps 5-6**).

4.2 Regularized Solutions of Newton Equations

By using regularized solutions of (1.8), we have the following regularized Newton method for finding a minimizer of $f(z)$.

Algorithm 4.2 (QSPLINE Method (I)) *Suppose that* P *is an* $r \times r$ *symmetric positive semidefinite matrix and* $f(z)$ *is defined in (1.1). For any given* $z^0 \in \mathbf{R}^r$, *generate a sequence of iterates* z^{k+1} *for* $k = 0, 1, 2, \ldots$ *as follows:*

Step 0: *choose an index subset* J_k *of* $\{1, \ldots, r\}$ *such that the principal submatrix of*

$$[P + Q^T D_s(J(z^k))Q]$$

corresponding to J_k *is nonsingular and has the same rank as*

$$[P + Q^T D_s(J(z^k))Q];$$

Step 1: *compute*

$$d^k := -\left(I - D_r(J_k) + D_r(J_k)\left[P + Q^T D_s(J(z^k))Q\right]\right)^{-1} \nabla f(z^k);$$

Step 2: *set* $t_k := \min\{1, \bar{t}_k\}$ *where* $\bar{t}_k > 0$ *satisfies* $\left(d^k\right)^T \nabla f\left(z^k + \bar{t}_k d^k\right) = 0$;

Step 3: *compute the next iterate* $z^{k+1} := z^k + t_k d^k$.

Remark. Note that d^k is a regularized solution of the Newton equation (1.8). Therefore, one can apply the Regularized Cholesky Factorization method to $A := \left[P + Q^T D_s(J(z^k))Q\right]$ and $b := -\nabla f(z^k)$ for computing d^k.

Theorem 4.3 *Suppose that* P *is an* $r \times r$ *symmetric positive semidefinite matrix, the eigenvalues of* $(P + Q^T Q)$ *are in the interval* $[0,1]$, *and* $f(z)$ *defined in (1.1) is bounded below on* $\mathbb{R}^r$. *If* $\{z^k\}$ *is the sequence generated by* **QSPLINE Method (I)**, *then there exist a positive integer* k_0 *and a minimizer* z^* *of* $f(z)$ *such that* $z^k = z^*$ *for* $k \geq k_0$. *That is,* **QSPLINE Method (I)** *finds a minimizer of* $f(z)$ *in finitely many iterations.*

Proof. Let $\mathcal{H}$ be the collection of matrices of the form:

$$\left(I - D_r(J) + D_r(J)\left[P + Q^T D_s(\hat{J})Q\right]\right), \tag{4.11}$$

where $\hat{J}$ is any subset of $\{1, \ldots, s\}$ and J is a subset of $\{1, \ldots, r\}$ such that the matrix in (4.11) is nonsingular. Since $\left[P + Q^T D_s(\hat{J})Q\right]$ is symmetric positive semidefinite and has eigenvalues between 0 and 1, by Lemma 4.1 and Theorem 4.2 we know that each matrix in $\mathcal{H}$ is positive definite. Since $\mathcal{H}$ is a finite set,

$$\gamma := \max_{H \in \mathcal{H}} \|H^{-1}\| < \infty.$$

Therefore,

$$\begin{aligned} \|d^k\| &= \left\|\left(I - D_r(J_k) + D_r(J_k)\left[P + Q^T D_s(J(z^k))Q\right]\right)^{-1} \nabla f(z^k)\right\| \\ &\leq \gamma \|\nabla f(z^k)\| \quad \text{for } k = 1, 2, 3, \ldots \end{aligned}$$

Moreover, by Theorem 4.1, d^k is a solution of (1.8) if (1.8) is consistent. That is, Algorithm 4.2 is a special case of **QSPLINE** method and Theorem 4.3 follows immediately from Theorem 3.1. ■

4.3 Regularized Solutions of Simplified Newton Equations

In this subsection, we assume that the Hessian matrices of $f(z)$ have the following special form:

$$P + Q^T D_s(J(z))Q = \hat{P} - \hat{P} D_r(\hat{J})\hat{P}, \tag{4.12}$$

where $\hat{P}$ is an $r \times r$ symmetric positive definite matrix with eigenvalues in the interval $(0, 1]$ and $\hat{J}$ is an index subset of $\{1, \ldots, r\}$ depending on z. Let $\hat{J}_k$ be an index set such that

$$P + Q^T D_s(J(z^k))Q = \hat{P} - \hat{P}D_r(\hat{J}_k)\hat{P}. \tag{4.13}$$

Then (1.8) can be simplified as

$$\left(\hat{P} - \hat{P}D_r(\hat{J}_k)\hat{P}\right) d = -\nabla f(z^k),$$

or

$$\left(I - D_r(\hat{J}_k)\hat{P}\right) d = -\hat{P}^{-1}\nabla f(z^k). \tag{4.14}$$

Even though the matrix $\left(I - D_r(\hat{J}_k)\hat{P}\right)$ is not symmetric, (4.14) always has a regularized solution. As a consequence, we have another regularized Newton method based on regularized solutions of (4.14).

Lemma 4.2 *Suppose that $J \subset \hat{J} \subset \{1, \ldots, r\}$, $(I - \hat{P})_{JJ}$ is nonsingular and has the same rank as $(I - \hat{P})_{\hat{J}\hat{J}}$. Let*

$$J_0 := J \cup \left(\{1, 2, \ldots, r\} \setminus \hat{J}\right).$$

Then the principal submatrix of $[I - D_r(\hat{J})\hat{P}]$ corresponding to J_0 is nonsingular and has the same rank as $[I - D_r(\hat{J})\hat{P}]$.

Proof. First consider the following homogeneous linear system:

$$[I - D_r(\hat{J})\hat{P}]_{J_0 J_0} x_{J_0} = 0. \tag{4.15}$$

Then $x_i = 0$ for $i \in J_0 \setminus \hat{J}$. Since $J = \hat{J} \cap J_0$, (4.15) implies

$$[I - D_r(\hat{J})\hat{P}]_{JJ} x_J = [I - D_r(\hat{J})\hat{P}]_{JJ_0} x_{J_0} = 0, \tag{4.16}$$

where the first equality follows from $x_i = 0$ for $i \in J_0 \setminus J = J_0 \setminus \hat{J}$. But $[I - D_r(\hat{J})\hat{P}]_{JJ} = (I - \hat{P})_{JJ}$ is nonsingular, (4.16) implies $x_i = 0$ for $i \in J$. Therefore, (4.15) has a unique solution $x_{J_0} = 0$. Thus, $[I - D_r(\hat{J})\hat{P}]_{J_0 J_0}$ is nonsingular.

Next consider the following homogeneous linear system:

$$[I - D_r(\hat{J})\hat{P}]_{J_0} x = 0. \tag{4.17}$$

Then $x_i = 0$ for $i \notin \hat{J}$. Since $J = \hat{J} \cap J_0$, (4.17) implies

$$(I - \hat{P})_{J\hat{J}} x_{\hat{J}} = (I - \hat{P})_J x = 0, \tag{4.18}$$

where the first equality follows from $x_i = 0$ for $i \notin \hat{J}$. Let rank(B) denote the rank of a matrix B. Then

$$\text{rank}[(I - \hat{P})_{\hat{J}\hat{J}}] = \text{rank}[(I - \hat{P})_{JJ}] \leq \text{rank}[(I - \hat{P})_{J\hat{J}}] \leq \text{rank}[(I - \hat{P})_{\hat{J}\hat{J}}],$$

where the first equality is the assumption of the lemma. As a consequence, $(I-\hat{P})_{J\hat{J}}$ has the same rank as $(I-\hat{P})_{\hat{J}\hat{J}}$, so (4.18) is equivalent to

$$(I-\hat{P})_{\hat{J}\hat{J}}x_{\hat{J}} = 0. \tag{4.19}$$

By $x_i = 0$ for $i \notin \hat{J}$, we can rewrite (4.19) as

$$(I - D_r(\hat{J})\hat{P})_{\hat{J}}x = 0. \tag{4.20}$$

By (4.17) and (4.20) we obtain

$$(I - D_r(\hat{J})\hat{P})x = 0. \tag{4.21}$$

Thus, we have proved that any solution of (4.17) is also a solution of (4.21), *i.e.*, the rank of $(I - D_r(\hat{J})\hat{P})$ is less than or equal to the number of indices in J_0, which is the rank of $[I - D_r(\hat{J})\hat{P}]_{J_0J_0}$. Hence, the principal submatrix of $[I - D_r(\hat{J})\hat{P}]$ corresponding to J_0 has the same rank as $[I - D_r(\hat{J})\hat{P}]$. This completes the proof of Lemma 4.2. ■

With regularized solutions of the simplified Newton equation (4.14), we can have the following version of the general **QSPLINE** method.

Algorithm 4.3 (QSPLINE Method (II)) *Suppose that the Hessian matrix of f has a simplified structure as given in (4.12). For any given $z^0 \in \mathbb{R}^r$, generate a sequence of iterates z^{k+1} for $k = 0, 1, 2, \ldots$ as follows:*

Step 0: *let $\hat{J}_k$ be an index subset of $\{1, \ldots, r\}$ such that*

$$P + Q^T D_s(J(z^k))Q = \hat{P} - \hat{P}D_r(\hat{J}_k)\hat{P}$$

and choose an index subset J_k of $\hat{J}_k$ such that $(I-\hat{P})_{J_kJ_k}$ is nonsingular and has the same rank as $(I-\hat{P})_{\hat{J}_k\hat{J}_k}$) (cf. Lemma 4.2);

Step 1: *compute $d^k := -(I - D_r(J_k)\hat{P})^{-1}\hat{P}^{-1}\nabla f(z^k)$;*

Step 2: *set $t_k := \min\{1, \bar{t}_k\}$ where $\bar{t}_k > 0$ satisfies $(d^k)^T \nabla f\left(z^k + \bar{t}_k d^k\right) = 0$;*

Step 3: *compute the next iterate $z^{k+1} := z^k + t_k d^k$.*

Theorem 4.4 *Suppose that $f(z)$ defined in (1.1) is bounded below on $\mathbb{R}^r$ and its Hessian matrix has a simplified structure as given in (4.12). If $\{z^k\}$ is the sequence generated by* **QSPLINE Method (II)**, *then there exist a positive integer k_0 and a minimizer z^* of $f(z)$ such that $z^k = z^*$ for $k \geq k_0$. That is,* **QSPLINE Method (II)** *finds a minimizer of $f(z)$ in finitely many iterations.*

Proof. Let $J_0 := J_k \cup \left(\{1, \ldots, r\} \setminus \hat{J}_k\right)$. Then

$$I - D_r(J_0) + D_r(J_0)[I - D_r(\hat{J}_k)\hat{P}] = I - D_r(J_k)\hat{P}. \tag{4.22}$$

By Lemma 4.2, d^k is a regularized solution of (4.14) corresponding to the index set J_0. If (1.8) is consistent, then (4.14) is consistent and d^k is a solution of (4.14) (cf. Theorem 4.1). Thus, d^k is also a solution of (1.8) whenever (1.8) is consistent.

Let $\mathcal{H}$ be a collection of positive definite matrices of the form $\hat{P}[I - D_r(J)\hat{P}]$, where J are index sets. Then $\mathcal{H}$ is a finite set.

It follows from (4.22) and Lemma 4.1 that $[I - D_r(J_k)\hat{P}]$ is a nonsingular matrix. Thus, $\hat{P}[I - D_r(J_k)\hat{P}]$ is nonsingular. Since $\hat{P}$ is symmetric positive definite with eigenvalues between 0 and 1, $\hat{P}[I - D_r(J_k)\hat{P}]$ is symmetric positive semidefinite. Since any nonsingular symmetric positive semidefinite matrix is positive definite, so $\hat{P}[I - D_r(J_k)\hat{P}] \in \mathcal{H}$ and

$$d^k = -\left(\hat{P}[I - D_r(J_k)\hat{P}]\right)^{-1} \nabla f(z^k).$$

Let

$$\gamma := \max_{H \in \mathcal{H}} \|H^{-1}\| < \infty.$$

Then

$$\|d^k\| \leq \gamma \|\nabla f(z^k)\| \quad \text{for } k = 1, 2, \ldots$$

That is, Algorithm 4.3 is a special case of **QSPLINE** method and Theorem 4.4 follows immediately from Theorem 3.1. ■

5 APPLICATIONS TO QUADRATIC PROGRAMMING

The motivation for this paper was the realization that certain types of quadratic programming problems could be reformulated as unconstrained minimization problems wherein the objective function is a convex quadratic spline. In this section we provide the details for three types of convex quadratic programming problems that lend themselves readily to unconstrained reformulations and indicate which **QSPLINE** method is applicable in each case.

5.1 Convex Quadratic Problems With Simple Bound Constraints

Consider the convex quadratic programming problem:

$$\min \left\{ \frac{1}{2} x^T M x - d^T x : l \leq x \leq u \right\}, \tag{5.1}$$

where M is an $n \times n$ symmetric positive semidefinite matrix, $x \in \mathbf{R}^n$ and l, u are vectors in $\mathbf{R}^n$. Assuming (5.1) is solvable, then, as shown in [20], solving (5.1) is equivalent to solving the following system of piecewise linear equations:

$$x = ((I - \alpha M)x + \alpha d)_l^u, \tag{5.2}$$

where $(z)_l^u$ denotes the vector whose components are given by

$$((z)_l^u)_i = \begin{cases} u_i & \text{if } z_i \geq u_i, \\ z_i & \text{if } l_i < z_i < u_i, \\ l_i & \text{if } z_i \leq l_i, \end{cases}$$

and $0 < \alpha < \frac{1}{\|M\|}$ (insuring that $I - \alpha M$ is positive definite). The system (5.2) can be rewritten as

$$\alpha Mx - \alpha d - (l - \alpha d - (I - \alpha M)\,x)_+ + ((I - \alpha M)\,x + \alpha d - u)_+ = 0. \quad (5.3)$$

Multiplying both sides of (5.3) by $I - \alpha M$ and writing the resulting system in block matrix form yield

$$\begin{aligned} &\alpha\left(M - \alpha M^2\right)x - \alpha\,(I - \alpha M)\,d \\ &\quad + ((\alpha M - I)\;\;(I - \alpha M))\left(\begin{pmatrix} \alpha M - I \\ I - \alpha M \end{pmatrix} x + \begin{pmatrix} l - \alpha d \\ \alpha d - u \end{pmatrix}\right)_+ \\ &= 0. \end{aligned} \quad (5.4)$$

Observe that the left hand side of (5.4) is the gradient of the convex quadratic spline

$$f(x) := \frac{1}{2}x^T Px + x^T p + \frac{1}{2}\left\|(Qx + q)_+\right\|^2,$$

where

$$P = \alpha(M - \alpha M^2),\; p = -\alpha\,(I - \alpha M)\,d,\; Q = \begin{pmatrix} \alpha M - I \\ I - \alpha M \end{pmatrix},\; q = \begin{pmatrix} l - \alpha d \\ \alpha d - u \end{pmatrix}.$$

Furthermore, from (5.2) and (5.3), it can be seen that the Hessian matrices of $f(x)$ actually fit the form (4.12) with $\hat{P} = I - \alpha M$ being a symmetric positive definite matrix. Thus **QSPLINE Method (II)** is applicable to produce a solution to (5.2), and consequently to (5.1), in a finite number of iterations.

5.2 Strictly Convex Quadratic Programs

Consider the strictly convex quadratic programming problem:

$$\min\left\{\frac{1}{2}x^T Mx - d^T x : Ax \geq b\right\}, \quad (5.5)$$

where M is an $n \times n$ symmetric positive definite matrix, $x \in \mathbf{R}^n$, A is an $m \times n$ matrix, and b is a vector in $\mathbf{R}^m$. It is shown in [20] that (5.5) can be solved by solving the piecewise linear system:

$$z - \left(\left(I - \alpha AM^{-1}A^T\right)z + \alpha\left(b - AM^{-1}d\right)\right)_+ = 0, \quad (5.6)$$

where $\alpha > 0$ is chosen so that $\alpha < \frac{1}{\|AM^{-1}A^T\|}$, thus insuring that $I - \alpha AM^{-1}A^T$ is positive definite. The solutions to (5.5) and (5.6) are related by

$$Mx = d + A^T z.$$

Since $I - \alpha AM^{-1}A^T$ is positive definite, we can rewrite (5.6) as follows:

$$\begin{aligned} &(I - \alpha AM^{-1}A^T)z \\ &\quad -(I - \alpha AM^{-1}A^T)\left(\left(I - \alpha AM^{-1}A^T\right)z + \alpha\left(b - AM^{-1}d\right)\right)_+ \\ &= 0. \end{aligned} \quad (5.7)$$

Solving (5.7) is equivalent to minimizing the convex quadratic spline

$$g(z) := \frac{1}{2}z^T(I-\alpha AM^{-1}A^T)z-\frac{1}{2}\left\|\left[\left(I-\alpha AM^{-1}A^T\right)z+\alpha\left(b-AM^{-1}d\right)\right]_+\right\|^2.$$

Since $(t)_+^2 = t^2 - (-t)_+^2$, we can rewrite $g(z)$ as follows:

$$g(z) := \frac{1}{2}z^TPz + \frac{1}{2}\|q\|^2 + z^Tp + \frac{1}{2}\left\|(Qz+q)_+\right\|^2,$$

where

$$Q = \alpha AM^{-1}A^T - I,\ q = \alpha\left(AM^{-1}d - b\right),\ P = Q^2 - Q,\ p = Q^Tq.$$

Let

$$f(z) := \frac{1}{2}z^TPz + z^Tp + \frac{1}{2}\left\|(Qz+q)_+\right\|^2.$$

Since $-Q$ is a symmetric positive definite matrix with eigenvalues in the interval $(0,1]$, P is a symmetric positive semidefinite matrix. As with the simple bound problem, the Hessian matrices of $f(z)$ have the special form (4.12) with $\hat{P} = I - \alpha AM^{-1}A^T$ and so **QSPLINE Method (II)** is applicable to produce a minimizer z^* of $f(z)$, which is also a minimizer of $g(z)$ or a solution of (5.6), in a finite number of iterations. As a consequence, we get the solution $x^* = M^{-1}(d + A^Tz^*)$ of (5.5) in finitely many iterations.

5.3 Least Distance Problems

Consider the least distance problem:

$$\min\left\{\frac{1}{2}\|x-d\|^2 : Ax = b,\ x \geq 0\right\}, \tag{5.8}$$

where $x \in \mathbf{R}^n$, A is an $m \times n$ matrix, $b \in \mathbf{R}^m$, and d is a given vector in $\mathbf{R}^n$. It is known that solving (5.8) is equivalent to solving the piecewise linear system [27, 28, 5, 6, 9, 18, 29, 14]:

$$A\left(A^Tz + d\right)_+ - b = 0. \tag{5.9}$$

The solutions to (5.8) and (5.9) are related by

$$x = \left(A^Tz + d\right)_+.$$

The left side of (5.9) is the gradient of

$$f(z) := \frac{1}{2}z^TPz + z^Tp + \frac{1}{2}\left\|(Qz+q)_+\right\|^2,$$

where

$$P = 0,\ p = -b,\ Q = A^T,\ q = d.$$

Thus **QSPLINE Method (I)** is applicable to produce a solution z^* to (5.9) and, consequently, we get the solution $x^* = (A^Tz^* + d)_+$ of (5.8) in a finite number of iterations.

References

[1] P. Berman, N. Kovoor and P. Pardalos, Algorithms for the least distance problem, in *Complexity in Numerical Optimization*, P. M. Pardalos (ed.), World Scientific, 1993, pp. 33-56,

[2] P. Brucker, An $O(n)$ algorithm for the quadratic knapsack problem, Oper. Res. Letters 3 (1984), 163-166.

[3] P. Calamai and J. Moré, Quasi-Newton updates with bounds, SIAM J. Numer. Anal. 24 (1987), 1434-1441.

[4] B. Chen and M. Pinar, On Newton's method for Huber's robust M-estimation problems in linear regression, BIT (to appear).

[5] C. Chui, F. Deutsch, and J. Ward, Constrained best approximation in Hilbert space, Constr. Approx. 6 (1990), 35-64.

[6] C. Chui, F. Deutsch, and J. Ward, Constrained best approximation in Hilbert space (II), J. Approx. Theory 71 (1992), 213-238.

[7] F. Clark, *Optimization and Nonsmooth Analysis*, John Wiley and Sons, New York, 1983.

[8] D. Clark and M. Osborne, Finite algorithms for Huber's M-estimator, SIAM J. Stat. Comput. 7 (1986), 72-85.

[9] F. Deutsch, W. Li, and J. Ward, A dual approach to constrained interpolation from a convex subset of Hilbert space, J. Approx. Theory 90 (1997), 385-414.

[10] M. Dyer, Linear time algorithms for two- and three-variable linear programs, SIAM Journal of Computing 13 (1984), 31-45.

[11] M. Dyer, On a multi-dimensional search technique and its application to the Euclidean one-center problem, SIAM J. Comput. 15 (1986), 725-738.

[12] H. Ekblom, A new algorithm for the Huber estimator in linear models, BIT 28 (1988), 123-132.

[13] P. Huber, *Robust Statistics*, John Wiley, New York, 1981.

[14] W. Li, Linearly convergent descent methods for unconstrained minimization of convex quadratic splines, J. Optim. Theory Appl. 86 (1995), 145-172.

[15] W. Li, A conjugate gradient method for unconstrained minimization of strictly convex quadratic splines, Math. Programming 72 (1996), 17-32.

[16] W. Li, Numerical algorithms for the Huber M-estimator problem, in *Approximation Theory VIII–Vol. 1: Approximation and Interpolation*, C. K. Chui and L. L. Schumaker, eds., World Scientific Publishing Co., Inc., New York, 1995, pp. 325-334.

[17] W. Li, Unconstrained minimization of quadratic splines and applications, in *Multivariate Approximation and Splines*, G. Nürnberger, J. W. Schmidt, and G. Walz (eds.), Birkhäuser, Basel, 1997, pp. 113-128.

[18] W. Li, P. Pardalos and C. G. Han, Gauss-Seidel method for least distance problems, J. Optim. Theory Appl. 75 (1992), 487-500.

[19] W. Li and J. Swetits, A Newton method for convex regression, data smoothing and quadratic programming with bounded constraints, SIAM J. Optim. 3 (1993), 466-488.

[20] W. Li and J. Swetits, A new algorithm for strictly convex quadratic programs, SIAM J. Optim. 7 (1997), 595-619.

[21] W. Li and J. Swetits, Linear ℓ_1 estimator and Huber M-estimator, SIAM J. Optim. 8 (1998), 457-475.

[22] K. Madsen and H. Nielsen, Finite algorithms for robust linear regression, BIT 30 (1990), 682-699.

[23] K. Madsen and H. Nielsen, A finite smoothing algorithm for linear l_1 estimation, SIAM J. Optim. 3 (1993), 223-235.

[24] K. Madsen, H. Nielsen, and M. Pinar, A new finite continuation algorithm for linear programming, SIAM J. Optim. 6 (1996), 600-616.

[25] K. Madsen, H. Nielsen, and M. Pinar, Bounded constrained quadratic programming via piecewise quadratic functions, Preprint, 1997.

[26] N. Megiddo, Linear programming in linear time when the dimension is fixed, J. ACM 31 (1984), 114-127.

[27] C. Micchelli, P. Smith, J. Swetits, and J. Ward, Constrained L_p-approximation, Constr. Approx. 1 (1985), 93-102.

[28] C. Micchelli and F. Utreras, Smoothing and interpolation in a convex subset of Hilbert space, SIAM J. Sci. Statist. Comput. 9 (1988), 728-746.

[29] C. Michelot and M. Bougeard, Duality results and proximal solutions of the Huber M-estimator problem, Appl. Math. Optim. 30 (1994), 203-221.

[30] J.-S. Pang and L. Qi, A globally convergent Newton method for convex SC^1 minimization problems, J. Optim. Theory Appl. 85 (1995), 633-648.

[31] P. Pardalos and N. Kovoor, An algorithm for a singly constrained class of quadratic programs subject to upper and lower bounds, Math. Programming 46 (1990), 321-328.

[32] D. Shanno and D. Rocke, Numerical methods for robust regression: linear models, SIAM J. Sci. Stat. Comp. 7 (1986), 86-97.

[33] J. Sun, On piecewise quadratic Newton and trust region problems, Math. Programming, 76 (1997), 451-468.

Reformulation: Nonsmooth, Piecewise Smooth,
Semismooth and Smoothing Methods, pp. 259–268
Edited by M. Fukushima and L. Qi

Regularized Linear Programs with Equilibrium Constraints [1]

Olvi L. Mangasarian[†]

Abstract We consider an arbitrary linear program with equilibrium constraints (LPEC) that may possibly be infeasible or have an unbounded objective function. We regularize the LPEC by perturbing it in a minimal way so that the regularized problem is solvable. We show that such regularization leads to a problem that is guaranteed to have a solution which is an exact solution to the original LPEC if that problem is solvable, otherwise it is a residual-minimizing approximate solution to the original LPEC. We propose a finite successive linearization algorithm for the regularized problem that terminates at point satisfying the minimum principle necessary optimality condition for the problem.

Key Words linear programs with equilibrium constraints, regularization, exact penalty, concave minimization.

1 INTRODUCTION

We consider the following linear program with equilibrium constraints:

$$
\begin{array}{rrrl}
 & \min\limits_{x,y} & c^Tx + d^Ty & \\
 & \text{s. t.} & Ax + By - b \geq 0 & \\
0 \leq y & \perp & Nx + My - q \geq 0 & \\
 & & x\ ,\ y \geq 0 &
\end{array}
\qquad (1.1)
$$

where $A \in R^{p\times n}$, $B \in R^{p\times m}$, $N \in R^{m\times n}$, and $M \in R^{m\times m}$, are given matrices, $c \in R^n$, $d \in R^m$, $b \in R^p$, and $q \in R^m$, are given vectors, and $\perp$

[1]This work was supported by National Science Foundation Grant CCR-9322479 and Air Force Office of Scientific Research Grant F49620-97-1-0326 as Mathematical Programming Technical Report 97-13, November 1997. Revised January 1998.

[†]Computer Sciences Department, University of Wisconsin, 1210 West Dayton Street, Madison, WI 53706, *olvi@cs.wisc.edu*.

denotes orthogonality, that is the scalar product of the two vectors appearing on either side of $\perp$ is zero. This is a special case of a mathematical program with equilibrium constraints [7] that has important applications in machine learning [9, 11, 3]. Fukushima and Pang [6] were the first to address the feasibility issue for a mathematical program with equilibrium constraints (MPEC) similar to those of our LPEC (1.1) and point out that it is a difficult problem in general. They gave sufficient conditions to ensure feasibility, which are not trivial to check, but did not address the question of handling possibly infeasible constraints. In the present work we wish to address this case as well as the more general problem of an unsolvable LPEC via an exact penalty regularization approach. We note that exact penalty approaches have been proposed for *solvable* LPECs in [9] and for *solvable* MPECs in [15].

We briefly outline the contents of the paper now. In Section 2 we give a general exact penalty result which shows that a fixed solution of a penalty problem, for an increasing sequence of penalty parameters tending to infinity, minimizes the penalty term. Such a solution also minimizes the objective function of the original unpenalized, possibly infeasible, problem over the set of minimizers of the penalty term. If the penalty term minimum is zero then the original unpenalized problem is feasible and solvable and a solution to it is obtained by the penalty function minimization. In Section 3 we apply the penalty result to regularize the LPEC (1.1) and show that solving the regularized penalty problem, which consists of minimizing a piecewise-linear concave function on a polyhedral set, for any value of the penalty parameter exceeding some threshold, leads to a residual-minimizing solution to the LPEC which is an exact solution if the LPEC is solvable. In Section 4 we propose a supergradient-based successive linearization algorithm, that terminates in a finite number of steps at a stationary point, for solving the regularized penalty problem for the LPEC. A similar supergradient-based algorithm has been proposed for a general linear complementarity problem [14], and successfully used on the NP-complete knapsack feasibility problem [10].

1.1 Notation and Background

A word about our notation and background material. All vectors will be column vectors unless transposed to a row vector by a prime superscript ^{T}. The scalar product of two vectors x and y in the n-dimensional real space R^n will be denoted by x^Ty. For a mathematical program $\min_{x\in X} f(x)$, where $f : R^n \longrightarrow R$, the notation $\arg\min_{x\in X} f(x)$, will denote the set of solutions of the mathematical program $\min_{x\in X} f(x)$, while arg vertex $\min_{x\in X} f(x)$ will denote the set of vertex solutions of the same problem when X is polyhedral. For $x \in R^n$ and $p \in [1,\infty)$, the norm $\|x\|_p$ will denote the p-norm: $(\sum_{i=1}^{n} |x_i|^p)^{\frac{1}{p}}$ and $\|x\|_\infty$ will denote $\max_{1\le i\le n} |x_i|$. For $x \in R^n$, $(x_+)_i = \max\ \{0, x_i\}$, $i = 1,\ldots,n$. For an $m \times n$ matrix A, A_i will denote row i of A and $A_{\cdot j}$ will denote column j of A. The identity matrix in a real space of arbitrary dimension will be denoted by I, while a column vector of ones of arbitrary dimension will be denoted by e and a column vector of zeros of arbitrary dimension will be denoted by 0. The symbol := will denote a definition of the term appearing to the left of the symbol by the term appearing to the right of the symbol. For a set $X \in R^n$ $\tilde{X}$ will denote its complement in R^n.

2 EXACT PENALTY FOR POSSIBLY INFEASIBLE PROBLEMS

We derive in this section an exact penalty formulation for a possibly infeasible mathematical program and show that such an exact penalty approach can yield an infeasibility-minimizing solution to the problem for the infeasible case and an exact solution for the solvable case. We state this result in the following theorem which may be useful for handling general infeasible linear and nonlinear programs other than infeasible LPECs.

Theorem 2.1 Exact Penalty for Inconsistent Problems. *Consider the possibly infeasible mathematical program* $\min_{x \in S \cap T} f(x)$ *where* $f : \emptyset \neq S \subset R^n \longrightarrow R$ *and* $T \subset R^n$. *Let the penalty* $Q : S \longrightarrow R$ *be defined such that:*

$$Q(x) = \left\langle \begin{array}{lll} > & 0 & on \quad S \cap \tilde{T} \\ \\ = & 0 & on \quad S \cap T \end{array} \right. , \tag{2.1}$$

and let

$$P(x, \alpha) := f(x) + \alpha Q(x), \quad \alpha \geq 0. \tag{2.2}$$

If for a sequence of positive numbers $\{\alpha^i\}_{i=0}^{\infty} \uparrow \infty$:

$$\bar{x} \in arg \min_{x \in S} P(x, \alpha^i), \quad i = 0, 1, \ldots, \tag{2.3}$$

then:

(i) $Q(\bar{x}) = \inf_{x \in S} Q(x) \geq 0$.

(ii) $\bar{x} \in arg \min_{x \in S} P(x, \alpha), \ \forall \alpha \geq \alpha^0$.

(iii) $\bar{x} \in arg \min_{x \in S} \{f(x) \mid Q(x) = \inf_{x \in S} Q(x)\}$.

(iv) If $\inf_{x \in S} Q(x) = 0$, *then* $\bar{x} \in arg \min_{S \cap T} f(x)$.

Proof

(i) Let $\bar{Q} = \inf_{x \in S} Q(x) \geq 0$. Suppose $Q(\bar{x}) > \bar{Q}$ and we will exhibit a contradiction and hence $Q(\bar{x}) = \bar{Q}$. To show a contradiction let:

$$\epsilon = \frac{Q(\bar{x}) - \bar{Q}}{4} > 0, \ x \in S, \text{ such that } Q(x) < \bar{Q} + \epsilon,$$

and choose $\alpha^{\bar{i}}$ such that:

$$\alpha^{\bar{i}} > \max\left\{\frac{f(x) - f(\bar{x})}{Q(\bar{x}) - \bar{Q} - 2\epsilon}, \alpha^0\right\}.$$

We then have the following contradiction:

$$\begin{array}{l} f(x) + \alpha^{\bar{i}}\bar{Q} + \alpha^{\bar{i}}\epsilon > f(x) + \alpha^{\bar{i}}Q(x) = P(x, \alpha^{\bar{i}}) \geq \\ P(\bar{x}, \alpha^{\bar{i}}) = f(\bar{x}) + \alpha^{\bar{i}}Q(\bar{x}) > f(x) + \alpha^{\bar{i}}\bar{Q} + 2\alpha^{\bar{i}}\epsilon, \end{array}$$

where the last inequality follows from the choice of $\alpha^{\bar{i}}$.

(ii) Since

$$f(x) - f(\bar{x}) + \alpha^0(Q(x) - Q(\bar{x})) \geq 0, \ \forall x \in S,$$

it follows, since $Q(x) - Q(\bar{x}) \geq 0, \forall x \in S$, that:

$$f(x) - f(\bar{x}) + \alpha(Q(x) - Q(\bar{x})) \geq 0, \ \forall x \in S, \forall \alpha \geq \alpha^0.$$

Hence

$$P(x, \alpha) \geq P(\bar{x}, \alpha), \ \forall x \in S, \forall \alpha \geq \alpha^0.$$

(iii) Suppose that $x \in S$ and $Q(x) = Q(\bar{x})$. Then

$$\begin{array}{l} f(x) = f(x) + \alpha^0(Q(x) - Q(x)) = P(x, \alpha^0) - \alpha^0 Q(x) \geq \\ P(x, \alpha^0) - \alpha^0 Q(\bar{x}) \geq P(\bar{x}, \alpha^0) - \alpha^0 Q(\bar{x}) = f(\bar{x}). \end{array}$$

Hence,

$$\bar{x} \in \arg\min_{x \in S}\{f(x) \mid Q(x) \leq Q(\bar{x}) = \inf_{x \in S} Q(x)\}.$$

(iv) This follows directly from *(iii)* and the definition (2.1) of $Q(x)$ which implies that $x \in T$ if $\inf_{x \in S} Q(x) = 0$. ∎

Remark 2.2 *For special classes of important problems that arise in machine learning and data mining [11, 2, 13] where $P(x, \alpha)$ is concave in x and S is a polyhedral set without straight lines going to infinity in both directions, such as problem (3.8) below, the point $\bar{x}$ of Theorem 2.1 is any repeated vertex solution of $\min_{x \in S} P(x, \alpha^i)$, $\{\alpha^i\}_{i=0}^{\infty} \uparrow \infty$. Although it is not easy to get such a globally optimal repeated vertex, a stationary repeated vertex can be easily obtained by successive linearization algorithms such as Algorithm 4.1 below. Such stationary vertices turn out to be very useful in machine learning and data mining applications [11, 2, 13].*

In a related weaker result, Tikhonov and Arsenin [19, Theorem 1, p 101] establish asymptotic convergence of an inexact quadratic penalty function solution for the very special case of projecting a point on a possibly empty solution set of a system of linear equations. They show convergence to a projection of the point onto the set of solutions of the normal equations of the linear system when the linear system is infeasible. In contrast for example, Part (iii) of the above theorem shows how the Big M Method of linear programming [17, pp 81-82] can terminate at an optimal solution over the set of infeasibility minimizers when a linear program is infeasible.

We turn our attention now to LPECs that may possibly be infeasible or whose objective may be unbounded.

3 LPEC REGULARIZATION

We consider now the LPEC (1.1) with no assumptions on feasibility or boundedness of the objective function to start with. We first regularize the underlying linear program to the LPEC (1.1), that is problem (1.1) with the complementarity condition removed, and make it solvable in case it is not. Hence if the underlying linear program to the LPEC (1.1) is solvable or regularized to be so, the LPEC (1.1) will have a bounded objective function if it is feasible. This is is so because the feasible region of the LPEC (1.1) is a subset of the feasible region of the underlying linear program, and both problems have the same objective function. This preliminary regularization can be achieved [12, Theorem 2.2] by appropriately modifying the underlying linear program by solving:

$$\min_{z} \|z - (z - Hz - h)_+\|_1, \tag{3.1}$$

which is a regularization of the linear complementarity problem associated with the linear program underlying LPEC (1.1):

$$0 \leq z \perp Hz + h \geq 0. \tag{3.2}$$

Here

$$H = \begin{bmatrix} 0 & 0 & -A^T & -N^T \\ 0 & 0 & -B^T & -M^T \\ A & B & 0 & 0 \\ N & M & 0 & 0 \end{bmatrix}, \quad h = \begin{bmatrix} c \\ d \\ -b \\ -q \end{bmatrix}, \quad z = \begin{bmatrix} x \\ y \\ u \\ v \end{bmatrix}. \tag{3.3}$$

We note that because of the skew symmetry of H, the regularization problem (3.1) can be rewritten as a linear program and its solution can be used to generate a solvable underlying linear program for LPEC (1.1) [12, Equation (9)]. For further details about this regularization we refer the interested reader to [12, Theorem 2.2]. Another and simpler regularization of the linear program underlying LPEC (1.1) is one where the coefficients c, d, b, q of LPEC (1.1) are replaced by $\bar{c}, \bar{d}, \bar{b}, \bar{q}$ as follows:

$$\bar{c} = c + \bar{r}_1, \ \bar{d} = d + \bar{r}_2, \ \bar{b} = b - \bar{s}_1, \ \bar{q} = q - \bar{s}_2, \tag{3.4}$$

where $(\bar{s}_1, \bar{s}_2)$ and $(\bar{r}_1, \bar{r}_2)$ are solutions of the following "Phase I" primal and dual linear programs:

$$(\bar{x}, \bar{y}, \bar{s}_1, \bar{s}_2) \in \arg \min_{x,y,s_1,s_2} \left\{ e^T s_1 + e^T s_2 \;\middle|\; \begin{array}{l} Ax + By + s_1 \geq b, \\ Nx + My + s_2 \geq q, \\ (x, y, s_1, s_2) \geq 0 \end{array} \right\}$$

$$(\bar{u}, \bar{v}, \bar{r}_1, \bar{r}_2) \in \arg \min_{u,v,r_1,r_2} \left\{ e^T r_1 + e^T r_2 \;\middle|\; \begin{array}{l} A^T u + N^T v - r_1 \leq c, \\ B^T u + M^T v - r_2 \leq d, \\ (u, v, r_1, r_2) \geq 0 \end{array} \right\} \tag{3.5}$$

We note that for LPECs arising in machine learning [9, 11, 3] there is no need for this preliminary regularization in as much as the underlying linear programs there are feasible and solvable. For the rest of the paper we shall assume that such a regularization as given by replacing c, d, b, q by $\bar{c}, \bar{d}, \bar{b}, \bar{q}$ of (3.4) has been either carried out or is un-necessary and for simplicity of notation we shall not replace c, d, b, q by $\bar{c}, \bar{d}, \bar{b}, \bar{q}$ as would be required if such a regularization takes place. Note that this preliminary regularization, which renders the underlying linear program solvable, does in no way ensure that the LPEC is solvable. It was pointed out in [6] that such feasibility of the underlying linear program does not ensure feasibility of the LPEC constraints. In fact one of the main objectives of this work is the ability to handle LPECs that may not have a solution. Hence starting with an LPEC (1.1) for which the underlying linear program is solvable, either through a preliminary regularization as described above or because it is naturally so, we define a regularized problem which is guaranteed to be solvable as follows. (We note in passing that we can forgo this preliminary regularization and pass on the regularization process in its entirety to one LPEC regularization process which would be considerably more complicated than the proposed one. For the sake of simplicity and because machine learning LPECs need no such preliminary regularization, we opt for the simpler regularization presented below.)

We define now two sets associated with the LPEC (1.1) which will be used in applying Theorem 2.1 to obtain our principal result.

$$\begin{array}{l} S := \{(x, y) \mid Ax + By \geq b, \; Nx + My \geq q, \; (x, y) \geq 0\} \\ T := \{(x, y) \mid y \perp Nx + My - q\}. \end{array} \tag{3.6}$$

We state our principal regularization result now.

Theorem 3.1 LPEC Regularization *Let S be nonempty, let $c^T x + d^T y$ be bounded below on S and consider the following penalty problem:*

$$\min_{(x,y) \in S} c^T x + d^T y + \alpha e^T \min\{y, \; Nx + My - q\}, \; \alpha \geq 0 \tag{3.7}$$

Then the following hold:

(i) The penalty problem (3.7) has a vertex solution for each $\alpha \geq 0$.

(ii) There exist $\bar{x}$ *and* $\alpha^0 \geq 0$ *such that for all* $\alpha \geq \alpha^0$

$$\bar{x} \in \textit{arg vertex} \min_{(x,y)\in S} c^T x + d^T y + \alpha e^T \min\{y,\ Nx + My - q\} \quad (3.8)$$

(iii) Furthermore, each $\bar{x}$ *that satisfies (ii) solves the following residual-minimizing problem:*

$$\min_{(x,y)} \{c^T x + d^T y \mid (x,y) \in \textit{arg} \min_{(x,y)\in S} e^T \min\{y,\ Nx + My - q\}\} \quad (3.9)$$

(iv) If $\min_{(x,y)\in S} e^T \min\{y,\ Nx + My - q\} = 0$ *then each* $\bar{x}$ *that satisfies (ii) is an exact solution of the LPEC (1.1).*

Proof

(i) Since the concave objective function of the penalty problem (3.7) is bounded below on the nonempty polyhedral region that contains no lines that go to infinity in both directions, it must by [18, Corollary 32.3.4] have a vertex solution.

(ii) Since S has a finite number of vertices, then there exists a sequence of positive numbers $\{\alpha^i\}_{i=0}^{\infty} \uparrow \infty$ such that:

$$\bar{x} \in \text{arg vertex} \min_{x\in S} c^T x + d^T y + \alpha e^T \min\{y,\ Nx + My - q\},\ \forall \alpha \in \{\alpha^i\}_{i=0}^{\infty} \uparrow \infty.$$

The result now follows from Theorem 2.1(i)-(ii).

(iii) Follows from Theorem 2.1(iii).

(iv) Follows from Theorem 2.1(iv). ∎

It is interesting to note that the penalty term that multiplies the parameter α in the penalty problem (3.7) is the natural error residual for the complementarity problem $0 \leq y \perp Nx + My - q \geq 0$ that appears in the constraints of LPEC (1.1) and constitutes a local error bound for the general linear complementarity problem, and a global error bound for $M \in R_0$, the class of matrices M for which 0 is the unique solution to the homogeneous linear complementarity problem: $0 \leq y \perp My \geq 0$ [8, 16].

We turn our attention now to solution methods for the regularized problem.

4 SUCCESSIVE LINEARIZATION ALGORITHM

We propose here a finite successive linearization algorithm introduced in [14] that utilizes a supergradient of the piecewise-linear concave objective function of the penalty problem (3.7) as follows.

Algorithm 4.1 Successive Linearization Algorithm *Choose $\alpha > 0$. Start with an arbitrary $(x^0, y^0) \in R^{n+m}$. Having (x^i, y^i) determine (x^{i+1}, y^{i+1}) as follows:*

$$(x^{i+1}, y^{i+1}) \in arg\ vertex \min_{(x,y)\in S} \left\{ \begin{array}{l} c^T(x-x^i)+d^T(y-y^i) \\ +\alpha(\partial_x\phi(x^i,y^i)^T(x-x^i)+\partial_y\phi(x^i,y^i)^T(y-y^i)) \end{array} \right\}, \quad (4.1)$$

where $\phi(x,y)$ is the penalty term of (3.7), that is:

$$\phi(x,y) := e^T \min\{y,\ Nx+My-q\}, \quad (4.2)$$

and $\partial_x\phi(x,y)$, $\partial_y\phi(x,y)$ are supergradients of $\phi(x,y)$ with respect to x and y respectively, that is:

$$\begin{array}{l} \partial_x\phi(x,y) \\ = \sum_{j=1}^m \left\langle \begin{array}{l} 0 \text{ if } y_j < N_jx+M_jy-q_j \\ (1-\lambda_j)0+\lambda_jN_j \text{ if } y_j = N_jx+M_jy-q_j,\ 0\le\lambda_j\le 1 \\ N_j \text{ if } y_j > N_jx+M_jy-q_j \end{array} \right\rangle \\ \partial_y\phi(x,y) \\ = \sum_{j=1}^m \left\langle \begin{array}{l} I_j \text{ if } y_j < N_jx+M_jy-q_j \\ (1-\lambda_j)I_j+\lambda_jM_j \text{ if } y_j = N_jx+M_jy-q_j,\ 0\le\lambda_j\le 1 \\ M_j \text{ if } y_j > N_jx+M_jy-q_j \end{array} \right\rangle \end{array} \quad (4.3)$$

Stop if (x^i, y^i) is a solution of (4.1).

By using [14, Theorem 3] finite termination of the above algorithm can be established as follows.

Theorem 4.2 Finite Termination *The Successive Linearization Algorithm 4.1 generates a finite sequence of points with strictly decreasing objective function values for the penalty problem (3.7). The sequence terminates at a point $(x^{\bar{i}}, y^{\bar{i}}) \in S$ that satisfies the minimum principle necessary optimality condition for (3.7):*

$$c^T(x-x^{\bar{i}})+d^T(y-y^{\bar{i}})+\alpha(\partial_x\phi(x^{\bar{i}},y^{\bar{i}})^T(x-x^{\bar{i}})+\partial_y\phi(x^{\bar{i}},y^{\bar{i}})^T(y-y^{\bar{i}})) \ge 0, \quad (4.4)$$

for all $(x,y) \in S$.

We note that the bilinear algorithm of [10] for solving the knapsack feasibility problem as a linear complementarity problem can be interpreted as a special case of Algorithm 4.1 with a *fixed* $\lambda_j = 0$. That bilinear algorithm solved 80 consecutive instances of the knapsack LCP ranging in size between 10 and 3000 without failure. This is an indication that the proposed Algorithm 4.1 may be effective for classes of LPECs.

5 SUMMARY AND FUTURE WORK

We have proposed an always-solvable regularization of a completely general LPEC. The regularized problem is a piecewise linear concave minimization problem on a polyhedral set that generates a residual-minimizing solution to the original LPEC which is an exact solution if the LPEC is solvable. A finite successive linearization algorithm is proposed for solving the regularized problem that terminates at a stationary point. A smoothing of the penalty problem (3.7) objective function using a smoothing of the plus-function proposed in [4, 5] can also be used and can be shown to lead to an exact solution the penalty problem (3.7) for a finite value of the penalty parameter as was done in [1]. The encouraging computational results of [10] on a special case of the proposed algorithm on an NP-hard problem is a possible indicator of the possible effectiveness of the proposed algorithm for solving the regularized LPEC.

References

[1] P. S. Bradley, O. L. Mangasarian, and J. B. Rosen. Parsimonious least norm approximation. Technical Report 97-03, Computer Sciences Department, University of Wisconsin, Madison, Wisconsin, March 1997. Computational Optimization and Applications, to appear. ftp://ftp.cs.wisc.edu/math-prog/tech-reports/97-03.ps.Z.

[2] P. S. Bradley, O. L. Mangasarian, and W. N. Street. Feature selection via mathematical programming. *INFORMS Journal on Computing*, 1998. To appear. Available at ftp://ftp.cs.wisc.edu/math-prog/tech-reports/95-21.ps.Z.

[3] E. J. Bredensteiner and K. P. Bennett. Feature minimization within decision trees. Department of Mathematical Sciences Math Report No. 218, Rensselaer Polytechnic Institute, Troy, NY 12180, 1995. Computational Optimizations and Applications. to appear.

[4] Chunhui Chen and O. L. Mangasarian. Smoothing methods for convex inequalities and linear complementarity problems. *Mathematical Programming*, 71(1):51–69, 1995.

[5] Chunhui Chen and O. L. Mangasarian. A class of smoothing functions for nonlinear and mixed complementarity problems. *Computational Optimization and Applications*, 5(2):97–138, 1996.

[6] M. Fukushima and J.-S. Pang. Some feasibility issues in mathematical programs with equilibrium constraints. Technical report, Department of Mathematics & Physics, Johns Hopkins University, January 1997. SIAM Journal on Optimization, to appear.

[7] Z.-Q. Luo, J.-S. Pang, and D. Ralph. *Mathematical Programs with Equilibrium Constraints*. Cambridge University Press, Cambridge, England, 1996.

[8] Z.-Q. Luo and P. Tseng. Error bound and convergence analysis of matrix splitting algorithms for the affine variational inequality problem. *SIAM Journal on Optimization*, 2:43–54, 1992.

[9] O. L. Mangasarian. Misclassification minimization. *Journal of Global Optimization*, 5:309–323, 1994.

[10] O. L. Mangasarian. The linear complementarity problem as a separable bilinear program. *Journal of Global Optimization*, 6:153–161, 1995.

[11] O. L. Mangasarian. Machine learning via polyhedral concave minimization. In H. Fischer, B. Riedmueller, and S. Schaeffler, editors, *Applied Mathematics and Parallel Computing - Festschrift for Klaus Ritter*, pages 175–188. Physica-Verlag A Springer-Verlag Company, Heidelberg, 1996. Available at ftp://ftp.cs.wisc.edu/math-prog/tech-reports/95-20.ps.Z.

[12] O. L. Mangasarian. The ill-posed linear complementarity problem. In Michael Ferris and Jong-Shi Pang, editors, *Complementarity and Variational Problems*, pages 226–233. SIAM, Philadelphia, PA, 1997. Available at ftp://ftp.cs.wisc.edu/math-prog/tech-reports/95-15.ps.Z.

[13] O. L. Mangasarian. Mathematical programming in data mining. *Data Mining and Knowledge Discovery*, 1(2):183–201, 1997. Available at ftp://ftp.cs.wisc.edu/math-prog/tech-reports/96-05.ps.Z.

[14] O. L. Mangasarian. Solution of general linear complementarity problems via nondifferentiable concave minimization. *Acta Mathematica Vietnamica*, 22(1):199–205, 1997. Available at ftp://ftp.cs.wisc.edu/math-prog/tech-reports/96-10.ps.Z.

[15] O. L. Mangasarian and J.-S. Pang. Exact penalty functions for mathematical programs with linear complementarity constraints. *Optimization*, 42:1–8, 1997. Available from: ftp://ftp.cs.wisc.edu/math-prog/tech-reports/96-06.ps.Z.

[16] O. L. Mangasarian and J. Ren. New improved error bounds for the linear complementarity problem. *Mathematical Programming*, 66:241–255, 1994.

[17] K. G. Murty. *Linear Programming.* John Wiley & Sons, New York, 1983.

[18] R. T. Rockafellar. *Convex Analysis.* Princeton University Press, Princeton, New Jersey, 1970.

[19] A. N. Tikhonov and V. Y. Arsenin. *Solutions of Ill-Posed Problems.* John Wiley & Sons, New York, 1977.

Reformulation: Nonsmooth, Piecewise Smooth, Semismooth and Smoothing Methods, pp. 269–291
Edited by M. Fukushima and L. Qi

Reformulations of a Bicriterion Equilibrium Model

Patrice Marcotte*

Abstract This paper is concerned with the formulation of an equilibrium model where each agent's objective is a linear combination of two criteria, and the ratio of the weights associated with each criterion is continuously distributed across the population. I present several reformulations of that model in different variable spaces, some finite, some infinite. While the various reformulations possess quite different theoretical properties, a common numerical procedure for their solution can be worked out. Indeed this algorithm has a very natural relationship with a subproblem common to each reformulation of the model.

Key Words variational inequalities, network equilibrium, bicriterion models

1 INTRODUCTION. THE BICRITERION MODEL

This paper is motivated by equilibrium models of trafic assignment, where uniform commuters are assigned to the (equilibrium) shortest routes of a congested transportation network that link their respective travel origin and travel destination. In its simplest form, the network equilibrium model consists in modelling the path choices of commuters in a transportation network subject to congestion. The network is composed of arcs and nodes. The demand d_{kl} for travel from an origin node k to a destination node l is assumed to be known. The travel time on an arc a of the network is expressed as $C_a(\bar{v})$ where $\bar{v}$ denotes the vector of arc flows, and C_a is a delay function that relates the travel time on arc a to the arc flow vector $\bar{v}$. The delay F_j on path j of the transportation network is obtained by summing the delays on the arcs along path j. Denoting by $\bar{x}$ the path flow vector, by A the arc-path incidence matrix and $F(\bar{x})$ the vector of path delays one obtains the relationships

$$v = Ax \tag{1.1}$$

*DIRO, Université de Montréal, C.P. 6128 Succursale Centre-Ville, Montréal, Canada H3C 3J7. email: marcotte@iro.umontreal.ca

$$F(x) = A^t C(Ax). \tag{1.2}$$

A path flow vector $\bar{x}$ is in equilibrium when it cannot be improved upon, with respect to the current path delay vector $F(\bar{x})$, i.e., $\bar{x}$ satisfies the variational inequality

$$\langle F(\bar{x}), \bar{x} - \bar{y} \rangle_2 \leq 0 \qquad \forall \bar{y} \in \bar{X}, \tag{1.3}$$

where $\bar{X}$ denotes the compact set of nonnegative and demand-feasible path flow vectors, and $\langle \cdot, \cdot \rangle_2$ the standard Euclidian scalar product. Whenever F is the gradient of some function f, $\bar{x}$ is a solution of the variational inequality (1.3) if and only if it satisfies the first-order optimality conditions associated with the mathematical program

$$\min_{\bar{y} \in \bar{X}} f(\bar{y}). \tag{1.4}$$

Various extensions of this basic model have been proposed since its introduction, and the interested reader could do much worse than consult the monograph by Patriksson [9] or the survey by Florian and Hearn [3]. In this paper, I focus on a variant of the traffic assignment model that involves two criteria, travel delay $F(\bar{x})$ and out-of-pocket travel cost $G(\bar{x})$, to fix ideas. Let α be a parameter that converts one time unit into one money unit. If this parameter, whose inverse is the value of one time unit, is uniform among the users of the network, then a time-money equilibrium corresponds to the solution of the variational inequality

$$\langle F(\bar{x}) + \alpha G(\bar{x}), \bar{x} - \bar{y} \rangle_2 \leq 0 \qquad \forall \bar{y} \in \bar{X}. \tag{1.5}$$

A novel situation occurs when the parameter α assumes distinct values across the population. Let the population be divided into p classes, each class c being endowed with its own conversion parameter α_c. If the distribution of users among the p classes is uniform throughout the network, [1] a multiclass vector $x = (x^c)_c$ is in equilibrium if and only if it satisfies:

$$x^c \in X^c = \{x^c : Ax^c = h^c b, x^c \geq 0\} = h^c \bar{X} \qquad c = 1, \ldots, p \tag{1.6}$$

$$\bar{x} = \sum_{c=1}^{p} x^c \tag{1.7}$$

$$\langle F(\bar{x}) + \alpha G(\bar{x}), x^c - y^c \rangle_2 \leq 0 \qquad \forall y^c \in X^c, \quad \forall c = 1, \ldots, p, \tag{1.8}$$

where $(h^c)_c$ is a mass function, i.e., a vector of nonnegative numbers summing to one. In this model, the functions F and G assume a very specific functional form; they are only dependent on the total flow vector $\bar{x}$. In other words, the influence of each user of the network on congestion and cost is the same. As we will see later, this has algorithmic implications.

The situation becomes all the more interesting when the conversion factor is continuously distributed among the population, i.e., the repartition of the

[1]While this assumption could easily be relaxed, I retain it for ease of notation.

parameter α is characterized by a continuous density function $h(\alpha)$ rather than the mass vector $(h^c)_c$. The history of the model is well presented in the thesis of Leurent [4]. Its rigourous treatment has been the subject of recent papers, and algorithms for its numerical solution have been proposed ([5], [6], [7], [8]). The aim of the present paper is to analyze the respective merits of old and new formulations of this model, both from the theoretical and algorithmical points of view. In particular I will show that the proposed formulations, whether monotone or non monotone, finite-dimensional or infinite-dimensional, all suggest the same 'natural' procedure for determining an equilibrium solution.

To close this section, I mention that, while the model is presented within the framework of traffic assignment, it could be generalized to virtually any equilibrium situation involving two criteria. From now on, the reader not familiar with the traffic assignment problem can simply assume that this paper is about the analysis of the infinite-dimensional variational inequality (1.5).

2 FORMULATIONS OF THE BICRITERION MODEL

In this section, several formulations of the bicriterion (or 'multiclass') problem are presented. The first is a direct infinite-dimensional adaptation of the model involving but a finite number of classes, while the others are finite-dimensional and defined either in total flow space or in 'breakpoint space'. While the infinite-dimensional formulation is always valid, a technical assumption is required to allow the validity of finite-dimensional formulations. The consequences of this assumption is far-reaaching and will be discussed in detail.

I now introduce, for ease of reference, the main notations. Throughout the paper, the plain symbols x, y will denote vectors of functions in the functional space X, while the overlined symbols $\bar{x}$, $\bar{y}$ live in the finite-dimensional convex set $\bar{X}$.

α: inverse of the value of time; one refers to 'α' both as a parameter and as the 'class' whose value of time is $1/\alpha$.

$\alpha_{\max}$: maximum value of the parameter α.

h: nonnegative and continuous density function over α: $\int_0^{\alpha_{\max}} h(\alpha)\,d\alpha = 1$.

ϕ: cumulative distribution function: $\phi(\alpha) = \int_0^{\alpha} h(t)\,dt$.

ψ: $\psi(\alpha) = \int_0^{\alpha} th(t)\,dt$.

μ : $\mu = \psi \circ \phi^{-1}$.

ν: $\nu(v) = \int_0^{v} \phi^{-1}(t)\,dt$.

ν_L: $\nu_L(v) = \int_0^{v} (1/\phi^{-1}(t))\,dt$.

ξ: positive weighing function.
$x(\alpha)$: vector of flow densities.
x: $x = (x(\alpha))_{\alpha \in [0,\alpha_{max}]}$
$\bar{X}$: total flow polyhedron: $\bar{X} = \{\bar{x} \geq 0 : A\bar{x} = b\}$.
$X(\alpha)$: convex, compact feasible set for class α: $X(\alpha) = h(\alpha)\bar{X}$.
X: feasible set: $X = \prod_{\alpha \in [0,\alpha_{max}]} X(\alpha)$.
$\bar{x}$: total flow vector: $\bar{x} = \int_0^{\alpha_{max}} x(\alpha)\, d\alpha$.
$\bar{y}^i$: ith extreme point of the polyhedron $\bar{X}$, $i = 1, \ldots, I$.
$\hat{\alpha}$: vector of breakpoints: $\hat{\alpha} = (\alpha_0 = 0, \alpha_1, \ldots, \alpha_I = \alpha_{\max})$.
S: $S = \{\hat{\alpha} : 0 = \alpha_0 \leq \alpha_1 \leq \ldots \leq \alpha_I = \alpha_{\max}\}$.
F: Delay function.
G: Cost function.
$F^i(\bar{x})$: $F^i(\bar{x}) = \langle F(\bar{x}), \bar{y}^i \rangle_2$.
$G^i(\bar{x})$: $F^i(\bar{x}) = \langle G(\bar{x}), \bar{y}^i \rangle_2$.

Note that, whenever $h(\alpha)$ is positive, i.e. nonzero, there is a one-to-one correspondance between the extreme points of $X(\alpha)$ and those of $\bar{X}$. Namely, every vertex of $X(\alpha)$ is of the form $h(\alpha)\bar{y}^i$ for some vertex $\bar{y}^i$ of $\bar{X}$. As mentioned in a previous footnote, the general form $X(\alpha) = h(\alpha)\bar{X}$ can be somewhat relaxed. For instance, in the case of the traffic assignment problem, the density functions h may vary from one origin-destination couple to the other.

2.1 An infinite dimensional formulation

The substitution of $h(\alpha)$ for h^c in (1.6)–(1.8) yields the characterization of an equilibrium as a solution of the infinite system:

$$x(\alpha) \in X(\alpha) \tag{2.1}$$

$$\bar{x} = \int_0^{\alpha_{max}} x(\alpha)\, d\alpha \tag{2.2}$$

$$\langle F(\bar{x}) + \alpha G(\bar{x}), x(\alpha) - y(\alpha) \rangle_2 \leq 0 \quad \forall y(\alpha) \in X(\alpha), \quad \forall \alpha \in [0, \alpha_{\max}]. \tag{2.3}$$

The previous conditions are equivalent to stating that, with respect to a given total flow vector $\bar{x}$, each vector $x(\alpha)$ is in equilibrium with respect to the cost vector $F(\bar{x}) + \alpha G(\bar{x})$ or, equivalently, is an optimal solution of the linear program

$$\min_{y(\alpha) \in X(\alpha)} \langle F(\bar{x}) + \alpha G(\bar{x}), y(\alpha) \rangle. \tag{2.4}$$

Solving the above linear program is of course equivalent, due to the special form of the set $X(\alpha)$, to solving the linear program

$$\min_{\bar{y} \in \bar{X}} \langle F(\bar{x}) + \alpha G(\bar{x}), \bar{y} \rangle. \tag{2.5}$$

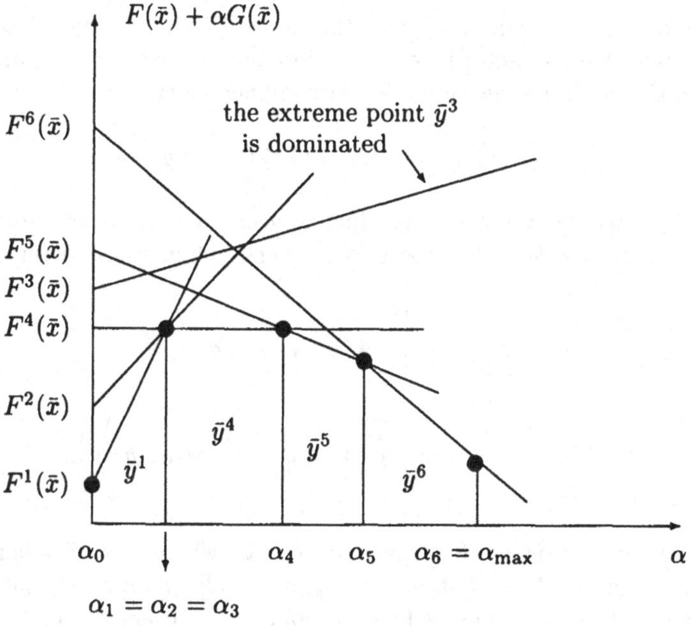

Figure 1.1 Solution of the parametric subproblem.

The solution of this linear program will be of the generic form $x(\alpha) = h(\alpha)\bar{y}^i$, where $\bar{y}^i$ is an extreme point of the polyhedron $\bar{X}$. This solution might be non unique, for instance if $F^i(\bar{x}) + \alpha G^i(\bar{x}) = F^j(\bar{x}) + \alpha G^j(\bar{x})$ for some distinct extreme points $\bar{y}^i$ and $\bar{y}^j$ of $\bar{X}$. If the numbers $G^i(\bar{x})$ and $G^j(\bar{x})$ are distinct whenever $i \neq j$, the previous equality can only occur for finitely many values of the parameter α.

A functional vector x will thus be in equilibrium if each of its values $x(\alpha)$ solves (2.4), i.e., x solves the parametric linear program (2.4). It is clear that one can restrict one's attention to 'standard' solutions of the form:

$$y(\alpha)(\bar{x}) = h(\alpha)\bar{y}^i \qquad \alpha \in [\alpha_{i-1}, \alpha_i) \qquad i = 1, \dots I-1 \tag{2.6}$$

$$y(\alpha)(\bar{x}) = h(\alpha)\bar{y}^I \qquad \alpha \in [\alpha_{I-1}, \alpha_I] \tag{2.7}$$

where $\alpha_0 = 0 \leq \alpha_1 \leq \dots \leq \alpha_I = \alpha_{\max}$ and each extreme point $\bar{y}^i$ is optimal for the linear program (2.5) over the interval $[\alpha_{i-1}, \alpha_i)$. These *breakpoints*, or values of α corresponding to changes in the optimal basis of the linear program, play a key role in the formulations considered in the paper. Implicit in the breakpoint ordering is the assumption that the slopes $G^i(\bar{x})$ are sorted in decreasing order. If either the numbers $F^i(\bar{x})$ or $G^i(\bar{x})$ are distinct for all extreme points $\bar{y}^i$ of $X(\alpha)$, then $y(\alpha)$ is uniquely determined, except at the breakpoints α_i. This situation is illustrated in Figure 1.1.

Integrating condition (2.3) over the interval $[0, \alpha_{\max}]$, one obtains that a feasible flow density vector is in equilibrium almost everywhere if and only if it satisfies the infinite-dimensional variational inequality:

$$\langle F(\bar{x}) + \alpha G(\bar{x}), x - y \rangle \leq 0 \qquad \forall y \in X, \tag{2.8}$$

where $\langle \cdot, \cdot \rangle$ denotes the scalar product of vector functions whose components are square integrable with respect to the Lebesgue measure, defined as:

$$\begin{aligned}\langle \Phi, \Psi \rangle &= \int_0^{\alpha_{\max}} \langle \Phi(\alpha), \Psi(\alpha) \rangle_2 \, d\alpha \\ &= \left\langle \int_0^{\alpha_{\max}} \Phi(\alpha)\, d\alpha, \int_0^{\alpha_{\max}} \Psi(\alpha)\, d\alpha \right\rangle_2 .\end{aligned}$$

This characterization was first given, in the case where F is a gradient mapping and the function G is constant, by Dafermos [2]. It is clear that (2.8) only implies that the conditions (2.1)–(2.3) hold almost everywhere. Marcotte and Zhu [8] proved the existence of at least one solution to the above variational inequality and gave a sufficient condition for monotonicity of the functional $F + \alpha G$ and/or uniqueness of the solution. A simpler proof of existence, to be given in the next subsection, shows that the system (2.1)–(2.3) actually admits a solution for *every* value of the parameter α, in contrast with an 'almost everywhere' solution.

In the case where F happens to be the gradient of some function f and G is constant (as shown by Marcotte and Zhu [8], this assumption is also necessary), x is an equilibrium solution if and only if it satisfies the first-order optimality conditions associated with the infinite-dimensional mathematical program

$$\min_{x \in X} f(\bar{x}) + \int_0^{\alpha_{\max}} \alpha \langle G, x(\alpha) \rangle \, d\alpha. \tag{2.9}$$

2.2 Finite-dimensional fixed point formulations

The rest of the paper will be devoted to finite-dimensional formulations of the bicriterion problem. Since the convex set $\bar{X}$ is also compact, any of its feasible points can be expressed as a convex combination of the extreme points $\bar{y}^i$. Therefore, from an existence point of view, it is equivalent to work in the set $\bar{X}$ or in its simplicial representation. Without loss of generality from the theoretical point of view, and for ease of notation, I will assume that the set $\bar{X}$ is the unit simplex. In this case, $G^i(\bar{x})$ and $F^i(\bar{x})$ reduce to $G_i(\bar{x})$ and $F_i(\bar{x})$, respectively, and the extreme points $\bar{y}^i$ are equivalent to the unit vectors e^i, although we will retain the former notation.

To each total flow vector $\bar{x}$ one can associate the point-to-set mapping

$$T_1(\bar{x}) = \left(\arg \min_{y(\alpha)\in X(\alpha)} \langle F(\bar{x}) + \alpha G(\bar{x}), y(\alpha)\rangle_2\right)_\alpha, \tag{2.10}$$

where the right-hand-side of the equality denotes the set of vector functions y whose values $y(\alpha)$ are optimal for the linear program

$$\min_{y(\alpha)\in X(\alpha)} \langle F(\bar{x}) + \alpha G(\bar{x}), y(\alpha)\rangle_2.$$

There holds (almost everywhere) the identity:

$$T_1(\bar{x}) = \arg\min_{y\in X} \langle F(\bar{x}) + \alpha G(\bar{x}), y\rangle.$$

Next, one associates with each functional vector $y(\alpha)$ the total flow vector

$$T_2(y) = \int_0^{\alpha_{\max}} y(\alpha)\, d\alpha. \tag{2.11}$$

The total flow vector $\bar{x}$ will be in equilibrium if:

$$\bar{x} \in (T_2 \circ T_1)(\bar{x}).$$

The previous fixed point mapping $T_2 \circ T_1$ does not involve explicitly the flow densities $x(\alpha)$. Indeed the sole knowledge of an equilibrium total flow vector $\bar{x}$ does not allow, without additional assumptions, to recover the vector $x(\alpha)$. Such a condition would be that the density function h is positive over the interval $(0, \alpha_{\max})$; if this were the case, there would be a one-to-one correspondance between the set of breakpoint vectors $\hat{\alpha}$ and the set of total flow vectors $\bar{x}$. If this condition is not fulfilled, it is yet possible to construct around the breakpoint vector $\hat{\alpha}$ a finite-dimensional fixed point problem whose solution yields complete information about an equilibrium solution. To this effect I introduce the mappings

$$U_1(\hat{\alpha}) = \bar{x} \qquad \text{with} \qquad \bar{x} = \sum_{i=1}^{I} \left(\int_{\alpha_{i-1}}^{\alpha_i} h(\alpha)\, d\alpha\right) \bar{y}^i \tag{2.12}$$

and U_2, which yields, for a given total flow vector $\bar{x}$, the breakpoints obtained by solving the parametric linear program (2.4). The domain of U_1 is the compact, convex set S of feasible breakpoint vectors $\hat{\alpha}$ defined as:

$$S = \{\hat{\alpha} : 0 = \alpha_0 \leq \alpha_1 \leq \ldots \leq \alpha_I = \alpha_{\max}\}, \tag{2.13}$$

and its range is a closed, convex subset of $\bar{X}$.

We have that $\hat{\alpha}$ is an equilibrium breakpoint vector if and only if it is a fixed point of the mapping $U_2 \circ U_1$. Since $U_2 \circ U_1$ involves the exact same operations as the mapping $T_2 \circ T_1$, it possesses a fixed point $\hat{\alpha}$ that corresponds to a fixed point $\bar{x}$ of $T_2 \circ T_1$ if and only $T_2 \circ T_1$ has a fixed point. Both finite-dimensional fixed point problems involve the solution of a parametric linear program whose solution yields infinite-dimensional information about the variational inequality (1.3).

Theorem 2.1 *Assume that the inequality $G^i(\bar{x}) \geq G^{i+1}(\bar{x})$ holds for all indices i and all $\bar{x}$ in $\bar{X}$. Then the solution set of (2.1)-(2.3) is nonempty.*

Proof. To prove the existence of a solution to the bicriterion problem, I will show that the mapping $U_2 \circ U_1$, which maps the convex compact set S onto itself, admits a fixed point. From Kakutani's fixed point theorem, it is sufficient to show that $U_2 \circ U_1$ is closed or simply, since U_1 is continuous, that U_2 is closed, i.e.,

$$\left.\begin{array}{l} \bar{x}^n \to \bar{x} \\ \hat{\alpha}^n \to \hat{\alpha} \\ \hat{\alpha}^n \in U_2(\bar{x}^n) \end{array}\right\} \Rightarrow \hat{\alpha} \in U_2(\bar{x}). \tag{2.14}$$

Assume that the result does not hold. This implies that there exists an interval $[\beta_1, \beta_2] \subset [\alpha_{k-1}, \alpha_k]$ $(\beta_1 < \beta_2)$ within which the extremal point $\bar{y}^k$ is not optimal (rather, $\bar{y}^j$ is), i.e.,

$$F^j(\bar{x}) + \alpha G^j < F^k(\bar{x}) + \alpha G^k \qquad \forall \alpha \in [\alpha_{k-1}, \alpha_k].$$

By continuity of F we have, for n sufficiently large:

$$[\beta_1, \beta_2] \subset [\alpha_{k-1}^n, \alpha_k^n] \tag{2.15}$$

and, by continuity of F again:

$$F^j(\bar{x}^n) + \alpha G^j < F^j(\bar{x})^n + \alpha G^j \qquad \forall \alpha \in [\alpha_{k-1}, \alpha_k], \tag{2.16}$$

which implies that $\bar{y}^k$ is not optimal within the range $[\beta_1, \beta_2]$, in contradiction with the very definition of $[\alpha_{k-1}^n, \alpha_k^n]$. (See Figure 1.2.) ∎

I now propose a family of finite-dimensional formulations involving the breakpoint vector $\hat{\alpha}$. This formulation will require that the ordering of the elements of the vector $\hat{\alpha}$ does not vary with the total flow vector $\bar{x}$. To ensure this property I will simply assume, from now on, that the function G is constant. This assumption has important consequences. Indeed, as shown in [8], it is necessary and sufficient to ensure that the variational inequality (1.3) be monotone. If F is a gradient mapping, it is also a necessary and sufficient for the variational inequality (1.3) to be reducible to the more tractable optimization form (1.4).

Let us consider Figure 1.3, where the vector $\hat{\alpha}$ does not coincide with the intersections of the straight lines $F^i(\bar{x}) + \alpha G^i$, clearly a nonequilibrium situation. In this situation, the vector $\hat{\alpha}$ does not correspond to an optimal solution of the parametric linear program (2.4). For instance, for values of the parameter α located in the interval (α_1, β_1), it is more efficient to adopt the extreme point

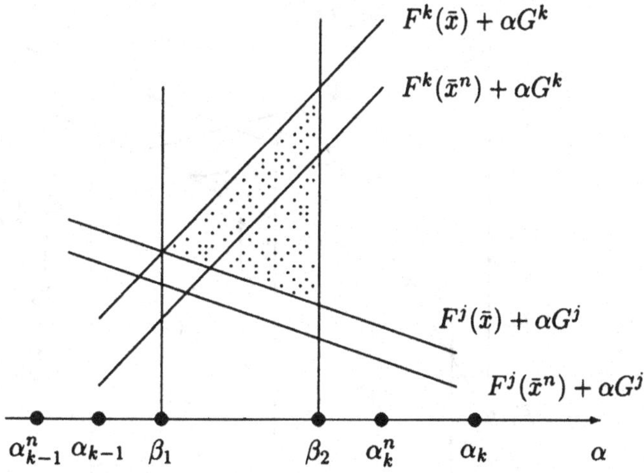

Figure 1.2 Closedness of the mapping U_2.

$\bar{y}^1$ than the extreme point $\bar{y}^2$. Actually, $\hat{\alpha}$ will be 'optimal' if and only if [2] the area

$$\sum_{i=1}^{I} \int_{\alpha_{i-1}}^{\alpha_i} (F^i(\bar{x}) + \alpha G^i)\, d\alpha = \sum_{i=1}^{I} [F^i(\bar{x})(\alpha_i - \alpha_{i-1}) + \frac{1}{2} G^i (\alpha_i - \alpha_{i-1})^2] \quad (2.17)$$

is minimized over all vectors $\hat{\alpha}$ in the set $S = \{\hat{\alpha} : 0 = \alpha_0 \leq \alpha_1 \leq \ldots \leq \alpha_I = \alpha_{\max}\}$. The minimum area will then be equal to the area under the curve defined by the value function of the parametric linear program (2.4), i.e.,

$$\min_{\alpha \in S} \sum_{i=1}^{I} \int_{\alpha_{i-1}}^{\alpha_i} (F^i(\bar{x}) + \alpha G^i)\, d\alpha = \int_0^{\alpha_{\max}} \min_{1 \leq i \leq I} (F^i(\bar{x}) + \alpha G^i)\, d\alpha. \quad (2.18)$$

This corresponds to driving to zero the areas of Δ_1, Δ_2 and Δ_3, the dotted areas in Figure 1.3. The minimization of the area under the value curve can also be achieved by minimizing, for some Lebesgue integrable, almost everywhere positive and bounded function ξ defined over the interval $[0, \alpha_{\max}]$, the weighted area:

$$\sum_{i=1}^{I} \int_{\alpha_{i-1}}^{\alpha_i} [F^i(\bar{x}) + \alpha G^i] \xi(\alpha)\, d\alpha. \quad (2.19)$$

[2]The 'only if' part holds if the density function h is positive over the open interval $(0, \alpha_{\max})$. Obviously, the value of $F^i(\bar{x}) + \alpha G^i$ is only relevant at points α where $h(\alpha)$ is positive.

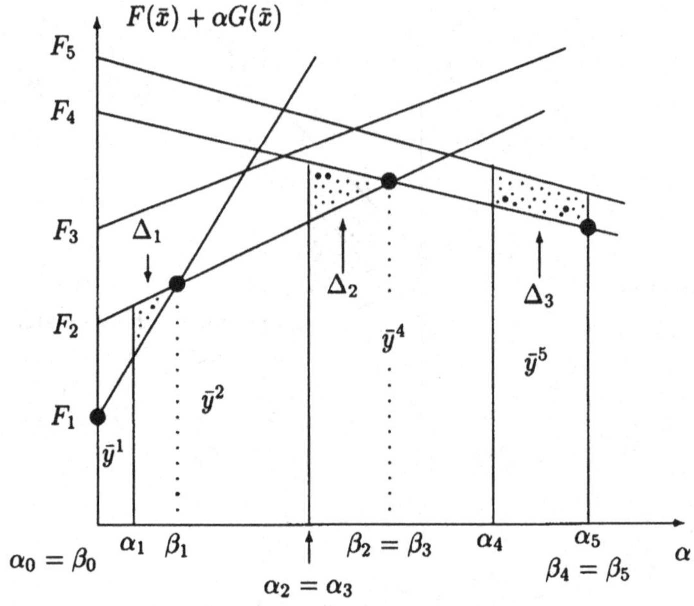

Figure 1.3 A nonequilibrium vector $\hat{\alpha}$.

Let us set $\Xi(\alpha) = \int_0^\alpha \xi(t)\,dt$ and $\chi(\alpha) = \int_0^\alpha t\xi(t)\,dt$. Let us consider again the mathematical program

$$\min_{\alpha \in S} q(\hat{\alpha}) = \sum_{i=1}^{I} \int_{\alpha_{i-1}}^{\alpha_i} [F^i(\bar{x}) + \alpha G^i]\xi(\alpha)\,d\alpha \tag{2.20}$$

$$= \sum_{i=1}^{I} [F^i(\bar{x})(\Xi(\alpha_i) - \Xi(\alpha_{i-1})) + G^i(\chi(\alpha_i) - \chi(\alpha_{i-1}))] \tag{2.21}$$

$$= \sum_{i=0}^{I-1} [(F^i(\bar{x}) - F^{i+1}(\bar{x}))\Xi(\alpha_i) + (G^i - G^{i+1})\chi(\alpha_i)]. \tag{2.22}$$

A vector $\bar{x}$ will be an equilibrium vector if, as noted earlier, it is a fixed point of the mapping $T_2 \circ T_1$, i.e., one has:

$$\bar{x}_i(\hat{\alpha}) = (\phi(\alpha_i) - \phi(\alpha_{i-1}))_{i=1,\dots,I}{}^3 \tag{2.23}$$

[3]Recall that I assumed that the set $\bar{X}$ was the unit simplex. In the general case of a convex, compact polyhedron set $\bar{X}$, one would have $\bar{x}(\hat{\alpha}) = \sum_{i=1}^{I} \lambda_i(\hat{\alpha})\bar{y}^i$ with $\lambda_i = \phi(\alpha_i) - \phi(\alpha_{i-1})$. The following derivations could be performed in λ-space by suitably redefining the numbers G^i and the functions F^i.

in accordance with the breakpoint vector that minimizes (2.19). Consequently, a breakpoint vector $\hat{\alpha}$ will be an equilibrium vector if and only if:

(i) $\hat{\alpha}$ is *globally* optimal for the the optimization problem:

$$\min_{\hat{\alpha}\in S} q(\hat{\alpha}) = \sum_{i=1}^{I} \int_{\alpha_{i-1}}^{\alpha_i} [F^i(\bar{x}) + \alpha G^i]\xi(\alpha)\, d\alpha; \tag{2.24}$$

(ii) $\bar{x}_i(\hat{\alpha}) = (\phi(\alpha_i) - \phi(\alpha_{i-1})) \qquad i = 1, \ldots, I.$

The first condition will be met only if $\hat{\alpha}$ satisfies the first-order optimality condition of (2.24), i.e., $\hat{\alpha}$ is solution of the variational inequality

$$\langle \nabla q(\hat{\alpha}), \hat{\alpha} - \bar{\beta} \rangle \leq 0 \qquad \forall \bar{\beta} \in S. \tag{2.25}$$

The gradient of the function q takes the form:

$$(\nabla q(\hat{\alpha}))_i = (F^i(\bar{x}) - F^{i+1}(\bar{x}))\xi(\alpha) + (G^i - G^{i+1})\alpha_i \xi(\alpha_i) \qquad i = 1, \ldots, I, \tag{2.26}$$

where $F^{I+1}(\bar{x})$ is set to zero and, breaking for the first and next-to-last time my earlier convention that the G^i's are sorted in decreasing order, I also set G^{I+1} to zero.

Replacing $\bar{x}$ by its expression (2.23) in terms of $\hat{\alpha}$ I obtain that $\hat{\alpha}$ must satisfy the variational inequality VI(Q, S) where the function Q is defined as:

$$\begin{aligned} Q_i(\hat{\alpha}) \quad = \quad & \left[F^i \begin{pmatrix} \phi(\alpha_1) \\ \phi(\alpha_2) - \phi(\alpha_1) \\ \vdots \\ \phi(\alpha_I) - \phi(\alpha_{I-1}) \end{pmatrix} - F^{i+1} \begin{pmatrix} \phi(\alpha_1) \\ \phi(\alpha_2) - \phi(\alpha_1) \\ \vdots \\ \phi(\alpha_I) - \phi(\alpha_{I-1}) \end{pmatrix} \right] \xi(\alpha_i) \\ & + (G^i - G^{i+1})\alpha_i \xi(\alpha_i), \qquad i = 1, \ldots, I. \end{aligned} \tag{2.27}$$

This equilibrium characterization of the vector $\hat{\alpha}$ will only be valid inasmuch as the first-order condition (2.26) is sufficient for ensuring *global* optimality of the optimization problem (2.24). For instance, if we set ξ to be constant and equal to one over the interval $[0, \alpha_{\max}]$, then the objective (2.19) is convex (actually strictly convex), as it is expressed as the sum of linear terms and single-variable quadratic terms with positive coefficients. [4] For other choices of the function ξ, the objective might fail to be convex if either Ξ or χ is nonconvex. However the first-order condition is yet sufficient to ensure global optimality. Indeed:

Theorem 2.2 *Let the weighing function ξ be positive, almost everywhere. Then any stationary point of q is a global solution of the mathematical program*

$$\min_{\hat{\alpha}\in S} q(\hat{\alpha}). \tag{2.28}$$

[4] Recall that $G^i - G^{i+1} \geq 0$ since the G^i's are sorted in decreasing order.

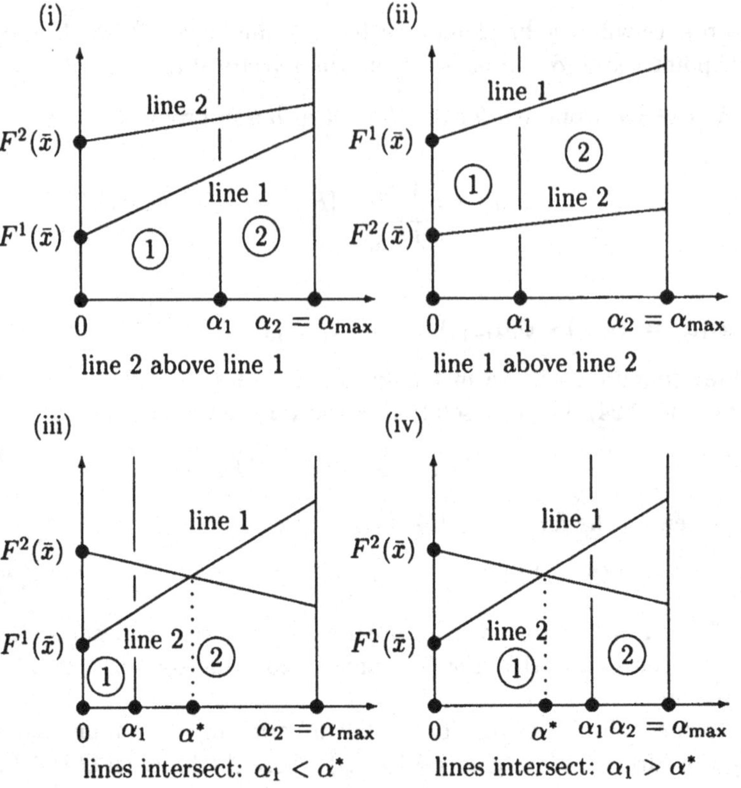

Figure 1.4 Induction basis: $I = 2$

Proof. Assume that $\hat{\alpha}$ does not achieve the minimum value of q over S. I will construct, inductively, a descent direction for q at $\hat{\alpha}$ that reduces the value of $F^i(\bar{x}) + \alpha G^i$ for each value of the state parameter α. This clearly implies that this direction is a descent direction for the weighted integral that defines the function q, whatever the choice of the weighting factor ξ. Actually, the descent direction is independent of ξ.

• INDUCTION BASIS: $I = 2$

In this case, the only free variable is α_1. If it is not optimal, one of four cases can occur, as illustrated in Figure 1.4. In cases (i) and (iii), increasing α_1 decreases the objective while, in cases (ii) and (iv), decreasing α_1 achieves the same aim.

• INDUCTION STEP

Assume that the theorem holds for the unit simplex of R^{I-1}, and let us insert into the problem an Ith extreme point, i.e, an Ith path in the realm of the traffic assignment problem. Departing from my earlier ordering convention, I set:

$$G^{k-1} \geq G^I \geq G^k \tag{2.29}$$

for some index k. This implies that $\alpha_{k-1} \leq \alpha_I \leq \alpha_k$. I consider all possible geometrical configurations in turn.

I: $\boxed{\alpha_{k-1} = \alpha_k}$ i.e.: $\alpha_{k-1} = \alpha_I = \alpha_k$.

Let $\hat{\alpha}^I$ denote the vector $\hat{\alpha}$ with respect to the I-dimensional simplex and $\hat{\alpha}^{I-1}$ the vector $\hat{\alpha}$ with respect to the $(I-1)$-dimensional simplex.

I.1 $\hat{\alpha}^{I-1}$ is nonoptimal: let d be a descent direction for $\hat{\alpha}^{I-1}$ in the $(I-1)$-dimensional simplex and make it a descent direction for $\hat{\alpha}^{I-1}$ in the I-dimensional simplex by imposing $d_I = d_{k-1}$.

I.2 $\hat{\alpha}^{I-1}$ is optimal: remove the kth value α_k and the corresponding line $F^k(\bar{x}) + \alpha G^k$ which coincides with the $(k-1)$st line. By the induction hypothesis, find a descent direction for the remaining vector $\hat{\alpha} = \hat{\alpha}^{I-1} \backslash \{\alpha_k\} \cup \{\alpha_I\}$ and make it into a descent direction in the I-dimensional simplex by imposing $d_I = d_{k-1}$, as in case I.1 above.

II: $\boxed{\alpha_{k-1} \neq \alpha_I = \alpha_k}$ i.e.: $\alpha_{k-1} < \alpha_k$ and $\alpha_{k-1} \leq \alpha_I \leq \alpha_k$.

II.1 $\boxed{\alpha_{k-1} = \alpha_I}$

II.1.1 $\hat{\alpha}^{I-1}$ is not optimal: see I.1.

II.1.2 $\hat{\alpha}^{I-1}$ is optimal; since $\hat{\alpha}^I$ is not optimal and $G^I \leq G^k$, one of the two situations illustrated in Figure 1.5 occurs. In both cases, increasing α_I decreases the objective q.

II.2 $\boxed{\alpha_I = \alpha_k}$

II.2.1 $\hat{\alpha}^{I-1}$ is not optimal: see I.1.

II.2.2 $\hat{\alpha}^{I-1}$ is optimal: increase all α_j's $(j \neq I)$ such that $\alpha_j = \alpha_k$ at the same rate (see Figure 1.6), or decrease α_I.

II.3 $\boxed{\alpha_{k-1} < \alpha_I < \alpha_k}$

II.3.1 line I lies above line k on the interval $[\alpha_{k-1}, \alpha_k]$: increase α_I.

II.3.2 line I lies below line k on the interval $[\alpha_{k-1}, \alpha_k]$: decrease α_I.

II.3.3 lines I and k coincide: line I can be deleted. The situation reduces to that of case I.2.

II.3.4 lines I and k intersect at a point of abscissa $\tilde{\alpha}$ in the interval (α_{k-1}, α_k):

II.3.4.1 $\alpha_I < \tilde{\alpha}$: increase α_I.

II.3.4.2 $\alpha_I > \tilde{\alpha}$: decrease α_I.

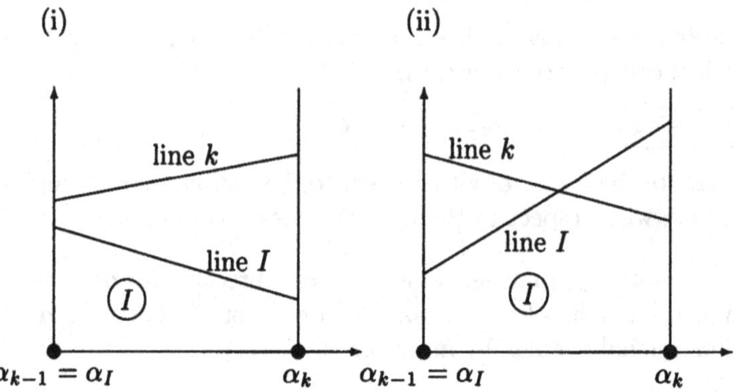

Figure 1.5 Case II.1.2

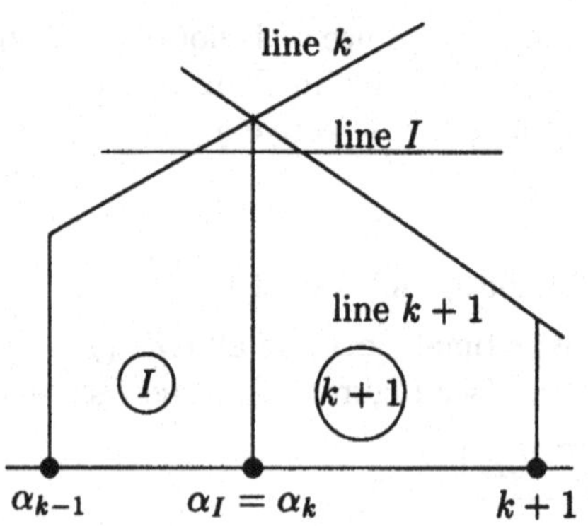

Figure 1.6 Case II.2.2

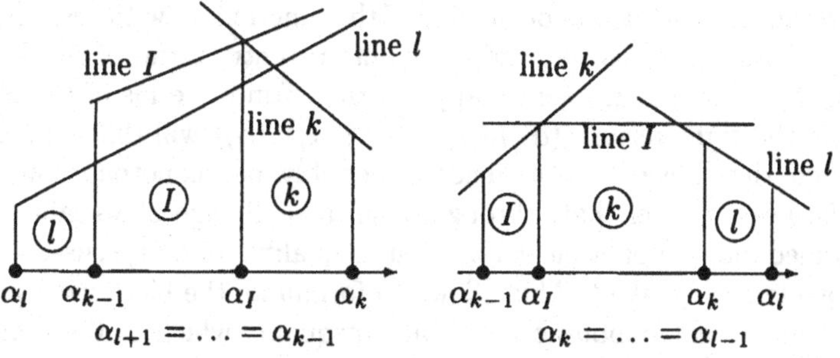

Figure 1.7 Case II.3.4.3

II.3.4.3 $\alpha_I = \tilde{\alpha}$:

Assume that $\hat{\alpha}^I$ is optimal on the interval (α_{k-1}, α_k) but suboptimal to the left of the breakpoint α_k. In this case, the subvector $(\alpha_1, \ldots, \alpha_{k-1})$ can be improved without resorting to breakpoints of indices larger than I. By the induction hypothesis, there exists a descent direction for this subvector which is readily extended to a feasible descent direction for the entire vector $\hat{\alpha}^I$. A descent direction can symmetrically be constructed if $\hat{\alpha}^I$ is suboptimal to the right of α_k.

Finally, if $\hat{\alpha}^I$ is optimal outside (α_{k-1}, α_k) but suboptimal within that interval, then there exists a line of index l which lies below the minimum of the lines I and k within the interval (α_{k-1}, α_k):

$$\exists \alpha \in (\alpha_{k-1}, \alpha_k): \quad F^l(\bar{x}) + \alpha G^l < \min\{F^k(\bar{x}) + \alpha G^k, F^I(\bar{x}) + \alpha G^I\}. \tag{2.30}$$

If the index l is less than $k-1$, increasing at a uniform rate the values of the breakpoints α_j, $j = l, \ldots, k-1$ will decrease the objective q (see Figure 1.7, left case). If l is larger than k, decreasing at a uniform rate the values of the breakpoints α_j, $j = k, \ldots, l$ will decrease the objective q (see Figure 1.7, right case).

∎

I now turn our attention to the mathematical properties of the variational inequality VI(Q, S). For algorithmic purposes, we would like the function Q to inherit the 'gradient mapping property' and the monotonicity property from F. Let us consider the first property. If F is continuously differentiable over $\bar{X}$ we derive, for $i \neq j$:

$$\frac{\partial Q_i}{\partial \alpha_j} = (F_j^i(\bar{x}) - F_{j+1}^i(\bar{x}) - F_j^{i+1}(\bar{x}) + F_{i+1}^{j+1}(\bar{x}))\xi(\alpha_i)h(\alpha_j) \tag{2.31}$$

$$\frac{\partial Q_j}{\partial \alpha_i} = (F_i^j(\bar{x}) - F_{i+1}^j(\bar{x}) - F_i^{j+1}(\bar{x}) + F_{j+1}^{i+1}(\bar{x}))\xi(\alpha_j)h(\alpha_i), \tag{2.32}$$

where F_j^i stands for the partial derivative of the function F^i with respect to the variable x_j. These terms correspond to symmetric entry terms of the Jacobian matrix $\nabla Q(\hat{\alpha})$. If F is a gradient mapping, the leftmost terms of the product agree, while the right terms $\xi(\alpha_i)h(\alpha_j)$ and $\xi(\alpha_j)h(\alpha_i)$ will differ unless $\xi \equiv h$. Although the choice $\xi = h$ yields a possibly nonmonotone mapping, I show in the next section that, through a suitable change of variable, it can be transformed into a monotone variational inequality. In that, sense, VI(Q, S) is 'monotone transformable'. This allows to formulate the bicriterion problem as a finite-dimensional monotone variational inequality which, in the symmetric case where F is a gradient, reduces to a finite-dimensional convex optimization problem.

As for the monotonicity property, let us consider:

$$\frac{\partial Q_i}{\partial \alpha_i} = (F_i^i(\bar{x}) - F_{i+1}^i(\bar{x}) - F_i^{i+1}(\bar{x}) + F_{i+1}^{i+1}(\bar{x}))\xi(\alpha_i)h(\alpha_i) \quad (2.33)$$

$$+(F^i(\bar{x}) - F^{i+1}(\bar{x}))\xi'(\alpha^i) + (G^i - G^{i+1})(\alpha_i\xi(\alpha_i))'. \quad (2.34)$$

The second term of the sum may make this diagonal term of the Jacobian negative, unless the weighing function ξ is constant. Moreover the third term could turn out negative as well, unless $\xi \equiv h$.

2.3 A finite-dimensional total flow formulation

If the density function h is positive over the interval $(0, \alpha_{\max})$, which implies that the cumulative distribution function ϕ and its inverse ϕ^{-1} are strictly increasing over that interval, one can derive a finite-dimensional formulation of the multiclass problem that only involves the total flow vector $\bar{x}$.

Theorem 2.3 *Let ϕ be continuous and positive over the interval $(0, \alpha_{\max})$ and set $\nu = \int \phi^{-1}$. Then $\bar{x} \in \bar{X}$ is an equilibrium total flow vector if and only if it satisfies the variational inequality VI($F + \nabla g, \bar{X}$) where g is a convex function defined by the expression:*

$$g(\bar{x}) = \sum_{i=1}^{I}[(G^i - G^{i+1})\nu(\sum_{j=1}^{i}\bar{x}_j)].$$

It follows that an optimal breakpoint vector is obtained by setting:

$$\alpha_i(\bar{x}_i) = \phi^{-1}(\bar{x}_1 + \ldots + \bar{x}_i) \qquad i = 1, \ldots, I.$$

Proof. Let us perform the substitution:

$$\bar{x}_i(\hat{\alpha}) = \phi(\alpha_i) - \phi(\alpha_{i-1}) \qquad i = 1, \ldots, I, \quad (2.35)$$

whose inverse is

$$\alpha_i(\bar{x}_i) = \phi^{-1}(\bar{x}_1 + \ldots + \bar{x}_i) \qquad i = 1, \ldots, I. \quad (2.36)$$

Through this transformation, the variational inequality:

$$\bar{x} \in S \tag{2.37}$$
$$\langle Q(\hat{\alpha}), \hat{\alpha} - \bar{\beta}\rangle_2 \leq 0 \qquad \forall \bar{\beta} \in S \tag{2.38}$$

becomes

$$\bar{x} \in \bar{X} \tag{2.39}$$
$$\langle J(\bar{x})^t Q(\hat{\alpha}(\bar{x})), \bar{x} - \bar{y}\rangle_2 \leq 0 \qquad \forall \bar{y} \in \bar{X}, \tag{2.40}$$

where $J(\bar{x})$ is the Jacobian matrix of the homeomorphism $\hat{\alpha}$ defined by (2.36), and whose elements are given by the expression:

$$[J(\bar{x})]_{ij} = \frac{\partial \hat{\alpha}_i(\bar{x})}{\partial \bar{x}_j} = \begin{cases} \dfrac{1}{\phi'(\phi^{-1}(\bar{x}_1 + \ldots + \bar{x}_j))} = \dfrac{1}{h(\alpha_i)}, & \text{if } j \leq i; \\ 0, & \text{otherwise.} \end{cases} \tag{2.41}$$

Let $H(\bar{x}) = J(\bar{x})^t Q(\hat{\alpha}(\bar{x}))$ and set, for notational purposes, $F^{I+1}(\bar{x}) = 0$. If one selects $\xi \equiv h$, the first term of H, that corresponds to the first term (2.27) of Q, is equal to

$$\begin{bmatrix} \frac{1}{h(\alpha_1)} & 0 & \cdots & 0 \\ \frac{1}{h(\alpha_2)} & \frac{1}{h(\alpha_2)} & \cdots & 0 \\ \vdots & \vdots & \ddots & \vdots \\ \frac{1}{h(\alpha_I)} & \frac{1}{h(\alpha_I)} & \cdots & \frac{1}{h(\alpha_I)} \end{bmatrix}^t \begin{bmatrix} h(\alpha_1)F^1(\bar{x}) - h(\alpha_1)F^2(\bar{x}) \\ h(\alpha_2)F^2(\bar{x}) - h(\alpha_2)F^3(\bar{x}) \\ \vdots \\ h(\alpha_I)F^I(\bar{x}) - h(\alpha_I)F^{I+1}(\bar{x}) \end{bmatrix} =$$

$$\begin{bmatrix} (F^1(\bar{x}) - F^2(\bar{x})) + \ldots + (F^I(\bar{x}) - F^{I+1}(\bar{x})) \\ (F^2(\bar{x}) - F^3(\bar{x})) + \ldots + (F^I(\bar{x}) - F^{I+1}(\bar{x})) \\ \vdots \\ (F^I(\bar{x}) - F^{I+1}(\bar{x})) \end{bmatrix} = \begin{bmatrix} F^1(\bar{x}) \\ F^2(\bar{x}) \\ \vdots \\ F^I(\bar{x}) \end{bmatrix}$$

$$= F(\bar{x})$$

since $F^{I+1}(\bar{x}) = 0$. The functional form of this first term should not be unexpected. It corresponds to the finite-dimensional term that makes the first part of the cost function in the infinite-dimensional formulation (2.8).

The second term of H is given by the expression:

$$\begin{bmatrix} \frac{1}{h(\alpha_1)} & 0 & \cdots & 0 \\ \frac{1}{h(\alpha_2)} & \frac{1}{h(\alpha_2)} & \cdots & 0 \\ \vdots & \vdots & \ddots & \vdots \\ \frac{1}{h(\alpha_I)} & \frac{1}{h(\alpha_I)} & \cdots & \frac{1}{h(\alpha_I)} \end{bmatrix} \begin{bmatrix} (G^1 - G^2)\alpha_1 h(\alpha_1) \\ (G^2 - G^3)\alpha_2 h(\alpha_2) \\ \vdots \\ (G^I - G^{I+1})\alpha_I h(\alpha_I) \end{bmatrix}$$

$$= \begin{bmatrix} (G^1 - G^2)\alpha_1 + (G^2 - G^3)\alpha_2 + \ldots + (G^I - G^{I+1})\alpha_I \\ (G^2 - G^3)\alpha_2 + \ldots + (G^I - G^{I+1})\alpha_I \\ \vdots \\ (G^I - G^{I+1})\alpha_I \end{bmatrix} \tag{2.42}$$

$$= \begin{bmatrix} (G^1 - G^2)\phi^{-1}(\bar{x}_1) + (G^2 - G^3)\phi^{-1}(\bar{x}_1 + \bar{x}_2) + \ldots + \\ (G^I - G^{I+1})\phi^{-1}(\sum_{i=1}^{I} \bar{x}_i) \\ (G^2 - G^3)\phi^{-1}(\bar{x}_1 + \bar{x}_2) + \ldots + (G^I - G^{I+1})\phi^{-1}(\sum_{i=1}^{I} \bar{x}_i) \\ \vdots \\ (G^I - G^{I+1})\phi^{-1}(\sum_{i=1}^{I} \bar{x}_i) \end{bmatrix} . \tag{2.43}$$

This latter term is the gradient with respect to the vector $\bar{x}$ of the convex function

$$g(\bar{x}) = \sum_{i=1}^{I} [(G^i - G^{i+1})\nu(\sum_{j=1}^{i} \bar{x}_j)] \tag{2.44}$$

where $\nu(v) = \int_0^v \phi^{-1}(t)\, dt$. The (strict) convexity of g follows from the fact that the function ϕ^{-1} is (strictly) increasing. Note also that, if the density ϕ is differentiable, one obtains:

$$\nu''(v) = \frac{1}{\phi'(\phi^{-1}(v))} = \frac{1}{h(\phi^{-1}(v))}, \tag{2.45}$$

which is nonnegative over the interval $[0, 1]$, and positive on the open interval $(0, 1)$. ∎

If one assumes furthermore that F is monotone, the variational inequality $\text{VI}(J^t Q, \bar{X})$ is monotone as well. Also, if F is the gradient of some function f, not necessarily convex, then this variational inequality can be solved by determining first-order solutions of the finite-dimensional mathematical program:

$$\min_{\bar{x} \in \bar{X}} f(\bar{x}) + \sum_{i=1}^{I} [(G^i - G^{i+1})\nu(\sum_{j=1}^{i} \bar{x}_j)]. \tag{2.46}$$

Actually, this miminization program could have been derived in a more straightforward manner by rewriting the objective in the infinite-dimensional formulation (2.9) as follows:

$$\begin{aligned} f(\bar{x}) + \int_0^{\alpha_{\max}} \alpha \langle G, x(\alpha) \rangle \, d\alpha &= f(\bar{x}_1, \ldots, \bar{x}_I) + \sum_{i=1}^{I} G^i \int_{\alpha_{i-1}}^{\alpha_i} \alpha x_i(\alpha)\, d\alpha \\ &= f(\bar{x}) + \sum_{i=1}^{I} G^i (\psi(\alpha_i) - \psi(\alpha_{i-1})), \end{aligned}$$

and performing the substitution (2.36) to obtain the mathematical program:

$$\min_{\hat{\alpha} \in S} f(\bar{x}) + \sum_{i=1}^{I} G^i[\psi(\phi^{-1}(\bar{x}_1 + \ldots + \bar{x}_i) - \psi(\phi^{-1}(\bar{x}_1 + \ldots + \bar{x}_{i-1}))]$$

$$= f(\bar{x}) + \sum_{i=1}^{I} G^i[\mu(\bar{x}_1 + \ldots + \bar{x}_i) - \mu(\bar{x}_1 + \ldots + \bar{x}_{i-1})], \tag{2.47}$$

$$= f(\bar{x}) + \sum_{i=1}^{I} [(G^i - G^{i+1})\mu(\sum_{j=1}^{i} \bar{x}_j)] \tag{2.48}$$

which is identical to (2.46), up to an additive constant, since

$$\begin{aligned} \mu' &= (\psi \circ \phi^{-1})' \\ &= (\phi^{-1})'\psi'(\phi^{-1}) \\ &= \frac{1}{\phi'(\phi^{-1})}\phi^{-1}h(\phi^{-1}) \\ &= \frac{1}{h(\phi^{-1})}\phi^{-1}h(\phi^{-1}) \\ &= \phi^{-1}. \end{aligned}$$

We close this section by relating the previous result to the finite-dimensional formulation of Leurent [5], where equilibrium conditions are stated *between* extreme points (paths) of distinct G-values and *within* extreme points having identical G-values. This author worked with a density of the value-of-time parameter, which is the inverse of the parameter α used in formulation (1.3). Let ϕ_L denote the distribution function associated with the value-of-time parameter $1/\alpha$ and define

$$\nu_L(v) = \int_0^v (1/\phi_L^{-1}(t))\, dt. \tag{2.49}$$

Let $x = \phi_L(y) = 1 - \phi(1/y)$. I obtain the derivative of ν_L in the following way:

$$\phi(1/y) = 1 - x \quad \Rightarrow \quad y = \frac{1}{\phi^{-1}(1-x)} = \phi_L^{-1}(x) \tag{2.50}$$

$$\Rightarrow \quad \nu_L'(x) = \frac{1}{\phi_L^{-1}(x)} = \phi^{-1}(1-x). \tag{2.51}$$

Assuming that f is a gradient mapping, and using my own notation, Leurent obtained the optimization characterization:

$$\min_{\bar{x}} f(\bar{x}) + \sum_{i=1}^{I} (G^{I-i+1} - G^{I-i})\nu_L(\bar{x}_I + \ldots + \bar{x}_{I-i+1}). \tag{2.52}$$

Denoting by $g_L(\bar{x})$ the summation on the right-hand-side of (2.52) one obtains:

$$g_L(\bar{x}) \;=\; \sum_{i=1}^{I} (G^i - G^{i-1})\nu_L(\bar{x}_i + \ldots + \bar{x}_I)$$

$$= \sum_{i=1}^{I}(G^i - G^{i-1})\nu_L(1 - (\bar{x}_1 + \ldots + \bar{x}_{i-1}))$$
$$= \sum_{i=0}^{I-1}(G^{i+1} - G^i)\nu_L(1 - (\bar{x}_1 + \ldots + \bar{x}_i))$$
$$= \sum_{i=1}^{I}(G^{i+1} - G^i)\nu_L(1 - (\bar{x}_1 + \ldots + \bar{x}_i)) + (G^1 - G^0)\nu_L(1).$$

Now, using (2.51) there comes:

$$\frac{\partial g_L}{\partial x_j} = \sum_{i \geq j} -(G^{i+1} - G^i)\phi^{-1}(\bar{x}_1 + \ldots + \bar{x}_i) = \frac{\partial g}{\partial x_j}. \qquad (2.53)$$

This confirms the equivalence of the optimization programs (2.52) and (2.44).

3 ALGORITHMIC CONSIDERATIONS

Throughout this short paper, I have considered several formulations of a bicriterion problem, some finite-dimensional, some infinite-dimensional, some monotone, some non monotone. The differences among these formulations are however superficial. A natural approach to solving the bicriterion problem is to use the information provided by the linear program resulting from fixing the total flow vector $\bar{x}$ to its current value. As shown in [8], the resulting algorithm achieves a linear rate of convergence whenever the quantities G^i are distinct and the function F is strongly monotone over $\bar{X}$. For all the proposed formulations, the algorithm involves solving, at each iteration, the same parametric linear program (2.4) and performs a linesearch in the corresponding feasible direction. In this respect, all formulations are equivalent. In the particular instance of traffic assignment problems, the solution of this parametric program could prove costly. Indeed, performing a pivot to leave one optimal basis for another one is costly for transportation problems and actually not much faster than re-solving for shortest paths from scratch. This suggests to solve the parametric shortest path problem arising from network equilibrium problems approximately, by selecting in a suitable manner values of the parameter α for which the linear program

$$\min_{x(\alpha)\in X(\alpha)} \langle F(\bar{x}) + \alpha G, x(\alpha)\rangle_2 \qquad (3.1)$$

will be solved. As shown in [7], the choice of a fixed number (less than twenty) equidistant values of α yields solutions that are virtually indistinguishable from the solution obtained from an exact solution of the parametric linear program, and this at a much lower computational cost. From the theoretical point of view, it is possible to implement an inexact version of the linear approximation method where the values of α are selected such as to guarantee a given error bound measured in the 'integral sense'. The subproblem is required to provide both an underestimation and an overestimation of the value function, and is stopped as soon as the integral error bound is inferior to some threshold factor ϵ_k that may depend on the iteration index k. Since the value function is concave, the underestimation is simply the piecewise linear function L going through the points

$$P_i = (\alpha_i, \min_{\bar{x} \in \bar{X}} \langle F(\bar{x}) + \alpha_i G, \bar{x} \rangle)_{i=1,\dots,I}. \tag{3.2}$$

On the other hand, each evaluation of the value function at the point α_i yields a supergradient G^i and an upper bound on the value function (3.1) in the form of the affine function

$$f_i(\alpha) = P_i + G^i(\alpha - \alpha_i). \tag{3.3}$$

It follows that the piecewise affine function

$$U(\alpha) = \min_i f_i(\alpha) \tag{3.4}$$

is an overestimation of the value function. The area $\int_0^{\alpha_{\max}} (U(\alpha) - L(\alpha))\, d\alpha$ provides the required error bound, which corresponds to the dotted area in Figure 1.8. The optimal computation of the evaluation points α_i can be achieved via a dynamic programming algorithm that is reminiscent of the one proposed by Bellman and Dreyfus [1] for determining the optimal locations of points where to evaluate a convex function whose zero is sought. I expect this procedure to be more efficient than indicated by the worst-case complexity resulting from the optimal approximation of a half-circle by $n+1$ equal chords, yielding the breakpoints

$$\alpha_k = \frac{\alpha_{\max}}{2}[1 + \cos(1 - \frac{k}{n})\pi] \qquad k = 0, 1, \dots, n. \tag{3.5}$$

Indeed, from elementary trigonometry, one obtains the error

$$\frac{{\alpha_{\max}}^2}{8}(\pi - n \sin\frac{\pi}{n}),$$

while the error estimate obtained from lower and upper bounds is equal to

$$\frac{{\alpha_{\max}}^2}{8}(2n \tan\frac{\pi}{2n} - n \sin\frac{\pi}{n}).$$

Both the exact error and its estimate are $O(1/n^2)$. Yet this worst-case complexity, coupled with the linear convergence achieved if one uses the true value function, could compare favourably to the $O(1/n)$ overall complexity of the Frank-Wolfe based algorithms which are currently used for solving network equilibrium problems. This leads me to believe that a careful implementation of an approximation scheme could allow to solve bicriterion assignment problems as efficiently as single-criterion problems.

Remark: Two key properties of the cost mapping G have been used in my analysis. The first is that the ordering of the numbers $G_i(\bar{x})$ be unchanged over

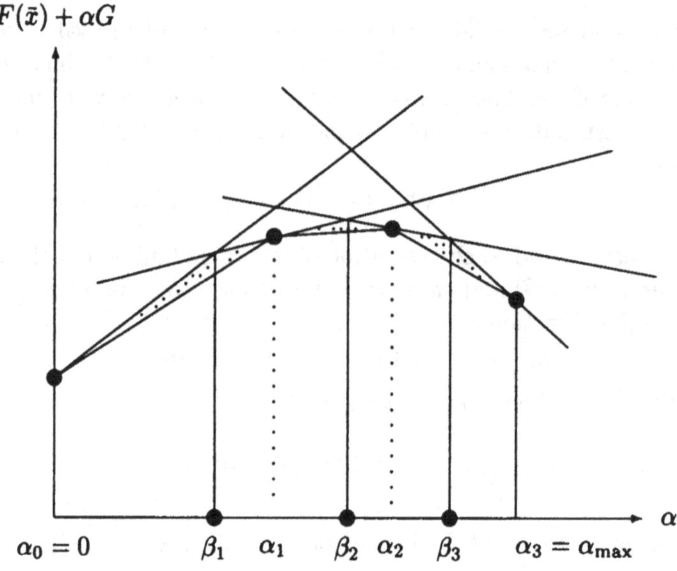

Figure 1.8 Approximation of the value function.

$\bar{X}$, the second that G_i be constant. The first property allows to develop a simple existence proof and to obtain a finite-dimensional formulation of the multiclass problem. Indeed, although the finite formulations VI(Q, S) and VI($J^t Q, \bar{X}$) have been obtained under the second condition, they also hold under the weaker first condition. It is an open question whether such results can be obtained without this assumption. As to the stronger second condition, it is necessary (and sufficient, together with the obvious conditions on the mapping F) for the multiclass problem to be reducible to a monotone variational inequality or a convex optimization problem.

Acknowledgements

This research was supported by grant A5789 from NSERC (Canada) and by FCAR (Québec).

References

[1] Bellman, R. and Dreyfus, S., Applied Dynamic Programming, Princeton University Press, Princeton, New-Jersey, 1962.

[2] Dafermos, S. C., *A multicriteria route-mode choice traffic equilibrium model*, unpublished manuscript, Lefschetz Center for Dynamical Systems, June 1981.

[3] Florian, M. and Hearn, D. W., Network Equilibrium Models and Algorithms, in: *Handbooks in Operations Research and Management Science: Network Routing*, North-Holland, Amsterdam, 1995.

[4] Leurent, F., Analyse et mesure de l'incertitude dans un modèle de simulation. Les principes, une méthode et l'exemple de l'affectation bicritère du trafic, Thèse de doctorat de l'Ecole Nationale des Ponts et Chaussées, France, 1997.

[5] Leurent, F. *Cost versus time equilibrium over a network*, European Journal of Operational Research **71** (1993) 205–221.

[6] Leurent, F. *The practice of a dual criteria assignment model with continuously distributed values-of-time*, Proceedings of the 23rd European Transport Forum: Transportation Planning Methods, London (1995) 117–128.

[7] Marcotte, P., Nguyen, S. and Tanguay, K., "Implementation of an efficient algorithm for the multiclass traffic assignment problem", Proceedings of the *13th International Symposium on Transportation and Traffic Theory*, Lyon, 24–26 juillet 1996, Jean-Baptiste Lesort, ed., Pergamon (1996) 217–236.

[8] Marcotte, P. and Zhu, D. L., "Equilibria with infinitely many differentiated classes of customers", in: Complementarity and Variational Problems. State of the Art, Proceedings of the *13th International Conference on Complementarity Problems*, Jong-Shi Pang and Michael Ferris, eds., SIAM, Philadelphia (1997) 234–258.

[9] Patriksson, M., The Traffic Assignment Problem. Models and Methods, VSP Press, Utrecht, The Netherlands, 1994.

[illegible] M. and [illegible] Network Equilibrium [illegible] *[illegible] in Operations Research and Management Science* [illegible] North-Holland, Amsterdam [illegible]

[4] [illegible] de [illegible] méthode [illegible] de l'École [illegible]

[5] [illegible] of Optimization Theory and Applications 71 (1992) [illegible]

[6] [illegible] d'une [illegible]

[7] [illegible] juillet 1986 [illegible]

[8] [illegible] and [illegible] with [illegible] State of the [illegible] SIAM, Philadelphia [illegible]

[9] [illegible] The Netherlands [illegible]

Reformulation: Nonsmooth, Piecewise Smooth, Semismooth and Smoothing Methods, pp. 293–316
Edited by M. Fukushima and L. Qi

A Smoothing Function and Its Applications [1]

Ji-Ming Peng[†]

Abstract The function $H(x) = \max\{x_1, x_2, \ldots, x_n\}$ plays an important role in mathematical programming since many problems can be reformulated via this function. In this paper, we reconsider a smooth approximation $p(x,t)$ to $H(x)$ and study its applications. The function $p(x,t)$ goes back to the exponential penalty function first proposed by Kort and Bertsekas and is also related to the network smooth function by Chen and Mangasarian. It is shown that linear and convex inequalities can be cast as smooth convex unconstrained minimization by means of this smoothing function. By reformulating the VNCP as an equation, we show under certain conditions that there exists a unique solution to the VNCP, and the norm of the function defining the equation provides a global error bound for the considered VNCP. Moreover, the smoothing function proposed in this paper is used to reformulate the VNCP as a system of parameterized smooth equations. Some properties of the smoothing path are explored. The box constrained variational inequalities can also be reformulated via the smoothing function $p(x,t)$.

Key Words smoothing function, smoothing path, vertical nonlinear complementarity problem, variational inequality problem, linear and convex inequalities.

1 INTRODUCTION

Given a point $x = (x_1, x_2, \ldots, x_n)^T \in \Re^n$, let

$$H(x) = \max\{x_1, x_2, \ldots, x_n\}. \tag{1.1}$$

The function $H(x)$ plays a fundamental role in the field of mathematical programming because many problems can be reformulated by using this function.

[1]This work was supported by Project 19601035 of NSFC in China. Current address: Faculty of Technical Mathematics and Informatics, Delft University of Technology P.O.Box 5031, 2600 GA Delft, The Netherlands. Email:j.peng@twi.tudelft.nl

[†]Institute of Computational Mathematics and Scientific Computing, Chinese Academy of Sciences, Beijing, P.O. Box 2719, 100080, China. Email: pjm@LSEC.cc.ac.cn

For example, if we consider a system of inequalities such that

$$g_1(x) \leq 0,\ g_2(x) \leq 0,\ \ldots,\ g_m(x) \leq 0, \tag{1.2}$$

where $g_j(x) : \Re^n \to \Re$ are continuous functions, then by defining

$$g(x) = \max\{g_1(x), g_2(x), \ldots, g_m(x)\}, \tag{1.3}$$

one can easily verify that (1.2) is equivalent to $g(x) \leq 0$. Let us consider the vertical nonlinear complementarity problem [11] (or VNCP for short)

$$x \geq 0,\ F_1(x) \geq 0,\ \ldots, F_m(x) \geq 0,\ x_i \prod_{j=1}^{m} F_j^i(x) = 0, \quad x \in \Re^n, \tag{1.4}$$

where $F_j(x),\ j = 1, \ldots, m$ are functions from $\Re^n$ into itself and $F_j^i(x)$ denotes the i-th component of $F_j(x)$. If we denote $F_0(x) = x$, then the VNCP (1.4) is equivalent to the following

$$\begin{aligned} & \min(F_0^i(x), F_1^i(x), \ldots, F_m^i(x)) \\ = \;& -\max(-x_i, -F_1^i(x), \ldots, -F_m^i(x)) = 0, \quad i = 1, 2, \ldots, n, \end{aligned} \tag{1.5}$$

where the minimum (or maximum) is taken componentwise. In the special case of $m = 1$, (1.4) reduces to a nonlinear complementarity problem (or NCP for short)

$$x \geq 0,\ F_1(x) \geq 0,\ x^T F_1(x) = 0.$$

An equivalent form widely used in the NCP literature is the the following system of equations $\min(x, F_1(x)) = -\max(-x, -F_1(x)) = 0$.

In recent years, smoothing methods for solving NCPs have attracted much attention and many results have been reported, see [4, 5, 6, 7, 8, 17, 18, 19] and the references therein. The first step of smoothing methods is to transform the original problem into a system of smooth equations or a smooth optimization problem via so-called smoothing functions. Then one solves the corresponding equations (or optimization problems). In [5, 6], Chen and Mangasarian introduced a class of smoothing functions and approximated NCP via a system of parameterized smooth equations. Their approach was further studied by Chen and Xiu [7], and others [4, 8, 17, 18]. In most cases, the solution set of these parameterized smooth equations forms a path as the parameter tends to zero. This path is usually called the smoothing path (or the homotopy path). All the above mentioned papers focus on LCPs and NCPs; in this paper, we consider smoothing methods for VNCP (1.4).

We consider the smooth function

$$p(x,t) = t \log \sum_{i=1}^{n} \exp^{x_i/t}, \quad t > 0. \tag{1.6}$$

It is easy to verify that $\lim_{t \to 0^+} p(x,t) = H(x)$, so we may define $p(x,0) := H(x)$. The function (1.6) was first introduced by Kort and Bertsekas [20] as a penalty function for constrained minimization. Bertsekas further studied this exponential penalty function [1, 2, 3]. Kort and Bertsekas's results were rediscovered by Li [21]. In [26], Tang and Zhang also studied the properties of the function (1.6).

We used the function (1.6) in [24] to reformulate the vertical linear complementarity problems (VLCP) as a system of parameterized smooth equations and showed that the function (1.6) is closely related to the neural networks smooth function considered by Chen and Mangasarian [5, 6]. A noninterior continuation method was also proposed to follow the smoothing path of the parameterized equations. We showed under certain conditions that the method is well defined, and provides a good approximation to the solution set of GLCP. Some interesting properties of the function (1.6) are also presented in [24].

The main motivation of this paper is to further explore the properties of the function (1.6). We study its applications in different settings, such as linear and convex inequalities, VNCPs and box constrained variational inequalities (or BVI for short). Our contributions are as follows: First, as in [5], a different reformulation of the system of linear and convex inequalities as a smooth unconstrained optimization problem is proposed. Some properties of the unconstrained optimization problem are also explored. Second, we study the equation (1.5) and discuss the existence and uniqueness of the solution. under certain conditions. Some sufficient conditions are given under which the norm of the function defining the equation (1.5) provides a global error bound for the considered VNCP. Then, by using the smooth function (1.6), we approximate VNCP as a system of parameterized smooth equations. Several interesting properties of the smoothing path are reported. Third, we cast BVI as a system of parameterized smooth equations. It is shown that one of our choices reduces to the same reformulation for BVI as considered by Gabriel and Morè [12], Chen and Ye [9].

The paper is organized as follows: In Section 2, we mainly review some properties of the function $p(x,t)$, but some new properties are also presented. Linear and convex inequalities are considered in Section 3. In Section 4, we discuss an equivalent system of VNCP and use the function $p(x,t)$ to smooth VNCP as a system of parameterized smooth equations. Some results about BVI are presented in Section 5. Finally we conclude this paper by some remarks.

Few words about the notations in the paper: Throughout this paper, we denote $\Re^n_+$, $\Re^n_-$ and $\Re^n_{--}$ the nonnegative, nonpositive and negative orthant respectively; $x_+ = \max\{x, 0\}$ where the maximum is taken componentwise; $\|\cdot\|$ denote the 2-norm in $\Re^n$; and for any function $\chi : \Re \to \Re$ and $F : \Re^n \to \Re^n$, $\chi(F)$ denote the composite function defined by $\chi(F)^i = \chi(F^i)$ for all $i = 1, \ldots, n$.

2 BASIC PROPERTIES OF THE SMOOTH FUNCTION

For any $x \in \Re^n$, let

$$B(x) = \{i \in \{1, 2, \ldots, n\} | x_i = H(x)\}, \tag{2.1}$$

and let κ_B be the number of elements in $B(x)$. Further, let $\partial p(x,0)$ be the generalized gradient of $p(x,0)$ at x in the sense of Clarke [10], and let $\text{dist}(X,Y)$ denote the minimum distance between two sets X and Y defined by

$$\text{dist}(X,Y) = \min_{x\in X, y\in Y} \|x-y\|.$$

The following lemma summarizes some basic properties of the function (1.6).

Lemma 2.1 *Suppose $p(x,t)$ is defined by (1.6). Then we have:*
(i) *For any fixed $t>0$, $p(x,t)$ is continuously differentiable and strictly convex in x.*
(ii)

$$\frac{\partial p(x,t)}{\partial x_i} = \lambda_i(x,t); \quad \lambda_i(x,t) := \frac{\exp(x_i/t)}{\sum_{j=1}^n \exp(x_j/t)} \in (0,1), \quad \sum_{i=1}^n \lambda_i(x,t) = 1.$$

(iii) *For any $x\in\Re^n$, $p(x,t)$ is continuously differentiable, increasing and convex with respect to $t>0$. Furthermore, we have*

$$p_t'(x,0^+) = \lim_{t\to 0^+} p_t'(x,t) = \log\kappa_B,$$

and

$$\lim_{t\to 0^+} \frac{p_t'(x,t)-\log\kappa_B}{t} = 0.$$

(iv) *For any $x\in\Re^n$ and $t>0$, $\log\kappa_B \le p_t'(x,t) \le \log n$.*
(v) *For any fixed $x\in\Re^n$, there exists a constant $C>0$ such that*

$$t\|\nabla_x^2 p(x,t)\| \le C, \quad \forall t>0.$$

If $\kappa_B = 1$ then

$$\lim_{t\to 0^+} \|\nabla_x^2 p(x,t)\| = 0.$$

(vi) *For any fixed $x\in\Re^n$,*

$$\nabla_x p(x,0^+) := \lim_{t\to 0^+} \nabla_x p(x,t) = \sum_{i\in B(x)} \frac{1}{\kappa_B} e_i$$

where e_i denotes the ith column of the unit matrix. Furthermore, we have

$$|\lambda_i(x,t)-\lambda_i(x,0^+)| \le \frac{t}{\kappa_B|x_i-p(x,0)|}, \quad \forall i\notin B(x),$$

and

$$|\lambda_i(x,t)-\lambda_i(x,0^+)| \le \sum_{j\notin B(x)} \frac{t}{\kappa_B|x_j-p(x,0)|}, \quad \forall i\in B(x).$$

In particularly, if $\kappa_B = 1$ then $p(x,0)$ (or $H(x)$) is differentiable at x.
(vii) *If $\lim_{k\to\infty} x^k = x$ and $\lim_{k\to\infty} t^k = 0$, then*

$$\lim_{k\to\infty} dist\{\nabla_x p(x^k,t^k), \partial p(x,0)\} = 0.$$

Proof. The results (i)-(v) have been proven in [24], we need only to prove statements (vi) and (vii). Denote $\lambda(x,0^+) = \lim_{t\to 0^+} \lambda(x,t)$. From the definition of $p(x,t)$ we get

$$\lambda_i(x,0^+) = \frac{1}{\kappa_B}, \quad \forall i \in B(x);$$

and

$$\lambda_i(x,0^+) = 0, \quad \forall i \notin B(x).$$

For $i \notin B(x)$, we have

$$\begin{aligned} 0 &\le \lambda_i(x,t) - \lambda_i(x,0^+) \\ &= \frac{\exp^{x_i/t}}{\sum_{j=1}^n \exp^{x_j/t}} \le \frac{\exp^{(x_i - p(x,0))/t}}{\kappa_B} \\ &= \frac{1}{\kappa_B \exp^{(p(x,0)-x_i)/t}} \le \frac{t}{\kappa_B |x_i - p(x,0)|}. \end{aligned}$$

For $i \in B(x)$, it is easy to verify that

$$\begin{aligned} 0 &\le \lambda_i(x,0^+) - \lambda_i(x,t) \\ &= \frac{\sum_{j\notin B(x)} \exp^{x_j/t}}{\sum_{j=1}^n \exp^{x_j/t}} = \sum_{j\notin B(x)} \lambda_j(x,t) - \lambda_j(x,0) \\ &\le \sum_{j\notin B(x)} \frac{t}{\kappa_B |x_j - p(x,0)|}. \end{aligned}$$

This completes the proof of (vi). Finally (vii) follows from the definition of $\partial p(x,0)$ and (vi). ∎

One can easily show by using the statements (iv) and (v) of Lemma 2.1 that for any $x \in \Re^n, t \ge 0$

$$p(x,0) \le H(x) + t\log\kappa_B \le p(x,t) \le H(x) + t\log n. \tag{2.2}$$

Since $\kappa_B \ge 1$, the above inequality and Lemma 2.1 imply the following result.

Lemma 2.2 *1. If $p(x,t) = 0$ for some $t > 0$ and $n > 1$, then $x \in \Re^n_{--}$.*
2. Let $n > 1$. Then for any $x \in \Re^n_-$, there is a unique $t \ge 0$ such that $p(x,t) = 0$. Furthermore, for any $x \in \Re^n$, the level set $\ell_c(t) = \{t : |p(x,t)| \le c\}$ is compact.

3. If $p(x,t) = a$ for $x \in \Re^n$ and $a \in \Re$, and $\bar{x} = (x_1 - a, x_2 - a, \ldots, x_n - a)^T$, then $p(\bar{x}, t) = 0$.
4. For any $x, y \in \Re^n$ and $t \geq 0$, $p(x,t) - p(-y,0) \leq p(x+y,t) \leq p(x,t) + p(y,0)$.
5. For any $\alpha \geq 1$, $p(\alpha x, t) \leq \alpha p(x,t)$.

Proof. Result 1 follows directly from (2.2). Statement (iv) of Lemma 2.1 and the fact that $\lim_{t \to \infty} p(x,t) = \infty$ yield 2. Results 3 and 4 follow easily from the definition of $p(x,t)$. To prove 5, we observe that

$$p(\alpha x, t) = \alpha p(x, t/\alpha) \leq \alpha p(x,t),$$

where the inequality follows from statement (i) of Lemma 2.1 and the fact $\alpha \geq 1$. This completes the proof of the lemma. ∎

In [24], we have compared the function $p(x,t)$ with Chen-Mangasarian's smooth functions for NCP [4, 5, 6, 7, 8] and showed that they share some common properties. An interrelation between the function $p(x,t)$ and the neural networks smooth function was also presented in [24]. For ease of reference, we denote the function $p(x,t)$ in the case of $n = 2$ by $p_2(x,t)$. Then the neural networks smooth function [5] is defined by $\bar{p}(\alpha, t) = \alpha + t\log(1 + \exp(-\alpha/t)) = \alpha + p_2((\alpha, 0)^T, t)$. Hence the function $p(x,t)$ can be viewed as a generalization of the neural networks smooth function.

In [8] and [9], the authors showed that for all $v = (v_1, v_2)^T \in \Re^2$ and $t > 0$, if

$$\begin{aligned} & v_1 - \bar{p}(v_1 - v_2, t) \\ = \; & v_2 - t\log\left(1 + \exp\left(\tfrac{v_2 - v_1}{t}\right)\right) \\ = \; & -t\log\left(\exp(-v_1/t) + \exp(-v_2/t)\right) \\ = \; & -p(v,t) = 0, \end{aligned} \tag{2.3}$$

then $v_1 v_2 \leq \zeta t^2$ with an appropriate constant $\zeta > 0$. One might conjecture that for any $x \in \Re^n$ and $t > 0$, if $p(x,t) = 0$ then there exists a constant $\zeta_1 > 0$ such that

$$\prod_{i=1}^{n} |x_i| \leq \zeta_1 t^n. \tag{2.4}$$

The following example shows that this conjecture is false if $n \geq 3$.

Example 2.1 *Consider*

$$\begin{aligned} X \;=\; & \{x = (x_1, x_2, x_3)^T : \; x_1 = t\log(1 - \tfrac{1}{\alpha} - \tfrac{1}{4}), \; x_2 = -t\log 4, \\ & x_3 = -t\log\alpha, \; t > 0, \; \alpha > 0\}. \end{aligned}$$

It is easy to see that $p(x,t) = 0$ for all $x \in X$. However, we have

$$|x_1 x_2 x_3| = t^3 \log\alpha \log 4 \log\frac{4\alpha}{3\alpha - 4}.$$

When α goes to infinity, the right side of the above equality also goes to infinity. One can similarly construct examples in $\Re^n$ for $n > 3$. This means that (2.4) is not true for $n \geq 3$.

On the other hand, if $p(x,t) = 0$ and $t > 0$, then it follows from (2.2) that

$$-t\log n \le H(x) \le -t\log \kappa_B, \tag{2.5}$$

which implies that $|H(x)| \le t\log n$.

3 LINEAR AND CONVEX INEQUALITIES

Consider the system of linear inequalities

$$Ax \le b, \tag{3.1}$$

where $A \in \Re^{m\times n}$ and $b \in \Re^m$ are given. In [5], Chen and Mangasarian used the neural networks smooth function to reformulate system (3.1) as an unconstrained minimization problem. The optimization problem considered in [5] is defined by

$$\min \sum_{i=1}^{m} p_2((a_i x - b_i, 0), t),$$

or

$$\min \sum_{i=1}^{m} p_2^2((a_i x - b_i, 0), t),$$

where a_i denotes the ith row of A. Clearly (3.1) is solvable if and only if the optimal values of these minimization problems are zero. Some further interesting properties of the optimization problems are presented in [5].

In this section, we find an approximation to (3.1) by solving the following unconstrained optimization problem

$$\min_{x\in\Re^n} f_1(x,t) \tag{3.2}$$

for an appropriate small $t > 0$, where

$$f_1(x,t) = p(Ax - b, t). \tag{3.3}$$

It is easy to see that the function $f_1(x,t)$ is convex in x, and is strictly convex in x if the matrix A has full column rank. Following the proof of Theorem 2.1 in [5], one can easily verify the following result.

Theorem 3.1 *Let $A \in \Re^{m\times n}$, $b \in \Re^m$ and let $f_1(x,t)$ be defined by (3.3). Then problem (3.2) has a solution if and only if $0 \ne Ax \le 0$ has no solution.*

Similar to Theorem 2.2 in [5], we can also give some conditions which ensure the uniqueness of the optimal solution of problem (3.2). Let $\ell_\nu(f) = \{x \in \Re^n | f(x) \le \nu\}$ denote the level set of $f(x)$.

Theorem 3.2 *Let $A \in \Re^{m\times n}, b \in \Re^m$, and let $f_1(x,t)$ be defined by (3.3). The following statements are equivalent:*
(i) For some $\nu \in \Re$, $\ell_\nu(f_1)$ is compact and nonempty.

(ii) For all $\nu \in \Re$, $\ell_\nu(f_1)$ is compact and nonempty.
(iii) $Ax \leq 0$, $x \neq 0$ has no solution.
(iv) Problem (3.2) has a unique optimal solution.
Furthermore, if the solution set X_1 of system (3.1) is not empty, then each of the above is equivalent to
(v) The solution set X_1 is nonempty and bounded.

Proof. The proofs of $(i) \Rightarrow (ii)$, $(ii) \Rightarrow (iii)$, $(iv) \Rightarrow (i)$ and $(iii) \Leftrightarrow (v)$ are similar to the proofs of Theorem 2.2 in [5], so we omit them here.

$(iii) \Rightarrow (iv)$. Since $Ax \leq 0$, $x \neq 0$ has no solution, the matrix A is of full column rank and $f_1(x,t)$ is strictly convex in x. So if the problem (3.2) has a solution, it is unique. Suppose (3.2) has no solution. Since $f_1(x,t)$ is continuous and convex, there exists an unbounded sequence $\{x^k\}$ such that

$$\lim_{k\to\infty} \|x^k\| = +\infty, \quad \lim_{k\to\infty} f_1(x^k,t) = \inf_{x\in\Re^n} f_1(x,t), \tag{3.4}$$

where $\inf_{x\in\Re^n} f_1(x,t)$ is either finite or $-\infty$. Hence there exists a constant $\nu > 0$ such that $f_1(x^k,t) \leq \nu$. Let $\bar{x}$ be an accumulation point of the sequence $\{\bar{x}^k = x_k/\|x^k\|\}$, i.e., there exists a subsequence $\{\bar{x}^{k_i}\}$ converging to $\bar{x}$. It follows that

$$\begin{aligned} \|x^{k_i}\|p(A\bar{x}^{k_i},0) &= p(Ax^{k_i},0) \leq p(Ax^{k_i}-b,0)+p(b,0) \\ &\leq p(Ax^{k_i}-b,t)+p(b,0) = f_1(Ax^{k_i},t)+p(b,0) \\ &\leq \nu + p(b,0), \end{aligned} \tag{3.5}$$

where the first inequality is given by result 4 of Lemma 2.2, and the second inequality follows from (2.2). Hence

$$p(A\bar{x}^{k_i},0) \leq \frac{\nu + p(b,0)}{\|x^{k_i}\|}. \tag{3.6}$$

Taking limits ($k_i \to \infty$) at both sides of the above inequality, we get $p(A\bar{x},0) \leq 0$. Thus $Ax \leq 0, x \neq 0$ has a nonzero solution which contradicts to (iii). The proof of the theorem is finished. ∎

By Lemma 2.3 in [5] (see also Hoffman [16]), if the solution set X_1 of system (3.1) is nonempty, then there exists a constant $\tau > 0$ such that for any $x \in \Re^n$, there is an $\bar{x} \in X_1$ satisfying

$$\|x - \bar{x}\| \leq \tau\|(Ax-b)_+\|. \tag{3.7}$$

For any $i \in \{1,2,\ldots,m\}$ and $t \geq 0$, the definition of $p(x,t)$ and (2.2) imply

$$(Ax-b)_i \leq p(Ax-b,0) \leq p(Ax-b,t). \tag{3.8}$$

Combining the above two inequalities we have that for any $x \in \Re^n$, there exists an $\bar{x} \in X_1$ such that

$$\|x-\bar{x}\| \leq \sqrt{m}\tau(p(Ax-b,t))_+. \tag{3.9}$$

Let $x(t)$ be a solution of (3.2). It follows that if the solution set X_1 of the inequality system (3.1) is nonempty, then there exists an $\bar{x} \in X_1$ such that

$$\|x(t) - \bar{x}\| \leq \sqrt{m}\tau(p(Ax(t) - b, t))_+ \leq \sqrt{m}\tau(p(A\bar{x} - b, t))_+ \leq t\tau\sqrt{m}\log m, \tag{3.10}$$

where the last inequality follows from (2.2). Hence, if the parameter $t > 0$ is sufficiently small, $x(t)$ can approximate the solution set X_1 to any desirable accuracy. Our next result says that if the solution set X_1 of system (3.1) has a nonempty interior, then the solution $x(t)$ of (3.2) solves the linear inequality system (3.1) exactly provided the parameter t is chosen sufficiently small.

Theorem 3.3 *Suppose the solution set X_1 of (3.1) has a nonempty interior and let $x(t)$ denote a solution of (3.2). Then there exists a constant $\bar{t} > 0$ such that for all $0 \leq t \leq \bar{t}$, one has $x(t) \in X_1$.*

Proof. Since X_1 has a nonempty interior, i.e., there exists an $\hat{x} \in X_1$ and $\epsilon_0 > 0$ such that $A\hat{x} - b \leq -\epsilon_0\bar{e}$ (here $\bar{e}$ denote the vector whose components are one). Let $\bar{t} = \epsilon_0/\log m$. Then for all $0 \leq t \leq \bar{t}$ and $i \in \{1, 2, \ldots, m\}$, we have

$$\begin{aligned} [Ax(t) - b]_i &\leq p(Ax(t) - b, 0) \leq p(Ax(t) - b, t) \leq p(A\hat{x} - b, t) \\ &\leq p(-\epsilon_0\bar{e}, t) \leq p(-\epsilon_0\bar{e}, \bar{t}) \leq 0, \end{aligned} \tag{3.11}$$

where the first inequality follows from the definition of $p(x, 0)$, the second and fifth inequalities follow from the result (iv) of Lemma 2.1, the third inequality is derived from the definition of $x(t)$, the fourth is given by the definition of $\hat{x}$ and the last one by the choice of $\bar{t}$. Therefore, $x(t) \in X_1$. ■

Next we consider the inequality system (1.2) in the case that all the functions $g_i(x), i = 1, \ldots, m$ are convex and continuous. Denote $f_2(x, t) = p(\bar{g}(x), t)$ with $\bar{g}(x) = (g_1(x), \ldots, g_m(x))^T \in \Re^m$. Consider the following unconstrained minimization

$$\min_{x \in \Re^n} f_2(x, t). \tag{3.12}$$

Similarly to the linear case, one can easily prove the following theorem.

Theorem 3.4 *Suppose that the solution set of (3.12) is nonempty and $\bar{g}(x)$ satisfy the Slater constraints qualification (i.e. $\bar{g}(\hat{x}) < 0$). Then there exists a constant $\bar{t} > 0$ such that for any $0 \leq t \leq \bar{t}$, all solutions of the problem (3.12) solve the system (1.2) exactly.*

If all $g_i(x)$ are twice continuously differentiable, then for $t > 0$, the function $f_2(x, t)$ is also twice continuously differentiable. Furthermore, if the following condition

$$\nabla\bar{g}(x)y = 0, y \neq 0 \Rightarrow y^T(\sum_{i=1}^{m} \nabla^2 g_i(x))y > 0$$

is satisfied, then one can easily verify that $f_2(x, t)$ is twice continuously differentiable and strictly convex in $\Re^n$ (see [26]).

4 VERTICAL NONLINEAR COMPLEMENTARITY PROBLEMS

In this section, we apply the function (1.6) to the vertical nonlinear complementarity problem (1.4). The section is divided into three parts: In the first subsection, we mainly study the relations between a matrix set and its row representatives. An equivalent system of VNCP is discussed in the second subsection. In the last subsection, we approximate VNCP as a system of parameterized smooth equations and discuss the properties of the smoothing path.

4.1. Matrix Set and Its Representatives

Let us first introduce some definitions which will be used in this section.

Definition 4.1 *[23] A mapping $F : \Re^n \to \Re^n$ is said to be is said to be a P_0-function if*

$$\max_{1\le i\le n, x_i\ne y_i} (x_i - y_i)(F^i(x) - F^i(y)) \ge 0 \quad \forall\, x, y \in \Re^n, x \ne y;$$

a P-function if

$$\max_{1\le i\le n} (x_i - y_i)(F^i(x) - F^i(y)) > 0 \quad \forall\, x, y \in \Re^n, x \ne y;$$

and a uniform P-function (with modulus μ) if

$$\max_{1\le i\le n} (x_i - y_i)(F^i(x) - F^i(y)) \ge \mu\|x - y\|^2, \quad \forall\, x, y \in \Re^n, x \ne y.$$

A matrix $M \in \Re^{n\times n}$ is a P_0-matrix if

$$\max_{x_i\ne 0, 1\le i\le n} x_i(Mx)_i \ge 0, \quad \forall\, x \ne 0 \in \Re^n;$$

and is a P-matrix if

$$\max_{1\le i\le n} x_i(Mx)_i > 0, \quad \forall\, x \ne 0 \in \Re^n.$$

It is known that ([23], Theorem 5.8) that if a $P_0(P)$-function is differentiable, then its Jacobian matrices are $P_0(P)$-matrices.

Let

$$\Gamma = \{M_1, \ldots, M_m\} \tag{4.1}$$

be a set of m matrices in $\Re^{n\times n}$. For any matrix M, let $M_{.j}$ $(M_{j.})$ be the j-th column (row) of M. A matrix $R \in \Re^{n\times n}$ is called a column representative of Γ if

$$R_{.j} \in \{(M_1)_{.j}, \ldots, (M_m)_{.j}\} \quad j = 1, 2, \ldots, n \tag{4.2}$$

and a row representative of Γ if

$$R_{j.} \in \{(M_1)_{j.}, \ldots, (M_m)_{j.}\} \quad j = 1, 2, \ldots, n. \tag{4.3}$$

The relationships between the matrix set Γ and its row (or column) representatives have been studied extensively by Sznajder and Gowda [25]. Their results are based on the so-called W (or W_0) properties of the matrix set Γ. Here we give a slightly different result about Γ and its representatives.

Lemma 4.1 *Suppose that every row representative of Γ is a P_0(or P)-matrix and that the matrices $D_1, \ldots, D_m$ are all nonnegative diagonal. If $\sum_{i=1}^m D_i$ is positive definite, then the matrix $\sum_{i=1}^m D_i M_i$ is a P_0 (or P)-matrix.*

Proof. We only prove the statement of the lemma in the case that each row representative of Γ is a P-matrix. The P_0 case follows similarly. Denote $D_i = \mathrm{diag}(d_{i,1}, d_{i,2}, \ldots, d_{i,n})$. Since D_i is nonnegative, we have $d_{i,j} \geq 0$ for all $i = 1, \ldots, m; j = 1, 2, \ldots, n$. Suppose to the contrary that the matrix $\sum_{i=1}^m D_i M_i$ is not a P-matrix, , i.e., there exists $x \neq 0 \in \Re^n$ such that

$$x_j \left(\sum_{i=1}^m D_i M_i x \right)_j \leq 0, \quad \forall x_j \neq 0,$$

which implies

$$\sum_{i=1}^m x_j d_{i,j} (M_i)_{j.} x \leq 0, \quad \forall x_j \neq 0. \tag{4.4}$$

Since all $d_{i,j}$ are nonnegative, from the assumption that $\sum_{i=1}^m D_i$ is positive definite we get $\bar{d}_j = \sum_{i=1}^m d_{i,j} > 0$. It follows from (4.4) that there exists an index $j_i \in \{1, 2, \ldots, m\}$ such that

$$x_j (M_{j_i})_{j.} x \leq 0, \quad \forall x_j \neq 0. \tag{4.5}$$

Let R be the matrix defined by

$$R_{j.} = \begin{cases} (M_{j_i})_{j.} & j_i \in \{i : x_j (M_i)_{j.} x \leq 0,\ i = 1, 2, \ldots, m\} \quad \text{if } x_j \neq 0; \\ (M_i)_{j.} & i \in \{0, 1, \ldots\} \quad \text{if } x_j = 0. \end{cases} \tag{4.6}$$

Clearly R is a row representative of Γ. By the assumption in the lemma, R is a P-matrix. So for any $s \neq 0 \in \Re^n$, there is an index j such that

$$s_j (Rs)_j = s_j (M_{j_i})_{j.} s > 0, \quad s_j \neq 0, \tag{4.7}$$

which contradicts to (4.5). This proves the statement of our lemma. ∎

It follows immediately from the above lemma

Corollary 4.1 *Suppose every row representative of Γ is a P_0-matrix. Then for any positive diagonal matrices $D_0, D_1, \ldots, D_m$, the matrix $D_0 + \sum_{i=1}^m D_i M_i$ is a P-matrix.*

Unfortunately, the assumptions in Lemma 4.1 seems to be too strict. Even all the matrices M_i are positive definite, there may exist a row representative R

of Γ which is not a P- (or P_0) matrix. This can be verified via the following example:

Example 4.2 *Let α be some constant larger than one. Define*

$$M_1 = \begin{pmatrix} 8\alpha & 7 \\ 7 & \frac{8}{\alpha} \end{pmatrix}; \quad M_2 = \begin{pmatrix} \frac{8}{\alpha} & 7 \\ 7 & 8\alpha \end{pmatrix}.$$

Since $\alpha > 1$, both M_1 and M_2 are positive definite. However, the row representative of $\{M_1, M_2\}$

$$R = \begin{pmatrix} \frac{8}{\alpha} & 7 \\ 7 & \frac{8}{\alpha} \end{pmatrix}$$

is not a P_0-matrix for all $\alpha > \frac{8}{7}$. If we choose $D_0 = 0$, $D_1 = diag(\frac{1}{\alpha^2}, 1 - \frac{1}{\alpha^2})$, $D_2 = diag(1 - \frac{1}{\alpha^2}, \frac{1}{\alpha^2})$, then

$$D_0 + D_1M_1 + D_2M_2 = \begin{pmatrix} \frac{16}{\alpha} - \frac{8}{\alpha^3} & 7 \\ 7 & \frac{16}{\alpha} - \frac{8}{\alpha^3} \end{pmatrix}.$$

It is easy to see that the matrix $D_0 + D_1M_1 + D_2M_2$ is singular when $\alpha = 2$.

On the other hand, if all M_k satisfy the following condition

$$(M_k)_{ii} \geq \sum_{j=1, j\neq i}^{n} |(M_k)_{ij}|, \quad k = 1, 2, \ldots, m, \tag{4.8}$$

then one can easily show that each row representative G of Γ is a P_0-matrix. If the inequality (4.8) holds strictly, then each row representative G of Γ is a P-matrix. Our next example gives two matrices satisfying (4.8).

Example 4.3 *Let*

$$M_1 = \begin{pmatrix} 0 & 0 & 0 \\ -1 & 2 & 0 \\ -1 & 0 & 1 \end{pmatrix}; \quad M_2 = \begin{pmatrix} 0 & 0 & 0 \\ 0 & 2 & 1 \\ -1 & -1 & 3 \end{pmatrix}.$$

Obviously, both M_1 and M_2 satisfy (4.8). But they are not positive semi-definite. One can easily verify that each row representative of $\{M_1, M_2\}$ is a P_0-matrix.

4.2. An Equivalent System of VNCP

For the mapping set $\Gamma_F = \{F_1, \ldots, F_m\}$, we say $G : \Re^n \to \Re^n$ is a representative of Γ_F if for all $1 \leq i \leq n$, $G_i = F_j^i$ for some $1 \leq j \leq m$. The following assumptions about VNCP (1.4) will be used in the rest of this paper.

Assumptions

(A1): Each representative of Γ_F is continuously differentiable and a uniformly P-function.

(A2): Each representative G of Γ_F is uniformly Lipschitz continuous.
Since there are finitely many representatives of Γ_F, we can assume that for any representative G, there exist two constants $\mu, L > 0$ such that

$$\max_{(x-y)_i \neq 0} (x-y)_i (G(x) - G(y))_i \geq \mu \|x-y\|^2, \quad \forall x, y \in \Re^n, \tag{4.9}$$

if (A1) is true, and that

$$\|G(x) - G(y)\| \leq L\|x-y\| \quad \forall x, y \in \Re^n \tag{4.10}$$

whenever (A2) is true.

Remark. In the case of NCP, the above assumptions reduce to the uniformly P-property and Lipschitz continuity of the function involved in NCP. On the other hand, since $F_0 = x$, it is easy to see that the assumptions (A1) and (A2) are equivalent to that each representative of the function set $\bar{\Gamma}_F = \{F_0, F_1, F_2, \ldots, F_m\}$ is continuously differentiable, Lipschitz continuous and a uniformly P-function.

For any diagonal matrix $D = \mathrm{diag}(d_{11}, d_{22}, \ldots, d_{nn})$ and any mapping F from $\Re^n$ into itself, let us denote the function

$$DF = (d_{11}F^1(x), d_{22}F^2(x), \ldots, d_{nn}F^n(x))^T.$$

Then by following the proof of Lemma 4.1, one can similarly show that

Lemma 4.2 *Suppose that every representative of Γ_F is P_0 (or P)-function and that the matrices $D_1, \ldots, D_m$ are all nonnegative diagonal. If the matrix $\sum_{i=1}^m D_i$ is diagonal positive definite, then the function $\sum_{i=1}^m D_i F_i$ is a P_0 (or P)-function.*

Now we turn our attention to VNCP (1.4). Define

$$\phi_i(x) = -(F_0^i(x), F_1^i(x), \ldots, F_m^i(x))^T \in \Re^{m+1}, \quad i = 1, 2, \ldots, n.$$

We approximate the VNCP (1.4) as the following system of equations

$$\Psi(x,t) = -\begin{pmatrix} p(\phi_1(x), t) \\ p(\phi_2(x), t) \\ \vdots \\ p(\phi_n(x), t) \end{pmatrix} = 0. \tag{4.11}$$

This system has been extensively studied for VLCP by the authors [24] under certain assumptions. The rest of this subsection is devoted to study (4.11) for VNCP in the limit case $t = 0$. Let us denote $\Psi(x, 0) = \lim_{t \to 0^+} \Psi(x, t)$. Then for any fixed $x \in \Re^n$ we have

$$\Psi_i(x, 0) = \min(F_0^i(x), F_1^i(x), \ldots, F_m^i(x)), \quad i = 1, 2, \ldots, n.$$

Our next result considers the uniqueness of the solution of VNCP (1.4).

Theorem 4.1 *Suppose that the equation $\Psi(x,0) = 0$ is solvable and that each representative of Γ_F is a P-function. Then the solution of the equation $\Psi(x,0) = 0$ (or VNCP) is unique.*

Proof. The proof of this lemma is provided by Lin [22]. For any $v^1, v^2 \in \Re^{m+1}$, from the definition of $H(v)$ it is easy to see that

$$H(v^1 + v^2) \leq H(v^1) + H(v^2). \tag{4.12}$$

The above inequality implies

$$-H(v^2 - v^1) \leq H(-v^2) - H(-v^1) \leq H(v^1 - v^2), \quad \forall v^1, v^2 \in \Re^{m+1}. \tag{4.13}$$

Assume x^* and y^* are two different solutions of the system $\Psi(x,0) = 0$, we have

$$-H(-\phi_i(x^*)) = -H(-\phi_i(y^*)) = 0, \quad i = 1, 2, \ldots, n. \tag{4.14}$$

Let I be the index set $I := \{i : x_i^* \neq y_i^*,\ i \in \{1, 2, \ldots, n\}\}$. Clearly for any $i \in I$, $\phi_i(x^*) \neq \phi_i(y^*)$. From (4.13) we conclude that

$$H(\phi_i(x^*) - \phi_i(y^*)) \geq 0; \tag{4.15}$$

and

$$-H(\phi_i(y^*) - \phi_i(x^*)) \leq 0 \tag{4.16}$$

for all $i \in I$. (4.15) implies for any $i \in I$ satisfying $x_i^* < y_i^*$, there are some indexs $j_i \in \{1, \ldots m\}$ such that

$$F_{j_i}^i(x^*) - F_{j_i}^i(y^*) \geq 0.$$

If $i \in I$ with $x_i^* > y_i^*$, from (4.16) we know that there exist some $j_i \in \{1, \ldots m\}$ such that

$$F_{j_i}^i(x^*) - F_{j_i}^i(y^*) \leq 0.$$

Let G be a representative of the function set $\bar{\Gamma}_F$ defined by

$$G_i = \begin{cases} F_j^i & j \in \{0, 1, \ldots, m\} & \text{if } x_i^* = y_i^*; \\ F_j^i & j \in J_1^i = \{j : F_j^i(x^*) - F_j^i(y^*) \geq 0, j = 1, 2, \ldots, m\} & \text{if } x_i^* < y_i^*; \\ F_j^i & j \in J_2^i = \{j : F_j^i(x^*) - F_j^i(y^*) \leq 0, j = 1, 2, \ldots, m\} & \text{if } x_i^* > y_i^*. \end{cases}$$

It follows

$$(x_i^* - y_i^*)(G_i(x^*) - G_i(y^*)) \leq 0, \quad \forall i \in I.$$

This contradicts the assumption that G is a P-function. So we have $x^* = y^*$. ∎

Let

$$\ell_c(x) = \{x \in \Re^n \,|\, \|\Psi(x,0)\| \leq c\}. \tag{4.17}$$

We have

Lemma 4.3 *Suppose that the assumption (A1) holds. Then the level set $\ell_c(x)$ is bounded.*

Proof. Since (A1) is true, for any $x \neq y \in \Re^n$ and any representative G of Γ_F, we have

$$\|G(x) - G(y)\| \geq \mu\|x - y\|. \tag{4.18}$$

Suppose the lemma is false, so there exists a sequence $\{x^k\}$ such that

$$\lim_{k\to\infty} \|x^k\| = +\infty, \quad \|\Psi(x^k, 0)\| \leq c.$$

Since there are only finitely many representatives of Γ_F, by choosing a subsequence if necessary, we can assume that there exists an index subset K and a representative G_K of Γ_F such that

$$\Psi(x^k, 0) = G_K(x^k), \quad \forall k \in K.$$

This relation and (A1) imply that for any $k \in K$,

$$\begin{aligned} &\|\Psi(x^k,0)\|^2 = \|G_K(x^k)\|^2 \\ \geq\ &\|G_K(x^k) - G_K(0)\|^2 - 2\|G_K(0)\| \cdot \|G_K(x^k) - G_K(0)\| \end{aligned}$$

holds. Since $\lim_{k\in K, k\to\infty} \|x^k\| = +\infty$, from (4.18) we get

$$\lim_{k\in K, k\to\infty} \|G_K(x^k) - G_K(0)\| = +\infty.$$

So there exists a number $\bar{k}$ such that

$$\|\Psi(x^k, 0)\|^2 > c^2, \quad \forall k \geq \bar{k}, k \in K, \tag{4.19}$$

which contradicts the assumption $x^k \in \ell_c(x)$. This completes the proof. ∎

Denote the unique solution of the system $\Psi(x, 0) = 0$ by x^*. Our next theorem studys the growth behavior of the norm $\|\Psi(x, 0)\|$.

Theorem 4.2 *Suppose that assumption (A1) holds and that x^* is the unique solution of VNCP (1.4). Then there exists a constant $\underline{L} > 0$ such that*

$$\|\Psi(x, 0)\| \geq \underline{L}\|x - x^*\| \quad \forall x \in \Re^n. \tag{4.20}$$

Proof. We first show that there exists a neighbourhood $\delta(x^*, \epsilon) = \{x : \|x - x^*\| < \epsilon\}$ of x^* such that

$$\|\Psi(x, 0)\| \geq \mu\|x - x^*\| \quad \forall x \in \delta(x^*, \epsilon). \tag{4.21}$$

Since assumption (A1) holds, each representative of Γ_F is locally Lipschitz continuous. Because there are only finitely many representatives, we can assume without loss of generality that for any representative G of Γ_F, there exist constants $\epsilon_1, L_1 > 0$ such that

$$\|G(x) - G(y)\| \leq L_1\|x - y\|, \quad \forall x, y \in \delta(x^*, \epsilon_1).$$

Let us define for all $x \in \Re^n$,

$$B_i(x) = \{j \in \{0,1,\ldots,n\} | \; F_j^i(x) = \Psi_i(x,0)\}.$$

Then for all $i = 1,2,\ldots,n$, we have

$$F_j^i(x^*) = 0, \quad \forall j \in B_i(x^*),$$

and

$$F_j^i(x^*) > 0, \quad \forall j \notin B_i(x^*).$$

Now, let

$$\underline{F} = \min\{F_j^i(x^*), \; i = 1,2,\ldots,n; \; j \notin B_i(x^*)\}.$$

Obviously we have $\underline{F} > 0$. If we choose

$$\epsilon = \min\{\frac{\underline{F}}{2L_1}, \epsilon_1\},$$

then for any $x \in \delta(x^*,\epsilon)$, we have

$$F_j^i(x) \le F_j^i(x^*) + L_1\|x - x^*\| < \frac{\underline{F}}{2}, \quad \forall j \in B_i(x^*); \tag{4.22}$$

and

$$F_j^i(x) \ge F_j^i(x^*) - L_1\|x - x^*\| \ge \frac{\underline{F}}{2}, \quad \forall j \notin B_i(x^*). \tag{4.23}$$

Inequalities (4.22) and (4.23) imply

$$F_{j_1}^i(x) < F_{j_2}^i(x), \quad \forall j_1 \in B_i(x^*), \; j_2 \notin B_i(x^*); \; i = 1,2,\ldots,n. \tag{4.24}$$

This means that for all $1 \le i \le n$, $B_i(x) \subset B_i(x^*)$. Hence

$$\Psi(x,0) = G(x), \quad \forall x \in \delta(x^*,\epsilon)$$

where G is a representative of Γ_F satisfying $G(x^*) = 0$. It follows that for any $x \in \delta(x^*,\epsilon)$, we have

$$\|\Psi(x,0)\| = \|G(x) - G(x^*)\| \ge \mu\|x - x^*\|,$$

where the last inequality is given by assumption (A1). This completes the proof of inequality (4.21).

We next claim that for all $x \in \bar{\delta}(x^*,r) = \{x : \|x - x^*\| \le r\}$ with $r \ge \epsilon$, there exists a constant $\mu_r > 0$ such that

$$\|\Psi(x,0)\| \ge \mu_r\|x - x^*\|. \tag{4.25}$$

Let

$$\mu_1 = \min_{\epsilon \le \|x - x^*\| \le r} \frac{\|\Psi(x,0)\|}{\|x - x^*\|}.$$

By the continuity of function and the fact that x^* is a unique solution of the equations $\Psi(x,0)=0$, we have $\mu_1>0$. As a result, inequality (4.25) follows directly by the choice $\mu_r=\min\{\mu,\mu_1\}$.

Now we are ready to prove the theorem. Suppose to the contrary that the statement of the theorem is not true, i.e., there exists a sequence $\{x^k\}$ such that

$$\|\Psi(x^k,0)\| \le \frac{1}{k}\|x^k-x^*\|. \tag{4.26}$$

The inequalities (4.21) and (4.25) imply that $\{x^k\}$ must be an unbounded sequence. By choosing a subsequence if necessary, we can assume that

$$\lim_{k\to\infty}\|x^k\|=+\infty. \tag{4.27}$$

Since there are only finitely many representatives of Γ_F, there exist an index subset K_1 and a representative G_{K_1} such that

$$\Psi(x^k,0)=G_{K_1}(x^k), \quad \forall k\in K_1.$$

It follows from (4.18) and (4.27) that

$$\begin{aligned}
\lim_{k\in K_1,k\to\infty}\frac{\|\Psi(x^k,0)\|}{\|x^k-x^*\|} &= \lim_{k\in K_1,k\to\infty}\frac{\|G_{K_1}(x^k)\|}{\|x^k-x^*\|}\\
&\ge \lim_{k\in K_1,k\to\infty}\frac{\|G_{K_1}(x^k)-G_{K_1}(x^*)\|-\|G_{K_1}(x^*)\|}{\|x^k-x^*\|}\\
&= \lim_{k\in K_1,k\to\infty}\frac{\|G_{K_1}(x^k)-G_{K_1}(x^*)\|}{\|x^k-x^*\|}\\
&\ge \mu,
\end{aligned}$$

which contradicts (4.26). This completes the proof. ∎

Now we give a result about the function $\Psi(x,0)$ when assumption (A2) holds.

Theorem 4.3 *Suppose that x^* is a solution of VNCP (1.4) and that assumption (A2) holds. Then for any $x\in\Re^n$,*

$$\|\Psi(x,0)\|\le L\|x-x^*\|. \tag{4.28}$$

Proof. We consider each component of $\Psi(x,0)$. For any fixed $i\in\{1,2,\ldots,n\}$, we first consider the case that $\Psi_i(x,0)<0$. Let $F^i_{j_i}=\Psi_i(x,0)$. Since $\Psi_i(x^*,0)=0$, we have $F^i_{j_i}(x^*)\ge 0$. So we have

$$\Psi_i(x,0)-F^i_{j_i}(x^*)\le\Psi_i(x,0)-0<0, \tag{4.29}$$

which implies $|\Psi_i(x,0)|\le|F^i_{j_i}(x)-F^i_{j_i}(x^*)|$. If $\Psi_i(x,0)\ge 0$, we can choose an index $j_i\in\{0,1,\ldots,m\}$ such that $\Psi_i(x^*,0)=F^i_{j_i}(x^*)=0$. Thus

$$F^i_{j_i}(x)-F^i_{j_i}(x^*)\ge\Psi_i(x,0)-F^i_{j_i}(x^*)\ge 0, \tag{4.30}$$

which yields $|\Psi_i(x,0)| \le |F^i_{j_i}(x) - F^i_{j_i}(x^*)|$. Let G be such a representative of Γ_F that $G_i(x) = F^i_{j_i}(x)$ where j_i was chosen by the rule stated above. Then we have

$$\|\Psi(x,0)\| \le \|G(x) - G(x^*)\| \le L\|x - x^*\|, \tag{4.31}$$

where the last inequality follows from assumption (A2). The proof of the theorem is completed. ∎

4.3. Properties of the Smoothing Path

In the previous subsection, we studied an equivalent system $\Psi(x,0) =$ of VNCP and gave some results about the mapping $\Psi(x,0)$. From Theorem 4.2 and Theorem 4.3 it is easy to see that if the assumptions (A1) and (A2) hold, then $\|\Psi(x,0)\|$ and $\|x-x^*\|$ have the same order. Hence we can apply function (1.6) to the equations system (1.5). In this case, one needs to estimate how large is the distance between a solution of (1.5) (or $\Psi(x,0) = 0$) and a solution of (4.11). This is done in our next result.

Theorem 4.4 *Suppose that for $t > 0$, $x(t)$ is a solution of (4.11) and that x^* is the unique solution of VNCP (1.4). If the assumptions (A1) is true, then*

$$\|x(t) - x^*\| \le \frac{\sqrt{n}t}{\underline{L}} \log(m+1). \tag{4.32}$$

Proof. Since the assumption (A1) is true, Theorem 4.2 implies

$$\|x(t) - x^*\| \le \frac{1}{\underline{L}}\|\Psi(x(t),0)\|. \tag{4.33}$$

By item (v) of Lemma 2.1 (or (2.2)), we have

$$\Psi_i(x(t),0) \le \Psi_i(x(t),t) \le \Psi_i(x(t),0) + t\log(m+1), \quad i \in \{1,2,...,n\}. \tag{4.34}$$

the last inequality and $\Psi(x(t),t) = 0$ imply

$$|\Psi_i(x(t),0)| \le t\log(m+1). \tag{4.35}$$

The statement of the theorem follows by combining (4.33) and (4.35). ∎

Before proving the existence and the uniqueness of the smoothing path, we give a simple result about the composite function $\chi(F)$. By the definition of the composite function $\chi(F)$, one can easily verify the following result.

Lemma 4.4 *Suppose the function $\chi(\cdot)$ is strictly increasing in $\Re$ and F is a P_0 (or P)-function. Then the function $\chi(F)$ is a P_0 (or P)-function.*

Now we have

Theorem 4.5 *Suppose that assumption (A1) holds. Then for any $t > 0$, there exists a unique point $x(t) \in \Re^n_+$ such that $\Psi(x,t) = 0$. Furthermore,*

$$\lim_{t\to 0^+} x(t) = x^*, \tag{4.36}$$

where x^* is the unique solution of VNCP (1.4).

Proof. We first discuss the existence of the smoothing path. Consider the following unconstrained minimization problem

$$\min_{x\in\Re^n} \|\Psi(x,t)\|^2. \tag{4.37}$$

We know from assumption (A1) that the level set $\ell_c(x)$ is compact. It follows from (2.2) that for any fixed $t > 0$, the level set $\ell_{ct}(x) = \{x : \|\Psi(x,t)\| \le c\}$ is also bounded. Therefore problem (4.37) is well defined. If the optimal value of (4.37) is zero , i.e., $\|\Psi(x(t),t)\| = 0$, then $x(t)$ is a solution of (4.11). Otherwise, there is a stationary point $x = x(t)$ of problem (4.37) such that

$$\nabla^T\Psi(x,t)\Psi(x,t) = (\bar{D}_0 + \bar{D}_1\nabla F_1 + \cdots + \bar{D}_m\nabla F_m)^T\Psi(x,t) = 0, \tag{4.38}$$

where $\bar{D}_i = \mathrm{diag}(\bar{\lambda}_{i,1}, \bar{\lambda}_{i,2}, \ldots, \bar{\lambda}_{i,n})$ with

$$\bar{\lambda}_{i,j} = \frac{\exp(-F_i^j/t)}{\sum_{k=0}^m \exp(-F_k^j/t)}, \quad i = 0,1,\ldots,m;\ j = 1,2,\ldots,n.$$

It is easy to see that all the matrices $\bar{D}_i$ are positive diagonal matrices and that $\sum_{i=0}^m \bar{D}_i = I$. By assumption (A1), we know each representative G of Γ_F is a continuously differentiable unifromly P-function and that each representative G of Γ_F has a Jacobian ∇G which is a P-matrix. We derive from the definition of G that ∇G is a row representative of the matrix set $\{\nabla F_1, \ldots, \nabla F_m\}$. This implies that each row representative of the matrix set $\{\nabla F_1, \ldots, \nabla F_m\}$ is a P-matrix, thus each row representative of the matrix set $\{I, \nabla F_1, \nabla F_2, \ldots, \nabla F_m\}$ is also a P-matrix. Now, recalling Lemma 4.1, we know that $\nabla\Psi(x,t)$ is nonsigular, hence $\Psi(x(t),t) = 0$. The first statement of Lemma 2.2 implies $x(t) \in \Re^n_+$.

Next we consider the uniqueness of the solution $x(t)$. Suppose to the contrary that there exist two different solution $x^1(t) \ne x^2(t)$ (for simplity we sometimes use x^1, x^2) such that $\Psi(x^1(t),t) = \Psi(x^2(t),t) = 0$. By the definiton of $\Psi(x,t)$, we have

$$\begin{aligned} x^1(t) &= -t\log\left(1 - \textstyle\sum_{j=1}^m \exp(-F_j(x^1)/t)\right), \\ x^2(t) &= -t\log\left(1 - \textstyle\sum_{j=1}^m \exp(-F_j(x^2)/t)\right). \end{aligned} \tag{4.39}$$

Since each representative of Γ_F is a uniformly P-function, from Lemma 4.4 we derive that every representative of the functions set

$$\Gamma_{\exp F} = \{-\exp(-F_1/t), -\exp(-F_2/t), \ldots, -\exp(-F_m/t))\}$$

is also a P-function for all fixed $t > 0$. By Lemma 4.2, we conclude that the function $-\sum_{j=1}^{m} \exp(-F_j/t)$ is a P-function. Hence for any $s^1 \neq s^2 \in \Re^n$, there exists an index $i \in \{1, 2, \ldots, n\}$ such that

$$(s_i^1 - s_i^2)(\sum_{j=1}^{m} \exp(-F_j^i(s^2)/t) - \sum_{j=1}^{m} \exp(-F_j^i(s^2)/t)) > 0, \tag{4.40}$$

which contradicts to (4.39). Therefore we have $x^1(t) = x^2(t)$.

Now we turn to the last statement of the theorem. From (2.2) we derive

$$\|\Psi(x,t) - \Psi(x,0)\| \leq \sqrt{n}t \log(m+1),$$

which implies

$$\|\Psi(x(t),0)\| \leq \sqrt{n}t \log(m+1)$$

if $x(t)$ is on the smoothing path. This relation shows that when t goes to zero, any accumulation point of $x(t)$ is a solution of VNCP (1.4). On the other hand, we know from Lemma 4.3 that the level set $\ell_c(x)$ is compact under the assumption (A1). So for any constant sequence $t^k \to 0^+$, the point sequence $\{x(t^k)\}$ is bounded and has at least an accumulation point x^* which solves VNCP (1.4). Since assumption (A1) holds, Theorem 4.1 implies that the solution of VNCP is unique. This fact and the conclusion that any accumulation point x^* of $x(t^k)$ solves VNCP imply the accumulation point of the sequence $x(t^k)$ is also unique. Hence we have

$$\lim_{t^k \to 0^+} x(t^k) = x^*, \tag{4.41}$$

which yields (4.36). The proof of the theorem is finished. ∎

Following the proof of Theorem 4.5, one can show the following result.

Theorem 4.6 *Suppose the following assumptions hold.*
(A3): *(i) Each representative of* Γ_F *is a continuously differentiable* P_0*-function.*
(ii) $\lim_{\|x\| \to \infty} \|\Psi(x,0)\| = \infty$.
Then for any fixed $t > 0$*, there exists a unique* $x(t) \in \Re_+^n$ *such that* $\Psi(x(t),t) = 0$*. Furthermore, any accumulation point of* $x(t)$ *(as* t *tends to zero) is a solution of VNCP (1.4).*

Remark. In the NCP case, conditions (i) and (ii) reduce to the so-called P_0 and R_0-conditions of the mapping involved in NCP, respectively. The above theorem implies that the problem (1.4) satisfying the assumption (A3) is solvable, and we can approximate such VNCP by solving the smooth system (4.11) with sufficiently small parameter. However, if the assumption (A1) is not satisfied, we can not ensure the whole smoothing path converges to a unique solution point of VNCP as t tends to zero. In such case, we also do not know whether the norm $\|\Psi(x,0)\|$ provides a global error bound for the problem considered.

Remark. After this paper is finished, we know a recent work [14] of Gowda and Tawhid. In [14], by using different approaches, the authors showed that if each representative of Γ_F is a $P_0(P)$-function, then for any fixed $t > 0$, the function $\Psi(x,t)$ is also a $P_0(P)$-function. Their results coincides with our conclusions here.

5 BOX CONSTRAINED VARIATIONAL INEQUALITIES

Let us consider box constrained variational inequalities, i.e., find an $x \in [l, u]$ such that

$$(y - x)^T F(x) \geq 0, \quad \forall y \in [l, u], \tag{5.1}$$

where $[l, u]$ is a box in $\Re^n$ with $l \leq u$. It is easy to see that the above problem can be writen as

$$x - \text{mid}(l, u, x - F(x)) = \text{mid}(x - l, x - u, F(x)) = 0, \tag{5.2}$$

where $\text{mid}(x^1, x^2, x^3) = x^1 + x^2 + x^3 - \min(x^1, x^2, x^3) - \max(x^1, x^2, x^3)$. In this section, we first consider the problem of finding a solution of

$$\text{mid}(F_1(x), F_2(x), F_3(x)) = 0, \tag{5.3}$$

where F_1, F_2 and F_3 are functions from $\Re^n$ into itself. This can be viewed as a generalization of BVI. Using the smoothing function $p(x, t)$, we approximate problem (5.3) by the following parameterized smooth equation system

$$\overline{\Psi}(x, t) =$$

$$\begin{pmatrix} \sum_{i=1}^3 F_i^1(x) - t \log \sum_{i=1}^3 \exp\left(\frac{F_i^1(x)}{t}\right) + t \log \sum_{i=1}^3 \exp\left(\frac{-F_i^1(x)}{t}\right) \\ \vdots \\ \sum_{i=1}^3 F_i^n(x) - t \log \sum_{i=1}^3 \exp\left(\frac{F_i^n(x)}{t}\right) + t \log \sum_{i=1}^3 \exp\left(\frac{-F_i^n(x)}{t}\right) \end{pmatrix} \tag{5.4}$$

$$= 0.$$

The Jacobian of this equation is defined by

$$\nabla \overline{\Psi}(x, t) = \hat{D}_1 \nabla F_1 + \hat{D}_2 \nabla F_2 + \hat{D}_3 \nabla F_3, \tag{5.5}$$

where $\hat{D}_i = \text{diag}(\hat{\lambda}_i^j), i = 1, 2, 3, \; j = 1, 2, \ldots, n$ with

$$\hat{\lambda}_i^j = 1 - \frac{\exp(F_i^j(x)/t)}{\sum_{k=1}^3 \exp(F_k^j(x)/t)} - \frac{\exp(-F_i^j(x)/t)}{\sum_{k=1}^3 \exp(-F_k^j(x)/t)}.$$

It is easy to verify that for all $t > 0$, $\hat{D}_i$ are positive diagonal. Following the proof of Theorem 4.5, one can easily show that if the mappings F_1, F_2 and F_3 satisfy to assumption (A1), then the Jacobian $\overline{\Psi}(x, t)$ is nonsingular, and the solution set of equation system (5.4) is nonempty. In this case we can approximate a solution of (5.3) by solving (5.4).

For BVI (5.1), we define $F_1(x) = x - l$, $F_2(x) = x - u$, and $F_3(x) = F(x)$. In this case assumption (A1) reduces to the condition that the mapping $F(x)$ is a continuously differentiable unifromly P-function. Observing that in this case, it holds $F_1(x) \geq F_2(x)$ which implies that $\max\{F_1(x), F_2(x), F_3(x)\} = \max\{F_1(x), F_3(x)\}$ and $\min\{F_1(x), F_2(x), F_3(x)\} = \min\{F_2(x), F_3(x)\}$. Thus by using the smooth function $p(.,t)$, the BVI (5.1) can be reformulated as the following system of equations

$$
\begin{aligned}
&2x_i - l_i - u_i + F^i - t\log\left(\exp\left(\tfrac{x_i - l_i}{t}\right) + \exp\left(\tfrac{F^i}{t}\right)\right) \\
&+t\log\left(\exp\left(\tfrac{u_i - x_i}{t}\right) + \exp\left(\tfrac{-F^i}{t}\right)\right) = 0,
\end{aligned}
$$

for all $i = 1, \ldots, n$. It is easy to see that the above equations are exactly the same as those considered by Gabriel amd Moré [12], Chen and Ye [9] via the neural networks function.

6 CONCLUDING REMARKS

In this paper, we have studied a smoothing function and explored some of its appealing properties. This smooth function is a generalization of the neural networks smooth function [5, 6], and essentially related to the exponential penalty function first proposed by Kort and Bertsekas [20].

We also discussed some applications of this function, and showed that, via this function systems of linear and convex inequalities can be reformulated as unconstrained minimization problems. Some properties of the solution set are derived under certain conditions. Using this smooth function, we approximate an equivalent system of VNCP as a system of parameterized smooth equations. Under some assumptions, the smoothing path of the parameterized equations can approximate the solution set of VNCP to any accuracy as the parameter goes to zero. A new problem that includes the BVI was also discussed. It is shown that we can reformulate this new problem as a system of parameterized equations defined by our smooth function. In the BVI case, a special choice of such reformulations leads to the same system what was considered by Gabrial and Moré [12], Chen and Ye [9].

A natural way to extend these results is to design efficient methods to follow the smooth paths of these parameterized equations which is the main topics of many recent works (see [4, 5, 6, 7, 8, 9, 12, 24] and the references therein). On the other hand, the smooth function in this paper is actually a generalization of the neural networks smooth function and applicable to VNCP. An interesting topic for further studies is to examine if we can get some new smoothing functions for VNCP based on so called Chen-Harker-Kanzow-Smale functions.

Acknowledgements

This work was completed when the author visited Delft University of Technology. The author would like to thank Prof. Roos and Prof. Terlaky for their help, particularly thanks Prof. Terlaky for his reading an early version of the paper. He also thanks the anonymous referees for their useful comments.

References

[1] D.P. Bertsekas, "A new algorithm for solution of nonlinear resistive

networks involving diodes", *IEEE Transactions on Circuit Theory*, Vol. CAS23, PP. 599-608,1976.

[2] D.P. Bertsekas, "Minimax Methods Based on Approximation", *Proceedings of the 1976 Johns Hopkins Conference on Information Sciences and Systems*, Baltimore, 1976.

[3] D.P. Bertsekas, "Approximation procedures based on the method of multipliers", *J. Optimization Theory and Applications*, 23 (1977) 487-510.

[4] B. Chen and P.T. Harker, "Smooth approximations to nonlinear complementarity problems", *SIAM Journal Optimization* 7 (1997) 403-420.

[5] C. Chen and O.L. Mangasarian, "Smoothing methods for convex inequalities and linear complementarity problems", *Mathematical Programming* 71 (1995), 51-69.

[6] C. Chen and O.L. Mangasarian, "A class of smoothing functions for nonlinear and mixed complementarity problems", *Computational Optimization and Applications* 5 (1996), 97-138.

[7] B. Chen and N. Xiu, "A global linear quadratic noninterior continuation method for nonlinear complementarity problems based on Chen-Mangasarian smoothing functions", to appear in *SIAM Journal on Optimization.*

[8] B. Chen and X. Chen, "A global and local superlinear continuation-smoothing method for $P_0 + R_0$ and monotone NCP", to appear in *SIAM Journal on Optimization.*

[9] X. Chen and Y. Ye, "On homotopy-smoothing methods for variational inequalities", to appear in *SIAM Journal on Control and Optimization.*

[10] F.H. Clarke, *Optimization and Nonsmooth Analysis*, Wiley, New York, 1983.

[11] R.W. Cottle and G.B. Dantzig, "A generalization of the linear complementarity problem", *Journal of Combinatorial Theory* 8 (1970) 79-90.

[12] S.A. Gabrial and J. Moré, "Smoothing of mixed complementarity problems," *Complementarity and Variational Problems: State of The Art*, Edited by M.C. Ferris and J.S. Pang, SIAM Publications, Philadephia, 1997, 105-116.

[13] M.S. Gowda and R. Sznajder, "The generalized order linear complementarity problem", *SIAM Journal Matrix Analysis and Applications* 159 (1994) 779-795.

[14] M.S. Gowda and M.A. Tawhid, "Existence and Limiting behavior of tracjectories associated with P_0-equations", Research Report No. 97-15, Department of mathematics, University of Maryland, November, 1997.

[15] P.T. Harker and J.S. Pang, "Finite-dimensional variational inequality and nonlinear complementarity problems: A survey of theory, algorithms and applications", *Mathematical Programming* 48 (1990), 339-357.

[16] A.J. Hoffman, "On approximate solutions of systems of linear inequalities", *Journal of Research of the National Bureau of Standards* 49 (1952) 263-265.

[17] C. Kanzow, "Some noninterior continuation methods for linear complementarity problems", *SIAM Journal Matrix Anal. Appl.*, 17 (1996) 851-868.

[18] C. Kanzow and H. Jiang, "A continuation method for the solution of monotone variational inequality problems", *Mathematical Programming*, 81 (1998) 103–125.

[19] M. Kojima, M. Megiddo and S. Mizuno, "A general framework of continuation methods for complementarity problems", *Mathematics of Operations Research*, 18 (1993) 945-963.

[20] B.W. Kort and D.P. Bertsekas, "A new penalty function algorithm for constrained minimization", *Proceedings of the 1972 IEEE Conference on Decision and Control*, New Orleans, Louisiana, 1972.

[21] X.S. Li, "An aggregate function method for nonlinear programming", *Science in China (A)*, 34 (1991) (12) 1467-1473.

[22] Z.H. Lin, *Private communication.*

[23] J.J. Moré, W.C. Rheinboldt, "On P- and S-Functions and related classes of n-dimensional nonlinear mappings", *Linear Algebra and Its Application*, 6(1973), pp. 45-68.

[24] J.M. Peng and Z.H. Lin, "A non-interior continuation method for generalized linear complementarity problems", Technical Report, Institute of Computational Mathematics and Scientific/Engineering Computing, Chinese Academy of Sciences, Beijing, China, 1997.

[25] R. Sznajder and M.S. Gowda, "Generalizations of P_0- and P-properties; extended vertical and horizontal linear complementarity problems", *Linear Algebra and its Applications* 223/224 (1995) 695-715.

[26] H. Tang and L. Zhang, "A maximum entropy method for convex programming", *Chinese Science Bulletin (Chinese)*, 39 (8) (1994) 682-684.

Reformulation: Nonsmooth, Piecewise Smooth, Semismooth and Smoothing Methods, pp. 317–334
Edited by M. Fukushima and L. Qi

On the Local Super–Linear Convergence of a Matrix Secant Implementation of the Variable Metric Proximal Point Algorithm for Monotone Operators

Maijian Qian* and James V. Burke†

Abstract Interest in the variable metric proximal point algorithm (VMPPA) is fueled by the desire to accelerate the local convergence of the proximal point algorithm without requiring the divergence of the proximation parameters. In this paper, the local convergence theory for matrix secant versions of the VMPPA is applied to a known globally convergent version of the algorithm. It is shown under appropriate hypotheses that the resulting algorithms are locally super–linearly convergent when executed with the BFGS and the Broyden matrix secant updates. This result unifies previous work on the global and local convergence theory for this class of algorithms. It is the first result applicable to general monotone operators showing that a globally convergent VMPPA with bounded proximation parameters can be accelerated using matrix secant techniques. This result clears the way for the direct application of these methods to constrained and non–finite–valued convex programming. Numerical experiments are included illustrating the potential gains of the method and issues for further study.

Key Words convex programming, optimization algorithms, maximal monotone operator, proximal point algorithm, variable metric proximal point algorithm, matrix secant methods

*Department of Mathematics, California State University, Fullerton, CA 92634, mqian@Exchange.FULLERTON.EDU

†This author's research is supported by the National Science Foundation Grant No. DMS-9303772, Department of Mathematics, Box # 354350, University of Washington, Seattle, Washington 98195–4350, burke@math.washington.edu

1 INTRODUCTION

In [4], we introduced the variable metric proximal point algorithm (VMPPA) for general monotone operators. The VMPPA builds on the classical proximal point algorithm and can be viewed as a Newton–like method for solving inclusions of the form

$$0 \in T(z)$$

where T is a maximal monotone operator on $\mathbb{R}^n$. In [4], we establish conditions under which the VMPPA is globally linearly convergent. In [3], we focus on the finite dimensional setting and more closely examine the local behavior of the algorithm under the assumption that the iterates converge linearly. In particular, we considered two matrix secant updating strategies for generating the Newton–like iterates: the BFGS and Broyden updates. The BFGS update is employed when it is known that the *derivative* (see Definition 3.1) of the operator T^{-1} at the origin is symmetric. This *symmetric* case occurs in applications to convex programming where the operator T is taken to be the subdifferential of a convex function. We show that if the sequence generated by the VMPPA is known to be linearly convergent, then it is also super–linearly convergent when the appropriate matrix secant update is employed: BFGS in the symmetric case and Broyden in the general case. In [3, Section 4], these results are applied to establish the local super–linear convergence of a variation on the Chen–Fukushima variable metric proximal point algorithm for convex programming.

In this paper, we show that the local theory developed in [3] can also be applied to the more general algorithm described in [4] and thereby obtain conditions under which the BFGS and Broyden updates can accelerate the convergence of the VMPPA applied to general monotone operators. This yields the first super-linear convergence result for the VMPPA applicable beyond the context of finite–valued convex programming. However, this extension comes at the cost of a more complicated statement of the algorithm. In particular, as given, the algorithm can only be implemented when further knowledge about the operator T is known, e.g. if the operator is strongly monotone and a lower bound on the modulus of strong monotonicity is known (see Part (a) of Lemma 3.1). The additional complexities of the algorithm can be traced back to the absence of an underlying objective function to which a line–search can be applied. On the other hand, the foundations laid in [4] and [3] clear the way for a very straightforward and relatively elementary proof of the superlinear convergence of the method.

In the case of finite–valued, finite dimensional convex programming, the VMPPA has recently received considerable attention [1, 6, 7, 9, 10, 11, 12, 17, 18]. In convex programming, the goal is to derive a variable metric method for minimizing the Moreau–Yosida regularization of a convex function $f : \mathbb{R}^n \to \mathbb{R} \cup \{+\infty\}$:

$$f_\lambda(x) := \min_{u \in \mathbb{R}^n} \{\lambda f(u) + \frac{1}{2}|u - x|^2\} \tag{1.1}$$

(in the finite–valued case, f cannot take the value $+\infty$). It is well known that the set of points x yielding the minimum value of f and f_λ coincide, and that the function f_λ is continuously differentiable with Lipschitz continuous derivative even if the function f is neither differentiable nor finite–valued. The challenge is to derive a super–linearly convergent method that does not require precise values for either f_λ or its derivative, and does not require excessively strong smoothness hypotheses on the function f. Detailed comparisons of the various contributions in this direction can be found in the introductions to the papers [4] and [3]. Here we only note that the references [3, 12] and the present contribution are the only papers containing local super–linear convergence results for the VMPPA when applied with only approximate values for f_λ and its derivative. The only such results for the case of general monotone operators are found in [3].

In Section 2, we review the VMPPA. The algorithm is motivated by considering the application to convex programming. Convergence results are presented in Section 3. In Section 4 we present the three applications of the VMPPA to convex programming, with numerical results.

2 THE VARIABLE METRIC PROXIMAL POINT ALGORITHM

An operator $T : \mathbb{R}^N \rightrightarrows \mathbb{R}^N$ (here the double arrows $\rightrightarrows$ are used to signify the fact that T is multi-valued) is said to be a monotone operator if $\langle z - z', w - w'\rangle \geq 0$ whenever $w \in T(z), w' \in T(z')$. It is said to be maximal monotone if, in addition, its graph $gph(T) := \{(z, w) \in \mathbb{R}^N \times \mathbb{R}^N | w \in T(z)\}$ is not properly contained in the graph of any other monotone operator. Monotone operators arise naturally in number of applications [20, 21]. Perhaps the most well–known of these is the ***subdifferential*** mapping of a closed proper convex function (see Minty [14] and Moreau [15]).

In most applications involving monotone operators, the central issue is the determination of those points z satisfying the inclusion $0 \in T(z)$, where $T : \mathbb{R}^N \rightrightarrows \mathbb{R}^N$. The proximal point algorithm is designed to solve inclusions of precisely this type. It does so by generating sequences $\{z^k\}$ satisfying the approximation rule

$$z^{k+1} \approx (I + c_k T)^{-1}(z^k)$$

for a given sequence of positive scalars $\{c_k\}$.

In the case of convex programming, the proximal point iteration has the form

$$z^{k+1} = z^k + w^k, \quad \text{where } w^k \approx -\nabla f_{c_k}(z^k)$$

and f_{c_k} is the Moreau–Yosida resolvent for f associated with the proximation parameter $\lambda = c_k$. That is, it is the method of steepest descent with unit step size applied to the function f_{c_k} with c_k varying between iterations.

Using the fact that

$$\nabla f_{c_k} = [I - (I + c_k T)^{-1}],$$

one can formally derive the algorithm for a general maximal monotone operator T by replacing ∂f with T and $-\nabla f_{c_k}$ with the operator

$$D_k = [(I + c_k T)^{-1} - I]. \tag{2.1}$$

The operator D_k will play a central role in the analysis to follow.

With these definitions, the proximal point algorithm takes the form

$$z^{k+1} = z^k + w^k, \quad \text{where} \quad w^k \approx D_k(z^k).$$

A Newton-like variation on this iteration yields the VMPPA.

The Variable Metric Proximal Point Algorithm:
Let $z^0 \in I\!R^N$ and $c_0 \geq 1$ be given. Having z^k, set

$$z^{k+1} := z^k + H_k w^k \quad \text{where} \quad w^k \approx D_k(z^k)$$

and choose $c_{k+1} \geq 1$.

The matrices H_k should be thought of as approximations to the inverse of the *derivative* of D_k at a solution to equation $D_k(z) = 0$, or equivalently, at a solution to the inclusion $0 \in T(z)$. The condition $c_k \geq 1$ is required to obtain the global convergence as in [4].

Explicit conditions on the accuracy of the approximation $w^k \approx D_k(z^k)$ are key to the convergence analysis. As in [4] and [3], we employ the following approximation criteria:

$$(\mathcal{G}) \qquad |w^k - D_k(z^k)| \leq \epsilon_k \;\text{ with } \sum_{k=0}^{\infty} \epsilon_k < \infty$$

and

$$(\mathcal{L}) \qquad |w^k - D_k(z^k)| \leq \delta_k |w^k| \;\text{ with } \sum_{k=0}^{\infty} \delta_k < \infty\,.$$

Criteria $(\mathcal{G})$ is used to establish global convergence while criteria $(\mathcal{L})$ is used to obtain local rates of convergence.

3 SUPER-LINEAR CONVERGENCE

3.1 Differentiability Hypotheses

In [4, Theorem 19], the global and linear convergence of the VMPPA is established by assuming that the operator T^{-1} has a certain smoothness at the origin. This smoothness property allows us to use matrix secant techniques to approximate ∇D_k at a unique solution to the inclusion $0 \in T(z)$. These hypotheses are reminiscent of those used in Newton's method for establishing rapid local convergence.

Definition 3.1 *We say that an operator $\Psi : \mathbb{R}^N \rightrightarrows \mathbb{R}^N$ is differentiable at a point $\bar{w}$ if $\Psi(\bar{w})$ consists of a single element $\bar{z}$ and there is a matrix $J \in \mathbb{R}^{N\times N}$ such that for some $\delta > 0$,*

$$\emptyset \neq \Psi(w) - \bar{z} - J(w - \bar{w}) \subset o(|w - \bar{w}|)\mathbb{B} \quad \text{whenever } |w - \bar{w}| \leq \delta\,. \tag{3.1}$$

We then write $J = \nabla\Psi(\bar{w})$. We say that the operator Ψ satisfies the quadratic growth condition at a point $\bar{w} \in \mathbb{R}^n$ if Ψ is differentiable at $\bar{w}$ and there are constants $C \geq 0$ and $\epsilon > 0$ such that

$$\Psi(w) - \Psi(\bar{w}) - \nabla\Psi(\bar{w})(w - \bar{w}) \subset C|w - \bar{w}|^2\mathbb{B} \quad \text{whenever } |w - \bar{w}| \leq \epsilon\,. \tag{3.2}$$

Remarks.

1) This notion of differentiability corresponds to the usual notion of differentiability when Ψ is single–valued.

2) In [4, Example 7], we show that it is possible to choose a convex function f so that ∂f^{-1} is differentiable at the origin, but does not satisfy the quadratic growth condition there.

3) In the context of convex programming, the strong second–order sufficiency condition implies that ∂f^{-1} satisfies the quadratic growth condition at the origin where f is the essential objective function ([19, Proposition 2] and [4, Theorem 8]).

4) Further discussion of these notions of differentiability as they relate to monotone operators can be found in [16] and ([4], Corollaries 12 and 13).

3.2 Conditions for Global Linear Convergence

We assume that the operator T^{-1} is differentiable at the origin. This implies the differentiability of the operators D_k at the unique global solution $\bar{z} = T^{-1}(0)$, with

$$\nabla D_k(\bar{z}) = -[I + \frac{1}{c_k}\nabla[T^{-1}](0)]^{-1} \tag{3.3}$$

[4, Proposition 9]. The matrices H_k appearing in the VMPPA are chosen to approximate the matrix $(-\nabla D_k(\bar{z}))^{-1} = [I + \frac{1}{c_k}\nabla[T^{-1}](0)]$. The accuracy of the approximation

$$H_k \approx [I + \frac{1}{c_k}\nabla[T^{-1}](0)]$$

determines both the global and local rates of convergence.

Observe that the smaller the value of c_k, the less the matrix $[I + \frac{1}{c_k}\nabla[T^{-1}](0)]$ looks like the identity, and therefore, the less the method looks like the classical proximal point algorithm. In the context of convex programming, this deviation from the direction of steepest descent is easily compensated for by appropriately *damping* the search direction with a line–search routine. This assures the global convergence of the method. However, in the operator setting, there is no objective function to which a line-search can be applied. For this reason, the global convergence analysis developed in [4] requires that the matrices H_k do not deviate from the identity too much. This allows us to extend the global linear convergence result for the classical proximal point algorithm to the variable metric setting. Specifically, we require that the matrices $\{H_k\}$ satisfy the condition

$$|(H_k - I)D_k(z^k)| \leq \gamma_k |D_k(z^k)| \quad \text{for all } k, \tag{3.4}$$

where

$$\gamma_k := \frac{|D_k(z^k)|}{2|z^k - \bar{z}| + 3|D_k(z^k)|} .$$

Of course, it is essential to know whether or not this condition can reasonably be achieved without requiring knowledge of $\bar{z}$ and $|D_k(z^k)|$ or that $H_k = I$ on all iterations. In this regard, we recall the following facts from [4].

Lemma 3.1 *[4, Lemma 14 and 15] Suppose $T^{-1}(0)$ is nonempty.*

(a) If the operator T is strongly monotone with modulus κ, then $T^{-1}(0) = \{\bar{z}\}$,

$$|z^k - \bar{z}| \leq (1 + \frac{1}{\kappa c_k})|D_k(z^k)| ,$$

and $\gamma_k \geq \frac{1}{5+\frac{2}{\kappa c_k}} \geq \frac{1}{5+2/\kappa}$ for all k.

(b) If T^{-1} is differentiable at the origin with derivative J, then there is a $\delta > 0$ such that $\gamma_k \geq \frac{1}{5+3\frac{|J|}{c_k}}$ for all k satisfying $|D_k(z^k)| \leq \delta$.

(c) Let $\xi, \hat{\gamma}_k, \delta_k \in I\!R_+$ be such that

$$0 \leq \xi < 1, \ \delta_k \leq \min\{1, |H_k|^{-1}\}\frac{3}{7}(1-\xi)\hat{\gamma}_k, \ \text{and } \hat{\gamma}_k \leq 1/3. \tag{3.5}$$

If $z^k, w^k \in I\!R^N$ satisfy

$$|(I - H_k)w^k| \leq \xi\hat{\gamma}_k|w^k| \text{ and } |w^k - D_k(z^k)| \leq \delta_k|w^k|, \tag{3.6}$$

then $|(I - H_k)D_k(z^k)| \leq \hat{\gamma}_k|D_k(z^k)|$. Therefore, if criterion $(\mathcal{L})$ is satisfied, and if ξ and the sequence $\{(\hat{\gamma}_k, \delta_k)\} \subset I\!R^2$ satisfy (3.5), with $\hat{\gamma}_k \leq \gamma_k$ for all k, then hypothesis (3.4) is satisfied.

Part (a) of the lemma says that the γ_k's are bounded below by a global positive constant if T is strongly monotone. Part (b) says that the γ_k's are locally bounded below by a positive constant under a differentiability hypothesis on T^{-1}. Part (c) says that condition (3.4) can be achieved by requiring a similar condition involving the computed quantities w^k.

Observe that the condition (3.4) is easily satisfied by taking $H_k = I$. In addition, from relation (3.3), we have $(\nabla D_k(\bar{z}))^{-1} \to -I$ as $c_k \uparrow \infty$. These observations motivate the choice of updating strategy given in the formal statement of the algorithm below. However, our practical experience indicates that the condition (3.4) can be significantly relaxed. We return to this issue in Section 4.2 where we discuss our numerical results.

3.3 The Algorithms and Their Convergence

We now give explicit statements of the algorithms. The somewhat unconventional form in which these algorithms are stated is a result of our desire to meld the global convergence theory in [4] with the local theory in [3].

BFGS Updating: Choose $0 \leq \xi_0 < 1$ and $\tilde{\gamma}_0$ as an estimate of γ_0. Choose $H_0 = \hat{H}_0 = I$. For $k \geq 0$, set $d^k = w^k - w^{k+1}$, $s^k = z^{k+1} - z^k$, and

$$\hat{H}_{k+1} = \hat{H}_k + \frac{(s^k - \hat{H}_k d^k){s^k}^T + s^k(s^k - \hat{H}_k d^k)^T}{\langle d^k, s^k\rangle} - \frac{\langle s^k - \hat{H}_k d^k, d^k\rangle s^k {s^k}^T}{\langle d^k, s^k\rangle^2}$$

if ${d^k}^T s^k > 0$; otherwise, set $\hat{H}_{k+1} = \hat{H}_k$. Set $H_{k+1} = \hat{H}_{k+1}$. Compute an estimate $\tilde{\gamma}_{k+1}$ of γ_{k+1} satisfying $0 \leq \tilde{\gamma}_{k+1} \leq \gamma_{k+1}$, and choose $0.95 \leq \xi_{k+1} < 1$. If $\|(I - H_{k+1})w^{k+1}\| > \xi_{k+1}\tilde{\gamma}_{k+1}\|w^{k+1}\|$, then reset $H_{k+1} = I$.

Remark. Note that the inverse Hessian approximations $\hat{H}_k$ are never restarted, even when they are not being used. This unusual updating strategy is required for our proof of super-linear convergence.

Broyden Updating: Choose $0 \leq \xi_0 < 1$ and $\tilde{\gamma}_0$ as an estimate of γ_0. Choose $H_0 = \hat{H}_0 = I$. For $k \geq 0$, set $d^k = w^k - w^{k+1}$, $s^k = z^{k+1} - z^k$. If $\langle s^k, H_k d^k\rangle \neq 0$, set

$$\hat{H}_{k+1} = H_k + \frac{(s^k - H_k d^k){s^k}^T H_k}{\langle s^k, H_k d^k\rangle} ;$$

otherwise, set $\hat{H}_{k+1} = I$. Set $H_{k+1} = \hat{H}_{k+1}$. Compute an estimate $\tilde{\gamma}_{k+1}$ of γ_{k+1} satisfying $0 \leq \tilde{\gamma}_{k+1} \leq \gamma_{k+1}$, and choose $0.95 \leq \xi_{k+1} < 1$. If $\|(I - H_{k+1})w^{k+1}\| > \xi_{k+1}\tilde{\gamma}_{k+1}\|w^{k+1}\|$, then reset $H_{k+1} = I$.

We require the following hypotheses in our convergence analysis:

(H1) The operator T^{-1} satisfies the quadratic growth condition at the origin with $J := \nabla T^{-1}(0)$ and $T^{-1}(0) = \{\bar{z}\}$.

(H2) The approximation criteria $(\mathcal{G})$ and $(\mathcal{L})$ are satisfied.

(H3) The parameters δ_k, ξ_k, and $\tilde{\gamma}_k$ satisfy $\delta_k \leq \min\{1, \|H_k\|^{-1}\}\frac{3}{7}(1-\xi_k)\tilde{\gamma}_k$ for all k such that $H_k \neq I$.

(H4) There exists $k_1 > 0$ such that $c_k \equiv \lambda > 6\|J\|$ and $1/6 \leq \xi_k \tilde{\gamma}_k$ for all $k \geq k_1$.

Remarks.

1. The global linear convergence of the iterates is insured by hypotheses (H3) and (H4) (See remarks below). Linear convergence in conjunction with hypotheses (H1), (H2), and (H4) allow us to apply [3, Theorem 3 and 4] to establish the local super–linear convergence of the iterates..

2. Hypotheses (H3) and (H4) concern the updating procedure for both c_k and $\tilde{\gamma}_k$. The parameters are related via Lemma 3.1, and the inequality

$$\|(I - H_k)w^k\| \leq \xi_k \tilde{\gamma}_k \|w^k\| \tag{3.7}$$

which must be satisfied or else H_k is reset to the identity. First observe that Lemma 3.1 indicates that $\gamma_k > K/(5K+3)$ whenever z^k is sufficiently close to $\bar{z}$ and $c_k = \lambda > K\|J\|$. Thus, for $K \geq 6$, $\tilde{\gamma}_k = (1/6)(.95)^{-1}$ is an acceptable lower bound for γ_k and hypothesis (H4) is satisfied. Therefore, we need only make $c_k = \lambda$ sufficiently large.

3. As is typical in the selection of penalization parameters, establishing whether c_k is large enough is not an easy matter in general. Indeed, we do not know of a general technique for this purpose. Therefore, we do not provide an explicit rule for updating the c_k's. However, there are some crude rules for recognizing when c_k is too small. For example, if inequality (3.7) fails to be satisfied, then c_k is probably too small and should be increased. In order to derive an effective strategy for updating the c_k's, more information on the structure of the operator T is required. For example, if it is known that T is strongly monotone with modulus κ, then $\|J\| \leq \kappa^{-1}$. In this case, Part (a) of Lemma 3.1 indicates that we can set $c_k = 6\kappa^{-1}$ and $\hat{\gamma}_k = \frac{1}{5+3\kappa^{-1}}$ for all k.

Theorem 3.1 *Let $\{z^k\}$ be any sequence generated by the variable metric proximal point algorithm using the BFGS updating scheme and suppose that hypotheses (H1)-(H4) are all satisfied. If J is symmetric, then there is a positive integer k_0 such that*

(i) ${d^k}^T s^k > 0$ for all $k \geq k_0$,

(ii) the sequences $\{\|H_k\|\}$and $\{\|H_k^{-1}\|\}$ are bounded,

(iii) $H_k = \hat{H}_k$ for all $k \geq k_0$, and

(iv) the sequence $\{z^k\}$ converges to $\bar{z}$ at a super–linear rate.

Proof. Hypothesis (H1)-(H4), Lemma 3.1, and [4, Theorem 19] guarantee that all hypotheses in [3, Theorem 3] are satisfied. We now show that $H_k = \hat{H}_k$ for all k large. Under hypotheses (H1)-(H4), [3, Lemma 8], and [3, Theorem 9] imply that $\{|\hat{H}_k|\}$ and $\{|\hat{H}_k|^{-1}\}$ are bounded and

$$\frac{|(\hat{H}_k^{-1} - (I + \frac{1}{\lambda}J)^{-1})s^k|}{|s^k|} \to 0 ,$$

or equivalently,

$$\frac{|(I - (I + \frac{1}{\lambda}J)\hat{H}_k^{-1})s^k|}{|s^k|} \to 0 .$$

Hence, the boundedness of $\{\hat{H}_k\}$ implies that

$$\frac{|\hat{H}_k s^k - s^k + \frac{1}{\lambda}\hat{H}_k J \hat{H}_k^{-1} s^k|}{|s^k|} \le \frac{|\hat{H}_k||(I - (I + \frac{1}{\lambda}J)\hat{H}_k^{-1})s^k|}{|s^k|} \to 0 .$$

Therefore, there is a sequence $\zeta_k \to 0$ such that

$$|\hat{H}_k s^k - s^k + \frac{1}{\lambda}\hat{H}_k J \hat{H}_k^{-1} s^k| \le \zeta_k |s^k| ,$$

which in turn implies that

$$\begin{aligned} |(I - \hat{H}_k)s^k| &\le (\frac{1}{\lambda}|\hat{H}_k J \hat{H}_k^{-1}| + \zeta_k)|s^k| \\ &= (\frac{1}{\lambda}|J| + \zeta_k)|s^k| \le \frac{1}{6}|s^k| \le \xi_k \gamma_k |s^k| \end{aligned} \tag{3.8}$$

for all k sufficiently large since $\lambda > 6|J|$. Now $H_k \neq \hat{H}_k$ implies that $s^k = w^k$ and

$$|(I - \hat{H}_k)s^k| > \xi_k \gamma_k |s^k| .$$

By (3.8) this cannot occur for k sufficiently large, therefore eventually $H_k = \hat{H}_k$.

The super–linear convergence of the iterates now follows from [3, Theorem 3].

■

Theorem 3.2 *Let $\{z^k\}$ be any sequence generated by the variable metric proximal point algorithm using the Broyden's updating scheme and suppose that the hypotheses (H1)-(H4) are all satisfied. If in addition, $\lambda > 16|J|$, then*

(i) there is a positive integer $\hat{k}$ such that ${s^k}^T H^k d^k \neq 0$ and H_k is updated using Broyden's formula for all $k \ge \hat{k}$,

(ii) the sequences $\{|H_k|\}$ and $\{|H_k^{-1}|\}$ are bounded, and

(iii) the sequence $\{z^k\}$ converges to $\bar{z}$ at a super–linear rate.

Proof. Hypothesis (H1)-(H4), Lemma 3.1, and [4, Theorem 19] guarantee that all hypotheses in [3, Theorem 4] are satisfied. We now show that the condition

$$\|(I - H_k)w^k\| < \xi_k \tilde{\gamma}_k \|w^k\| \tag{3.9}$$

is satisfied for all k large.

Set $A_k = \hat{H}_k^{-1}$. By [3, Lemma 11], for any matrix G we have

$$\|A_{k+1} - G\| \leq \|A_0 - G\| + \sum_{j=0}^{k} \frac{\|y^j - Gs^j\|}{\|s^j\|} . \tag{3.10}$$

Set $G = (I + \frac{1}{\lambda}J)^{-1}$. Then, by the Banach Lemma,

$$\|I - G\| \leq \sum_{i=1}^{\infty} (\frac{1}{\lambda}\|J\|)^i < \frac{1}{15} . \tag{3.11}$$

By [3, Lemma 8], there is a k_0 such that

$$\sum_{j=k_0}^{\infty} \frac{\|y^j - Gs^j\|}{\|s^j\|} \leq (\frac{1}{7} - 2\|I - G\|) . \tag{3.12}$$

We need to show that there exists a $\hat{k} \geq k_0$ such that (3.9) is satisfied for all $k \geq \hat{k}$. If we cannot take $\hat{k} = k_0$, then there is a $\hat{k} > k_0$ such that $H_{\hat{k}} = A_{\hat{k}} = I$. Now for all $k \geq \hat{k}$

$$\begin{aligned} \|A_k - I\| &\leq \|I - G\| + \|A_k - G\| \\ &\leq 2\|I - G\| + \sum_{j=\hat{k}}^{k} \frac{\|y^j - Gs^j\|}{\|s^j\|} < \frac{1}{7} \end{aligned}$$

by (3.10), (3.12), and (3.11). Therefore,

$$\|\hat{H}_k - I\| = \|A_k^{-1} - I\| \leq \|A_k^{-1}\|\|A_k - I\| \leq \frac{1/7}{1 - 1/7} = \frac{1}{6} \leq \xi_k \tilde{\gamma}_k ,$$

for all $k > \hat{k}$. Therefore (3.9) is satisfied for all $k > \hat{k}$.

The super–linear convergence of the iterates now follows from [3, Theorem 4].

■

4 APPLICATION TO CONVEX PROGRAMMING

In this section we study the realizations of the VMPPA in optimization problems.

Let C be a nonempty closed convex subset of $I\!R^n$, and for $i = 0, 1, \ldots, m$ let $f_i : C \to I\!R$ be a lower semi-continuous convex function. We consider the convex programming problem

$$\text{minimize}\{f_0(x) | x \in C, f_i(x) \leq 0, i = 1, 2, \ldots, m\} \qquad (P)$$

under the assumption that (P) is solvable.

In [19], Rockafellar presented theoretical results on the convergence of three approaches to solve (P). All three of these approaches are realizations of the general proximal point algorithm for maximal monotone operators. Each algorithm replaces (P) by a sequence of approximating minimization problems. Here we consider variable metric versions of these three approaches.

The first algorithm is the primal application, which we call the *variable metric proximal minimization algorithm (VPA)* for (P). The second algorithm is the dual application, which we call the *variable metric proximal dual algorithm (VDA)* for the dual problem associated with (P):

$$\max_y\{g_0(y) | y \in I\!R^m_+\} \text{ where } g_0(y) := \inf_{x \in C}\{f_0(x) + y_1 f_1(x) + \cdots + y_m f_m(x)\}. \quad (D)$$

The third method, the *variable metric proximal method of multipliers (VPM)*, applies the general variable metric proximal point algorithm to the mini-max problem associated with (P):

$$\min_x \max_y\{l(x, y) := f_0(x) + y_1 f_1(x) + \cdots + y_m f_m(x) | x \in C, y \in I\!R^m_+\}. \quad (L)$$

For all three applications, detailed discussions of the approximation criteria in solving the subproblems can be found in [19].

4.1 Three Algorithms

Applying the variable metric version to the first algorithm in [19] using the BFGS update yields the VPA Algorithm:

The VPA Algorithm: *Given x^0 feasible, for $k = 0, 1, \ldots$, choose $c_k \geq 1$, then:*

1°. *Set*

$$w^k \approx \arg\min_{w \in I\!R^n} \phi_k^{(P)}(w) \qquad (4.1)$$

where $\phi_k^{(P)}$ is the closed convex function on $I\!R^n$ defined by

$$\phi_k^{(P)}(w)) := \begin{cases} f_0(x^k + w) + \frac{1}{2c_k} w^T w & \text{if } x \text{ is feasible} \\ +\infty & \text{otherwise.} \end{cases} \qquad (4.2)$$

The subproblem (4.1) is solved iteratively, with the feasible starting point $w^{0(0)} = 0$ and $w^{k(0)} = x^{k-1} - x^k + w^{k-1}$ for $k \geq 1$.

2°. *For $k = 0$, set $H_k = I$; For $k \geq 1$, use the BFGS updating procedure introduced in Section 3.3 to obtain $H_k \in I\!R^{n \times n}$, then set*

$$x^{k+1} := x^k + H_k w^k. \qquad (4.3)$$

Remark. A detailed discussion of the approximation criteria in solving the subproblem (4.1) for w^k can be found in [19, Section 3].

In order to apply the VMPPA to the dual problem, we cite the definition of the *augmented Lagrangian* as given in [19]:

$$L(x,y,c) := \begin{cases} f_0(x) + \sum_{j=1}^{m} \psi(f_i(x), y_i, c) & \text{if } x \in C \\ +\infty & \text{if } x \notin C \end{cases} \tag{4.4}$$

for all $y \in \mathbb{R}^m$ and $c > 0$, where

$$\psi(f_i, y_i, c) := \begin{cases} y_i f_i + \frac{c}{2} f_i^2 & \text{if } cf_i \geq -y_i \\ -\frac{1}{2c} y_i^2 & \text{if } cf_i \leq -y_i \ . \end{cases} \tag{4.5}$$

We now introduce the VDA Algorithm:

The VDA Algorithm: *Given $y^0 \in \mathbb{R}^m_+$, for $k = 0, 1, \ldots$, we choose $c_k \geq 1$, then*

1°. *Set*

$$x^{k+1} \approx \arg \min_{x \in \mathbb{R}^n} L(x, y^k, c_k) \ . \tag{4.6}$$

2°. *Set u^k be such that*

$$u_i^k := \max\{-y_i^k, c_k f_i(x^{k+1})\}. \tag{4.7}$$

3°. *For $k = 0$, set $H_k = I$; For $k \geq 1$, using $d^{k-1} = u^{k-1} - u^k$ and $s^{k-1} = x^k - x^{k-1}$, apply the non-symmetric updating procedure to determine*

$$H_k \in \mathbb{R}^{m \times m}$$

and set

$$y^{k+1} = y^k + H_k u^k \ . \tag{4.8}$$

Remarks.

1) A detailed discussion of the approximation criteria in solving the subproblem (4.6) for x^{k+1} can be found in [19, Section 4].

2) Although the $y^{k'}s$ are not necessary non-negative, the sequence $\{y^k + u^k\}$ is in $\mathbb{R}^m_+$ and has the same converges behavior as $\{y^k\}$.

We now apply the variable metric proximal point algorithm to the minimax problem and introduce the VPM Algorithm:

The VPM Algorithm *Given $\begin{pmatrix} x^0 \\ y^0 \end{pmatrix}$ with $y^0 \in \mathbb{R}^m_+$, for $k = 0, 1, \ldots$, we choose $c_k \geq 1$, then:*

1°. *Set*

$$v^k \approx \arg\min_{v\in \mathbb{R}^n} L(x^k+v, y^k, c_k) + \frac{1}{2}v^T v\,. \tag{4.9}$$

2°. *Set* u^k *be such that*

$$u_i^k := \max\{-y_i^k, c_k f_i(x^k+v^k)\}. \tag{4.10}$$

3°. *For* $k=0$*, set* $H_k = I$*; For* $k \geq 1$*, using*

$$d^{k-1} = \begin{pmatrix} v^{k-1} - v^k \\ u^{k-1} - u^k \end{pmatrix} \quad \text{and} \quad s^{k-1} = \begin{pmatrix} x^k - x^{k-1} \\ y^k - y^{k-1} \end{pmatrix},$$

apply the non-symmetric updating procedure to determine

$$H_k \in \mathbb{R}^{(n+m)\times(n+m)}$$

and set

$$\begin{pmatrix} x^{k+1} \\ y^{k+1} \end{pmatrix} = \begin{pmatrix} x^k \\ y^k \end{pmatrix} + H_k \begin{pmatrix} v^k \\ u^k \end{pmatrix}. \tag{4.11}$$

Remark. A detailed discussion of the approximation criteria in solving the subproblem (4.9) for v^k can be found in [19, Section 5].

4.2 Numerical Results

We test the three algorithms on nine test problems. All problems are convex programs. The first four of the test problems are from [8] (#43, #49, #50, and #100). The fifth problem is from [13] and the last four problems are from [5] (section 3.3). We test each of the three types of the problems from [5] with four variables, and combine the three problems to form the last test problem with twelve variables. Of these problems, only Problem # 49 [8] does not satisfy the second-order sufficiency condition. Therefore, in all but Problem # 49 [8], the inverse of the monotone operator for the primal, dual, and minimax formulations satisfies the second–order growth condition (3.2) at the origin.

The three algorithms are applied to these test problems to be compared with the three corresponding algorithms suggested in [19], namely the proximal minimization algorithm (PPA), the method of multipliers (MM), and the proximal method of multipliers (PMM). Since problems #49 and #50[8] only involve equality constrains, they are used for the primal version only. The problem from [13] only has one constraint, hence it is omitted for the dual version. The MATLAB routine "constr" is employed for solving the constrained subproblems in the primal cases, and the MATLAB routine "fminu" is employed for solving the unconstrained subproblems in the dual and mini-max cases. Both routines are in the MATLAB Optimization Toolbox [2].

Selection of Parameters:
There are four control parameters in the each of the algorithms described in Section 4.1: δ, δ_k, c_k, and $\xi_k\hat{\gamma}_k$ (here we take $\xi = \xi_k\hat{\gamma}_k$ as a single parameter choice).

The Global Stopping Criteria δ:

Primal Problems, The VPA Algorithm: $\|x^{k+1} - x^k\| \leq \delta = 10^{-7}$.

Dual Problems, The VDA Algorithm: $\|y^{k+1} - y^k\| \leq \delta = 10^{-5}$.

Minimax Problems, The VPM Algorithm: $\|(x^{k+1}, y^{k+1}) - (x^k, y^k)\| \leq \delta = 10^{-5}$.

The Subproblem Stopping Criteria δ_k:

MATLAB subroutine stopping tolerance are set to $\delta_k = \max\{0.2\delta_{k-1}, \delta\}$ for $k \geq 1$ with $\delta_0 = 0.1$.

The Proximation Parameters c_k:

Since we focus on local analysis, this parameter is set to be a constant, $c_k = \lambda$ for all $k = 0, 1, 2, \ldots$. The specific choice of λ depends on the problem to be solved.

The Matrix Secant Updating Criteria $\xi_k\hat{\gamma}_k$:

The product $\xi_k\tilde{\gamma}_k$ employed in the matrix updating conditions is set to the constant value $\xi = 0.5$ for all problems.

When the pairs of the algorithms (classic vs. variable metric) are applied to each problem, all parameters are identical. The only difference is in the matrix secant updating formula. The classic algorithms use the identity matrices, while the variable metric algorithms use the matrices updated by the Broyden and BFGS formulas. The numerical results are shown in the following three tables. The second column lists the dimensions of the problem (n denotes the number of variables and m denotes the number of constraints) and the value of parameter λ. For each problem, we list the number of iterations required, the norm of $\|x^k - x^*\|$ at termination where x^* is the known optimal solution, and the numbers of both the function and the gradient evaluations.

Problem	n, m & λ	PPA				VPA			
		iter	$\|x^k - x^*\|$	f eval	g eval	iter	$\|x^k - x^*\|$	f eval	g eval
#43[8]	4,3, 8	17	$4 \cdot 10^{-8}$	108	72	13	$9 \cdot 10^{-9}$	83	55
#49[8]	5,2, 5	10	$2 \cdot 10^{-8}$	73	67	9	$7 \cdot 10^{-9}$	67	61
#50[8]	5,3, 5	20	$2 \cdot 10^{-7}$	95	75	18	$7 \cdot 10^{-8}$	89	66
#100[8]	7,4, 10	25	$7 \cdot 10^{-7}$	420	140	20	$7 \cdot 10^{-7}$	287	98
[13]	2,1, 0.5	19	$4 \cdot 10^{-9}$	137	103	14	$8 \cdot 10^{-9}$	91	65
[5]type I	4,6, 0.5	11	$3 \cdot 10^{-10}$	62	52	9	10^{-9}	41	35
[5] type II	4,6, 5	10	$4 \cdot 10^{-8}$	32	32	3	$2 \cdot 10^{-16}$	12	12
[5] type III	4,6,10	10	$7 \cdot 10^{-8}$	40	40	9	$6 \cdot 10^{-16}$	29	29
[5] comb.	12,18, 4	15	$3 \cdot 10^{-7}$	69	69	9	$3 \cdot 10^{-7}$	33	33

Problem	n, m & λ	MM				VDA			
		iter	$\|z^k - z^*\|$	f eval	g eval	iter	$\|z^k - z^*\|$	f eval	g eval
#43[8]	4,3,10	10	$7 \cdot 10^{-7}$	169	54	6	$4 \cdot 10^{-7}$	124	39
#100[8]	7,4, 6	11	$8 \cdot 10^{-6}$	333	108	7	$8 \cdot 10^{-6}$	319	102
[5] type I	4,6, 12	14	10^{-9}	59	23	6	$7 \cdot 10^{-10}$	34	12
[5] type II	4,6, 12	8	10^{-8}	52	19	4	0	35	12
[5] type III	4,6, 20	14	$4 \cdot 10^{-7}$	273	90	11	$5 \cdot 10^{-8}$	228	74
[5] comb.	12,18,2.5	10	$3 \cdot 10^{-7}$	166	55	8	$8 \cdot 10^{-7}$	136	45

Problem	n, m & λ	PMM				VPM			
		iter	$\|z^k - z^*\|$	f eval	g eval	iter	$\|z^k - z^*\|$	f eval	g eval
#43[8]	4,3, 8	9	$3 \cdot 10^{-6}$	162	51	7	$2 \cdot 10^{-6}$	133	41
#100[8]	7,4, 6	11	$7 \cdot 10^{-6}$	158	49	7	10^{-6}	138	43
[13]	2,1, 0.5	13	$2 \cdot 10^{-6}$	149	49	8	$4 \cdot 10^{-7}$	103	34
[5] type I	4,6, 3	11	$6 \cdot 10^{-7}$	116	40	8	$4 \cdot 10^{-6}$	93	31
[5] type II	4,6, 3	10	$2 \cdot 10^{-6}$	95	35	7	$3 \cdot 10^{-12}$	69	23
[5] type III	4,6, 20	12	$2 \cdot 10^{-6}$	213	69	10	$6 \cdot 10^{-6}$	152	50
[5] comb.	12,18, 2	16	$3 \cdot 10^{-7}$	260	85	12	$6 \cdot 10^{-7}$	207	68

Clearly, the choice of the parameters $\lambda = c_k$ and $\xi = \xi_k \hat{\gamma}_k$ has a direct impact on how often the matrix secant updates are employed. In turn, this impacts the performance of the algorithm. Our experience indicates that for this set of test problems the parameters should be chosen to encourage the use of the matrix secant updates. The results of two numerical experiments are included to illustrate the relationship between the choice of these parameters and the performance of the VPA algorithm.

In the first experiment, we compare performance of the PPA and VPA algorithms for different values of the proximation parameter λ. We do this by applying the algorithms to the last problem for λ varying between 2.5 and 28. The results are shown in the following table. The entries in this table give the number of function evaluations plus the dimension (which in this case is 12) times the number of gradient evaluations before termination. The last row shows the difference of the combined evaluations between the two methods.

λ	2.5	3	4	5	6	8	10	12	14	17	20	23	28
PPA	1444	1262	897	689	637	598	533	494	416	390	442	429	377
VPA	1444	718	429	390	429	481	416	429	377	364	403	403	377
Diff.	0	544	468	299	208	117	117	65	39	26	39	26	0

Observe that for both small and large values of λ the PPA and VPA algorithms are comparable. Real gains in performance only occur in the middle range near $\lambda = 5$. This behavior is typical of all the test problems: poor performance for extreme values of λ with improved performance in some middle range of values. To gain insight into this behavior, recall the relation

$$H_k \approx (-\nabla D_k(\bar{z}))^{-1} = [I + \frac{1}{c_k} \nabla [T^{-1}](0)]$$

from Section 3.2. This relationship indicates that if c_k is too small, then $(-\nabla D_k(\bar{z}))^{-1}$ is most likely very difficult to approximate in which case the variable metric update is probably rejected. This observation is borne out by our experiments. On the other hand, if c_k is large, then $H_k \approx I$ so there is little difference between the PPA and VPA iterates.

In the second experiment, we varied the value of ξ for $\lambda = 3$, 5, and 7. Again the VPA was applied to the last problem. For each value of λ, we discovered what appeared to be a break–point value for ξ. For all values below the break–point the matrix secant updates were almost never employed and the number of combined function and gradient evaluations remained constant. On the other hand, for all values above the break–point the matrix secant updates were almost always employed, and again the number of combined function and gradient evaluations remained constant, but at a significantly reduced level. The results are shown in the following table.

λ	Fun. and Grad. Eval.	
3	1262 for $\xi < 0.5$	507 for $\xi > 0.5$
5	689 for $\xi < 0.3$	390 for $\xi > 0.3$
7	598 for $\xi < 0.25$	468 for $\xi > 0.25$

These experiments indicate that the variable metric proximal point algorithm can be used successfully to improve the performance of the classical proximal point algorithm. However, a number of practical issues remain open. The foremost of these are implementable strategies for updating the proximation parameters c_k and the acceptance criteria $\xi_k\hat{\gamma}_k$ for the matrix secant updates. Our simple experiments indicate that the choice of proximation parameters has a much more dramatic effect on the performance of the method than does the choice of $\xi_k\hat{\gamma}_k$ for $1 > \xi_k\hat{\gamma}_k \geq 0.5$. This corresponds to practical experience with the classical PPA where the choice of proximation parameters is known to critically impact performance.

References

[1] J.F. Bonnans, J.C. Gilbert, C. Lemaréchal, and C. Sagastizábal. A family of variable metric proximal point methods. *Mathematical Programming*, 68:15—47, 1995.

[2] M. A. Branch and A. Grace. *Optimization Toolbox*. The Math Works, Inc., Natick, MA (1996).

[3] J.V. Burke and M. Qian. On the super–linear convergence of the variable metric proximal point algorithm using Broyden and BFGS matrix secant updating. Submitted to *Mathematical Programming*, August 1996.

[4] J.V. Burke and M. Qian. A variable metric proximal point algorithm for monotone operators. To appear in *SIAM J. Control and Optimization*, 1998.

[5] P.H. Calamai, L.N. Vicente, and J.J. Judice. A new technique for generating quadratic programming test problems. *Mathematical Programming*, 61:215—231, 1993.

[6] X. Chen and M. Fukushima. Proximal quasi-Newton methods for nondifferentiable convex optimization. Technical Report AMR 95/32, Dept. of Applied Math., University of New South Wales, Sydney, Australia, 1995.

[7] M. Fukushima and L. Qi. A globally and superlinearly convergent algorithm for nonsmooth convex minimization. *SIAM J. Optim.*, 30:1106—1120, 1996.

[8] W. Hock. Test Examples for Nonlinear Programming Codes. Springer-Verlag, New York, 1981

[9] C. Lemaréchal and C. Sagastizábal. An approach to variable metric bundle methods. In J. Henry and J.P. Yuan, editors, *IFIP Proceedings, Systems Modeling and Optimization*, pages 144—162. Springer, Berlin, 1994.

[10] C. Lemaréchal and C. Sagastizábal. Variable metric bundle methods: from conceptual to implementable forms. Preprint, INRIA, BP 105, 78153 Le Chesnay, France, 1995.

[11] R. Mifflin. A quasi-second-order proximal bundle algorithm. *Mathematical Programming*, 73:51—72, 1996.

[12] R. Mifflin, D. Sun, and L. Qi. Quasi-Newton bundle-type methods for nondifferentiable convex optimization. *SIAM J. Optimization*, 8:563–603, 1998.

[13] H. Mine, K. Ohno, and M. Fukushima. A conjugate interior penalty method for certain convex programs. *SIAM J. Control and Optimization*, 15:747—755, 1977.

[14] G.J. Minty. Monotone (nonlinear) operators in Hilbert space. *Duke Math. J.*, 29:341—346, 1962.

[15] J.J. Moreau. Proximité et dualité dans un espace Hilbertien. *Bull. Soc. Math. France*, 93:273—299, 1965.

[16] L. Qi. Second-order analysis of the Moreau-Yosida regularization of a convex function. Technical Report AMR 94/20, Dept. of Applied Math., University of New South Wales, Sydney, Australia, 1994.

[17] L. Qi and X. Chen. A preconditioning proximal Newton method for nondifferentiable convex optimization. *Mathematical Programming*, 76:411—429, 1997.

[18] M. Qian. The Variable Metric Proximal Point Algorithm: Theory and Application. Ph.D., University of Washington, Seattle, WA, 1992.

[19] R.T. Rockafellar. Augmented Lagrangians and applications of the proximal point algorithm in convex programming. *Math. of Operations Research*, 1:97—116, 1976.

[20] E. Zeidler. *Nonlinear Functional Analysis and its Applications: II/A, Linear Monotone Operators*. Springer-Verlag, New York, 1990.

[21] E. Zeidler. *Nonlinear Functional Analysis and its Applications: II/B, Nonlinear Monotone Operators.* Springer–Verlag, New York, 1990.

Reformulation: Nonsmooth, Piecewise Smooth, Semismooth and Smoothing Methods, pp. 335–354
Edited by M. Fukushima and L. Qi

Reformulation of a Problem of Economic Equilibrium

Alexander M. Rubinov* and Bevil Glover†

Abstract We show that equilibrium prices within the classical Arrow-Debreu equilibrium model are a solution of a special nonsmooth optimization problem. The objective function of this problem is the difference of two marginal functions. Differential properties of these marginal functions are studied.

Key Words Arrow-Debreu equilibrium, directional differentiability, piecewise smooth optimization

1 INTRODUCTION

Mathematical models of economic equilibrium occupy an important place in mathematical economics. The existence of equilibria can be proved by applying fixed point theorems. Usually, methods based on a search of fixed points of special set-valued mappings are used for the calculation of equilibria. A different approach was proposed in [7, 8], for the calculation of equilibria for certain special classes of models. For such equilibrium models a special family of nonsmooth and nonconvex extremal problems which depend on a parameter is constructed. It was shown that the set of global minimizers of each of these extremal problems coincides with the set of vectors of equilibrium prices for the equilibrium model under consideration. The optimal values of these extremal problems are known (in fact are they are equal to zero). This approach is closely related to the least square methods for a solution of a system of equations. In particular, the set of all solutions of the system $f_j(x) = 0$, $j = 1, \ldots, m$, $x \in X$, coincides with the set of global minimizers for the following optimization prob-

*This research has been supported by the Australian Research Council Grant A69701407, School of Information Technology and Mathematical Sciences, The University of Ballarat, Vic 3353, Australia, amr@ballarat.edu.au
†School of Mathematics and Statistics, Curtin University of Technology, Perth, WA, Australia, rgloverb@alpha2.curtin.edu.au

lems which depend on parameters $\rho_j > 0$:

$$\sum_j \rho_j f_j^2(x) \longrightarrow \min \quad \text{subject to} \quad x \in X.$$

The optimal value each of these problems is known (and equal to zero). This follows by assuming that the system is consistent.

In this paper we extend the approach proposed in [7, 8] to the classical Arrow-Debreu model of economic equilibrium (see, for example [4, 5]) including not only consumers but also producers. We show that the problem of finding the equilibria can be reformulated as the minimization of a function H, represented as the difference of two special marginal functions H_1 and H_2, subject to very simple constraints. The objective function H can be chosen depending on a parameter so we again have a family of optimization problems with the same set of solutions. The optimal value of each of these problems is again known (and equal to zero).

Here we also study differential properties of the functions H_1 and H_2. Each of them is the maximum of a concave function subject to a special (and essentially simple) set-valued mapping constraint defined by linear functions. The differential properties of such marginal functions are usually studied, in a very general setting, by assuming that the objective function is smooth and by applying appropriate Lagrange multiplier results (see, for example [2, 3] and references therein). In the simple case under consideration in this paper it is more convenient to apply a direct approach exploiting a special notion of *first order approximation* for set-valued mappings (see [1]). This approach permits the existence of directional derivatives even for nonsmooth concave objective functions and allows the calculation these derivatives in a very convenient way. Some preliminary results in this direction have been obtained in [8].

Applying recent results by D. Ralph and S. Dempe (see [6]) we show that for an exchange model, where only consumers are present and under some other very natural assumptions, the marginal functions H_1 and H_2 are differentiable with piecewise smooth gradient mappings.

The structure of this paper is as follows. In section 2 we briefly recall the Arrow-Debreu equilibrium model. In section 3 we reformulate the equilibrium problem as a special optimization problem. Section 4 provides a discussion on directional derivatives of the objective function of this problem and finally in section 5 we consider an exchange model and discuss differential properties of the objective function for this model.

2 THE ARROW-DEBREU EQUILIBRIUM MODEL

Let I be a finite set of indices. We shall use the following notation:

- $\mathbb{R}^I$ is the space of all vectors $(x^i)_{i \in I}$;
- x^i is the i-th coordinate of a vector $x \in \mathbb{R}^I$;
- if $x, y \in \mathbb{R}^I$ then $x \geq y \iff x^i \geq y^i \quad \text{for all} \quad i \in I$;

- if $x, y \in \mathbb{R}^I$ then $x \gg y \iff x_i > y^i$ for all $i \in I$;
- $\mathbb{R}^I_+ = \{x = (x^i) \in \mathbb{R}^I : x \geq 0\}$;
- $\mathbb{R}^I_{++} = \{x = (x^i) \in \mathbb{R}^I : x \gg 0\}$.

If I consists of n elements we will also use the notation $\mathbb{R}^n$, $\mathbb{R}^n_+$ and $\mathbb{R}^n_{++}$ instead of $\mathbb{R}^I$, $\mathbb{R}^I_+$ and $\mathbb{R}^I_{++}$ respectively.

In this section we describe the version of Arrow-Debreu model that will be investigated in the paper. There are two kinds of economic agents in this model, consumers and producers. We denote the set of consumers by J and the set of producers by K. It is assumed that both J and K are finite sets, $J = \{1, \ldots, m\}$, $K = \{1, \ldots, M\}$. Let also $I = \{1, \ldots, n\}$ be a finite set of products in the economy.

Description of a consumer. A consumer $j \in J$ is a pair (U_j, ω_j) where U_j is a utility function and $\omega_j \in \mathbb{R}^I_+$ is a vector of initial endowments. It is assumed that U_j is concave and increasing ($x_1 \gg x_2$ implies $U(x_1) > U(x_2)$) continuous function defined on $\mathbb{R}^I_+$. As a rule we assume that U_j is a smooth function defined on some open set containing $\mathbb{R}^I_+$. A state of the consumer $j \in J$ is an arbitrary vector $x \in \mathbb{R}^I_+$.

Description of a producer. The producer $k \in K$ is described by a compact convex set $Z_k \subset \mathbb{R}^I$. It is assumed that $0 \in Z_k$ and $Z_k \cap \mathbb{R}^I_+ = \{0\}$. A state of the producer k is a vector $z_k \in Z_k$. (The negative coordinates of the vector z_k are considered as an input of resources and positive coordinates as an output of produced products with respect to the technological process z_k). We denote by Z the Cartesian product of the sets Z_k that is $Z = \prod_{k \in K} Z_k$.

States of the economy. A state of the economy is a vector $v = (x, z)$ with $x = (x_j)_{j \in J}$ and $z = (z_k)_{k \in K}$ such that $x_j \in \mathbb{R}^I_+$, $j \in J$ and $z \in Z$. We denote the set of all states of the economy by V. Thus $V = (\mathbb{R}^n_+)^m \times Z$. A state v is called feasible if it belongs to the following set Ω:

$$\Omega = \{v = (x, z) \in V : x = (x_j)_{j \in J}, z = (z_k)_{k \in K}; \sum_j x_j \leq \sum_k z_k + \sum_j \omega_j\} \tag{2.1}$$

Prices and budget functions. A price vector (prices) is a vector $p \in \mathcal{P}$ where $\mathcal{P} = \{p \in \mathbb{R}^I_+ : \sum_{i \in I} p^i = 1\}$. With a given price vector producers can determine their profit: namely the profit of a technological process $z_k \in Z_k$ is $[p, z_k]$ (the inner product of vectors p and z_k). Let $\beta_k(p)$ be the maximal profit of the producer k under prices p:

$$\beta_k(p) = \max_{z_k \in Z_k} [p, z_k], \qquad (k \in K) \tag{2.2}$$

A budget $\alpha_j(p)$ of the consumer j under prices p are defined by the following formula:

$$\alpha_j(p) = [p, \omega_j] + \sum_k \theta_{kj}\beta_k(p) \tag{2.3}$$

where

$$\theta_{kj} \geq 0 \quad \text{and} \quad \sum_j \theta_{kj} = 1 \quad \text{for all} \quad k \in K. \tag{2.4}$$

The following equality, which is called the Walras Law, follows directly from (2.3) and (2.4):

$$\sum_j \alpha_j(p) = \sum_j [p, \omega_j] + \sum_k \beta_k(p). \tag{2.5}$$

Let $p \in \mathcal{P}$. The set

$$B_j(p) = \{x_j \in \mathbb{R}^I_+ : \; [p, x_j] \leq \alpha_j(p)\} \tag{2.6}$$

is called the budget set of the consumer $j \in J$. If $p \gg 0$ then the budget set is compact, if there exists i such the that i-th coordinate p^i of the vector p is equal to zero, then this set is unbounded. The mapping $p \mapsto B_j(p)$ is called the budget mapping of the consumer j.

We shall denote the model under consideration by $\mathcal{M}$:

$$\mathcal{M} = \{I, J, K, (U_j)_{j \in J}, (\omega_j)_{j \in J}, (\theta_{j,k})_{j \in J, k \in K}, (Z_k)_{k \in K}\}. \tag{2.7}$$

Semi-equilibrium and equilibrium. A semi-equilibrium of the model $\mathcal{M}$ is a pair $(\bar{p}, \bar{v}) \in \mathcal{P} \times \Omega$ with $\bar{v} = (\bar{x}, \bar{z})$ where $\bar{x} = (\bar{x}_j)_{j \in J})$, $\bar{z} = (\bar{z}_k)_{k \in K}$, such that

1) for all $j \in J$ the vector $\bar{x}_j$ is a solution of the problem:

$$U_j(x_j) \longrightarrow \max \quad \text{subject to} \quad x_j \in B_j(\bar{p}). \tag{2.8}$$

2) for all $k \in K$ the vector $\bar{z}_k$ is a solution of the problem

$$[\bar{p}, z_k] \longrightarrow \max \quad \text{subject to} \quad z_k \in Z_k \tag{2.9}$$

A semi-equilibrium $(\bar{p}, \bar{v})$ is called an equilibrium if

$$\sum_j \bar{x}_j = \sum_k \bar{z}_k + \sum_j \omega_j. \tag{2.10}$$

The following Proposition is well known. We give its proof for the sake of completeness.

Proposition 2.1 *A semi-equilibrium $(\bar{p}, \bar{v})$ with a strictly positive price vector $\bar{p} \gg 0$ is an equilibrium.*

Proof: Let $\bar{v} = ((\bar{x}_j)_{j\in J}, (\bar{z}_k)_{k\in K})$. Then

$$\sum_j \bar{x}_j \leq \sum_k \bar{z}_k + \sum_j \omega_j \tag{2.11}$$

and

$$U_j(\bar{x}_j) = \max\{U(x_j) : [\bar{p}, x_j] \leq \alpha_j(\bar{p})\} \quad \text{for all} \quad j \in J, \tag{2.12}$$

$$[\bar{p}, \bar{z}_k] = \max_{z_k \in Z_k} [\bar{p}, z_k] \quad \text{for all} \quad k \in K. \tag{2.13}$$

Assume that the equality does not hold in (2.11). Since $\bar{p} \gg 0$ we can conclude by applying (2.13) and the Walras Law (2.5) that

$$\sum_j [\bar{p}, \bar{x}_j] < \sum_k [\bar{p}, \bar{z}_k] + \sum_j [\bar{p}, \omega_j] = \sum_j \beta_j(\bar{p}) + \sum_j [p, \omega_j] = \sum_j \alpha_j(\bar{p}).$$

So there exists an index j' such that $[p, \bar{x}_{j'}] < \alpha_{j'}(\bar{p})$. Let $\tilde{x} \gg \bar{x}_{j'}$ be a vector such that $[p, \tilde{x}] = \alpha_{j'}(p)$. Then $\tilde{x} \in B_{j'}(\bar{p})$ and $U_{j'}(\tilde{x}) > U_{j'}(\bar{x}_{j'})$. We thus have a contradiction with (2.12) and the desired result follows. ∎

It can be shown (see, for example, [4, 5]) that a semi-equilibrium exists if the utility functions U_j $(j \in J)$ and production sets Z_k $(k \in K)$ possess the properties described above and the following additional assumption holds: $\alpha_j(p) > 0$ for all $p \in \mathcal{P}$ and $j \in J$. These inequalities are valid if, for example, $\omega_j \gg 0$ for all $j \in J$.

We shall assume in the sequel that these conditions hold and therefore a semi-equilibrium exists. We shall assume also that the model under consideration possesses a semi-equilibrium $(\bar{p}, \bar{v})$ with a strictly positive price vector $\bar{p}$ (therefore, an equilibrium is guaranteed to exist by the above).

An equilibrium model which describes only the interaction of consumers is called an *exchange model.* There is no production activity in such a model, in other words, an exchange model is an Arrow-Debreu model (2.7) with $Z_k = \{0\}$ for all $k \in K$. We can also suppose that $K = \emptyset$. The budget function $\alpha_j(p)$ has the following form for an exchange model: $\alpha_j(p) = [p, \omega_j]$.

3 EQUILIBRIUM AS A SOLUTION OF AN OPTIMIZATION PROBLEM

We now reformulate the equilibrium problem as a special optimization problem. More precisely we will consider a family of optimization problems with the same set of optimal solutions.

Consider an equilibrium model

$$\mathcal{M} = \{I, J, K, (U_j)_{j\in J}, (\omega_j)_{j\in J}, (\theta_{j,k})_{j\in J, k\in K}, (Z_k)_{k\in K}\}.$$

Let

$$\mathcal{M}' = \{I, J, K, (\rho_j U_j)_{j\in J}, (\omega_j)_{j\in J}, (\theta_{j,k})_{j\in J,k\in K}, (Z_k)_{k\in K}\}.$$

where ρ_j ($j \in J$) are positive numbers. It is straightforward to show that the models $\mathcal{M}$ and $\mathcal{M}'$ have the same equilibria. Thus we can exploit the model $\mathcal{M}'$ with arbitrary $\rho_j > 0$ in order to find an equilibrium $(\bar{p}, \bar{v})$.

Let $\mathcal{P}_o = \{p \gg 0 : \sum_i p^i = 1\} = \mathcal{P} \cap \{p : p \gg 0\}$. If $p \in \mathcal{P}_o$ the budget sets $B_j(p)$ ($j \in J$), defined by (2.6), are compact and so the following maximization operation is well-defined:

$$H_1(p) = \sum_j \max_{x_j \in B_j(p)} \rho_j U_j(x_j) + \sum_k \beta_k(p) \tag{3.1}$$

(For simplicity we omit ρ in this notation).

We now consider the following functions:

$$F_j(p) = \max_{x_j \in B_j(p)} \rho_j U_j(x_j) \qquad (p \in \mathcal{P}_o), \tag{3.2}$$

and

$$\tilde{F}(p) = \sum_k \beta_k(p) \qquad (p \in \mathcal{P}_o). \tag{3.3}$$

It follows directly from the definition that

$$H_1(p) = \sum_j F_j(p) + \tilde{F}(p). \tag{3.4}$$

In order to give a suitable economic interpretation of the function $H_1(p)$ we consider the following set-valued mapping with compact convex values:

$$A_*(p) = \{x = ((x_j)_{j\in J} \in (\mathbb{R}^n)^m : x_j \in B_j(p),\ (j \in J)\} \equiv \prod_j B_j(p), \tag{3.5}$$

We also consider the following function:

$$U_*(p, x) = \sum_j \rho_j U_j(x_j) + \sum_k \beta_k(p) \quad (p \in \mathcal{P},\ x = ((x_j) \in (\mathbb{R}^n)^m\} \tag{3.6}$$

Let

$$Z_k(p) = \arg\max{}_{z'_k \in Z_k}[p, z_k] \quad (k \in K), \qquad Z(p) = \prod_k Z_k(p) \tag{3.7}$$

The quantity $U_*(p, x)$ expresses the total weighted welfare of all agents, if they use the price vector p and producers choose the vector $z(p) = (z_k(p))_{k\in K} \in Z(p)$.

It is straightforward to show that $H_1(p)$ can be represented in the following form:

$$H_1(p) = \max_{x \in A_*(p)} U_*(p, x) \tag{3.8}$$

We also consider the following function:

$$U(p,v) = \sum_j \rho_j U_j(x_j) + \sum_k [p, z_k] \quad (p \in \mathcal{P}, v = ((x_j)_{j\in J}, (z_k)_{k\in K})) \tag{3.9}$$

It is clear that $U_*(p,x) = \max_{z\in Z} U(p,(x,z))$ and

$$H_1(p) = \max_{v=(x,z);\, x\in A_*(p),\ z\in Z} U(p,v) \tag{3.10}$$

The quantity $H_1(p)$ can be interpreted as the maximal total welfare (weighted by the vector $\rho = (\rho_1, \ldots, \rho_m) \gg 0$) for both consumers and producers which they can obtain under the application of the price vector p, if the feasibility conditions are ignored. In order to take into account these conditions we shall consider the following set-valued mapping $A(p)$:

$$\begin{aligned} A(p) \quad = \quad & \{v = (x,z) : \ x = (x_j)_{j\in J} \in A_*(p),\ z = (z_k)_{k\in K} \in Z(p) \ : \\ & \sum_j x_j \le \sum_k z_k + \sum_j \omega_j\}. \end{aligned} \tag{3.11}$$

The domain $\operatorname{dom} A = \{p \in \mathcal{P}_o : \ A(p) \neq \emptyset\}$ of the mapping A coincides with the set

$$\mathcal{P}_1 = \{p \in \mathcal{P} : \ p \gg 0,\ \exists z_k \in Z_k(p) \quad \text{such that} \quad \sum_k z_k + \sum_j \omega_j \ge 0\}.$$

Thus, if we assume that for all $p \gg 0$ there exists $z_k \in Z_k$ with the property $\sum_k z_k \ge -\sum_j \omega_j$ then $\mathcal{P}_1 = \mathcal{P}_o$. In particular, $\mathcal{P}_1 = \mathcal{P}_o$ for an exchange model. Note that the set $\mathcal{P}_1$ contains the set of all positive equilibrium prices.

Let

$$H_2(p) = \max_{v\in A(p)} U(p,v) \qquad (p \in \mathcal{P}_1) \tag{3.12}$$

Clearly we can represent the quantity $H_2(p)$ in the following form:

$$H_2(p) = \sum_k \beta_k(p) + \max\{\sum_j \rho_j U_j(x_j) :$$

$$x_j \in B_j(p),\ \exists z_k \in Z_k(p) \quad \text{such that} \quad \sum_j x_j \le \sum_k z_k + \sum_j \omega_j\} \tag{3.13}$$

The quantity $H_2(p)$ is the maximal total welfare (weighted by ρ), which agents can obtain if they consider only feasible states. Thus the quantity $H_1(p)$ is the maximal welfare which agents (consumers and producers) desire to obtain (ignoring feasibility conditions) and $H_2(p)$ is the maximal welfare which they are able to obtain (by considering the feasibility conditions). We now consider the difference between this desired state and the available state:

$$H(p) = H_1(p) - H_2(p) \qquad (p \in \mathcal{P}_1). \tag{3.14}$$

It is clear that $A_*(p) \times Z \supseteq A(p)$ for all $p \in \mathcal{P}_1$ so it follows from (3.10) that

$$H(p) \geq 0 \quad \text{for all} \quad p \in \mathcal{P}_1.$$

Lemma 3.1 *Let $\bar{p} \in \mathcal{P}_1$. The equality $H(\bar{p}) = 0$ holds if and only if $\bar{p}$ is a vector of equilibrium prices (that is there exists $\bar{v} \in V$ such that $(\bar{p}, \bar{v})$ is an equilibrium).*

Proof: Let $H(\bar{p}) = 0$ and $\bar{v} = (\bar{x}, \bar{z}) \in A(\bar{p})$ be a vector such that $H_2(\bar{p}) = U(\bar{p}, \bar{v})$. Let $\bar{x} = (\bar{x}_j)_{j \in J}, \bar{z} = (\bar{z}_k)_{k \in K}$. Then $\bar{z} \in Z(\bar{p})$, that is $[\bar{p}, \bar{z}_k] = \beta_k(\bar{p})$ for all k, and

$$\sum_j \max_{x_j \in B_j(\bar{p})} \rho_j U_j(x_j) + \sum_k \beta_k(\bar{p}) = H_1(\bar{p}) = H_2(\bar{p}) = U_*(\bar{p}, \bar{x}).$$

Since

$$U_*(\bar{p}, \bar{x}) = \sum_j \rho_j U_j(\bar{x}_j) + \sum_k \beta_(\bar{p}),$$

we have that

$$\sum_j \rho_j U_j(\bar{x}_j) = \sum_j \rho_j \max_{x_j \in B_j(\bar{p})} U_j(x_j). \tag{3.15}$$

Since $\bar{v} \in A(\bar{p})$ it follows that $\bar{x}_j \in B_j(\bar{p})$. Therefore

$$U_j(\bar{x}_j) \leq \max_{x_j \in B_j(\bar{p})} U_j(x_j) \qquad (j \in J).$$

These inequalities and (3.15) show that

$$U_j(\bar{x}_j) = \max_{x'_j \in B_j(\bar{p})} U_j(x_j), \quad (j \in J).$$

Since $\bar{v} \in A(\bar{p})$ it follows that $\bar{v}$ is a feasible state. Hence, by Proposition 2.1, $(\bar{p}, \bar{v})$ is an equilibrium.

We now prove the reverse assertion. Let $\bar{p} \gg 0$ be equilibrium prices. We need only show that $H(\bar{p}) \leq 0$ and the result will follow. Let $(\bar{p}, \bar{v})$ with $\bar{v} = (\bar{x}, \bar{z})$ be an equilibrium. Let $\bar{x} = (\bar{x}_j)_{j \in J}$, $\bar{z} = (\bar{z}_k)_{k \in K}$. Then

$$[\bar{p}, \bar{z}_k] = \beta_k(\bar{p}) \quad (k \in K), \qquad \bar{x}_j \in B_j(\bar{p}) \quad (j \in J)$$

and $\bar{v}$ is a feasible state. Thus $\bar{v} \in A(\bar{p})$. We also have

$$H_1(p) = \sum_j \rho_j U_J(\bar{x}_j) + \sum_k \beta_k(\bar{p}) = U_*(\bar{p}, \bar{x}) = U(\bar{p}, \bar{v}).$$

Since $\bar{v} \in A(p)$ it follows that $U(\bar{p}, \bar{v}) \leq H_2(\bar{p})$. Thus $H(p) = H_1(p) - H_2(p) \leq 0$. ∎

We now denote the function $H = H_1 - H_2$ by H_ρ. Here $\rho = (\rho_1, \dots, \rho_m) \gg 0$. Assume for the sake of simplicity that $\mathcal{P}_1 = \mathcal{P}_o \equiv \{p \in \mathcal{P} : p \gg 0\}$. Thus strictly positive equilibrium prices can be found by solving the following extremal problem (and by simultaneously assuming that this problem has a strictly positive solution):

$$H_\rho(p) \longrightarrow \min \quad \text{subject to} \quad p \in \mathcal{P}$$

If there exists a vector of strictly positive equilibrium prices, then it follows from Lemma 3.1, that the value of this problem, for each $\rho \gg 0$, is equal to zero. The problem of locating a global minimum by changing the parameters ρ requires special attention and will be the focus of future research. However, in order to use known methods to search for a minimum we need to study the differential properties of the objective function H. This will be the focus of the remainder of the paper.

4 THE DIRECTIONAL DERIVATIVE OF THE FUNCTION H

We shall prove in this section that the function $H = H_1 - H_2$ defined in the previous section is directionally differentiable. We shall apply an approach based on a special first order approximation for set-valued mappings (see [1] and the references therein).

Let Y_1, Y_2 be finite dimensional spaces and a be a set-valued mapping defined on an open subset T of the space Y_1 and mapping into the collection of all nonempty subsets of the space Y_2. Let $x \in T$, $y \in a(x)$ and $g \in Y_1$. The following definitions (see [1]) will be useful in the sequel.

Definition 4.1 *The set*

$$\mathcal{K}(x,y,g) = \{u \in Y_2 : \ y + \gamma u + o(\gamma) \in a(x + \gamma g)\}$$

is called the set of admissible directions of the mapping a at the points x, y in the direction g.

It can be shown that $\mathcal{K}(x,y,g)$ is a closed set for all x, $y \in a(x)$ and $g \in Y_1$.

Definition 4.2 *The mapping a has a first order approximation at the points $x \in T$, $y \in a(x)$ in the direction g if for any sequence of positive numbers $\gamma^{(l)}$ and vectors $y^{(l)}$ such that $\gamma^{(l)} \to 0^+$, $y^{(l)} \in a(x + \gamma^{(l)} g)$, $y^{(l)} \to y$ there exists a sequence $y_*^{(l)}$ such that*

$$y_*^{(l)} \in \mathcal{K}(x,y,g) \quad \textit{and} \quad y^{(l)} = y + \gamma^{(l)} y_*^{(l)} + o(\gamma^{(l)}).$$

Let $f(x,y)$ be a real function of two variables and

$$\psi(x) = \max\{f(x,y) : y \in a(x)\} \tag{4.1}$$

be a marginal function. Let $R(x) = \{y \in a(x) : \psi(x) = f(x,y)\}$. The following theorem (see [1]) will be useful in the sequel.

Theorem 4.1 *Let $x \in T$ and assume the following conditions be satisfied:*

1) *the set-valued mapping a is closed and bounded in some neighborhood* $\mathcal{N}$ *of a point* x *(that is* $\bigcup_{x\in\mathcal{N}} a(x)$ *is bounded);*

2) *the mapping a has a first order approximation at the points* x, y *with* $y \in R(x)$*, in a direction* g*;*

3) *the function* $f(x,y)$ *is concave jointly in its variables in some convex neighborhood of the set* $\{(x,y) : y \in R(x)\}$*.*

Then the function ψ *defined by (4.1) is directionally differentiable at* x *in the direction* g *and its directional derivative* $\psi'(x,g)$ *has the following form:*

$$\psi'(x,g) = \sup_{y\in\mathbb{R}(x)} \sup_{u\in\mathcal{K}(x,y,g)} f'((x,y),(g,u))$$

where $f'((x,y),(g,u))$ *is the directional derivative of the concave function* f *at the point* (x,y) *in the direction* (g,u)*.*

We shall apply this theorem in order to describe the directional derivative of the function $H = H_1 - H_2$ introduced in the previous section.

For the sake of simplicity we shall assume the domain, dom A, of the mapping $A(p)$ is equal to $\mathcal{P}_o$ (see (3.11) and the discussion after this formula). We suppose that the mappings B_j $(j \in J)$ and A are defined by (2.6) and (3.11) respectively for all $p \gg 0$, not only for $p \in \mathcal{P}$. Then the functions H_1 and H_2 are defined for all $p \gg 0$ by (3.1) and (3.13) respectively.

First, we shall describe the set of admissible directions $\mathcal{K}_j(p,x,g)$ for the mapping $B_j(p)$ defined by (2.6):

$$B_j(p) = \{x_j \geq 0 : [p, x_j] \leq \alpha_j(p)\}.$$

where α_j is defined by (2.3). It follows from (2.3) that α_j is a sublinear function and, therefore, is directionally differentiable. For $p \gg 0$, $x \in B_j(p)$ and $q \in \mathbb{R}^l$ consider the set

$$\Delta_j(p,x,q) = \{u \in \mathbb{R}^n : \ [q,x] + [p,u] \leq \alpha_j'(p,q);\ u^i \geq 0 \ \text{ if } \ x^i = 0\}. \quad (4.2)$$

Proposition 4.1 *Let* $p \gg 0$, $x \in B_j(p)$ *and* $[p,x] = \alpha_j(p)$*. Then* $\mathcal{K}_j(p,x,q) = \Delta_j(p,x,q)$ *for all* $q \in \mathbb{R}^n$*.*

Proof: Let $u \in \mathcal{K}_j(p,x,q)$. Then

$$[p+\gamma q, x+\gamma u + o(\gamma)] \leq \alpha_j(p+\gamma q) = \alpha_j(p) + \gamma\alpha_j'(p,q) + o(\gamma).$$

Since $[p,x] = \alpha_j(p)$ we have $[q,x] + [p,u] \leq \alpha_j'(p,q)$. Since $x + \gamma u + o(\gamma) \geq 0$ it follows that $u^i \geq 0$ if $x^i = 0$. Thus the inclusion $\mathcal{K}_j(p,x,q) \subseteq \Delta_j(p,x,q)$ has been verified. We now prove the reverse inclusion.

First we show that

$$\tilde{\Delta}_j \subseteq \mathcal{K}_j(p,x,q) \quad (4.3)$$

where

$$\tilde{\Delta}_j(p,x,q) = \{u : [q,x] + [p,u] \le \alpha_j'(p,q);\ u^i > 0 \quad \text{if} \quad x^i = 0\}.$$

Let $u \in \tilde{\Delta}d_j(p,x,q)$. Then $[q,x]+[p,u] \le \alpha_j'(p,q)$. We also have $[p,x] = \alpha_j(p)$. So

$$[p+\gamma q, x+\gamma u] = [p,x] + \gamma([q,x]+[p,u]) + \gamma^2[q,u] \le$$
$$\alpha_j(p) + \gamma\alpha_j'(p,q) + o(\gamma) = \alpha_j(p+\gamma q) + \xi(\gamma)$$

where $\xi(\gamma)/\gamma \to 0$ as $\gamma \to 0^+$. Take an index $i \in I$ such that $x^i > 0$ and put

$$\tilde{x}(\gamma) = x + \gamma u - \frac{\xi(\gamma)}{p_i + \gamma q_i} e_i$$

where e_i it the i-th unit vector of the space $\mathbb{R}^I$. Then

$$\tilde{x}(\gamma) = x + \gamma u + o(\gamma), \quad [p+\gamma q, \tilde{x}(\gamma)] \le \alpha_j(p+\gamma q)$$

Since $u \in \tilde{\Delta}_j(p,q,x)$ it follows that $u^i > 0$ for i such that $x^i = 0$. Then $\tilde{x}(\gamma) \ge 0$ for sufficiently small γ. Thus $\tilde{x}(\gamma) \in B_j(p+\gamma q)$ and therefore $u \in \mathcal{K}(p,x,q)$. The inclusion (4.3) has been proved. Since $\mathcal{K}_j(p,x,q)$ is closed and the closure of $\tilde{\Delta}_j(p,x,q)$ contains $\Delta_j(p,x,q)$ it follows that $\Delta_j(p,x,q) \subseteq \mathcal{K}_j(p,x,q)$. ∎

Lemma 4.1 *Let $p \gg 0$, $x \in B_j(p)$ and $[p,x] = \alpha_j(p)$. Let further $\gamma^{(l)} \to 0^+$, $x^{(l)} \in B_j(p+\gamma^{(l)}q)$ and $x^{(l)} \to x$. Then $(1/\gamma^{(l)})(x^{(l)} - x) \in \Delta_j(p,x,q)$.*

Proof: It follows from (2.6) that

$$[p+\gamma^{(l)}q, x^{(l)}] \le \alpha_j(p+\gamma^{(l)}q) = \alpha_j(p) + \gamma^{(l)}\alpha_j'(p,q) + o(\gamma^{(l)}). \tag{4.4}$$

Let $u^{(l)} = (1/\gamma^{(l)})(x^{(l)} - x)$. Then

$$[p+\gamma^{(l)}q, x+\gamma^{(l)}u^{(l)}] = [p,x] + \gamma^{(l)}([q,x]+[p,u]) + o(\gamma^{(l)})$$

so, applying (4.4) and the equality $[p,x] = \alpha(p)$, we have $[q,x] + [p,u^{(l)}] \le \alpha_j'(p,q)$. It also follows from the definition of $u^{(l)}$ that $(u^{(l)})_i \ge 0$ if $x^i = 0$. Thus $u^{(l)} \in \Delta_j(p,x,q)$. ∎

We now check that the mapping B_j has a first order approximation.

Proposition 4.2 *Let $p \gg 0$, $x \in B_j(p)$ and $[p,x] = \alpha_j(p)$. Then the mapping B_j has a first order approximation at the points p, x in each direction q.*

Proof: This result follows immediately from Proposition 4.1 and Lemma 4.1. ∎

Consider the set-valued mapping

$$Z_k(p) = \arg\max_{z_k \in Z_k} [p, z_k].$$

We denote by $\Gamma_k(p, z_k, q)$ the set of admissible directions for the mapping Z_k at the points p, z_k in the direction q. Thus for $z_k \in Z_k(p)$ we have:

$$\Gamma_k(p, z_k, q) = \{s_k \in \mathbb{R}^n : \ z_k + \gamma s_k + o(\gamma) \in Z_k(p + \gamma q)\}. \tag{4.5}$$

Proposition 4.3 *Let $z_k \in Z_k(p)$. Then*

$$\Gamma_k(p, z_k, q) = \{s_k \in \mathbb{R}^n : \ \beta_k'(p, q) = [q, z_k] + [p, s_k]\}. \tag{4.6}$$

where β_k is the support function of the convex compact set Z_k.

Proof: It follows from definition that $s_k \in \Gamma_k(p, z_k, q)$ if and only if

$$[p + \gamma q, z_k + \gamma s_k + o(\gamma)] = \max_{z_k' \in Z_k} [p + \gamma q, z_k'] = \beta_k(p + \gamma q).$$

We can represent this equality in the following form:

$$[p, z_k] + \gamma[q, z_k] + \gamma[p, s_k] + o(\gamma) = \beta_k(p) + \gamma \beta_k'(p, q) + o(\gamma). \tag{4.7}$$

Since $[p, z_k] = \beta_k(p)$ it follows from (4.7) that $s_k \in \Gamma_k(p, z_k, q)$ if and only if (4.6) holds. ∎

We now describe the set of admissible directions $\mathcal{K}_A(p, v, q)$ for the mapping $A(p)$ defined by (3.11).

Proposition 4.4 *Let $p \gg 0$ and $v = ((x_j)_{j\in J}, (z_k)_{k\in K}) \in A(p)$. Assume that*

$$[p, x_j] = \alpha_j(p) \quad \text{for all} \quad j \in J \tag{4.8}$$

where α_j is defined by (2.3). Assume also that

$$\sum_j x_j = \sum_k z_k + \sum_j \omega_j \tag{4.9}$$

Then, for all $q \in \mathbb{R}^l$, $\mathcal{K}_A(p, v, q)$ coincides with the following set

$$\mathcal{D}(p, v, q) = \{u = ((r_j)_{j\in J}, (s_k)_{k\in K}) :$$

$$\sum_j r_j + \sum_k s_k \le 0, \ r_j \in \Delta_j(p, q, x_j) \ (j \in J), \ [p, s_k] = c_k \ (k \in K)\}.$$

where Δ_j is defined by (4.2) and $c_k = \beta_k'(p, q) - [q, z_k]$.

Proof: Let $u = ((r_j)_{j\in J}, (s_k)_{k\in K}) \in \mathcal{K}_A(p, v, q)$. Then $v + \gamma u + o(\gamma) \in A(p + \gamma q)$ that is

$$\sum_j x_j + \gamma \sum_j r_j \le \sum_k z_k + \gamma \sum_k s_k + \sum_j \omega_j + o(\gamma) \tag{4.10}$$

and

$$x_j + \gamma r_j + o(\gamma) \in B_j(p + \gamma q) \qquad (j \in J), \tag{4.11}$$

$$z_k + \gamma s_k + o(\gamma) \in Z_k(p + \gamma q) \qquad (k \in K) \tag{4.12}$$

It easily follows from (4.10) and (4.9) that $\sum_j r_j + \sum_k s_k \leq 0$. We also can conclude, by applying (4.11), (4.8) and Proposition 4.1, that $r_j \in \Delta_j(p, q, x_j)$. It follows immediately from (4.12) that $s_k \in \Gamma_k(p, z_k, q)$. Therefore, by applying Proposition 4.3, we can conclude that $[p, s_k] = c_k$. Thus $\mathcal{K}_A(p, v, q) \subseteq \mathcal{D}(p, v, q)$.

We now check that $\mathcal{D}(p, v, q) \subseteq \mathcal{K}_A(p, v, q)$. Let $u = ((r_j)_{j \in J}, (s_k)_{k \in K}) \in \mathcal{D}(p, v, q)$. Then $r_j \in \Delta_j(p, q, x_j)$ for all $j \in J$, therefore (see Proposition 4.1) (4.11) holds. Since $s_k \in \Gamma_k(p, z_k, q)$ it follows that (4.12) holds. Applying (4.9) and the inequality $\sum_j r_j + \sum_k s_k \leq 0$ we have

$$\sum_j x_j + \gamma \sum_j r_j \leq \sum_k z_k + \gamma \sum_k s_k + \sum_j \omega_j + o(\gamma).$$

Thus $v + \gamma u + o(\gamma) \in A(p + \gamma q)$ and therefore $u \in \mathcal{K}(p, v, q)$. ∎

We now prove that the mapping A has a first order approximation.

Proposition 4.5 *Let $p \gg 0$ and a vector $v = ((x_j)_{j \in J}, z = (z_k)_{k \in K}) \in A(p)$ such that (4.9) and (4.8) hold. Then the mapping A has a first order approximation at the points p, v for each direction q.*

Proof: Let $\gamma^{(l)} \to 0^+$ and $v^{(l)} \to v$, $v^{(l)} \in A(p+\gamma^{(l)} q)$. Let $u^{(l)} = (1/\gamma^{(l)})(v^{(l)} - v)$ where $u^{(l)} = ((r_j^{(l)})_{j \in J}, ((s_k^{(l)})_{k \in K})$. Since

$$v^{(l)} = v + \gamma^{(l)} u^{(l)} \in A(p + \gamma^{(l)} q)) \tag{4.13}$$

we easily conclude, by applying (4.9), that $\sum_j r_j^{(l)} + \sum_k s_k^{(l)} \leq 0$. It also follows from (4.13) that $x_j^{(l)} \in B_j(p+\gamma^{(l)} q)$ and $r_j^{(l)} = (1/\gamma^{(l)}(x_j^{(l)} - x_j)$. So by applying Lemma 4.1 we conclude that $r_j^{(l)} \in \Delta_j(p, x_j, q)$. The relation $z_k + \gamma^{(l)} s_k \in Z_k(p + \gamma_{(l)} q)$, which also follows from (4.13), shows that $s_k \in \Gamma_k(p, z_k, q)$. Thus $u^{(l)} \in \mathcal{D}(p, v, q) = \mathcal{K}(p, v, q)$. ∎

We now calculate the directional derivative of the function H_1 and H_2 defined in the previous section. Let $p \gg 0$ and $q \in \mathbb{R}^n$. It follows from (3.4) that

$$H_1'(p, q) = \sum_j F_j'(p, q) + \tilde{F}'(p, q)$$

where F_j and $\tilde{F}$ are defined by (3.2) and (3.3) respectively.

Proposition 4.6 *Let $p \gg 0$ and $q \in \mathbb{R}^n$. Then*

$$F_j'(p, q) = \sup_{x_j \in R_j(p)} \sup_{u \in \Delta_j(p, x_j, q)} \rho_j U_j'(x_j, u)) \tag{4.14}$$

where $R_j(p) = \{x_j \in B_j(p) : \; F_j(p) = U_j(x_j)\}$.

Proof: Since U_j is increasing it follows that $[p, x_j] = \alpha_j(p)$ for $x_j \in R_j(p)$. Hence we can apply Proposition 4.1 which shows that $\mathcal{K}_j(p, x_j, q) = \Delta_j(p, x_j, q)$. The desired result follows directly from Theorem 4.1 and Proposition 4.2. ∎

Proposition 4.7 *Let $p \gg 0$ and $q \in \mathbb{R}^n$. Then*

$$\tilde{F}'(p,q) = \sum_k \beta_k'(p,q)$$

where β_k is the support function of the compact convex set Z_k.

Proof: This follows immediately from (3.3). ∎

Corollary 4.1 *The function H_1 is directionally differentiable at the point $p \gg 0$ in the direction $q \in \mathbb{R}^n$ and*

$$H_1'(p,q) = \sum_j \sup_{x_j \in R_j(p)} \max_{u \in \Delta_j(p,x_j,q)} \rho_j U_j'(x_j, u) + \sum_k \beta_k'(p,q)$$

Proposition 4.8 *The function H_2 is directional differentiable at the point $p \gg 0$ in the direction $q \in \mathbb{R}^n$ and*

$$H_2'(p,q) = \sup_{v \in R_A(p)} \sup_{u \in \mathcal{D}(p,v,q)} U'((p,v),(q,u)). \tag{4.15}$$

where $R_A(p) = \{v \in A(p) : \ H_2(p) = U(p,v)\}$.

Proof: First we verify that

$$\sum_j x_j = \sum_k z_k + \sum_j \omega_j \tag{4.16}$$

and

$$[p, x_j] = \alpha_j(p) \quad \text{for all} \quad j \in J. \tag{4.17}$$

for $v = ((x_j)_{j \in J}, z = (z_k)_{k \in K} \in R_A(p)$. Assume $\sum_j x_j \neq \sum_k z_k + \sum_j \omega_j$. Since v is a feasible element and $p \gg 0$ we can conclude, by applying the Walras Law (2.5)

$$\sum_j [p, x_j] < [p, \sum_k z_k] + [p, \sum_j \omega_j] = \sum_j \alpha_j(p), \tag{4.18}$$

since $z_k \in Z_k(p)$. Thus it follows from (4.18) that there exists $j' \in J$ such that $[p, x_{j'}] < \alpha_{j'}(p)$. Clearly, there exists a vector $\tilde{x}_{j'} \gg x_{j'}$ such that $[p, \tilde{x}_{j'}] = \alpha_{j'}(p)$. It follows from monotonicity of the function $U_{j'}$ that $U_{j'}(\tilde{x}_{j'}) > U_{j'}(x_{j'})$ which is a contradiction. Thus (4.16) holds. It follows immediately from (4.16) and the Walras Law that

$$\sum_j [p, x_j] = [p, \sum_k z_k] + [p, \sum_j \omega_j] = \sum_j \alpha_j(p, v) = \sum_j \alpha_j(p)$$

Since $x_j \in B_j(p)$, we conclude that (4.17) holds. It follows from (4.16) and (4.17), by applying Proposition 4.4, that $\mathcal{K}_A(p,v,q) = \mathcal{D}(p,v,q)$. Thus the desired result follows from Proposition 4.5 and Theorem 4.1. ∎

Remark. The inner suprema in both formulae (4.14) and (4.15) are calculated by maximization of a superlinear function over a polyhedron. It is possible to simplify the calculation of these suprema by exploiting the special structure of the constraints. We will discuss this question for exchange models with smooth strictly concave utility functions in the next section.

5 EXCHANGE MODELS

We shall consider in this section an exchange model, in other words the following will be assumed:

Assumption 5.1 $Z_k = \{0\}$ *for all* $k \in K$

Thus $\beta_k = 0$ for all $k \in K$ and $\alpha_j(p) = [p, \omega_j]$ for all $j \in J$.

We shall also require the following assumption:

Assumption 5.2 *Utility functions* U_j *are twice continuously differentiable and*

$$[\nabla^2 U_j(x_j)y_j, y_j] < 0 \quad \text{for all} \quad x_j \in \mathbb{R}^n_+, \quad \text{and} \quad y_j \neq 0 \qquad (j \in J). \tag{5.1}$$

It follows from (5.1) that U_j is strictly concave on $\mathbb{R}^n_+$. The function U defined by (3.9) has now the following form:

$$U(\rho, v) \equiv U(x) = \sum_j \rho_j U_j(x_j) \qquad (v \equiv x = (x_j)_{j \in J}),$$

hence is strictly concave. Let $p \gg 0$. The function U_j attains its maximum over the set $B_j(p)$ at the unique point and the function U attains its maximum over the set $A(p)$ at the unique point. For $p \gg 0$ let

$$x_j(p) = \arg\max{}_{x_j \in B_j(p)} U_j(x_j) \tag{5.2}$$

and

$$\tilde{x}(p) = (\tilde{x}_1(p), \ldots, \tilde{x}_m(p)) = \arg\max{}_{x \in A(p)} U(x). \tag{5.3}$$

Thus $R_j(p) = \{x_j(p)\}$ and $R_A(p) = \{\tilde{x}(p)\}$.

A function $g(p)$ defined on an open set is said to be PC^1 at a point p^* if g is continuous and there is a finite family of C^1 functions $g^1(p), \ldots, g^N(p)$ defined on a neighborhood of p^*, such that $g(p) \in \{g^1(p), \ldots, g^N(p)\}$ for each p in that neighborhood. A PC^1 function is locally Lipschitz.

We shall use the following theorem which is a special case of a general result from [6]

Theorem 5.1 *Let $x^*(p)$ be a solution of the following parametric convex programming problem:*

$$T(x) \longrightarrow \max$$

subject to

$$x \in \mathbb{R}^n, \quad [a_j, x] \le b_j, \quad (j \in J_1), \ [p, A_j x - b_j] \le 0, \quad (j \in J_2). \qquad (5.4)$$

where T is a twice continuously differentiable concave function defined on an open set X such that

$$[\nabla^2 T(x^*(p))y, y] < 0 \quad \text{for all} \quad y \ne 0;$$

$a_j \, (j \in J_1)$ are vectors and $A_j \, (j \in J_2)$ are matrices.
Assume that the Mangasarian-Fromovitz (MF) constraint qualification holds; that is there exists $y \in \mathbb{R}^n$ such that

$$[a_j, y] < 0 \ \text{if} \ [a_j, x^*(p)] = b_j \, (j \in J_1), \quad [p, A_j y] < 0 \ \text{if} \ [p, A_j x - b_j] = 0 \, (j \in J_2).$$

Then $x^*(\cdot)$ is a PC^1 function.

Remark. The PC^1 property of the solution function (from the parametrized family of optimization problems) is proved in [6] in a general setting. As it is noted in [6] this result holds due to the constant rank constraint qualification (CRCQ). In the case under consideration CRCQ has the following form:
there exists a neighbourhood $\mathcal{N}$ of $x^*(p)$ such that for any subsets J_1^* of the set $\{j \in J_1 : [a_j, x^*(p)] = b_j\}$ and J_2^* of the set $\{j \in J_2 : [p, A_j x(p) - b_j)] = 0\}$, the family of gradients (with respect to x) of constraints has the same rank (depending on J_1^* and J_2^*) for all vectors $x \in \mathcal{N}$.

In the case under consideration the gradients of the constraint functions are vectors $\{a_j : j \in J_1^*\}$ and $\{pA_j : j \in J_2\}$. Since these vectors do not depend on x the CRCQ trivially holds in this case.

Proposition 5.1 *The functions $x_j(p) \, (j \in J)$ and $\tilde{x}(p)$, defined by (5.2) and by (5.3) respectively are PC^1.*

Proof: We can easily represent both mappings B_j and $A(p)$ in the form (5.4) such that the MF constraint qualification holds. Thus the desired result follows directly from Theorem 5.1. ∎

Corollary 5.1 *The function $H(p)$ is PC^1.*

Proof: The function F_j is PC^1 as it is the composition of the PC^1 mapping $x_j(p)$ and the smooth function $U_j, j \in J$. Hence the function H_1 is PC^1. Since the function H_2 is also PC^1 it follows that H is PC^1. ∎

Consider the following index sets for $p \gg 0$:

$$I_j^1(p) = \{i : x_j^i(p) = 0\}, \quad I_j^2(p) = \{i : x_j^i(p) > 0\}.$$

Clearly $I_j^2(p) \neq \emptyset$ for all $p \gg 0$.

Actually the function H_1 is differentiable. In order to show this we shall prove the following statement.

Proposition 5.2 *The function F_j is Frechet differentiable at $p \gg 0$ and*

$$\nabla F_j(p) = \rho_j\left(\frac{1}{p_i}\frac{\partial U_j(x_j(p))}{\partial x^i}\right)(\omega_j - x_j(p)). \tag{5.5}$$

where i is an arbitrary index belonging to $I_j^2(p)$.

Proof: Let us calculate directional derivative $F_j'(p,q)$ for $p \gg 0$ and $q \in \mathbb{R}^n$. It follows from Proposition 4.6 and (4.2) that

$$F_j'(p,q) = \rho_j \sup\{[\nabla U_j(x_j(p), u] : [p,u] \le [q, \omega_j - x_j(p)], u^i \ge 0 \quad \text{if} \quad x^i(p) = 0\}.$$

Proposition 5.1 shows that the function $x_j(p)$ is locally Lipschitz so $F_j(p) = U_j(x_j(p))$ is also locally Lipschitz and therefore the directional derivative $F_j'(p,q)$ is finite. Let

$$c_j = [q, \omega_j - x_j(p)], \quad \rho_j \nabla U_j(x_j(p)) = l.$$

Then $F_j'(p,q)$ is the value of the following linear programming problem:

$$(P_j) \qquad [l,u] \longrightarrow \max \quad \text{subject to} \quad [p,u] \le c_j,\ u^i \ge 0, \quad (i \in I_j^1(p)).$$

The dual problem to the problem P_j has the following form:

$$(D_j) \qquad c_j t \longrightarrow \min \quad \text{subject to}$$

$$tp^i \ge l^i \quad (i \in I_j^1(p)), \qquad tp^i = l^i, \quad (i \in I_j^2(p)), \qquad t \ge 0. \tag{5.6}$$

Since the value of the problem P_j is finite it follows that there exists a number t such that (5.6) holds. Since the set $I_j^2(p)$ is nonempty it follows that $t = \min_i(l^i/p^i)$. Thus the value of the problem D_j is equal to $c_j t$ and therefore

$$F_j'(p,q) = c_j t = [q, \omega_j - x_j(p)]t. \tag{5.7}$$

Since $F_j'(p,q)$ is a linear function of the direction q and F_j is locally Lipschitz it follows that F_j is Frechet differentiable. The equality (5.5) immediately follows from (5.7). ∎

Corollary 5.2 *The function H_1 is Frechet differentiable with a PC^1 gradient mapping.*

Proof: This follows directly from Proposition 5.2. ∎

We now show that the function H_2 is also Frechet differentiable with a PC^1 gradient mapping under rather mild additional assumptions.

Let $p \gg 0$ and let $\tilde{x}(p) = (\tilde{x}_1(p), \ldots, \tilde{x}_m(p))$ be the point defined by (5.3). We shall use the following notation:

$$\tilde{I}_j^1(p) = \{i : \tilde{x}_j^i(p) = 0\} \qquad \tilde{I}_j^2(p) = \{i : \tilde{x}_j^i(p) > 0\}. \tag{5.8}$$

Proposition 5.3 *Let $p \gg 0$ and the set*

$$\tilde{I}(\tilde{x}(p)) = \{i : \tilde{x}^i_j(p) > 0 \quad \text{for all} \quad j \in J\} \equiv \cap_j \tilde{I}^2_j(p) \tag{5.9}$$

be non-empty. Then the function H_2 is Frechet differentiable at the point p and

$$\nabla H_2(p) = \frac{1}{p^{i_o}} \sum_j \frac{\partial U_j(\tilde{x}_j(p)}{\partial x^{i_o}} (\omega_j - \tilde{x}_j(p)). \tag{5.10}$$

where i_o is an arbitrary index belonging to the set $\tilde{I}(\tilde{x}(p))$. The mapping ∇H_2 is PC^1 at the point p.

Proof: First we calculate the directional derivative $H_2'(p,q)$ in a direction $q \in \mathbb{R}^n$. It follows from Propositions 4.8 and 4.4 that

$$H_2'(p,q) = \sup_{u=(u_j)_{j \in J} \in \mathcal{D}(p,\tilde{x}(p),q)} \sum_j [l_j, u_j]$$

where $l_j = \nabla U_j(\tilde{x}_j)$ and

$$\mathcal{D}(p, \tilde{x}(p), q)$$

$$= \{(u_j)_{j \in J} : \sum_j u_j \leq 0, [q, \tilde{x}_j(p)] + [p, u_j] \leq [q, \omega_j], \; u^i_j \geq 0, \; (j \in J, \; i \in \tilde{I}^1_j(p))\},$$

(Here $\tilde{I}^1_j$ is defined by (5.8)). Let

$$\tilde{c}_j = [q, \omega_j - \tilde{x}_j(p)] \qquad (j \in J)$$

Thus $H_2'(p,q)$ is the value of the following linear programming problem:

$$(\tilde{P}) \qquad \sum_j [l_j, u_j] \longrightarrow \max$$

subject to

$$u^i_1 + \ldots + u^i_m \leq 0 \qquad (i = 1, \ldots, n)$$

$$p^1 u^1_j + \ldots + p^n u^n_j \leq \tilde{c}_j \qquad (j = 1, \ldots, m)$$

$$u^i_j \geq 0, \qquad j \in J, \, i \in \tilde{I}^1_j(p).$$

The dual problem to the problem $\tilde{P}$ has the following form

$$(\tilde{D}) \qquad \sum_j \tilde{c}_j \mu_j \longrightarrow \min$$

subject to

$$\lambda^i + p^i \mu_j \geq l^i_j, \qquad j \in J, \, i \in \tilde{I}^1_j(p).$$

$$\lambda^i + p^i \mu_j = l^i_j, \qquad j \in J, \, i \in \tilde{I}^2_j(p). \tag{5.11}$$

$$\lambda^i \geq 0, \quad i = 1, \dots, n; \qquad \mu_j \geq 0, \quad j = 1, \dots, m.$$

It follows from Proposition 5.1 that the function $H_2(p)$ is locally Lipschitz so the directional derivative $H_2'(p, q)$ is finite. Thus the set of admissible solutions of the dual problem is not empty and the value of dual problem is equal to $H_2'(p, q)$.

Let $((\lambda_i)_{i \in I}, (\mu_j)_{j \in J})$ be an admissible solution of the dual problem. Let $i_o \in \tilde{I}((\tilde{x}(p))$. It follows immediately from (5.11) that

$$l_j^{i_o} - p^{i_o}\mu_j = l_1^{i_o} - p^{i_o}\mu_1 \quad \text{for all} \quad j = 1, \dots, m.$$

Hence

$$\mu_j = \mu_1 = \frac{1}{p^{i_o}}(l_j^{i_o} - l_1^{i_o}) \quad \text{for all} \quad j = 1, \dots, m$$

and

$$\sum_j \tilde{c}_j \mu_j = \left(\sum_j \tilde{c}_j\right) \mu_1 + \frac{1}{p^{i_o}} \sum_j \tilde{c}_j l_j^{i_o} - \frac{1}{p^{i_o}} \left(\sum_j \tilde{c}_j\right) l_1^{i_o}.$$

It has been shown in the proof of Proposition 4.8 (see (4.16)) that $\sum_j \tilde{x}_j(p) - \sum_j \omega_j = 0$. Therefore

$$\sum_j \tilde{c}_j = [q, \sum_j (\omega_j - \tilde{x}_j(p)] = 0.$$

Thus

$$\sum_j \tilde{c}_j \mu_j = \frac{1}{p^{i_o}} \sum_j \tilde{c}_j l_j^{i_o}.$$

We have proved that the objective function of the dual problem is constant on the set of admissible solutions. Thus

$$H_2'(p, q) = \sum_j \tilde{c}_j \mu_j = \frac{1}{p^{i_o}} \sum_j \tilde{c}_j l_j^{i_o}$$

$$= \frac{1}{p^{i_o}} [q, \sum_j \frac{\partial U_j(\tilde{x}_j(p))}{\partial p^{i_o}} (\omega_j - \tilde{x}_j(p))].$$

Since $H_2'(p, q)$ is a linear function of q and H_2 is locally Lipschitz it follows that H_2 is Frechet differentiable. We have also obtained (5.10). Let $i_o \in \tilde{I}(\tilde{x}(p))$. It follows from the definition of the set $\tilde{I}(\tilde{x}(p))$ that $i_o \in \tilde{I}(\tilde{x}(p'))$ for all p' sufficiently close to p. Applying (5.10) we can conclude that the mapping ∇H_2 is PC^1 at the point p. ■

Theorem 5.2 *Let Assumptions 5.1 and 5.2 hold and assume that there is a point $p \gg 0$ such that (5.9) is valid. Then the function H is differentiable at the point p and the gradient mapping ∇H is PC^1 at this point.*

Proof: This follows from Corollary 5.2 and Proposition 5.3. ■

Acknowledgement

The authors are very grateful to Dr. Danny Ralph for many helpful discussions.

References

[1] V. F. Demyanov and A. M. Rubinov, *Constructive Nonsmooth Analysis*, Peter Lang, Frankfurt am Main, 1995.

[2] Z.-Q. Luo, J. S. Pang and D. Ralph, *Mathematical Programs with Equilibrium Constraints*, Cambridge University Press, Cambridge, 1996.

[3] L. I. Minchenko, O. F. Borisenko and C. P. Gritcai, *Set-valued analysis and perturbed problems of nonlinear programming*, Navuka i Technika, Minsk, 1993 (In Russian).

[4] V. L. Makarov, M. I. Levin and A. M. Rubinov, *Mathematical Economic Theory: Pure and Mixed Types of Economic Mechanisms*, Advanced Textbooks in Economics, 33, Elsevier, Amsterdam, 1995

[5] H. Nikaido, *Convex Structures and Economic Theory*, Academic Press, New York, 1969.

[6] D. Ralph and S. Dempe, Directional derivatives of the solution of a parametric nonlinear program, *Mathematical Programming*, 70(1995), 159-172.

[7] A. M. Rubinov, On some problems of nonsmooth optimization in economic theory, in: *Nonsmooth Optimization: Methods and Applications*, F. Giannessi (ed.). Gordon and Breach, Amsterdam, 1992, 379-391.

[8] A. M. Rubinov and A. A. Yagubov, Investigations of an exchange model by methods of nonsmooth analysis, *Izv. AN Azerb. SSR*, No 5, 1988, 13-20 (In Russian).

Reformulation: Nonsmooth, Piecewise Smooth, Semismooth and Smoothing Methods, pp. 355–369
Edited by M. Fukushima and L. Qi

A Globally Convergent Inexact Newton Method for Systems of Monotone Equations [1]

Michael V. Solodov[†] and Benav F. Svaiter[†]

Abstract We propose an algorithm for solving systems of monotone equations which combines Newton, proximal point, and projection methodologies. An important property of the algorithm is that the whole sequence of iterates is always globally convergent to a solution of the system without any additional regularity assumptions. Moreover, under standard assumptions the local superlinear rate of convergence is achieved. As opposed to classical globalization strategies for Newton methods, for computing the stepsize we do not use linesearch aimed at decreasing the value of some merit function. Instead, linesearch in the approximate Newton direction is used to construct an appropriate hyperplane which separates the current iterate from the solution set. This step is followed by projecting the current iterate onto this hyperplane, which ensures global convergence of the algorithm. Computational cost of each iteration of our method is of the same order as that of the classical damped Newton method. The crucial advantage is that our method is truly globally convergent. In particular, it cannot get trapped in a stationary point of a merit function. The presented algorithm is motivated by the hybrid projection-proximal point method proposed in [25].

Key Words nonlinear equations, Newton method, proximal point method, projection method, global convergence, superlinear convergence.

---

[1]Research of the first author is supported by CNPq Grant 300734/95-6 and by PRONEX–Optimization, research of the second author is supported by CNPq Grant 301200/93-9(RN) and by PRONEX–Optimization.
[†]Instituto de Matemática Pura e Aplicada, Estrada Dona Castorina 110, Jardim Botânico, Rio de Janeiro, RJ 22460-320, Brazil.
Email : solodov@impa.br and benar@impa.br.

1 INTRODUCTION

We consider the problem of finding solutions of systems of nonlinear equations

$$F(x) = 0, \tag{1.1}$$

where $F : \Re^n \to \Re^n$ is continuous and monotone, i.e.

$$\langle F(x) - F(y), x - y\rangle \geq 0 \quad \text{for all} \quad x, y \in \Re^n.$$

Note that under this assumption, the solution set of (1.1) is convex. Systems of monotone equations arise in various applications. One important example is subproblems in the generalized proximal algorithms with Bregman distances [9, 7].

Among numerous algorithms for solving systems of equations, the Newton method and its variants are of particular importance [16, 5, 19, 2]. Given a point x, the Newton method generates the next iterate by

$$x^{new} = x + d,$$

where d is a (possibly approximate) solution of the system of linear equations

$$\nabla F(x)d = -F(x),$$

with $\nabla F(x)$ being the Jacobian of F at the point x (assuming F is differentiable). If the starting point is sufficiently close to some solution $\bar{x}$ of (1.1) where $\nabla F(\bar{x})$ is nonsingular, the sequence generated by the Newton method converges superlinearly or quadratically, depending on further assumptions. Similar local results hold in the more general case of semismooth equations (see [20, 21, 18, 15, 10]). Having in mind the problem under consideration, it is worth to mention that monotonicity of F is not needed for such local analysis.

To enlarge the domain of convergence of the Newton method, some globalization strategy has to be used. The most common globalization strategy is the damped Newton method

$$x^{new} = x + \alpha d$$

which employs a linesearch procedure along the Newton direction d to compute the stepsize $\alpha > 0$. This linesearch is typically based on an Armijo-type [1] sufficient descent condition for some merit function, usually the squared 2-norm merit function

$$f(x) := \|F(x)\|^2.$$

Sometimes also the damped Gauss-Newton method is used, where the search direction d is obtained by solving the regularized linear system

$$(\nabla F(x)^\top \nabla F(x) + \mu I)d = -\nabla F(x)^\top F(x), \quad \mu > 0.$$

The latter method has the advantage that the search direction always exists, even if $\nabla F(x)$ is singular. Note that in the case of semismooth equations, some additional assumptions are needed to apply the linesearch globalization strategy (see [11]).

To motivate the development of our algorithm, we emphasize the following drawbacks of damped Newton and damped Gauss-Newton methods :

- Either method can only ensure that all accumulation points of the generated sequence of iterates are stationary points of the merit function $f(x)$. In the absence of regularity, there is no guarantee that these stationary points are solutions of the original problem (1.1).

- Moreover, even to ensure the existence of such accumulation points, some additional assumptions are needed (for example, on boundedness of the level sets of f and, hence, of the solution set of the system of equations).

- Without some regularity conditions, one cannot prove convergence of the *whole* sequence of iterates even assuming that the sequence of merit function values actually goes to zero (so that the iterates do approach the solution set of the problem).

Note that by itself, monotonicity of F does not help to resolve any of the above listed difficulties. For example, $f(x) = \|F(x)\|^2$ may still have stationary points which are not solutions of the system of equations $F(x) = 0$. In addition, if the solution set of (1.1) is not a singleton, regularity conditions do not hold.

In this paper, we present a Newton-type algorithm for solving systems of monotone equations which overcomes the above mentioned drawbacks. In particular, under the assumption of continuity and monotonicity of F, from any starting point the *whole* sequence of iterates converges to a solution of the problem, provided one exists (neither regularity nor boundedness of the solution set are needed for global convergence). Under the assumptions of differentiability and nonsingularity, we prove superlinear convergence of the inexact version of the algorithm, similar to the classical inexact Newton method [4].

Our algorithm is motivated, to some extent, by the hybrid projection-proximal point method proposed in [25] in the more general context of finding zeroes of set-valued maximal monotone operators in a Hilbert space. Let T be a maximal monotone operator on a real Hilbert space $\mathcal{H}$. And consider, for a moment, the problem of finding an $x \in \mathcal{H}$ such that $0 \in T(x)$.

Algorithm 1.1 (Hybrid Projection-Proximal Point Method) *[25]*
Choose any $x^0 \in \mathcal{H}$ and $\sigma \in [0, 1)$; set $k := 0$.

Inexact proximal step. *Choose $\mu_k > 0$ and find $y^k \in \mathcal{H}$ and $v^k \in T(y^k)$ such that*

$$0 = v^k + \mu_k(y^k - x^k) + \varepsilon^k,$$

where

$$\|\varepsilon^k\| \leq \sigma \max\{\|v^k\|, \mu_k\|y^k - x^k\|\}.$$

Stop if $v^k = 0$ or $y^k = x^k$. Otherwise,

Projection step. *Compute*

$$x^{k+1} = x^k - \frac{\langle v^k, x^k - y^k \rangle}{\|v^k\|^2} v^k.$$

Set $k := k + 1$*; and repeat.*

If problem $0 \in T(x)$ has a solution and the sequence $\{\mu_k\}$ is bounded above, then the generated sequence $\{x^k\}$ either is finite and terminates at a solution, or it is infinite and converges (weakly) to a solution (for complete properties of the method, see [25]). The idea of Algorithm 1.1 is to use an approximate proximal iteration to construct a hyperplane

$$H_k := \{x \in \Re^n \mid \langle v^k, x - y^k \rangle = 0\},$$

which separates the current iterate x^k from the solutions of $0 \in T(x)$. The last step of Algorithm 1.1 is equivalent to projecting x^k onto this hyperplane. Separation arguments show that the distance to the solution set for thus constructed sequence monotonically decreases, which essentially ensures global convergence of the algorithm. The advantage of Algorithm 1.1 over the classical proximal point method [22] (of course, we are talking about the general context of operator equations here) is that the condition imposed on ε^k in the inexact proximal step is significantly less restrictive (and more constructive) than the corresponding tolerance requirements in the standard proximal point settings (see [22, 3]) :

$$\|\varepsilon^k\| \le \sigma_k \mu_k, \quad \sum_{k=0}^{\infty} \sigma_k < \infty$$

or

$$\|\varepsilon^k\| \le \sigma_k \mu_k \|y^k - x^k\|, \quad \sum_{k=0}^{\infty} \sigma_k < \infty.$$

Note that in the classical approach one further sets $x^{k+1} := y^k$ to obtain the next iterate. For one thing, the right-hand-side of the corresponding tolerance inequality in Algorithm 1.1 involves the *largest* of two quantities, $\|v^k\|$ and $\mu_k \|y^k - x^k\|$. And more importantly, the tolerance parameter σ can be fixed, which means that the *relative error* in solving the proximal subproblems has to be merely bounded, while in the classical setting it has to be summable (hence tend to zero). Thus Algorithm 1.1 has better robustness features while preserving computational costs and convergence properties of the classical proximal point method (computational costs are actually expected to be reduced, thanks to the relaxed tolerance requirements). We refer the readers to [25] for complete analysis of Algorithm 1.1 and a more detailed comparison with the classical proximal point method. Another related work using inexact proximal iterations with relative error bounded away from zero is [24].

Now let us go back to our problem (1.1). To ensure global convergence to a solution, one can apply Algorithm 1.1, setting $T = F$. However, it is clear that this approach, without modifications, is not quite practical for the following reasons. The resulting proximal subproblem $F(x) + \mu_k(x - x^k) = 0$ is still a rather difficult nonlinear problem, albeit with better nondegeneracy properties compared to the original equation. Solving this nonlinear equation even with the more realistic error tolerance of Algorithm 1.1, can be quite difficult. Therefore in this paper, we propose an algorithm where an approximate proximal subproblem solution is obtained by a single Newton step. Moreover, the (linear) regularized Newton equation itself can be solved inexactly (see Algorithm 2.1). Because such step may fail to generate an appropriate separating hyperplane (i.e. the tolerance requirements of Algorithm 1.1 need not be met), a linesearch procedure similar to the one in [23] is employed. Finally, as in Algorithm 1.1, a projection step is made. This hybrid algorithm is globally convergent to a solution of the system of equations, provided one exists, under no assumptions on F other than continuity and monotonicity (see Theorem 2.1). When F is differentiable at this solution and the Jacobian is nonsingular, the rate of convergence is superlinear (see Theorem 2.2). Thus attractive global convergence properties of proximal-like methods are successfully combined with fast local convergence of Newton-type methods. Globalization is achieved via a linesearch and projection strategy. The total computational cost of one iteration of the algorithm is of the same order as that of the standard damped Newton method.

2 THE ALGORITHM AND ITS CONVERGENCE

We now describe our algorithm in detail. Given a current iterate x^k and a regularization parameter $\mu_k > 0$, consider the proximal point subproblem (see [22, 6, 8, 12, 13]),

$$0 = F(x) + \mu_k(x - x^k).$$

The first step consists of solving the linearization of this subproblem at the point x^k

$$0 = F(x^k) + G_k(x - x^k) + \mu_k(x - x^k),$$

where G_k is a positive semidefinite matrix (this is similar to standard Newton-proximal point approaches, for example, [9]). We allow this Newton-type linear equation to be solved approximately (see Algorithm 2.1), much in the spirit of inexact Newton methods (see [4, 26, 14, 17, 15]). This feature is of particular importance for large-scale problems.

Because a full step in the obtained Newton direction may not satisfy the tolerance conditions imposed on solving the proximal subproblems in Algorithm 1.1, we cannot immediately perform the projection step. Thus the second step of the method is a linesearch procedure in the Newton direction d^k (see Algorithm 2.1) which computes a point $y^k = x^k + \alpha_k d^k$ such that

$$0 < \langle F(y^k), x^k - y^k \rangle.$$

A similar linesearch technique was used in [23]. Note that by monotonicity of F, for any $\bar{x}$ such that $F(\bar{x}) = 0$ we have

$$0 \geq \langle F(y^k), \bar{x} - y^k \rangle.$$

Thus the hyperplane

$$H_k := \{x \in \Re^n \mid \langle F(y^k), x - y^k \rangle = 0\}$$

strictly separates the current iterate x^k from zeroes of the system of equations. Once the separating hyperplane is obtained, the next iterate x^{k+1} is computed by projecting x^k onto it.

We now formally state the algorithm.

Algorithm 2.1 *Choose any* $x^0 \in \Re^n$, $\beta \in (0,1)$ *and* $\lambda \in (0,1)$; *set* $k := 0$.

Inexact Newton step. *Choose a positive semidefinite matrix* G_k. *Choose* $\mu_k > 0$ *and* $\rho_k \in [0,1)$. *Compute* $d^k \in \Re^n$ *such that*

$$0 = F(x^k) + (G_k + \mu_k I)d^k + e^k, \tag{2.1}$$

where

$$\|e^k\| \le \rho_k \mu_k \|d^k\|.$$

Stop if $d^k = 0$. *Otherwise,*

Linesearch step. *Find* $y^k = x^k + \alpha_k d^k$, *where* $\alpha_k = \beta^{m_k}$ *with* m_k *being the smallest nonnegative integer* m *such that*

$$-\langle F(x^k + \beta^m d^k), d^k \rangle \ge \lambda(1 - \rho_k)\mu_k \|d^k\|^2. \tag{2.2}$$

Projection step. *Compute*

$$x^{k+1} = x^k - \frac{\langle F(y^k), x^k - y^k \rangle}{\|F(y^k)\|^2} F(y^k). \tag{2.3}$$

Set $k := k + 1$; *and repeat.*

We note that Algorithm 2.1 has computational costs per iteration comparable to those of damped (Gauss-) Newton method : solving a system of linear equations followed by a linesearch procedure (the projection step is explicit and hence computationally negligible). Algorithm 2.1 has the advantage that the whole sequence of iterates is globally convergent to a solution of the system of equations under no regularity assumptions (Theorem 2.1). When F is differentiable at this solution, ∇F is nonsingular, and the parameters of the algorithm are set properly, fast superlinear rate of convergence is obtained (Theorem 2.2).

We start with a preliminary result.

Lemma 2.1 *Let* F *be monotone and* $x, y \in \Re^n$ *be such that*

$$\langle F(y), x - y \rangle > 0.$$

Let

$$x^+ = x - \frac{\langle F(y), x - y \rangle}{\|F(y)\|^2} F(y).$$

Then for any $\bar{x} \in \Re^n$ such that $F(\bar{x}) = 0$, it holds that

$$\|x^+ - \bar{x}\|^2 \leq \|x - \bar{x}\|^2 - \|x^+ - x\|^2.$$

Proof. Let $\bar{x} \in \Re^n$ be any point such that $F(\bar{x}) = 0$. By monotonicity of F,

$$\langle F(y), \bar{x} - y\rangle \leq 0.$$

It follows from the hypothesis that the hyperplane

$$H := \{s \in \Re^n \mid \langle F(y), s - y\rangle = 0\}$$

strictly separates x from $\bar{x}$. It is also easy to verify that x^+ is the projection of x onto the halfspace $\{s \in \Re^n \mid \langle F(y), s - y\rangle \leq 0\}$. Since $\bar{x}$ belongs to this halfspace, it follows from the basic properties of the projection operator (see [19, p.121]) that $\langle x - x^+, x^+ - \bar{x}\rangle \geq 0$. Therefore

$$\begin{aligned} \|x - \bar{x}\|^2 &= \|x - x^+\|^2 + \|x^+ - \bar{x}\|^2 + 2\langle x - x^+, x^+ - \bar{x}\rangle \\ &\geq \|x - x^+\|^2 + \|x^+ - \bar{x}\|^2 \\ &= \left(\frac{\langle F(y), x - y\rangle}{\|F(y)\|}\right)^2 + \|x^+ - \bar{x}\|^2. \end{aligned}$$

■

We are now ready to prove our main global convergence result. Throughout we assume that the solution set of the problem is nonempty.

Theorem 2.1 *Suppose that F is continuous and monotone and let $\{x^k\}$ be any sequence generated by Algorithm 2.1.*

For any $\bar{x}$ such that $F(\bar{x}) = 0$, it holds that

$$\|x^{k+1} - \bar{x}\|^2 \leq \|x^k - \bar{x}\|^2 - \|x^{k+1} - x^k\|^2.$$

In particular, $\{x^k\}$ is bounded. Furthermore, it holds that either $\{x^k\}$ is finite and the last iterate is a solution, or the sequence is infinite and $0 = \lim_{k\to\infty} \|x^{k+1} - x^k\|$.

Suppose the sequence $\{x^k\}$ is infinite. Suppose that $1 > \limsup_{k\to\infty} \rho_k$ and there exist constants $C_1, C_2, C_3 > 0$ such that $\|G_k\| \leq C_1$ for all k and, starting with some index k_0, $C_2 \geq \mu_k \geq C_3\|F(x^k)\|$. Then $\{x^k\}$ converges to some $\bar{x}$ such that $F(\bar{x}) = 0$.

Proof. First note that if the algorithm terminates at some iteration k then $d^k = 0$, and consequently, $e^k = 0$. It follows from (2.1) that $F(x^k) = 0$, so that x^k is a solution. From now on, we assume that $d^k \neq 0$ for all k. We next show that the method is well-defined and an infinite sequence $\{x^k\}$ is generated.

Since G_k is positive semidefinite, the regularized Newton equation is always solvable. Hence the inexact Newton step (2.1) is well-defined.

We now show that the linesearch procedure (2.2) always terminates with a positive stepsize α_k. Suppose that for some iteration index k this is not the case. That is, for all integers m we have

$$-\langle F(x^k + \beta^m d^k), d^k\rangle < \lambda(1-\rho_k)\mu_k\|d^k\|^2. \tag{2.4}$$

We further obtain

$$\begin{aligned} -\lim_{m\to\infty}\langle F(x^k + \beta^m d^k), d^k\rangle &= -\langle F(x^k), d^k\rangle \\ &= \langle (G_k + \mu_k I)d^k + e^k, d^k\rangle \\ &\geq \mu_k\|d^k\|^2 - \|e^k\|\|d^k\| \\ &\geq (1-\rho_k)\mu_k\|d^k\|^2, \end{aligned}$$

where the second equality and the last inequality follow from (2.1), and the first inequality follows from positive semidefiniteness of G_k and the Cauchy-Schwarz inequality. Now taking the limits as $m \to \infty$ in both sides of (2.4) implies that $\lambda \geq 1$, which contradicts the choice of $\lambda \in (0,1)$. It follows that the linesearch step (and hence the whole algorithm) is well-defined.

By (2.2) we have

$$\langle F(y^k), x^k - y^k\rangle = -\alpha_k\langle F(y^k), d^k\rangle \geq \lambda(1-\rho_k)\mu_k\alpha_k\|d^k\|^2 > 0. \tag{2.5}$$

Let $\bar{x}$ be any point such that $F(\bar{x}) = 0$. By (2.3), (2.5) and Lemma 2.1, it follows that

$$\|x^{k+1} - \bar{x}\|^2 \leq \|x^k - \bar{x}\|^2 - \|x^{k+1} - x^k\|^2. \tag{2.6}$$

Hence the sequence $\{\|x^k - \bar{x}\|\}$ is nonincreasing and convergent, therefore the sequence $\{x^k\}$ is bounded, and also

$$0 = \lim_{k\to\infty}\|x^{k+1} - x^k\|. \tag{2.7}$$

By (2.1), the triangle inequality and the choice of μ_k for $k \geq k_0$, we have

$$\begin{aligned} \|F(x^k)\| &\geq \|(G_k + \mu_k I)d^k\| - \|e^k\| \\ &\geq (1-\rho_k)\mu_k\|d^k\| \\ &\geq (1-\rho_k)C_3\|F(x^k)\|\|d^k\|. \end{aligned}$$

It follows that the sequence $\{d^k\}$ is bounded, hence so is $\{y^k\}$. Now by continuity of F and condition $1 > \limsup_{k\to\infty}\rho_k$, there exists $C_4 > 0$ such that $\lambda(1-\rho_k)\|F(y^k)\|^{-1} \geq C_4$. Using (2.3) and (2.5), we thus obtain

$$\begin{aligned} \|x^{k+1} - x^k\| &= \frac{\langle F(y^k), x^k - y^k\rangle}{\|F(y^k)\|} \\ &\geq C_4\mu_k\alpha_k\|d^k\|^2. \end{aligned}$$

From the latter relation and (2.7) it follows that

$$0 = \lim_{k\to\infty} \mu_k \alpha_k \|d^k\|^2. \tag{2.8}$$

We consider the two possible cases :

$$0 = \liminf_{k\to\infty} \|F(x^k)\| \quad \text{and} \quad 0 < \liminf_{k\to\infty} \|F(x^k)\|.$$

In the first case, continuity of F implies that the sequence $\{x^k\}$ has some accumulation point $\hat{x}$ such that $F(\hat{x}) = 0$ (recall that $\{x^k\}$ is bounded). Since $\bar{x}$ was an *arbitrary* solution, we can choose $\bar{x} = \hat{x}$ in (2.6). The sequence $\{\|x^k - \hat{x}\|\}$ converges, and since $\hat{x}$ is an accumulation point of $\{x^k\}$, it must be the case that $\{x^k\}$ converges to $\hat{x}$.

Consider now the second case. From the choice of μ_k for $k \geq k_0$,

$$\liminf_{k\to\infty} \mu_k \geq C_3 \liminf_{k\to\infty} \|F(x^k)\| > 0.$$

Furthermore, by (2.1) and the triangle inequality, we have

$$\begin{aligned} \|F(x^k)\| &\leq \|(G_k + \mu_k I)d^k\| + \|e^k\| \\ &\leq (\|G_k\| + \mu_k + \rho_k\mu_k)\,\|d^k\| \\ &\leq (C_1 + 2C_2)\|d^k\|. \end{aligned}$$

Hence,

$$0 < \liminf_{k\to\infty} \|d^k\|.$$

Then, by (2.8), it must hold that

$$0 = \lim_{k\to\infty} \alpha_k.$$

The latter is equivalent to saying that $m_k \to \infty$. By the stepsize rule we have that (2.2) is not satisfied for the value of β^{m_k-1}, i.e.

$$-\langle F(x^k + \beta^{m_k-1}d^k), d^k\rangle < \lambda(1-\rho_k)\mu_k\|d^k\|^2.$$

Taking into account boundedness of the sequences $\{x^k\}$, $\{\mu_k\}$, $\{\rho_k\}$ and $\{d^k\}$, and passing onto a subsequence if necessary, as $k \to \infty$ we obtain

$$-\langle F(\hat{x}), \hat{d}\rangle \leq \lambda(1-\hat{\rho})\hat{\mu}\|\hat{d}\|^2,$$

where $\hat{x}$, $\hat{\mu}$, $\hat{\rho}$ and $\hat{d}$ are limits of corresponding subsequences. On the other hand, by (2.1) and already familiar argument,

$$-\langle F(\hat{x}), \hat{d}\rangle \geq (1-\hat{\rho})\hat{\mu}\|\hat{d}\|^2.$$

Taking into account that $\hat{\mu}\|\hat{d}\| > 0$ and $\hat{\rho} \leq \limsup_{k\to\infty} \rho_k < 1$, the last two relations are a contradiction because $\lambda \in (0,1)$. Hence the case $0 < \liminf_{k\to\infty} \|F(x^k)\|$ is not possible.

This completes the proof. ∎

Remark. It can be observed from the proof of Theorem 2.1 that conditions imposed on the choice of μ_k (even though they are not restrictive) can be replaced by more general ones. Let t_1 and t_2 be two nonnegative monotonically increasing functions (not necessarily continuous) such that $t_1(0) = t_2(0) = 0$ and $t_2(s) \geq t_1(s) > 0$ for all $s > 0$. If μ_k is such that

$$t_1(\|F(x^k)\|) \leq \mu_k \leq t_2(\|F(x^k)\|),$$

then the global convergence result still holds.

We now turn our attention to local rate of convergence analysis.

Theorem 2.2 *Let F be monotone and continuous on $\Re^n$. Let $\bar{x}$ be the (unique) solution of (1.1) at which F is differentiable with $\nabla F(\bar{x})$ nonsingular. Let ∇F be locally Lipschitz continuous around $\bar{x}$. Suppose that starting with some index k_0, $G_k = \nabla F(x^k)$. Suppose that*

$$0 = \lim_{k\to\infty} \mu_k = \lim_{k\to\infty} \rho_k = \lim_{k\to\infty} \mu_k^{-1}\|F(x^k)\|. \tag{2.9}$$

Then the sequence $\{x^k\}$ converges to $\bar{x}$ Q-superlinearly.

If starting with some index k_0, $\mu_k = \gamma_1\|F(x^k)\|^{1/2}$ and $\rho_k = \gamma_2\|F(x^k)\|^{1/2}$ for some positive constants γ_1 and γ_2, then the order of superlinear convergence is at least 1.5 .

Proof. By Theorem 2.1, we already know that the sequence $\{x^k\}$ converges to $\bar{x}$ (note that monotonicity of F and nonsingularity of $\nabla F(\bar{x})$ imply that $\bar{x}$ is the unique solution).

Since $\{x^k\}$ converges to $\bar{x}$ and $G_k = \nabla F(x^k)$ starting with some index k_0, nonsingularity of $\nabla F(\bar{x})$ implies that there exists a constant $C_5 > 0$ such that $\|G_k z\| \geq C_5\|z\|$ for all $z \in \Re^n$ and all k sufficiently large. By (2.1) and the triangle inequality, we have

$$\begin{aligned} \|F(x^k)\| &\geq \|(G_k + \mu_k I)d^k\| - \|e^k\| \\ &\geq \|G_k d^k\| - \|e^k\| \\ &\geq C_5\|d^k\| - \rho_k\mu_k\|d^k\|. \end{aligned}$$

Since $0 = \lim_{k\to\infty} \rho_k\mu_k$, it follows that for some $C_6 > 0$

$$\|d^k\| \leq C_6\|F(x^k)\|. \tag{2.10}$$

In particular, the sequence $\{d^k\}$ converges to zero. By Lipschitz continuity of ∇F in the neighbourhood of $\bar{x}$, it follows that there exists $C_7 > 0$ such that

$$\begin{aligned} &F(x^k + d^k) = F(x^k) + \nabla F(x^k)d^k + R^k, \\ &\|R^k\| \leq C_7\|d^k\|^2 \end{aligned} \tag{2.11}$$

for k large enough.

We next establish that for k large, the unit stepsize is always accepted. By (2.1) and (2.11) we have

$$F(x^k + d^k) = -\mu_k d^k - e^k + R^k. \tag{2.12}$$

Using the Cauchy-Schwarz inequality, for k large enough we obtain

$$\begin{aligned} -\langle F(x^k + d^k), d^k\rangle &= \langle \mu_k d^k + e^k - R^k, d^k\rangle \\ &\geq \mu_k \|d^k\|^2 - (\|e^k\| + \|R^k\|)\|d^k\| \\ &\geq (1-\rho_k)\mu_k\|d^k\|^2 - C_7\|d^k\|^3 \\ &= \left[1 - \frac{C_7\|d^k\|}{(1-\rho_k)\mu_k}\right](1-\rho_k)\mu_k\|d^k\|^2. \end{aligned}$$

Note that from (2.10) and (2.9), it follows that

$$\lim_{k\to\infty} \frac{C_7\|d^k\|}{(1-\rho_k)\mu_k} \leq C_7 C_6 \lim_{k\to\infty} \frac{\|F(x^k)\|}{(1-\rho_k)\mu_k} = 0.$$

Therefore (recall that $\lambda < 1$), for k sufficiently large it holds that

$$-\langle F(x^k + d^k), d^k\rangle > \lambda(1-\rho_k)\mu_k\|d^k\|^2,$$

i.e. condition (2.2) is satisfied for $m = 0$. Hence $\alpha_k = 1$ and $y^k = x^k + d^k$ for all indices k large enough.

From now on, we assume that k is large enough so that (2.11) and (2.12) hold, and $y^k = x^k + d^k$.

Denote

$$s^k = -(\nabla F(x^k) + \mu_k I)^{-1} F(x^k) \quad \text{and} \quad z^k = x^k + s^k.$$

By the triangle inequality,

$$\|x^{k+1} - \bar{x}\| \leq \|x^{k+1} - y^k\| + \|y^k - z^k\| + \|z^k - \bar{x}\|. \tag{2.13}$$

We proceed to analyze the three terms in the right-hand-side of the above inequality separately.

We start with the first term. Since the point x^{k+1} is the projection of x^k onto the hyperplane H_k (defined previously), and $y^k \in H_k$, the vectors $x^{k+1} - x^k$ and $x^{k+1} - y^k$ are orthogonal. Hence

$$\|x^{k+1} - y^k\| = \|y^k - x^k\| \sin\theta_k = \|d^k\| \sin\theta_k,$$

where θ_k is the angle between $x^{k+1} - x^k$ and $y^k - x^k$. Because $x^{k+1} - x^k = -t_k F(y^k)$ for a certain $t_k > 0$ and $y^k - x^k = d^k$, the angle between the vectors $F(y^k)$ and $-\mu_k d^k$ is also θ_k. Since $y^k = x^k + d^k$, by (2.12) we have

$$F(y^k) = -\mu_k d^k - e^k + R^k.$$

Given the above relation, sin of the angle between $F(y^k)$ and $-\mu_k d^k$ can be easily bounded:

$$\sin\theta_k \le \frac{\|-e^k + R^k\|}{\mu_k\|d^k\|}.$$

By the triangle inequality, we further obtain

$$\sin\theta_k \le \rho_k + C_7\frac{\|d^k\|}{\mu_k}.$$

Therefore,

$$\|x^{k+1} - y^k\| \le (\rho_k + C_7\frac{\|d^k\|}{\mu_k})\|d^k\|. \tag{2.14}$$

For the second term in (2.13) we have

$$\begin{aligned}
\|y^k - z^k\| &= \|d^k - s^k\| \\
&= \|(\nabla F(x^k) + \mu_k I)^{-1}e^k\| \\
&\le C_8\|e^k\| \\
&\le C_8\rho_k\mu_k\|d^k\|,
\end{aligned} \tag{2.15}$$

where C_8 is some positive constant.

We now consider the third term in the right-hand-side of (2.13). By the Cauchy-Schwarz and triangle inequalities, and Lipschitz continuity of ∇F in the neighbourhood of $\bar{x}$, we have

$$\begin{aligned}
\|z^k - \bar{x}\| &= \|x^k - \bar{x} - (\nabla F(x^k) + \mu_k I)^{-1}F(x^k)\| \\
&= \|(\nabla F(x^k) + \mu_k I)^{-1}[F(\bar{x}) - F(x^k) + (\nabla F(x^k) + \mu_k I)(x^k - \bar{x})]\| \\
&\le C_8\|F(\bar{x}) - F(x^k) - \nabla F(x^k)(\bar{x} - x^k)\| + C_8\mu_k\|x^k - \bar{x}\| \\
&\le C_9\|x^k - \bar{x}\|^2 + C_8\mu_k\|x^k - \bar{x}\|,
\end{aligned} \tag{2.16}$$

with some $C_9 > 0$.

By (2.10) and Lipschitz continuity of F around $\bar{x}$, we have

$$\|d^k\| \le C_6\|F(x^k)\| \le C_{10}\|x^k - \bar{x}\|, \tag{2.17}$$

where $C_{10} > 0$. Combing (2.13)-(2.16) and taking into account (2.17), we obtain

$$\|x^{k+1} - \bar{x}\| \le$$
$$\left(C_{10}(\rho_k + C_7C_6\tfrac{\|F(x^k)\|}{\mu_k} + C_8\rho_k\mu_k) + C_9\|x^k - \bar{x}\| + C_8\mu_k\right)\|x^k - \bar{x}\|. \tag{2.18}$$

From (2.9) it follows that the sequence $\{x^k\}$ converges to $\bar{x}$ Q-superlinearly.

If $\mu_k = \gamma_1\|F(x^k)\|^{1/2}$ and $\rho_k = \gamma_2\|F(x^k)\|^{1/2}$ for some positive constants γ_1 and γ_2, then (2.17) and (2.18) yield

$$\|x^{k+1} - \bar{x}\| \le C_{11}\|x^k - \bar{x}\|^{3/2}, \quad C_{11} > 0,$$

i.e. the order of superlinear convergence is at least 1.5 . ■

Remark. Inequality (2.18) also shows that if parameters μ_k and ρ_k are fixed and sufficiently small, then the linear rate of convergence is attained.

Remark. It can be observed that under the hypotheses of Theorem 2.2, for k large enough (so that the unit stepsize is always accepted), the steps of Algorithm 2.1 become a particular (practical) implementation of the more general Algorithm 1.1.

Remark. The superlinear convergence of order 1.5 implies at least 2-step Q-quadratic convergence.

Finally, we give one possible choice of parameters which satisfies the conditions of Theorems 2.1 and 2.2 :

$$\mu_k = \max\{C_3\|F(x^k)\|, \gamma_1\|F(x^k)\|^{1/2}\}$$

and

$$\rho_k = \min\{1/2, \gamma_2\|F(x^k)\|^{1/2}\}.$$

3 CONCLUDING REMARKS

A hybrid algorithm for solving systems of monotone equations was presented. The algorithm combines elements of Newton, proximal point and projection methods. An important property of the presented method is that it is truly globally convergent to a solution without any regularity assumptions. Under the assumption of differentiability and nonsingularity at the solution, locally superlinear rate of convergence was established.

References

[1] L. Armijo. Minimization of functions having Lipschitz continuous first partial derivatives. *Pacific Journal of Mathematics*, 16:1–3, 1966.

[2] D.P. Bertsekas. *Nonlinear programming.* Athena Scientific, Belmont, Massachusetts, 1995.

[3] J.V. Burke and Maijian Qian. The variable metric proximal point algorithm, I: Basic convergence theory, 1996. Department of Mathematics, University of Washington, Seattle, WA.

[4] R.S. Dembo, S.C Eisenstat, and T. Steihaug. Inexact Newton methods. *SIAM Journal of Numerical Analysis*, 19:400–408, 1982.

[5] J.E. Dennis and R.B. Schnabel. *Numerical Methods for Unconstrained Optimization and Nonlinear Equations.* Prentice-Hall, Englewood Cliffs, N.J., 1983.

[6] J. Eckstein and D.P. Bertsekas. On the Douglas-Rachford splitting method and the proximal point algorithm for maximal monotone operators. *Mathematical Programming*, 55:293–318, 1992.

[7] J. Eckstein and M.C. Ferris. Smooth methods of multipliers for complementarity problems. Technical Report RRR 27-96, Rutgers Center for Operations Research, Rutgers University, New Brunswick, New Jersey, August 1996. Revised February 1997.

[8] M.C. Ferris. Finite termination of the proximal point algorithm. *Mathematical Programming*, 50:359–366, 1991.

[9] A.N. Iusem and M.V. Solodov. Newton-type methods with generalized distances for constrained optimization. *Optimization*, 41:257–278, 1997.

[10] H. Jiang, L. Qi, X. Chen, and D. Sun. Semismoothness and superlinear copnvergence in nonsmooth optimization and nonsmooth equations. In G. Di Pillo and F. Giannessi, editors, *Nonlinear Optimization and Applications*, pages 197–212. Plenum Press, 1996.

[11] H. Jiang and D. Ralph. Global and local superlinear convergence analysis of Newton-type methods for semismooth equations with smooth least squares. Department of Mathematics, The University of Melbourne, Australia. July 1997.

[12] B. Lemaire. The proximal algorithm. In J.P. Penot, editor, *International Series of Numerical Mathematics*, pages 73–87. Birkhauser, Basel, 1989.

[13] F.J. Luque. Asymptotic convergence analysis of the proximal point algorithm. *SIAM Journal on Control and Optimization*, 22:277–293, 1984.

[14] J.M. Martínez. Local convergence theory for inexact Newton methods based on structural least-squares updates. *Mathematics of Computation*, 55:143-168, 1990.

[15] J.M. Martínez and L. Qi. Inexact Newton methods for solving nonsmoooth equations. *Journal of Computational and Applied Mathematics*, 60:127–145, 1995.

[16] J.M. Ortega and W.C. Rheinboldt. *Iterative Solution of Nonlinear Equations in Several Variables.* Academic Press, 1970.

[17] J.-S. Pang and S.A. Gabriel. An inexact NE/SQP method for solving the nonlinear complementarity problem. *Computational Optimization and Applications*, 1:67–92, 1992.

[18] J.-S. Pang and L. Qi. Nonsmooth equations : Motivation and algorithms. *SIAM Journal on Optimization*, 3:443–465, 1995.

[19] B.T. Polyak. *Introduction to Optimization.* Optimization Software, Inc., Publications Division, New York, 1987.

[20] L. Qi. Convergence analysis of some algorithms for solving nonsmooth equations. *Mathematics of Operations Research*, 18:227–244, 1993.

[21] L. Qi and J. Sun. A nonsmooth version of Newton's method. *Mathematical Programming*, 58:353–367, 1993.

[22] R.T. Rockafellar. Monotone operators and the proximal point algorithm. *SIAM Journal on Control and Optimization*, 14:877–898, 1976.

[23] M.V. Solodov and B.F. Svaiter. A new projection method for variational inequality problems. Technical Report B-109, Instituto de Matemática Pura e Aplicada, Estrada Dona Castorina 110, Jardim Botânico, Rio de Janeiro, RJ 22460, Brazil, November 1996. *SIAM Journal on Control and Optimization*, submitted.

[24] M.V. Solodov and B.F. Svaiter. Forcing strong convergence of proximal point iterations in a Hilbert space, 1997. *Mathematical Programming*, submitted.

[25] M.V. Solodov and B.F. Svaiter. A hybrid projection – proximal point algorithm. Technical Report B-115, Instituto de Matemática Pura e Aplicada, Estrada Dona Castorina 110, Jardim Botânico, Rio de Janeiro, RJ 22460, Brazil, January 1997. *Journal of Convex Analysis*, submitted.

[26] T.J. Ypma. Local convergence of inexact Newton methods. *SIAM Journal of Numerical Analysis*, 21:583–590, 1984.

Reformulation: Nonsmooth, Piecewise Smooth, Semismooth and Smoothing Methods, pp. 371–379
Edited by M. Fukushima and L. Qi

On the Limiting Behavior of the Trajectory of Regularized Solutions of a P_0-Complementarity Problem

Roman Sznajder* and M. Seetharama Gowda†

Abstract Given a continuous $\mathbf{P}_0$-function on R^n, we consider the nonlinear complementarity problem NCP(f) and the trajectory of regularized solutions $\{x(\varepsilon) : 0 < \varepsilon < \infty\}$ where $x(\varepsilon)$ is the unique solution of NCP(f_ε) with $f_\varepsilon(x) := f(x) + \varepsilon x$. Given a sequence $\{x(\varepsilon_k)\}$ with $\varepsilon_k \downarrow 0$, we discuss ($i$) the existence of a bounded/convergent subsequence in the affine case, (ii) a property of any subsequential limit x^*, and (iii) the convergence of the entire trajectory in the polynomial case.

Key Words $\mathbf{P}_0$-complementarity problem, trajectory, monotone, Pareto minimal, semi-algebraic

1 INTRODUCTION

Consider the nonlinear complementarity problem NCP(f) of finding a vector x in R^n such that

$$x \geq 0, \quad f(x) \geq 0 \quad \text{and} \quad \langle f(x), x\rangle = 0, \tag{1.1}$$

where $f : R^n \to R^n$ is continuous and satisfies the $\mathbf{P}_0$-property that for all $x, y \in R^n$ with $x \neq y$,

$$\max_{\{i: x_i \neq y_i\}} (x_i - y_i)[f_i(x) - f_i(y)] \geq 0.$$

*Department of Mathematics, Bowie State University, Bowie, Maryland 20715-9465, rsznajde@bowiestate.edu

†Research supported by the National Science Foundation Grant CCR-9307685, Department of Mathematics & Statistics, University of Maryland Baltimore County, Baltimore, Maryland 21250, gowda@math.umbc.edu

The importance of NCP and related problems is well documented in the literature, see e.g., [6].

For $\varepsilon > 0$, let $f_\varepsilon(x) := f(x) + \varepsilon x$ be the Tikhonov regularization of f. Extending earlier results of Megiddo and Kojima (Thm. 3.4, [12]), and Facchinei and Kanzow [5] (corresponding to a continuously differentiable f), Ravindran and Gowda proved (see Corollary 6, [15]) that NCP(f_ε) has a unique solution $x(\varepsilon)$, the mapping $\varepsilon \mapsto x(\varepsilon)$ is continuous on $(0,\infty)$, and when the solution set SOL(f) of NCP(f) is nonempty and bounded, $\text{dist}(x(\varepsilon), \text{SOL}(f)) \to 0$ as $\varepsilon \to 0$. The objective of this paper is to study the limiting behavior of the trajectory $\{x(\varepsilon) : 0 < \varepsilon < \infty\}$ as $\varepsilon \to 0$. Specifically, we consider the following questions:

(*i*) Given $\varepsilon_k \downarrow 0$, when does the sequence $\{x(\varepsilon_k)\}$ have a bounded/convergent subsequence (so that a limit of some subsequence is a solution of NCP(f))?

(*ii*) If $x(\varepsilon_k) \to x^*$ for some $\varepsilon_k \downarrow 0$, what can be said about x^*?

(*iii*) When does the entire trajectory converge?

When f is monotone, that is,

$$\langle f(x) - f(y), x - y\rangle \geq 0 \quad (\forall x, y \in R^n),$$

there is a nice and simple answer for all the questions: Either the trajectory diverges to infinity (in the norm) in which case NCP(f) has no solution or it converges to the least two-norm solution of NCP(f) [17]. (For convergence in the setting of maximal monotone operators via Yosida approximations, see Theorem 3.5.9 in [1].) The $\mathbf{P}_0$ situation is not so simple. Even in the case of affine f given by $f(x) = Mx + q$, the NCP(f) (which now becomes the linear complementarity problem LCP(M,q)) may have a solution, yet the trajectory could diverge to infinity. Regarding question (*i*), several answers are known for the LCP. When M is a $\mathbf{P}_0 \cap \mathbf{R}_0$-matrix, every sequence $\{x(\varepsilon_k)\}$ is bounded as $\varepsilon_k \downarrow 0$ [3], [7]. By considering complementary cones associated with the given sequence $\{x(\varepsilon_k)\}$, Ebiefung [4] describes sufficient conditions for the existence of a convergent subsequence. Venkateswaran [18], based on a condition (which essentially says that the problem has a unique solution) proves that for the LCP(M,q), the entire trajectory converges. Our contribution, here, is as follows. In the affine case, by considering a 'directional perturbation' we extend results of Ebiefung, in the general case (of a continuous $\mathbf{P}_0$-function f) we show that the limit x^* mentioned in question (*ii*) is a weak Pareto minimal element of the solution set of NCP(f), and that when f is a polynomial map (in particular, affine) the entire trajectory (as $\varepsilon \downarrow 0$) either diverges to infinity (in the norm) or converges to a solution of NCP(f). One interesting aspect of the proof of this last result is the applicability of a result from algebraic geometry, which, we think should be useful to researchers working in this area.

2 PRELIMINARIES

Our reference for linear complementarity problems is [3]. For the sake of completeness, we include two important definitions.

Definition 2.1 *A matrix $M \in R^{n\times n}$ is called*

1. *a* **P**-*matrix* (**P**$_0$-*matrix*) *if all its principal minors are positive (respectively, nonnegative);*
2. *an* **R**$_0$-*matrix if* SOL$(M,0) = \{0\}$, *i.e., the homogeneous* LCP *has only the trivial solution.*

We sometimes write the complementarity problem NCP(f) as

$$x \wedge f(x) = 0$$

where '$\wedge$' denotes the componentwise minimum of vectors involved. We say that a function
$g : R^n \to R^n$ is a **P**-function if for all $x, y \in R^n$ with $x \neq y$,

$$\max_{\{i: x_i \neq y_i\}} (x_i - y_i)[g_i(x) - g_i(y)] > 0.$$

Note that if f is a **P**$_0$-function, then $f + \varepsilon I$ is a **P**-function; moreover, $f(x) = Mx + q$ is a **P**$_0$-function if and only if M is a **P**$_0$-matrix.

3 EXISTENCE OF A CONVERGENT SUBSEQUENCE

In this section, we address the first question stated in the Introduction, namely, for a given sequence $\{x(\varepsilon_k)\}$ with $\varepsilon_k \downarrow 0$, when we can expect to have a bounded/convergent subsequence. For a general continuous **P**$_0$-function, the Ravindran-Gowda result mentioned in the Introduction (also, Facchinei-Kanzow [5] for C^1 **P**$_0$-function) gives an answer when the solution set of NCP(f) is nonempty and bounded. What happens when the solution set is unbounded? We do not have an answer for the general **P**$_0$ case and so we restrict our attention to affine functions. To this end, we consider $f(x) = Mx + q$ and assume that M is a **P**$_0$-matrix.

The following simple example shows what can go wrong even in this affine case.

Example 3.1 Let

$$M = \begin{bmatrix} 0 & 0 \\ 1 & 0 \end{bmatrix} \quad q = \begin{bmatrix} 0 \\ -1 \end{bmatrix}.$$

For every $\varepsilon > 0$, SOL$(M + \varepsilon I, q) = \{(0, \frac{1}{\varepsilon})\}$, while

$$\text{SOL}(M, q) = \{(1, x_2) : x_2 \geq 0\} \cup \{(x_1, 0) : x_1 \geq 1\}.$$

The trajectory diverges to infinity, yet keeping the same distance (one) from SOL(M, q).

Throughout this section, we fix a $\mathbf{P}_0$-matrix M, vectors p and q in R^n, and consider the problem LCP$(M+\varepsilon I, q+\varepsilon p)$ for $\varepsilon > 0$. This problem can be considered as a 'directional perturbation' of the problem LCP(M,q) in the direction of (I,p); the case $p=0$ reduces to the original problem LCP(M,q) and its perturbation LCP$(M+\varepsilon I, q)$. Since $M+\varepsilon I$ is a $\mathbf{P}$-matrix, LCP$(M+\varepsilon I, q+\varepsilon p)$ has a unique solution, say, $x(\varepsilon)$. We fix a sequence $\varepsilon_k \downarrow 0$ and seek conditions under which the sequence $\{x(\varepsilon_k)\}$ will have a convergent subsequence. Our first result relies on the boundedness of SOL(M,q).

Theorem 3.1 *Assume* $M \in \mathbf{P}_0$, $\{x(\varepsilon_k)\} = \text{SOL}\,(M+\varepsilon_k I, q+\varepsilon_k p)$, *and* SOL$\,(M,q) \neq \emptyset$ *and bounded. Then,* $\{x(\varepsilon_k)\}$ *is bounded and hence some subsequence of* $\{x(\varepsilon_k)\}$ *converges to a solution of* LCP$\,(M,q)$.

Proof. Since SOL$\,(M,q)$ is nonempty and bounded, it is stable (see Theorem 7, [8] or Theorem 1, [15]), that is, for all small $\varepsilon > 0$ and some bounded open set Ω in R^n, $\{x(\varepsilon)\} = \text{SOL}\,(M+\varepsilon I, q+\varepsilon p) \subseteq \Omega$. Hence, $\{x(\varepsilon_k)\}$ is bounded and some subsequence converges. ∎

When SOL(M,q) is unbounded, we can describe a condition (similar to the one in [4]) for the existence of a bounded subsequence of $\{x(\varepsilon_k)\}$ in terms of complementary cones. Recall that (Def. 1.3.2, [3]) for a given $M \in R^{n\times n}$ and $\alpha \subseteq \{1,\ldots,n\}$, the matrix $C_M(\alpha) \in R^{n\times n}$ defined by

$$C_M(\alpha)_{\cdot i} = \begin{cases} -M_{\cdot i} & \text{if } i \in \alpha, \\ I_{\cdot i} & \text{if } i \notin \alpha, \end{cases}$$

is called a complementary matrix of M (or a complementary submatrix of $(-M,I)$). The associated cone, *pos* $C_M(\alpha) := C_M(\alpha)(R^n{}_+)$, is called a *complementary cone* relative to M. Note that LCP(M,q) is solvable if and only if $q \in$ *pos* $C_M(\alpha)$ for some α.

We now fix M, p, q, and a sequence $\{\varepsilon_k\} \downarrow 0$. For each k, there is an α (depending on k) such that $q+\varepsilon_k p \in$ *pos* $(-(M+\varepsilon I)_\alpha, I_{\bar\alpha})$ where $\bar\alpha$ is the set of indexes not in α. Since these index sets α are finite in number, we can take a subsequence k_j so that the same α works for all k_j. Writing

$$C_{M+\varepsilon_{k_j}I}(\alpha) = \left(-(M+\varepsilon_{k_j}I)_\alpha, I_{\bar\alpha}\right) = (-M_\alpha, I_{\bar\alpha}) + \varepsilon_{k_j}(-I_\alpha, 0_{\bar\alpha}) = A + \varepsilon_{k_j}B,$$

we see that

$$q+\varepsilon_{k_j}p \in (A+\varepsilon_{k_j}B)(R^n{}_+) \quad (\forall j = 1,2,\ldots)$$

from which we can write $q+\varepsilon_{k_j}p = (A+\varepsilon_{k_j}B)(u(\varepsilon_{k_j}))$ where $u(\varepsilon_{k_j}) \geq 0$. It follows from a standard normalization argument that the sequence $\{u(\varepsilon_{k_j})\}$ is bounded if the following condition holds:

$$Au = 0,\ u \geq 0 \Longrightarrow u = 0. \tag{3.1}$$

Since $x(\varepsilon_{k_j})$ is identical to $u(\varepsilon_{k_j})$ in α components and zero in $\bar\alpha$ components, we see that (3.1) is a sufficient condition for the boundedness of $\{x(\varepsilon_{k_j})\}$.

We end this section with the monotone case. The result below and its proof are similar to the ones for the standard case (with $p = 0$) [17]. For the sake of completeness, we include a proof.

Theorem 3.2 *Let M be positive semidefinite, $p, q \in R^n$ and $\mathrm{SOL}(M,q) \neq \emptyset$. Let $\{x(\varepsilon)\} = \mathrm{SOL}(M + \varepsilon I, q + \varepsilon p)$. Then $\{x(\varepsilon)\}$ converges as $\varepsilon \downarrow 0$, to the (unique) element x^* in $\mathrm{SOL}(M,q)$ that is closest to $-p$ in the two-norm.*

Proof. Let $y(\varepsilon) = Mx(\varepsilon) + \varepsilon x(\varepsilon) + \varepsilon p + q$ and $v = Mu + q$ where $u \in \mathrm{SOL}(M,q)$. Then

$$0 \geq \langle y(\varepsilon)-v, x(\varepsilon)-u\rangle = \langle M(x(\varepsilon)-u)+\varepsilon(x(\varepsilon)+p), x(\varepsilon)-u\rangle \geq \varepsilon\langle x(\varepsilon)+p, x(\varepsilon)-u\rangle,$$

in view of the assumption that the matrix M is positive semidefinite. The above inequality shows that $\langle x(\varepsilon) + p, x(\varepsilon) - u\rangle \leq 0$. It follows that $\langle x(\varepsilon) + p, x(\varepsilon) + p - [u + p]\rangle \leq 0$. The Cauchy-Schwarz inequality gives $\|x(\varepsilon) + p\|^2 \leq \langle x(\varepsilon) + p, u + p\rangle \leq \|x(\varepsilon) + p\| \cdot \|u + p\|$. Thus,

$$\|x(\varepsilon) + p\| \leq \|u + p\|. \tag{3.2}$$

This proves the boundedness of $\{x(\varepsilon) : \varepsilon > 0\}$. To prove the convergence of the entire trajectory, let $x(\varepsilon_k) \to x^*$ for some $\varepsilon_k \to 0$. By continuity and complementarity, we get $x^* \in \mathrm{SOL}(M,q)$. From (3.2), we see that for all $u \in \mathrm{SOL}(M,q)$, $\|x^* + p\| \leq \|u + p\|$. Since $\mathrm{SOL}(M,q)$ is convex, x^* is the (unique) element of $\mathrm{SOL}(M,q)$ that is closest to $-p$. Since the sequence $\{\varepsilon_k\}$ is arbitrary, the entire trajectory must go to x^*. This completes the proof. ∎

4 WEAK PARETO MINIMAL PROPERTY

We now consider the second question raised in the Introduction: Given that $x(\varepsilon_k) \to x^*$, what is the nature of x^*? The following example shows that unlike the monotone case, in the $\mathbf{P_0}$ case, the limit x^* need not be the least two-norm solution of NCP(f).

Example 4.1 Let

$$M = \begin{bmatrix} 0 & 1 \\ 0 & 1 \end{bmatrix} \in \mathbf{P_0} \text{ and } q = \begin{bmatrix} -1 \\ -1 \end{bmatrix}.$$

Then

$$\mathrm{SOL}(M,q) = \{(x_1, 1) : x_1 \geq 0\},$$

$$\mathrm{SOL}(M + \varepsilon I, q) = \left\{\left(\frac{1}{1+\varepsilon}, \frac{1}{1+\varepsilon}\right)\right\},$$

and $x^* = (1,1)$.

For our next result, we need the following definition.

Definition 4.1 *Consider an element x^* of a nonempty set S. We say that x^* is a weak Pareto minimal element (Pareto minimal element) of S (with respect to the nonnegative orthant) if*

$$(x^* - int\, R^n_+) \cap S = \emptyset$$

(respectively, $(x^* - R^n_+) \cap S = \{x^*\}$). In other words, x^* is a weak Pareto minimal element of S if there is no element of S satisfying the inequality $s < x^*$ and is a Pareto minimal element of S if x^* is the only element of S satisfying the inequality $s \leq x^*$.

The above two notions appear in multi-objective optimization [11] and can be defined with respect to any cone/order.

Theorem 4.1 *Let f be a continuous* $\mathbf{P}_0$*-function and $\{x(\varepsilon)\} = \mathrm{SOL}\,(f_\varepsilon)$. Then every accumulation point of $\{x(\varepsilon)\}$ (as $\varepsilon \to 0$) is a weak Pareto minimal element of* SOL (f).

Proof. Suppose $x(\varepsilon_k) \to x^*$ as $\varepsilon_k \downarrow 0$. Suppose, if possible, that there exists $\overline{x} \in \mathrm{SOL}\,(f)$ such that $\overline{x} < x^*$. Since $0 \leq \overline{x} < x^* \in \mathrm{SOL}\,(f)$, by complementarity, $f(x^*) = 0$. Also, $x(\varepsilon_k) \to x^* > \overline{x} \geq 0$, so, for large k, $x(\varepsilon_k) > \overline{x} \geq 0$. Again, the complementarity condition implies

$$f(x(\varepsilon_k)) + \varepsilon_k x(\varepsilon_k) = 0. \tag{4.1}$$

Since $f(x) + \varepsilon_k x$ is a **P**-function and $x(\varepsilon_k) \neq \overline{x}$ (for large k), there exists $i \in \{1, \ldots, n\}$ (depending on k) such that

$$[x(\varepsilon_k) - \overline{x}]_i[f(x(\varepsilon_k)) + \varepsilon_k x(\varepsilon_k) - (f(\overline{x}) + \varepsilon_k \overline{x})]_i > 0. \tag{4.2}$$

By taking a subsequence, we may assume that the same i works for all $k \in N$. Letting $\varepsilon_k \downarrow 0$ in (4.2), we get $[x^* - \overline{x}]_i[f(x^*) - f(\overline{x})]_i \geq 0$. By expanding this inequality, and by using the complementarity conditions $x^* \wedge f(x^*) = 0 = \overline{x} \wedge f(\overline{x})$, we get the cross-complementarity conditions

$$x_i^*(f(\overline{x}))_i = 0 = \overline{x}_i(f(x^*))_i. \tag{4.3}$$

As $x_i^* > 0$, (4.3) implies $(f(\overline{x}))_i = 0$. By (4.1) and (4.2), $[x(\varepsilon_k) - \overline{x}]_i[-\varepsilon_k \overline{x}]_i > 0$, so $[x(\varepsilon_k) - \overline{x}]_i < 0$, which contradicts the earlier condition $x(\varepsilon_k) > \overline{x}$. This proves that x^* is a weak Pareto minimal element of SOL(f). ■

Remark Referring to Example 4.1, we notice that the trajectory converges to $x^* = (1, 1)$ and that x^* is a weak Pareto minimal element of the solution set. Unfortunately, every element of the solution set is a weak Pareto minimal element. This raises the following question: How does one identify x^* among all weak Pareto minimal elements of the solution set?

5 LIMITING BEHAVIOR OF THE ENTIRE TRAJECTORY

In this section, we consider the question of convergence of the entire trajectory. We restrict our attention to polynomial functions (these are functions from R^n to itself whose component functions are polynomials in n real variables) and appeal to a result from algebraic geometry. While proving the convergence of an 'interior point' trajectory in the context of monotone LCPs, Kojima, Megiddo, and Noma [10] use triangulation techniques and refer to a Referee's remark that such convergence can be established via a result from algebraic geometry. It is this algebraic geometry result that we use here. First a definition.

Definition 5.1 *A subset of R^n is called semi-algebraic if it is a finite union of sets of the form*

$$E = \{x \in R^n : p_j(x) \ \circ_j \ 0, \ j = 1, \ldots, L\}$$

where for each j, $p_j : R^n \to R$ is a polynomial and $\circ_j \in \{=, \leq, <\}$.

The following theorem emphasizes the key property of semi-algebraic sets needed in this section.

Theorem 5.1 *(Thm. 2.2.1, [2]) Every semi-algebraic set in R^n has a finite number of (connected) components.*

Theorem 5.2 *Consider a polynomial $\mathbf{P}_0$-function f. Let* SOL$(f_\varepsilon) = \{x(\varepsilon)\}$, *for $\varepsilon > 0$. Then the following alternative holds:*
As $\varepsilon \downarrow 0$, the trajectory $\{x(\varepsilon) : \varepsilon > 0\}$ either converges to an element of SOL(f) *or diverges to infinity in the norm.*

Proof. Suppose that the trajectory $\{x(\varepsilon) : \varepsilon > 0\}$ does not diverge to infinity in the norm, in which case we may assume that for some sequence $\varepsilon_k \downarrow 0$, $x(\varepsilon_k) \to x^*$. Clearly, $x^* \in$ SOL(f). We claim that $x(\varepsilon) \to x^*$ as $\varepsilon \to 0$. For any $\delta > 0$, let $B_\delta := \{x : -\delta \leq x_i - x_i^* \leq \delta, \ i = 1, 2, \ldots, n\}$ be a box around x^*. To prove the convergence $x(\varepsilon) \to x^*$, we show that for any such δ, there exists $\bar{\varepsilon} > 0$ such that for $0 < \varepsilon < \bar{\varepsilon}$, $x(\varepsilon) \in B_\delta$. Assume the contrary that for some $\delta > 0$ there exists a sequence $\{\bar{\varepsilon}_l\} \downarrow 0$ such that $x(\bar{\varepsilon}_l) \notin B_\delta$. By renaming the sequences, if necessary, we may assume that

(a) $x(\varepsilon_k) \in B_\delta$, $x(\bar{\varepsilon}_k) \notin B_\delta$ for all k and

(b) $1 > \varepsilon_1 > \bar{\varepsilon}_1 > \varepsilon_2 > \bar{\varepsilon}_2 > \varepsilon_3 > \cdots$.

Now consider the set

$$E := \{(x, \varepsilon) : \varepsilon > 0, \ x \geq 0, \ f(x) + \varepsilon x \geq 0, \ x_i(f_i(x) + \varepsilon x_i) = 0 \text{ for } i = 1, 2, \ldots, n\} \tag{5.1}$$

which is really the graph $\{(x(\varepsilon), \varepsilon) : \varepsilon > 0\}$. Since the function f is a polynomial, the sets E and $E \cap B_\delta \times (0, 1)$ are semi-algebraic.

We show that the semi-algebraic set $E \cap B_\delta \times (0,1)$ has infinitely many components, thus reaching a contradiction to Theorem 5.1. For each $j = 1, 2, \ldots$, let C_j be the (connected) component of the set $E \cap B_\delta \times (0,1)$ containing $(x(\varepsilon_j), \varepsilon_j)$. Since connected sets in R are intervals, C_j must be of the form $\{(x(\varepsilon), \varepsilon) \ : \ \varepsilon \in I_j\}$ where I_j is an interval in $(0,1)$ containing ε_j. Since $x(\bar{\varepsilon}_l) \notin B_\delta$ for all l, we see that no $\bar{\varepsilon}_l$ can be in any I_j. By the interlacing property (b) above, the intervals I_j $(j = 1, 2 \ldots)$ and hence the components C_j $(j = 1, 2, \ldots)$ are pairwise disjoint. Thus we see that there are infinitely many components in $E \cap B_\delta \times (0,1)$ contradicting Theorem 5.1. This proves that $x(\varepsilon) \to x^*$. ■

Remarks (1) In the above proof we relied on Theorem 5.1. Instead, we could have used the so called *curve selection theorem* (Prop. 2.6.19, [2]) which says that if X is a semi-algebraic set in R^n and x^* is an accumulation point of X, then there is a continuous curve in X that begins at x^*.
(2) For the linear complementarity problem, R. Stone [16] has given an elementary proof of the above theorem based on complementary cones and rational functions.

We now state our final result for linear complementarity problems without proof.

Corollary 5.1 *Let M be a* $\mathbf{P}_0$*-matrix,* $\mathrm{SOL}\,(M + \varepsilon I, q) = \{x(\varepsilon)\}$ *for* $\varepsilon > 0$. *Then as* $\varepsilon \downarrow 0$, *either the trajectory* $\{x(\varepsilon) : \varepsilon > 0\}$ *diverges to infinity in the norm or converges to a weak Pareto minimal point of* $\mathrm{SOL}\,(M, q)$.

Acknowledgements

During the 16th International Symposium on Mathematical Programming held August 24-29, 1997, in Lausanne, Switzerland, the second author asked Rick Stone about the possibility of proving Theorem 5.2 for the LCPs. As usual, Rick Stone spontaneously came up with a proof. It is a pleasure to thank him for stimulating conversations. We thank Osman Güler for his suggestions on the curve selection theorem and for providing appropriate references. We also thank Christian Kanzow for his comments and Hou-Duo Qi for sending his papers [13] and [14] where the boundedness of the trajectory $\{x(\varepsilon)\}$ is discussed.

References

[1] J.-P. Aubin and H. Frankowska, *Set valued analysis*, Birkhäuser, Boston, 1990.

[2] R. Benedetti and J.J. Risler, *Real algebraic and semi-algebraic sets*, Hermann, Éditeurs des Sciences et des Arts, Paris, 1990.

[3] R.W. Cottle, J.-S. Pang and R.E. Stone, *The linear complementarity problem*, Academic Press, Boston, 1992.

[4] A.A. Ebiefung, "New perturbation results for the linear complementarity problem with P_0-matrices," Appl. Math. Letters, forthcoming.

[5] F. Facchinei and C. Kanzow, "Beyond monotonicity in regularization methods for nonlinear complementarity problems," Universitàdi Roma "La Sapienza," Dipartimento di Informatica e Sistemistica, Via Buonarotti 12, 00185 Roma, Italy, May 1997.

[6] M.C. Ferris and J.-S. Pang (Eds.), *Complementarity and Variational problems: State of the art*, SIAM, Philadelphia, 1997.

[7] A. Gana, *Studies in the complementarity problem*, Ph.D. Dissertation, Department of Industrial and Operations Engineering, University of Michigan, Ann Arbor, 1982.

[8] M.S. Gowda and R. Sznajder, "The generalized order complementarity problem," SIAM J. Matrix Anal. Appl. 15 (1994) 779-795.

[9] G. Isac, "Tikhonov's regularization and the complementarity problem in Hilbert spaces," J. Math. Anal. Appl. 174 (1993) 53-66.

[10] M. Kojima, N. Megiddo, and T. Noma, "Homotopy continuation methods for nonlinear complementarity problems," Math. Operations Res., 16 (1991) 754-774.

[11] D.T. Luc, *Theory of vector optimization*, Springer-Verlag, Lecture Notes in Economics and Mathematical Systems, 319, New York, 1989.

[12] N. Megiddo and M. Kojima, "On the existence and uniqueness of solutions in nonlinear complementarity theory," Math. Programming, 12 (1977) 110-130.

[13] H.-D. Qi, "A note on regularization methods for a $\mathbf{P}_0$-box variational inequality problem," Research Report, Institute of Computational Mathematics and Scientific/Engineering Computing, Chinese Academy of Sciences, P.O. Box 2719, Beijing, China, November 6, 1997.

[14] H.-D. Qi and L.-Z. Liao, "Beyond boundedness and regularization methods for nonlinear complementarity problems and box variational inequalities," Research Report, Institute of Computational Mathematics and Scientific/Engineering Computing, Chinese Academy of Sciences, P.O. Box 2719, Beijing, China, September 22, 1997.

[15] G. Ravindran and M.S. Gowda, "Regularization of P_0-functions in Box Variational Inequality Problems," Research Report, Department of Mathematics and Statistics, University of Maryland Baltimore County, Baltimore, Maryland 21250, August 1997.

[16] R.E. Stone, Personal communication, August 1997.

[17] P.K. Subramanian, "A note on least two norm solutions of monotone complementarity problems," Appl. Math. Letters, 1 (1988) 395-397.

[18] V. Venkateswaran, "An algorithm for the linear complementarity problem with a P_0-matrix," SIAM J. Matrix Anal. Appl., 14 (1993) 967-977.

Reformulation: Nonsmooth, Piecewise Smooth, Semismooth and Smoothing Methods, pp. 381–404
Edited by M. Fukushima and L. Qi

Analysis of a Non-Interior Continuation Method Based on Chen-Mangasarian Smoothing Functions for Complementarity Problems [1]

Paul Tseng[†]

Abstract Recently Chen and Mangasarian proposed a class of smoothing functions for linear/nonlinear programs and complementarity problems that unifies many previous proposals. Here we study a non-interior continuation method based on these functions in which, like interior path-following methods, the iterates are maintained to lie in a neighborhood of some path and, at each iteration, one or two Newton-type steps are taken and then the smoothing parameter is decreased. We show that the method attains global convergence and linear convergence under conditions similar to those required for other methods. We also show that these conditions are in some sense necessary. By introducing an inexpensive active-set strategy in computing one of the Newton directions, we show that the method attains local superlinear convergence under conditions milder than those for other methods. The proof of this uses a local error bound on the distance from an iterate to a solution in terms of the smoothing parameter.

Key Words complementarity problem, smoothing, continuation method, global linear convergence, local superlinear convergence

1 INTRODUCTION

We consider the well-known complementarity problem (CP for short) of finding an $(x, y) \in \Re^n \times \Re^n$ satisfying

$$x \geq 0, \quad y \geq 0, \quad x^T y = 0, \quad F(x) - y = 0, \tag{1.1}$$

[1]This research is supported by National Science Foundation Grant CCR-9311621.

[†]Department of Mathematics, University of Washington, Seattle, Washington 98195, U.S.A., tseng@math.washington.edu

where $F = (F_1, ..., F_n)$ is a given continuously differentiable function from $\Re^n$ to $\Re^n$ [12, 14, 30]. In the case where F is affine, CP is called the linear complementarity problem or LCP [13, 26]. We denote by S the set of of solutions of CP, i.e., $S := \{(x,y) \in \Re^n \times \Re^n : (x,y) \text{ satisfy } (1.1)\}$, which we assume is nonempty. There has been proposed various approaches for solving CP and, of these, one that has generated much interest recently (and is the focus of this paper) is a non-interior smoothing approach proposed by Chen and Mangasarian [8, 9], based on an earlier work of Kreimer and Rubinstein. In this approach, we rewrite (1.1) as a set of nonsmooth equations $H_0(x,y) = 0$, where

$$H_0(x,y) := \begin{bmatrix} x - \max\{0, x-y\} \\ F(x) - y \end{bmatrix}, \tag{1.2}$$

and the max is taken componentwise, and then we approximate the nondifferentiable convex function $\max\{0, \tau\}$ by a smooth function of the form

$$\mu g(\tau/\mu), \tag{1.3}$$

where $\mu \in \Re_{++}$ and $g : \Re \to \Re$ is any convex continuously differentiable function with the properties that $\lim_{\tau \to -\infty} g(\tau) = 0$ and $\lim_{\tau \to \infty} g(\tau) - \tau = 0$ and $0 < g'(\tau) < 1$ for all $\tau \in \Re$. [We denote the class of such functions g by $\mathcal{CM}$. In the case where g is twice continuously differentiable, (1.3) may be written as the integral $\int_{-\infty}^{\tau} \int_{-\infty}^{\alpha} \frac{1}{\mu} d(\xi/\mu) d\xi d\alpha$, where $d := g''$ is interpreted as a probability density function. This is the form presented in [9] and used by others. We prefer the form (1.3) for its simplicity of analysis.] This yields a set of nonlinear equations of the form $H_\mu(x,y) = 0$, parameterized by the smoothing parameter μ, where

$$H_\mu(x,y) := \begin{bmatrix} x - \mu G((x-y)/\mu) \\ F(x) - y \end{bmatrix}, \qquad G(\tau_1, ..., \tau_n) := \begin{bmatrix} g(\tau_1) \\ \vdots \\ g(\tau_n) \end{bmatrix}. \tag{1.4}$$

[Thus, H_μ converges to H_0 pointwise as $\mu \to 0$.] Then we can use a continuation/homotopy solution approach whereby, starting with any $\mu \in \Re_{++}$ and $z \in \Re^{2n}$, we fix μ and apply a few Newton-type steps for $H_\mu(z) = 0$ to update z, and then we decrease μ and re-iterate. This approach has been widely used [2, 3, 7, 10, 11, 18, 20, 38] and is the one that we will consider. A related approach involves having a separate μ for each component and viewing it as a variable so that one instead works with $3n$ equations in $3n$ variables [18, 31] (in [32], $H_0(x,y) = 0$ is reformulated, say via $y := F(x)$ and $u := x - y$, as the normal equation $u + F(\max\{0,u\}) - \max\{0,u\} = 0$, to which smoothing is applied). Also, instead of applying Newton steps to $H_\mu(z) = 0$, one can fix μ and minimize $\|H_\mu(z)\|$, possibly inexactly, using standard methods for unconstrained minimization and then decrease μ [6, 8, 9, 15].

Examples of the smoothing function g include one proposed independently by Chen and Harker [4, 5], Kanzow [20, 21], and Smale [33]:

$$g(\tau) = (\sqrt{\tau^2 + 4} + \tau)/2, \tag{1.5}$$

and one obtained by integrating the sigmoid function $\tau \mapsto 1/(1+e^{-\tau})$ used in neural networks [8, 9]:

$$g(\tau) = \ln(e^{\tau} + 1). \tag{1.6}$$

See [8, 9] for other examples of g, and see [15] for extensions of (1.3) to the mixed complementarity problem. [The graph of $\zeta = g(\tau)$ given by (1.5) has the equation $-\tau = 1/\zeta - \zeta$. Other such g can be obtained from the general equation $-\tau = h(\zeta) - \zeta$, with $h : \Re_{++} \mapsto \Re_{++}$ any convex twice continuously differentiable function satisfying $\lim_{\zeta \to 0} h(\zeta) = \infty$ and $\lim_{\zeta \to \infty} h(\zeta) = 0$.] Also, the function (1.3) has been further generalized to the form $p(\tau, \mu)$ and properties of solution to the equation $H_{\mu}(z) = 0$ has been studied similarly [6]. Finally, we note the related works in [1, 19, 37], which analyze the global and local convergence rate of continuation methods and generalized Newton methods based on approximating $a - \max\{0, a - b\}$ by a smoothing function $a + b - \sqrt{a^2 + b^2 + 2\mu}$ proposed by Kanzow [20]. [In particular, [1] gave the first global linear convergence analysis of a non-interior continuation method for LCP.] This function cannot be written in the Chen-Mangasarian form of $a - \mu g((a-b)/\mu)$ for some $g \in \mathcal{CM}$, as neither can the function $ab - \mu$ used in interior-point methods [24, 25, 27, 35, 36].

As is mentioned earlier, we are interested in continuation methods and, in particular, ones that maintain (z, μ) to lie in a neighborhood of the "path" defined by $H_{\mu}(z) = 0$. As in the case of infeasible interior-point methods [35], such a path-following approach appears to be needed for achieving both global linear and local superlinear convergence on nonlinear problems [2, 7, 38]. We will use the following choice of neighborhood:

$$\mathcal{N}_{\beta} := \{ (z, \mu) \in \Re^{2n} \times \Re_{++} : \|H_{\mu}(z)\| \le \beta\mu \}, \tag{1.7}$$

where $\beta \in \Re_{++}$ is fixed. This choice, an analog of that used in [35] for an interior path-following method, was also used in [2] (see [7] for a similar choice and see [3] for a weighted version involving different μ for different components). Our method iteratively moves (z, μ) along the Newton direction $-(H'_{\mu}(z)^{-1}H_{\mu}(z), \sigma\mu)$ ($\sigma \in (0,1)$) while maintaining it to remain in $\mathcal{N}_{\beta}$ and, to accelerate convergence, a further move is made towards $(\hat{z}, \pi(\mu))$, where π is a superlinear function and $\hat{z}$ is any "guess" of a solution (see Algorithm 2.1). We show that the method converges globally and linearly under conditions similar to those required by other methods [2, 7] (see Proposition 3.1). Moreover, we show that these conditions are in some sense necessary (see Corollaries 3.1 and 3.2) and we give a condition for $(z, \mu) \in \mathcal{N}_{\beta}$ to be bounded when F is monotone that is simpler than existing conditions (see Lemma 3.4). Our proofs are also quite compact. Lastly, we propose a choice of $\hat{z}$ that generalizes $z - H'_{\mu}(z)^{-1}H_0(z)$ used in [2, 3, 7, 10] via an active-set strategy, and we show that the associated method attains local superlinear convergence under conditions milder than those required for other methods (see Proposition 4.1). Moreover, this choice of $\hat{z}$ can be computed relatively inexpensively. Our analysis uses in part a new local error bound on the distance from z to a solution

z^* in terms of μ (see Lemma 4.1). Some preliminary computational experience with the above method is reported in Section 5.

In our notation, all vectors are column vectors, $\Re^n$ denotes the space of n-dimensional real column vectors, $\Re^{n\times n}$ denotes the space of $n \times n$ matrices with real entries, ^{T} denotes transpose, $\Re_{++}$ denotes the positive reals, and $N := \{1, ..., n\}$. For any vector $x \in \Re^n$, we denote by x_i the ith component of x and, for any $I \subseteq N$, by x_I the vector obtained after removing from x those x_i with $i \notin I$. We also denote by $\|x\|_1$, $\|x\|$, and $\|x\|_\infty$ the 1-, 2- and ∞-norm of x, so $\|x\| = \sqrt{x^T x}$, etc. For any $M \in \Re^{n\times n}$ and any $I \subset N$ and $J \subset N$, we denote by M_{IJ} the submatrix of M obtained by removing all rows of M with indices outside of I and removing all columns of M with indices outside of J. Also, we denote by $|I|$ the cardinality of I and denote $I^c := N \backslash I$ and $\|M\| := \max_{x\in\Re^n:\|x\|=1} \|Mx\|$. For any continuously differentiable function $H = (H_1, ..., H_m)^T : \Re^m \mapsto \Re^m$, we denote its Jacobian by $H' = (H_1', ..., H_m')^T$, where H_i' denotes the gradient of H_i for $i = 1, ..., m$.

2 ALGORITHM DESCRIPTION

In this section we formally describe our method, parameterized by $\beta \in \Re_{++}$ and $g \in CM$. This method is patterned after an interior path-following method studied in [35] and, except for the choice of the Newton direction (2.1) and the neighborhood $\mathcal{N}_\beta$ (and the fact that an Armijo-Goldstein line search is used), is quite different from other non-interior path-following methods [2, 7, 38].

Algorithm 2.1 *Choose any* $(z^0, \mu^0) \in \mathcal{N}_\beta$, *any* ψ, σ *in* $(0,1)$ *satisfying*

$$\sqrt{n}g(0)\sigma < (1-\sigma)\beta,$$

and any continuous $\pi : \Re_{++} \mapsto \Re$ *satisfying* $0 < \pi(\mu) \le (1-\sigma)\mu$ *for all* $\mu \in \Re_{++}$, *and* $\pi(\mu)/\mu \to 0$ *as* $\mu \to 0$. *For* $t = 0, 1, ...,$ *we generate* (z^{t+1}, μ^{t+1}) *from* (z^t, μ^t) *as follows:*

Iteration t. *Let* $w^t \in \Re^{2n}$ *satisfy*

$$H'_{\mu^t}(z^t)w^t = -H_{\mu^t}(z^t). \tag{2.1}$$

Choose θ^t *to be the largest* $\theta \in \{1, \psi, \psi^2, ...\}$ *such that* $(z^t + \theta w^t, (1-\sigma\theta)\mu^t) \in \mathcal{N}_\beta$. *Choose any* $\hat{z}^t \in \Re^{2n}$ *and any* $\nu^t \in \{0, 1, ...\}$. *Choose* α^t *to be the largest* $\alpha \in \{1, \psi, \psi^2, ..., \psi^{\nu^t}, 0\}$ *such that*

$$((z^t + \theta^t w^t)(1-\alpha) + \hat{z}^t\alpha, (1-\sigma\theta^t)\mu^t(1-\alpha) + \pi(\mu^t)\alpha) \in \mathcal{N}_\beta, \tag{2.2}$$

and let

$$z^{t+1} := (z^t+\theta^t w^t)(1-\alpha^t)+\hat{z}^t\alpha^t, \quad \mu^{t+1} := (1-\sigma\theta^t)\mu^t(1-\alpha^t)+\pi(\mu^t)\alpha^t. \tag{2.3}$$

Roughly speaking, at iteration t of Algorithm 2.1, we first compute a Newton direction w^t by solving the linear equations (2.1), and next we move (z^t, μ^t) in the direction $(w^t, -\sigma\mu^t)$ by as "large" a stepsize θ^t as possible while remaining in the neighborhood $\mathcal{N}_\beta$ (this is done using an Armijo-Goldstein-type line search rule), and lastly we move the resulting pair as near to $(\hat{z}^t, \pi(\mu^t))$ as possible while remaining in $\mathcal{N}_\beta$. [If $\alpha^t = 1$, then $(z^{t+1}, \mu^{t+1}) = (\hat{z}^t, \pi(\mu^t))$. The integer ν^t controls the accuracy and the work in computing α^t.] This last move is designed to accelerate the convergence of the method (note that, by choice of π, $\pi(\mu^t)$ is always below $1-\sigma\theta^t$ and tends to zero superlinearly in μ^t) and, while $\hat{z}^t$ can be chosen arbitrarily without affecting the global convergence properties of the method, we would like $\hat{z}^t$ to be near the solution set S for reasons of practical efficiency and improved local convergence rate. In Section 4 we will consider a choice of $\hat{z}^t$ that is relatively cheap to compute and yields local superlinear convergence under assumptions milder than those for other methods. Also, in Proposition 3.1 of the next section, we will show that if $H'_{\mu^t}(z^t)$ is nonsingular (so w^t is uniquely defined), then θ^t is defined and positive (due to our choice of σ) and, since α^t is well defined ($\alpha = 0$ always satisfies (2.2)), hence (z^{t+1}, μ^{t+1}) is well defined and, by $\sigma \in (0,1)$ and $\pi(\mu^t) > 0$, μ^{t+1} is positive.

3 GLOBAL (LINEAR) CONVERGENCE ANALYSIS

In this section we show that Algorithm 2.1 attains global convergence provided that $H'_\mu(z)$ is nonsingular and $\|z\|$ is bounded for all $(z,\mu) \in \mathcal{N}_\beta$ with μ bounded, and if in addition $\|H'_\mu(z)^{-1}\|$ is bounded and F and g are sufficiently smooth, then linear convergence is also achieved. The proof, which uses ideas similar to those used for other path-following methods, is quite compact. Moreover, we show that existing sufficient condition for $H'_\mu(z)$ to be nonsingular (namely, F being a P_0-function) is in some sense also necessary, and similarly for the boundedness of $\|H'_\mu(z)^{-1}\|$ (see Corollaries 3.1(b), 3.2(b)). We also refine existing results on boundedness of $\|z\|$ when F is monotone and a feasible interior point exists (see Lemma 3.4).

First, we derive some basic properties of any $g \in \mathcal{CM}$. Since $g'(\tau) > 0$ for all $\tau \in \Re$ and $\lim_{\tau\to-\infty} g(\tau) = 0$, it must be that $g(\tau)$ is positive and strictly increasing with τ. Similarly, since $g'(\tau) - 1 < 0$ for all τ and $\lim_{\tau\to\infty} g(\tau) - \tau = 0$, it must be that $g(\tau) - \tau$ is positive and strictly decreasing with τ (so $g(\tau) > \max\{0, \tau\}$ for all $\tau \in \Re$). Using these properties, we have the following lemma showing that $\mu g(\tau/\mu)$ is Lipschitz continuous in μ with Lipschitz constant independent of τ. This result has been shown by Kanzow [20, Lemma 3.7] in the case g given by (1.5) and by Chen and Xiu [7, Proposition 1] in the case g is twice continuously differentiable and g'' is an even function with $\int_0^\infty \tau g''(\tau)d\tau < \infty$. Our proof seems simpler than that given in [7].

Lemma 3.1 *Fix any $g \in \mathcal{CM}$. For any $\tau \in \Re$ and any scalars $\mu > \nu > 0$, we have $0 < \nu g(\tau/\nu) - \mu g(\tau/\mu) \le g(0)(\mu - \nu)$.*

Proof. Fix any $\tau \in \Re$ and define $h : \Re_{++} \mapsto \Re$ by $h(\mu) := \mu g(\tau/\mu)$ for all $\mu \in \Re_{++}$. Then, h is continuously differentiable and $h'(\mu) = g(\tau/\mu) - g'(\tau/\mu)(\tau/\mu)$. Letting $a := \tau/\mu$ and using the properties that $g(a) > \max\{0, a\}$

and $0 < g'(a) < 1$, we see that $g(a) - g'(a)a > 0$. Also, since g is convex, we have $g(0) \geq g(a) - g'(a)a$. Thus, $0 < h'(\mu) \leq g(0)$ for all $\mu \in \Re_{++}$ so, for any scalars $\mu > \nu > 0$, we have from the mean-value theorem that

$$0 < h(\mu) - h(\nu) = h'(\bar{\mu})(\mu - \nu) \leq g(0)(\mu - \nu)$$

for some $\bar{\mu} \in [\nu, \mu]$. ∎

Lemma 3.1 also holds if $\nu = 0$, as was shown by Chen and Mangasarian, in the case of g given by (1.6) [8, Lemma 1.1]. Using this lemma, we obtain the following proposition, establishing the global (linear) convergence of Algorithm 2.1 under conditions similar to those used for related methods [2, 7]. The proof is, in part, adapted from that used in [35] for an interior path-following method.

Proposition 3.1 *Fix any $\beta \in \Re_{++}$ and $g \in \mathcal{CM}$. Assume $H'_\mu(z)$ is nonsingular for all $(z, \mu) \in \mathcal{N}_\beta$ with $\mu \leq \mu^0 \in \Re_{++}$. Then $\{(z^t, \mu^t, \theta^t)\}_{t=0,1...}$ generated by Algorithm 2.1 is well defined and satisfies $(z^t, \mu^t) \in \mathcal{N}_\beta$ and $\mu^{t+1} \leq (1 - \sigma\theta^t)\mu^t$ for all t, where σ and ψ are chosen in the method. Moreover, the following hold.*
(a) If $\{z^t\}$ has a convergent subsequence, then $\{\mu^t\} \to 0$ and the limit is in S.
(b) If there exist $\kappa, \lambda_1, \lambda_2 \in \Re_{++}$ and $\gamma > 1$ such that $\|H'_{\mu^t}(z^t)^{-1}\| \leq \kappa$ for all t and $\|G(r+s) - G(r) - G'(r)s\| \leq \lambda_1\|s\|^2$ and $\|F(x+u) - F(x) - F'(x)u\| \leq \lambda_2\|u\|^\gamma$ for all $r, s, x, u \in \Re^n$, then $\mu^{t+1} \leq c\mu^t$ for all t, where $c \in (0,1)$ depends on $\beta, \psi, \sigma, \sqrt{n}g(0), \lambda_1\kappa^2, \lambda_2\kappa^\gamma(\mu^0)^{\gamma-1}, \gamma$ only.

Proof. At the start of each iteration $t = 0, 1, ...$, we have $0 < \mu^t \leq \mu^0$ (since μ^t is monotonically decreasing with t) and $(z^t, \mu^t) \in \mathcal{N}_\beta$, so $H'_{\mu^t}(z^t)$ is nonsingular by assumption, implying w^t is well defined. We show below that θ^t is well defined and positive. Then, since α^t is well defined, so is (z^{t+1}, μ^{t+1}) given by (2.3). Moreover, our choice (2.2) of α^t ensures that $(z^{t+1}, \mu^{t+1}) \in \mathcal{N}_\beta$ and the property $\pi(\mu) \leq (1 - \sigma)\mu$ for all $\mu \in \Re_{++}$ ensures that $\mu^{t+1} \leq (1 - \sigma\theta^t)\mu^t$.

(a) By assumption, there is some subsequence T of $\{0, 1, ...\}$ and some $z^\infty = (x^\infty, y^\infty) \in \Re^{2n}$ such that $\{z^t\}_{t \in T} \to z^\infty$. Since μ^t is monotonically decreasing, $\{\mu^t\} \to$ some $\mu^\infty \geq 0$. Since $(z^t, \mu^t) \in \mathcal{N}_\beta$ for all $t \in T$ and, by properties of g, $\lim_{\mu \to 0, \tau \to \alpha} \mu g(\tau/\mu) = \max\{0, \alpha\}$ for any $\alpha \in \Re$, then if $\mu^\infty = 0$, we would have in the limit $x^\infty - \max\{0, x^\infty - y^\infty\}, F(x^\infty) - y^\infty = 0$, implying $z^\infty \in S$ (see (1.2)). Thus, it remains to consider the case $\mu^\infty > 0$. For any $(z, \mu) \in \mathcal{N}_\beta$ and $\theta \in [0, 1]$, $w := -H'_\mu(z)^{-1}H_\mu(z)$ and $z^+ := z + \theta w$ satisfy

$$\begin{aligned} \|H_\mu(z^+)\| &= \|H_\mu(z + \theta w) - H_\mu(z) - \theta H'_\mu(z)w + (1 - \theta)H_\mu(z)\| \\ &\leq \|H_\mu(z + \theta w) - H_\mu(z) - \theta H'_\mu(z)w\| + (1 - \theta)\|H_\mu(z)\| \\ &\leq r(z, \mu, \theta w) + (1 - \theta)\beta\mu, \end{aligned}$$

where we denote the 1st-order remainder $r(z, \mu, a) := \|H_\mu(z + a) - H_\mu(z) - H'_\mu(z)a\|$. Thus, for $\mu^+ := (1 - \sigma\theta)\mu$ and writing $z^+ = (x^+, y^+)$, we have

$$\|H_{\mu^+}(z^+)\| \leq \|H_\mu(z^+)\| + \|H_\mu(z^+) - H_{\mu^+}(z^+)\|$$

$$\begin{aligned} &= \|H_\mu(z^+)\| + \|\mu G((x^+ - y^+)/\mu) - \mu^+ G((x^+ - y^+)/\mu^+)\| \\ &\le r(z, \mu, \theta w) + (1-\theta)\beta\mu + \sqrt{n}g(0)(\mu - \mu^+), \end{aligned}$$

where the equality uses (1.4) and the last inequality also uses Lemma 3.1. Since the right-hand side is below $\beta\mu^+$ whenever

$$r(z, \mu, \theta w)/\theta + \sqrt{n}g(0)\sigma\mu \le (1-\sigma)\beta\mu,$$

which, by our choice of σ and the fact $r(z, \mu, \theta w)/\theta \to 0$ as $\theta \to 0$, occurs whenever θ is sufficiently small, it follows from our choice of θ^t that θ^t is well defined and positive for all t. Moreover, either $\theta^t = 1$ or else

$$r(z^t, \mu^t, (\theta^t/\psi)w^t)/(\theta^t/\psi) > ((1-\sigma)\beta - \sqrt{n}g(0)\sigma)\mu^t. \tag{3.1}$$

Since $\pi(\mu^t) \le (1-\sigma)\mu^t$ so that, by (2.3), $\mu^{t+1} \le (1-\sigma\theta^t)\mu^t$ for all t, we see from $\{\mu^t\} \to \mu^\infty > 0$ that $\{\theta^t\} \to 0$. Also, since $\{H'_{\mu^t}(z^t)\}_{t\in T}$ converges entrywise to $H'_{\mu^\infty}(z^\infty)$, which is nonsingular, and $\{H_{\mu^t}(z^t)\}_{t\in T}$ converges, (2.1) implies $\{w^t\}_{t\in T}$ is bounded. These two observations, together with $\{z^t\}_{t\in T} \to z^\infty$ and $\{\mu^t\} \to \mu^\infty > 0$ (and g, F being continuously differentiable), implies the left-hand side of (3.1) tends to zero as $t \to \infty$, $t \in T$. On the other hand, by our choice of σ and $\mu^t \ge \mu^\infty > 0$, the right-hand side of (3.1) is bounded away from zero for all t, a contradiction.

(b) By (1.4) and our assumptions on G and F,

$$\begin{aligned} r(z, \mu, a) &= \left\| \begin{bmatrix} -\mu(G(r+s) - G(r) - G'(r)s) \\ F(x+u) - F(x) - F'(x)u \end{bmatrix} \right\| \\ &\le \mu\|G(r+s) - G(r) - G'(r)s\| + \|F(x+u) - F(x) - F'(x)u\| \\ &\le \mu\lambda_1\|s\|^2 + \lambda_2\|u\|^\gamma \\ &\le 4\lambda_1\|a\|^2/\mu + \lambda_2\|a\|^\gamma, \end{aligned} \tag{3.2}$$

where for simplicity we write $z := (x, y), a := (u, v)$, $r := (x-y)/\mu, s := (u-v)/\mu$. Also, by (2.1) and $(z^t, \mu^t) \in \mathcal{N}_\beta$, we have

$$\|w^t\| = \|H'_{\mu^t}(z^t)^{-1} H_{\mu^t}(z^t)\| \le \|H'_{\mu^t}(z^t)^{-1}\| \|H_{\mu^t}(z^t)\| \le \kappa\beta\mu^t.$$

This together with (3.1) and (3.2) yields

$$\begin{aligned} 0 < (1-\sigma)\beta - \sqrt{n}g(0)\sigma &< r(z^t, \mu^t, (\theta^t/\psi)w^t)/(\mu^t\theta^t/\psi) \\ &\le 4\lambda_1(\theta^t/\psi)\|w^t\|^2/(\mu^t)^2 + \lambda_2(\theta^t/\psi)^{\gamma-1}\|w^t\|^\gamma/\mu^t \\ &\le 4\lambda_1(\kappa\beta)^2(\theta^t/\psi) + \lambda_2(\kappa\beta)^\gamma(\theta^t/\psi)^{\gamma-1}(\mu^t)^{\gamma-1}, \end{aligned}$$

from which we obtain that θ^t is bounded below by a positive constant depending on $\beta, \psi, \sigma, \sqrt{n}g(0), \lambda_1\kappa^2, \lambda_2\kappa^\gamma(\mu^0)^{\gamma-1}, \gamma$ only. Since $\mu^{t+1} \le (1-\sigma\theta^t)\mu^t$ for all t, the global linear convergence of $\{\mu^t\}$ follows. ∎

If in addition $\|\hat{z}^t - z^t\|$ is in the order of μ^t (such as when $\hat{z}^t$ is given by (4.1)–(4.2) and the assumptions of Proposition 4.1 hold), then the global linear convergence of μ^t in Proposition 3.1(b) yields, as a byproduct, that $\{z^t\}$ converges linearly in the root sense. Also, note that the convergence ratio c depends on $\lambda_1, \lambda_2, \kappa, \mu^0$ through their respective products only and, in the case where F is affine (so $\lambda_2 = 0$), c does not depend on μ^0.

We next have a lemma giving a characterization of P_0-matrices and an accompanying corollary showing that $F'(x)$ being a P_0-matrix is sufficient and in some sense necessary for $H'_\mu(x, y)$ to be nonsingular. The lemma, as is noted in [15, page 112], was first given by Sandberg and Willson and was rediscovered by Chen and Harker [4, Theorem 3.3] and De Luca et al. The sufficient part of the corollary is not new (see [4, Proposition 3.6], [21, Theorem 3.5], [20, Theorem 3.5] for the case of g given by (1.5) and see [6, Proposition 14], [15, page 112] for the general case) but the necessary part appears to be new (see [4, Proposition 3.6] for a related result). It is known [29, Corollary 5.3 and Theorem 5.8] that $F'(x)$ being a P_0-matrix for all x in an open box $X \subset \Re^n$ is equivalent to F being a P_0-function on X (i.e., for every $x, y \in X$ with $x \neq y$, there is an $i \in N$ with $x_i \neq y_i$ and $(x_i - y_i)(F_i(x) - F_i(y)) \geq 0$).

Lemma 3.2 *An $M \in \Re^{n\times n}$ is a P_0-matrix if and only if $D+M$ is nonsingular for all diagonal $D \in \Re^{n\times n}$ with positive diagonal entries.*

Corollary 3.1 *Fix any $g \in \mathcal{CM}$ and any $x \in \Re^n$. If $F'(x)$ is a P_0-matrix, then $H'_\mu(x,y)$ is nonsingular for all $\mu \in \Re_{++}$ and $y \in \Re^n$. If $F'(x)$ is not a P_0-matrix, then for every $\mu \in \Re_{++}$ there exists $y \in \Re^n$ such that $H'_\mu(x,y)$ is singular.*

Proof. Using (1.4), we have for every $\mu \in \Re_{++}$ and $y \in \Re^n$ that

$$\begin{aligned} \det\left[H'_\mu(x,y)\right] &= \det\begin{bmatrix} I-E & E \\ M & -I \end{bmatrix} \qquad (3.3) \\ &= \det\begin{bmatrix} I & -E \\ 0 & I \end{bmatrix} \cdot \det\begin{bmatrix} I-E+EM & 0 \\ M & -I \end{bmatrix} \\ &= (-1)^n \det\left[I - E + EM\right] \\ &= (-1)^n \det[E] \det\left[E^{-1} - I + M\right], \qquad (3.4) \end{aligned}$$

where $E := G'((x-y)/\mu)$ and $M := F'(x)$ (so E is a diagonal matrix with diagonal entries in $(0,1)$).

Assume M is a P_0-matrix. Since the diagonal entries of E are between 0 and 1, then $E^{-1} - I$ is a diagonal matrix with positive diagonal entries, so, by Lemma 3.2, $E^{-1} - I + M$ is nonsingular (in fact, a P-matrix) and hence, by (3.4), $H'_\mu(x,y)$ is nonsingular.

Assume M is not a P_0-matrix. By Lemma 3.2, there exists a diagonal $D \in \Re^{n\times n}$ with positive diagonal entries such that $D + M$ is singular. Then, for each $\mu \in \Re_{++}$, choose $y \in \Re^n$ so $G'((x-y)/\mu) = (I + D)^{-1}$, i.e., $y_i = x_i - \mu \cdot (g')^{-1}(1/(1 + D_{ii}))$ for $i \in N$. For this choice of y, (3.4) yields that $\det[H'_\mu(x,y)] = 0$. ∎

Next, we have the following lemma and an accompanying corollary showing that, in the case where F is a P_0-function, $\det[F'(\bar{x})_{II}]$ being positive for all $\{i \in N : \bar{x}_i > \bar{y}_i\} \subset I \subset \{i \in N : \bar{x}_i \geq \bar{y}_i\}$ is sufficient and in some sense necessary for $\|H'_\mu(x,y)^{-1}\|$ to be uniformly bounded (with an explicit bound given) near $(\bar{x},\bar{y})$. The sufficient part can also be inferred from the proof of [2, Proposition 5] while the explicit bound and the necessary part appears to be new.

Lemma 3.3 *Fix any $\omega \in (0,1)$ and any $M \in \Re^{n\times n}$.*
(a) If $\min_{J\subset N}\det[M_{JJ}] \geq 0$, then for every nonempty $I \subset N$ and every diagonal $E \in \Re^{n\times n}$ with diagonal entries in $[0,1]$ and satisfying $I = \{i \in N : E_{ii} \geq \omega\}$, we have $\left|\det\begin{bmatrix} I-E & E \\ M & -I\end{bmatrix}\right| \geq \det[M_{II}]\min\{\omega, 1-\omega\}^n$.
(b) If $\det[M_{II}] = 0$ for some nonempty $I \subset N$, then for the diagonal $E \in \Re^{n\times n}$ with $E_{II} = I$ and $E_{I^cI^c} = 0$, we have $\det\begin{bmatrix} I-E & E \\ M & -I\end{bmatrix} = 0$.

Proof. For any $M \in \Re^{n\times n}$ and any diagonal $E \in \Re^{n\times n}$ with diagonal entries in $[0,1]$, we have that

$$\begin{aligned} \left|\det\begin{bmatrix} I-E & E \\ M & -I\end{bmatrix}\right| &= \left|\det\begin{bmatrix} I & -E \\ 0 & I\end{bmatrix}\cdot\det\begin{bmatrix} I-E+EM & 0 \\ M & -I\end{bmatrix}\right| \\ &= \left|\det\left[I-E+EM\right]\right| \\ &= \left|\sum_{J\subset N}\det[I-E_{J^cJ^c}]\det[E_{JJ}]\det[M_{JJ}]\right|, \end{aligned} \quad (3.5)$$

where the third equality uses formula (2.2.1) in [13].

(a) Since $\det[E_{JJ}]$, $\det[I-E_{J^cJ^c}]$ and $\det[M_{JJ}]$ are nonnegative for all $J \subset N$ and, for $J = I$, we have $\det[E_{JJ}] \geq \omega^{|I|}$ and $\det[I-E_{J^cJ^c}] \geq (1-\omega)^{|I^c|}$, then the right-hand side of (3.5) is bounded below by $\det[E_{II}]\omega^{|I|}(1-\omega)^{|I^c|}$.

(b) For our particular choice of I and E, the formula above (3.5) yields that

$$\left|\det\begin{bmatrix} I-E & E \\ M & -I\end{bmatrix}\right| = |\det[I-E_{II}+E_{II}M_{II}]| = |\det[M_{II}]| = 0.$$

∎

Corollary 3.2 *Fix any $g \in \mathcal{CM}$ and any $\bar{z} = (\bar{x},\bar{y}) \in \Re^{2n}$. Let $\bar{I} := \{i \in N : \bar{x}_i > \bar{y}_i\}$, $\bar{J} := \{i \in N : \bar{x}_i < \bar{y}_i\}$.*
(a) If $\rho := \min_{\bar{I}\subset I\subset \bar{J}^c}\det[F'(\bar{x})_{II}] > 0$ and $\min_{J\subset N}\det[F'(x)_{JJ}] \geq 0$ for all x in some open set containing $\bar{x}$, then there exists $\delta \in \Re_{++}$ such that for every $z \in \Re^{2n}$ with $\|z-\bar{z}\| \leq \delta$ and every $\mu \in \Re_{++}$, we have

$$\left|\det\left[H'_\mu(z)\right]\right| \geq \rho\varpi/2, \qquad \|H'_\mu(z)^{-1}\| \leq 2\|\mathrm{adj}H'_\mu(z)\|/(\rho\varpi),$$

where $\varpi := \min\{g'(0), 1-g'(0)\}^n$ and $\mathrm{adj}A$ denotes the adjoint of $A \in \Re^{2n\times 2n}$.

(b) If $\det[F'(\bar{x})_{II}] = 0$ *for some* $\bar{I} \subset I \subset \bar{J}^c$, *then*

$$\lim \inf_{(z,\mu)\to(\bar{z},0)} \det\left[H'_\mu(z)\right] = 0, \qquad \lim \sup_{(z,\mu)\to(\bar{z},0)} \|H'_\mu(z)^{-1}\| = \infty.$$

Proof. (a) Fix any $\delta \in \Re_{++}$ small enough so that for every $z = (x, y) \in \Re^{2n}$ with $\|z - \bar{z}\| \le \delta$, we have $x_{\bar{I}} > y_{\bar{I}}$ and $x_{\bar{J}} < y_{\bar{J}}$ and $\min_{\bar{I}\subset I\subset \bar{J}^c} \det[F'(x)_{II}] \ge \rho/2$ and $\min_{J\subset N} \det[F'(x)_{JJ}] \ge 0$. Then, for any $\mu \in \Re_{++}$, by using (3.3) with $E := G'((x-y)/\mu)$ and $M := F'(x)$ and observing that E is a diagonal matrix with diagonal entries in $(0,1)$ and $I := \{i \in N : x_i \ge y_i\} = \{i \in N : E_{ii} \ge g'(0)\}$ satisfies $\bar{I} \subset I \subset \bar{J}^c$, it readily follows from Lemma 3.3(a) that

$$\left|\det\left[H'_\mu(z)\right]\right| = \left|\begin{bmatrix} I-E & E \\ M & -I \end{bmatrix}\right| \ge \det[M_{II}] \cdot \varpi \ge \rho\varpi/2.$$

Using the adjoint formula for matrix inverse [17, page 20], i.e., $A^{-1} = \mathrm{adj}A/\det[A]$ for any $A \in \Re^{2n\times 2n}$, completes the proof.

(b) For each $\mu \in \Re_{++}$, let $x_i := \bar{x}_i + \sqrt{\mu}$ and $y_i := \bar{y}_i$ for all $i \in I$ and let $x_i := \bar{x}_i$ and $y_i := \bar{y}_i + \sqrt{\mu}$ for all $i \in I^c$. Define E and M as above. Then, as $\mu \to 0$ (so $E_{II} \to I$ and $E_{I^cI^c} \to 0$), we have from (3.3) and Lemma 3.3(b) (and using the continuity property of determinant) that

$$\det[H'_\mu(x,y)] \;=\; \det\begin{bmatrix} I-E & E \\ M & -I \end{bmatrix} \to 0.$$

Since the entries of $H'_\mu(x, y)$ are bounded as $\mu \to 0$, using the aforementioned adjoint formula for matrix inverse completes the proof. ∎

Corollary 3.2(a), together with Proposition 3.1(b) and Corollary 3.1, shows that Algorithm 2.1 attains linear convergence provided that, in addition to F being a P_0-function, the ratio $\bar{\kappa} := \|F'(\bar{x})\|/\min_{\bar{I}\subset I\subset \bar{J}^c} \det[F'(\bar{x})_{II}]$ is uniformly bounded at all cluster points $\bar{x}$ of the iterate sequence, where $\bar{I}$ and $\bar{J}$ are as defined in Corollary 3.2, and F and g are sufficiently smooth (see [2, Propositions 4 and 5] for a similar result). Moreover, with some algebra, an explicit formula for the convergence ratio c in terms of the parameters in Proposition 3.1(b) and $\bar{\kappa}$, $g'(0)$ (replacing κ) can be derived. Corollary 3.2(b), together with Corollary 3.1, suggests that, as long as our linear convergence analyses rely on showing $\|H'_\mu(z)^{-1}\|$ to be bounded along the iterate sequence, we cannot hope to significantly improve upon the above sufficient condition. In particular, for any $M \in \Re^{n\times n}$ satisfying $\det[M_{II}] = 0$ for some $I \subset N$, we can choose any $(x^*, y^*) \in \Re^{2n}$ satisfying $x^*_I > y^*_I = 0$ and $y^*_{I^c} > x^*_{I^c} = 0$, and let $q := y^* - Mx^*$. Then, for $F(x) := Mx + q$, we have that $(x^*, y^*) \in S$ while, as $\mu \to 0$, $H_\mu(x^*, y^*) \to 0$ and, by Corollary 3.2(b), either $H'_\mu(x^*, y^*)$ is singular or $\|H'_\mu(x^*, y^*)^{-1}\| \to \infty$.

To ensure that the iterates z^t generated by Algorithm 2.1 remain bounded, we need to show that $(z, \mu) \in \mathcal{N}_\beta$ is bounded whenever μ is bounded. It is known that this holds if F is an R_0-function [7, Proposition 4] (for earlier results, see [4, Lemma 3.8], [21, Theorem 3.9], [20, Theorem 3.8] in the case of g given by (1.5), and see [9, Proposition 3.2] in the case where F is strongly monotone and Lipschitz continuous) or if F is monotone and β and μ are sufficiently small in some sense [2, Proposition 8] (see [38, Lemma 2.4] for a similar result in the case F is monotone affine and g is given by (1.5)). An example by Sun cited in [38, Example 2.5] ($n = 1$, $F(x) = 1$, $\beta > 1$, g given by (1.5)) suggests that, for monotone F, β and/or μ being sufficiently small is also necessary (see [23] for an analogous example, where boundedness of $\{x \in \Re^n : \|H_0(x, F(x))\| \leq \gamma\}$ requires γ to be sufficiently small). The analysis in [2] assumes g is twice continuously differentiable and g'' is an even function and $g''(\tau)$ decays to zero faster in some sense than $1/\tau^2$ as $|\tau| \to \infty$. We show in the lemma below that this assumption can be removed and the threshold on β and μ can be more simply estimated.

Lemma 3.4 *Assume F is monotone, i.e., $(x-y)^T(F(x) - F(y)) \geq 0$ for all $x, y \in \Re^n$. Also, assume there exists $(\bar{x}, \bar{y}) \in \Re^{2n}$ satisfying $\bar{x} > 0, \bar{y} > 0, \bar{y} = F(\bar{x})$. Then, for any $\beta \in \Re_{++}$ and $\mu^0 \in \Re_{++}$ satisfying $2\beta\mu^0 < \min_{i\in N} \bar{y}_i$ and $\beta\mu^0 < \min_{i\in N} \bar{x}_i$ and any $g \in \mathcal{CM}$, the set $\{(z, \mu) \in \mathcal{N}_\beta : 0 < \mu \leq \mu^0\}$ is bounded.*

Proof. Fix any $\epsilon \in \Re_{++}$ satisfying $(2\beta+\epsilon)\mu^0 < \min_i \bar{y}_i$ and $(\beta+\epsilon)\mu^0 < \min_i \bar{x}_i$. Since $g(\tau)$ and $g(\tau) - \tau$ monotonically decrease to 0 as $\tau \to -\infty$ and $\tau \to \infty$ respectively, there exist unique τ_1 and τ_2 satisfying $g(\tau_1) = \epsilon$ and $g(\tau_2) - \tau_2 = \epsilon$ with $0 < g(\tau) < \epsilon$ for all $\tau < \tau_1$ and $0 < g(\tau) - \tau < \epsilon$ for all $\tau > \tau_2$. Consider any $a, b \in \Re$ satisfying $|a - g(a-b)| \leq \beta$. If $a - b < \tau_1$ so that $|g(a-b)| < \epsilon$, then we have $|a| < \beta + \epsilon$. If $a - b > \tau_2$ so that $|a - b - g(a-b)| < \epsilon$, then we have $|b| < \beta + \epsilon$. Otherwise $\tau_1 \leq a - b \leq \tau_2$, implying $|a| \leq |a - g(a-b)| + |g(a-b)| \leq \beta + g(\tau_2)$ and $|b| \leq |a - g(a-b)| + |a - b - g(a-b)| \leq \beta + g(\tau_1) - \tau_1$. Also, $|a - g(a-b)| \leq \beta$ implies (since $g(a-b) > 0$) $a > -\beta$ and (since $g(a-b) - (a-b) > 0$) $b > -\beta$.

Fix any $(x, y, \mu) \in \mathcal{N}_\beta$ with $0 < \mu \leq \mu^0$. The monotonicity of F implies

$$
\begin{aligned}
0 &\leq (\bar{x} - x)^T(F(\bar{x}) - F(x)) \\
&= (\bar{x} - x)^T(\bar{y} - y + y - F(x)) \\
&= \bar{x}^T\bar{y} - x^T\bar{y} - \bar{x}^T y + x^T y + (\bar{x} - x)^T(y - F(x)) \\
&\leq \bar{x}^T\bar{y} - x^T\bar{y} - \bar{x}^T y + x^T y + \|\bar{x} - x\|_1 \beta\mu,
\end{aligned}
$$

where the last inequality uses (1.4) and (1.7). Thus,

$$x^T\bar{y} + \bar{x}^T y - x^T y - \|\bar{x} - x\|_1\beta\mu \leq \bar{x}^T\bar{y}. \tag{3.6}$$

Let $I := \{i \in N : (x_i - y_i)/\mu > \tau_2\}$, $J := \{i \in N : (x_i - y_i)/\mu < \tau_1\}$. Since $(x, y, \mu) \in \mathcal{N}_\beta$ so that $|x_i/\mu - g((x_i - y_i)/\mu)| \leq \beta$ for all $i \in N$, it follows from the above facts that $|y_i|/\mu < \beta + \epsilon$ for $i \in I$, $|x_i|/\mu < \beta + \epsilon$ for $i \in J$, and $|x_i|/\mu \leq \beta + g(\tau_2)$, $|y_i|/\mu \leq \beta + g(\tau_1) - \tau_1$ for $i \in (I \cup J)^c$. This, together with (3.6) and $\mu \leq \mu^0$, implies

$$\sum_{i\in I} x_i\bar{y}_i + \sum_{i\in J} \bar{x}_i y_i - \sum_{i\in I} x_i y_i - \sum_{i\in J} x_i y_i - \sum_{i\in I} |\bar{x}_i - x_i|\beta\mu^0 \leq \eta$$

for some constant η. Rewriting this as

$$\sum_{i\in I} x_i(\bar{y}_i - y_i - \beta\mu^0) + \sum_{i\in J}(\bar{x}_i - x_i)y_i \le \eta$$

and noting that, by the above facts and our choice of ϵ, $\bar{y}_i - y_i - \beta\mu^0 > \bar{y}_i - (\beta+\epsilon)\mu^0 - \beta\mu^0 > 0$ and $x_i > -\beta$ for all $i \in I$ and $\bar{x}_i - x_i > \bar{x}_i - (\beta+\epsilon)\mu^0 > 0$ and $y_i > -\beta$ for all $i \in J$, we see that x_i, $i \in I$, and y_i, $i \in J$, are bounded above by some constant. ∎

As was noted by one referee, Lemma 3.4 is related to [11, Prop. 2.4]. In particular, Lemma 3.4 equivalently says that the set $\{(x,y) \in \Re^{2n} : \|H_\mu(x,y)\| \le \gamma, 0 < \mu \le \mu^0\}$ is bounded for any $\mu^0 > 0$ and any $0 < \gamma < \sup_{\bar{x}>0, F(\bar{x})>0} \min_{i\in N} \{\bar{x}_i, F_i(\bar{x})/2\}$, and Proposition 2.4 in [11] equivalently says that the set $\{x \in \Re^n : \|H_0(x, F(x))\| \le \gamma\}$ is bounded for any $0 < \gamma < \sup_{\bar{x}>0, F(\bar{x})>0} \min_{i\in N} \{\bar{x}_i, F_i(\bar{x})\}$. Thus the two results are analogous to each other, with one for $H_\mu(x,y)$ $(\mu > 0)$ and the other for $H_0(x, F(x)) = \min\{x, F(x)\}$.

It is readily seen from the proof that Lemma 3.4 can be refined to show that, under its assumptions, the set

$$\{(x,y,\mu) \in \Re^{2n} \times (0,\mu^0] : \|x - \mu G((x-y)/\mu)\|_\infty \le \beta_1\mu, \|F(x) - y\|_\infty \le \beta_2\mu\}, \tag{3.7}$$

where $\beta_1, \beta_2, \mu^0 \in \Re_{++}$ and $g \in \mathcal{CM}$, is bounded whenever $(\beta_1 + \beta_2)\mu^0 < \min_{i\in N} \bar{y}_i$ and $\beta_1 < \min_{i\in N} \bar{x}_i$. In the case where F is affine, say $F(x) = Mx + q$ for some $M \in \Re^{n\times n}$ and $q \in \Re^n$, this sufficient condition is also close to being necessary. In particular, assume there exists $(\bar{x}, \bar{y}) \in \Re^{2n}$ satisfying $\min_{i\in N} |\bar{x}_i| > 0, \min_{i\in N} |\bar{y}_i| > 0, \bar{y} = M\bar{x} + q$ and assume M is not an R_0-matrix (otherwise the set (3.7) is bounded for any $\beta_1, \beta_2, \mu^0 \in \Re_{++}$), so there exists nonzero $(u,v) \in \Re^{2n}$ satisfying $u \ge 0, v = Mu \ge 0, u^T v = 0$. Let $I := \{i \in N : u_i > 0\}$ and $J := \{i \in N : v_i > 0\}$. Then, for any $\beta_1, \beta_2, \mu^0 \in \Re_{++}$ satisfying $\beta_1 > g(0)$ and

$$\beta_1\mu^0 > \max\{\|\bar{y}_I\|_\infty, \|\bar{x}_J\|_\infty\}, \quad (\beta_1 - g(0))\mu^0 \ge \max_{i\in(I\cup J)^c} |\min\{\bar{x}_i, \bar{y}_i\}|,$$

the set (3.7) is not bounded. To see this, for each $\zeta \in \Re_{++}$, let $x := \bar{x} + \zeta u$ and $y := \bar{y} + \zeta v$. Then $F(x) - y = 0$ and, by $\lim_{\tau\to\infty} g(\tau) - \tau = 0$, we have for all $i \in I$ that

$$\left|\frac{x_i}{\mu^0} - g\left(\frac{x_i - y_i}{\mu^0}\right)\right| = \left|\frac{\bar{x}_i + \zeta u_i}{\mu^0} - g\left(\frac{\bar{x}_i + \zeta u_i}{\mu^0} - \frac{\bar{y}_i}{\mu^0}\right)\right| \to \left|\frac{\bar{y}_i}{\mu^0}\right| < \beta_1 \text{ as } \zeta \to \infty$$

and, by $\lim_{\tau\to-\infty} g(\tau) = 0$, we have for all $i \in J$ that

$$\left|\frac{x_i}{\mu^0} - g\left(\frac{x_i - y_i}{\mu^0}\right)\right| = \left|\frac{\bar{x}_i}{\mu^0} - g\left(\frac{\bar{x}_i - \bar{y}_i - \zeta v_i}{\mu^0}\right)\right| \to \left|\frac{\bar{x}_i}{\mu^0}\right| < \beta_1 \text{ as } \zeta \to \infty$$

and for all $i \in (I \cup J)^c$ with $\bar{x}_i \ge \bar{y}_i$ that

$$\left|\frac{x_i}{\mu^0} - g\left(\frac{x_i - y_i}{\mu^0}\right)\right| = \left|\frac{\bar{x}_i}{\mu^0} - g\left(\frac{\bar{x}_i - \bar{y}_i}{\mu^0}\right)\right| \le \left|\frac{\bar{y}_i}{\mu^0}\right| + \left|\frac{\bar{x}_i - \bar{y}_i}{\mu^0} - g\left(\frac{\bar{x}_i - \bar{y}_i}{\mu^0}\right)\right|$$

$$\leq \quad \left|\frac{\bar{y}_i}{\mu^0}\right| + g(0) \leq \beta_1$$

and for all $i \in (I \cup J)^c$ with $\bar{x}_i < \bar{y}_i$ that

$$\begin{aligned} \left|\frac{x_i}{\mu^0} - g\left(\frac{x_i - y_i}{\mu^0}\right)\right| &= \left|\frac{\bar{x}_i}{\mu} - g\left(\frac{\bar{x}_i - \bar{y}_i}{\mu^0}\right)\right| \leq \left|\frac{\bar{x}_i}{\mu^0}\right| + \left|g\left(\frac{\bar{x}_i - \bar{y}_i}{\mu^0}\right)\right| \\ &\leq \left|\frac{\bar{x}_i}{\mu^0}\right| + g(0) \leq \beta_1. \end{aligned}$$

[The above argument does not depend on F being monotone.]

4 LOCAL SUPERLINEAR CONVERGENCE ANALYSIS

In this section we consider a special choice of $\hat{z}^t$ in Algorithm 2.1, based on active-set strategy, that can be computed using only one extra backsolve and for which local superlinear convergence can be shown without assuming the limit point is strictly complementary. In contrast, other non-interior continuation methods either require solving an extra $n \times n$ system of linear equations with different matrix coefficients [2, 3, 11] or assume the limit point is strictly complementary and nondegenerate [2, 3, 7]. The choice of $\hat{z}^t$ we consider is one that satisfies the following reduced system of linear equations:

$$H'_{\mu^t}(z^t)_{K^tK^t}(\hat{z}^t_{K^t} - z^t_{K^t}) = -H_0(z^t_{K^t}, 0)_{K^t}, \qquad \hat{z}^t_{(K^t)^c} = 0, \tag{4.1}$$

where

$$I^t := \{i \in N : x^t_i > \sqrt{\mu^t}\}, \;\; J^t := \{i \in N : y^t_i > \sqrt{\mu^t}\}, \;\; K^t := I^t \cup J^t. \tag{4.2}$$

[Here, for any $A \in \Re^{2n\times 2n}$ and $z \in \Re^{2n}$ and $K \subset N$, we denote by A_{KK} the matrix obtained by removing all rows and columns of A not indexed by K nor $n+K$, and by z_K the vector obtained by removing all components not indexed by K nor $n+K$.] The index sets I^t, J^t, K^t estimate, respectively,

$$I^* := \{i \in N : x^*_i > 0, y^*_i = 0\}, \;\; J^* := \{i \in N : x^*_i = 0, y^*_i > 0\}, \;\; K^* := I^* \cup J^*, \tag{4.3}$$

when $z^t = (x^t, y^t)$ is near an $z^* = (x^*, y^*) \in S$. In fact, we would have $I^t = I^*, J^t = J^*$ if μ^t is small and $\|z^t - z^*\|$ is in the order of μ^t. The choice of $\hat{z}^t$ given by (4.1) may be viewed as a generalization of that used in [2, 3, 7, 10] which corresponds to $K^t = N$. The idea of setting the smoothing parameter μ to zero on the right-hand side to achieve superlinear convergence can be traced to the interior predictor-corrector methods. Lastly, instead of $\sqrt{\mu}$, any sublinear function of μ can be used in (4.2).

An important feature of $\hat{z}^t$ given by (4.1)–(4.2) is that it can be computed with little extra effort. For example, if we compute w^t given by (2.1) using Gaussian elimination on the $n \times (n+1)$ matrix $[H'_{\mu^t}(z^t)\ H_{\mu^t}(z^t)]$, then we can simultaneously compute $\hat{z}^t$ by appending to this matrix the column $H_0(z^t_{K^t}, 0)$

and ordering the row reduction so that the rows indexed by K^t are reduced first (to, say, upper triangular form). The extra effort for computing $\hat{z}^t$ in this manner is in the order of $|K^t|^2$. Alternatively, we can compute the inverse of $H'_{\mu^t}(z^t)_{K^tK^t}$, represented in some suitable form, and use this inverse to compute the inverse of $H'_{\mu^t}(z^t)$ using an appropriate formula such as ones given in [13, page 76], [17, page 18].

We will show that if $\hat{z}^t$ is chosen according to (4.1)–(4.2) and z^t is sufficiently near an $z^* = (x^*, y^*) \in S$ that is nondegenerate in the sense that $F'(x^*)_{I^*I^*}$ is nonsingular and

$$\{F'(x^*)_{II}u_I = 0, F'(x^*)_{(I^c\setminus J^*)I}u_I \geq 0, u_{I\setminus I^*} \geq 0 \implies u_I = 0\}$$
$$\forall I^* \subset I \subset (J^*)^c, \tag{4.4}$$

where I^*, J^*, K^* are given by (4.3), then $z^t \to z^*$ and $\mu^t \to 0$ superlinearly (and "almost" quadratically if F and g are sufficiently smooth). Thus, our condition for local superlinear convergence is less restrictive than the widely used condition of z^* being a strictly complementary nondegenerate solution [2, 3, 7, 35, 36], which further requires $K^* = N$. Condition (4.4) is also less restrictive than the nonsingularity condition used in [2, Proposition 4], [3, 11] for hybrid non-interior methods, which further requires $F'(x^*)_{II}$ to be nonsingular for all $I^* \subset I \subset (J^*)^c$. In particular, for the example

$$n = 2, \quad F(x) = Mx, \quad M = \begin{bmatrix} 0 & 1 \\ -1 & 1 \end{bmatrix}, \quad x^* = y^* = \begin{bmatrix} 0 \\ 0 \end{bmatrix},$$

we have $I^* = J^* = \emptyset$ and $F'(x) = M$ is a P_0-matrix for all $x \in \Re^2$. Clearly $M_{I^*I^*}$ is nonsingular and it can be seen that (4.4) holds. However, M_{II} is singular for $I = \{1\}$.

The proof of our result is rather intricate since we need to show that superlinear convergence in μ^t can be achieved while maintaining $(z^t, \mu^t) \in \mathcal{N}_\beta$ for all t and eventually identifying the index sets (4.3) implicitly. [The index sets (4.3) cannot be identified explicitly since x^* cannot be known exactly.] We start with the following lemma showing that the distance from z to any $z^* = (x^*, y^*) \in S$ satisfying (4.4) is in the order of μ, for all $(z, \mu) \in \mathcal{N}_\beta$ near $(z^*, 0)$. This local error bound is reminiscent of one used in the superlinear convergence analysis of an interior path-following method for monotone LCP [34, Lemma 4.1].

Lemma 4.1 *Fix any $\beta \in \Re_{++}$, $g \in \mathcal{CM}$, and $z^* = (x^*, y^*) \in S$ satisfying (4.4), where I^*, J^*, K^* are given by (4.3). There exists $\delta_1 \in \Re_{++}$ such that for all $(z, \mu) = (x, y, \mu) \in \mathcal{N}_\beta$ satisfying $\|z - z^*\| + \mu \leq \delta_1$, we have $\|z - z^*\| \leq \mu/\delta_1$.*

Proof. Suppose our lemma is false so that, for every $\delta > 0$, there exists some $(z, \mu) = (x, y, \mu) \in \mathcal{N}_\beta$ such that $\|z - z^*\| + \mu \leq \delta$ but $\|z - z^*\| > \mu/\delta$. Since $(z, \mu) \in \mathcal{N}_\beta$, we have from (1.4) and (1.7) that $\beta \geq |x_i/\mu - g((x_i - y_i)/\mu)|$ for all $i \in N$. Thus, for each $i \in I := \{i \in N : x_i \geq y_i\}$,

$$\beta \geq |y_i/\mu| - |(x_i - y_i)/\mu - g((x_i - y_i)/\mu)| \geq |y_i/\mu| - g(0), \tag{4.5}$$

where the last inequality follows from $0 < g(\tau) - \tau \le g(0)$ for all $\tau \ge 0$. Similarly, for each $i \in J := \{i \in N : x_i < y_i\} = I^c$,

$$\beta \ge |x_i/\mu| - |g((x_i - y_i)/\mu)| \ge |x_i/\mu| - g(0), \tag{4.6}$$

where the last inequality follows from $0 < g(\tau) < g(0)$ for all $\tau < 0$. Also, we have $\|F(x) - y\| \le \beta\mu$ and $F(x^*) - y^* = 0$, implying

$$\begin{aligned}
&\beta\mu/\|z - z^*\| \\
&\ge \|F(x) - F(x^*) + y^* - y\|/\|z - z^*\| \\
&= \|M(x - x^*) + y^* - y\|/\|z - z^*\| \\
&= \left\| \begin{bmatrix} M_{II} & 0 \\ M_{JI} & -I \end{bmatrix} \begin{bmatrix} x_I - x_I^* \\ y_J - y_J^* \end{bmatrix} + \begin{bmatrix} M_{IJ}(x_J - x_J^*) - (y_I - y_I^*) \\ M_{JJ}(x_J - x_J^*) \end{bmatrix} \right\| / \|z - z^*\|
\end{aligned}$$

where we denote $M := \int_0^1 F'(x^* + \tau(x - x^*))d\tau$. As $\delta \to 0$, we obtain from $\|z - z^*\| + \mu \le \delta$ and $\|z - z^*\| > \mu/\delta$ that $\|z - z^*\| \to 0$ (so that M converges to $F'(x^*)$ entrywise and, for all δ sufficiently small, $I^* \subset I$ and $J^* \subset J$, implying $y_I^* = 0, x_J^* = 0$ and $x_i - x_i^* \ge y_i - y_i^*$ for all $i \in I \backslash I^*$, $x_i - x_i^* < y_i - y_i^*$ for all $i \in J \backslash J^*$) and $\mu/\|z - z^*\| \to 0$ (so that, by (4.5), $y_I/\|z - z^*\| \to 0$ and, by (4.6), $x_J/\|z - z^*\| \to 0$). By passing to a subsequence if necessary, we assume I and J are fixed. Then, the above inequality yields in the limit

$$0 \ge \left\| \begin{bmatrix} F'(x^*)_{II} & 0 \\ F'(x^*)_{JI} & -I \end{bmatrix} \begin{bmatrix} u_I \\ v_J \end{bmatrix} \right\|, \quad v_I = 0,\ u_J = 0,\ u_{I\backslash I^*} \ge v_{I\backslash I^*},\ u_{J\backslash J^*} \le v_{J\backslash J^*}, \tag{4.7}$$

where (u, v) is any cluster point of $(x - x^*, y - y^*)/\|z - z^*\|$. Thus, (4.7) yields $F'(x^*)_{II}u_I = 0$ and $F'(x^*)_{(J\backslash J^*)I}u_I = v_{J\backslash J^*} \ge 0$ and $u_{I\backslash I^*} \ge 0$. Moreover, we have $(u, v) \ne 0$, so either $u_I \ne 0$ or $v_J \ne 0$. Since, by (4.7), $F'(x^*)_{JI}u_I = v_J$ so that $v_J \ne 0$ implies $u_I \ne 0$, it must be that $u_I \ne 0$. Since $J = I^c$, this contradicts (4.4). ∎

In the case where F is affine, i.e., $F(x) = Mx + q$ for some $M \in \Re^{n \times n}$ and $q \in \Re^n$, and $g(0) \le \beta/\sqrt{n}$, the converse of Lemma 4.1 also holds. In particular, for any $z^* = (x^*, y^*) \in S$ not satisfying (4.4), so that there exist $I^* \subset I \subset (J^*)^c$ and nonzero $u_I \in \Re^{|I|}$ satisfying $M_{II}u_I = 0$ and $M_{(I^c\backslash J^*)I}u_I \ge 0$ and $u_{I\backslash I^*} \ge 0$, by letting $z := (x, y)$ with

$$x_I := x_I^* + \sqrt{\mu}u_I, \quad x_{I^c} := x_{I^c}^*, \quad y_I := y_I^*, \quad y_{I^c} := y_{I^c}^* + \sqrt{\mu}M_{I^cI}u_I,$$

we obtain $F(x) - y = 0$ and, as $\mu \to 0$, that $\|z - z^*\| \to 0$, $\|z - z^*\|/\mu \to \infty$ and $\|x/\mu - G((x - y)/\mu)\| \le \beta$. Thus, for any $\delta > 0$, we have $(z, \mu) \in \mathcal{N}_\beta$ and $\|z - z^*\| + \mu \le \delta$ but $\|z - z^*\| > \mu/\delta$, for all μ sufficiently small. [Instead of $\sqrt{\mu}$, we can use any function of μ that tends to 0 slower than μ, thus showing that $\|z - z^*\|$ cannot be bounded by any positive power of μ either.]

The next lemma shows that $\hat{z}^t$ given by (4.1) approaches an $z^* \in S$ superlinearly provided that $\|z^t - z^*\|$ is in the order of μ^t and $K^t = K^*$ (so the degenerate components are identified), and $F'(x^*)_{I^*I^*}$ is nonsingular, where I^* and K^* are given by (4.3).

Lemma 4.2 *Fix any $\beta \in \Re_{++}$, $g \in \mathcal{CM}$, and $z^* = (x^*, y^*) \in S$ such that $F'(x^*)_{I^*I^*}$ is nonsingular and there exists $\delta_1 \in \Re_{++}$ such that for all $(z, \mu) \in \mathcal{N}_\beta$ satisfying $\|z - z^*\| + \mu \leq \delta_1$ we have $\|z - z^*\| \leq \mu/\delta_1$, where I^*, J^*, K^* are given by (4.3). There exist $\delta_2 \in \Re_{++}$ and $\chi : \Re_{++} \mapsto \Re_{++}$ with $\lim_{\mu \to 0} \chi(\mu)/\mu = 0$ such that for all $(z, \mu) \in \mathcal{N}_\beta$ with $\|z - z^*\| + \mu \leq \delta_2$, we have that $H'_\mu(z)_{KK}$ is nonsingular and $\|\hat{z} - z^*\| \leq \chi(\mu)$, where $\hat{z}$ is given by*

$$H'_\mu(z)_{KK}(\hat{z}_K - z_K) = -H_0(z_K, 0)_K, \qquad \hat{z}_{K^c} = 0, \tag{4.8}$$

with $K = K^*$. If in addition $\limsup_{\tau \to \infty} \tau(1 - g'(\tau)) < \infty$, $\liminf_{\tau \to -\infty} \tau g'(\tau) > -\infty$, *and there exists $\lambda \in \Re_{++}$ such that $\|F(x+u) - F(x) - F'(x)u\| \leq \lambda \|u\|^2$ whenever $\|x - x^*\|$ and $\|u\|$ are sufficiently small, then we can further choose χ such that* $\limsup_{\mu \to 0} \chi(\mu)/\mu^2 < \infty$.

Proof. For simplicity, denote $I := I^*, J := J^*$ (so $K = I \cup J$). Consider the continuously differentiable mapping $\hat{H} : \Re^{2|K|} \mapsto \Re^{2|K|}$ given by

$$\hat{H}(z_K) = \hat{H}(x_K, y_K) := (y_I, x_J, F(x_K, 0)_K - y_K).$$

By taking δ_1 smaller if necessary, we can further assume that $x_I - y_I > x_I^*/2$ and $x_J - y_J < -y_J^*/2$ whenever $\|z - z^*\| \leq \delta_1$, in which case $\hat{H}(z_K) = H_0(z_K, 0)_K$. Now, for all $(z, \mu) \in \mathcal{N}_\beta$ satisfying $\|z - z^*\| + \mu \leq \delta_1$ we have

$$\begin{aligned} &H'_\mu(z)_{KK} - \hat{H}'(z_K) \\ &= \begin{bmatrix} I - E_{II} & 0 & E_{II} & 0 \\ 0 & I - E_{JJ} & 0 & E_{JJ} \\ M_{II} & M_{IJ} & -I & 0 \\ M_{JI} & M_{JJ} & 0 & -I \end{bmatrix} - \begin{bmatrix} 0 & 0 & I & 0 \\ 0 & I & 0 & 0 \\ \hat{M}_{II} & \hat{M}_{IJ} & -I & 0 \\ \hat{M}_{JI} & \hat{M}_{JJ} & 0 & -I \end{bmatrix} \end{aligned} \tag{4.9}$$

where $E = G'((x - y)/\mu)$, $M = F'(x)$ and $\hat{M} = F'(x_K, 0)$. Moreover, for $i \in I$, we have $0 < 1 - E_{ii} = 1 - g'((x_i - y_i)/\mu) < 1 - g'(x_i^*/(2\mu)) \to 0$ as $\mu \to 0$. Similarly, for $i \in J$, we have $0 < E_{ii} = g'((x_i - y_i)/\mu) < g'(-y_i^*/(2\mu)) \to 0$ as $\mu \to 0$. Also, as $\mu \to 0$ we have $x_{K^c} \to x_{K^c}^* = 0$ (since $\|z - z^*\| \leq \mu/\delta_1$) so that $\|M_{II} - \hat{M}_{II}\| = \|F'(x)_{II} - F'(x_K, 0)_{II}\| \to 0$. A similar argument shows that $\|M_{IJ} - \hat{M}_{IJ}\|$, $\|M_{JI} - \hat{M}_{JI}\|$, and $\|M_{JJ} - \hat{M}_{JJ}\|$ also tend to zero. Thus, the matrix difference (4.9) tends to zero entrywise as $\mu \to 0$. Since $\hat{H}(z_K^*) = 0$ and $\hat{H}(z_K) = H_0(z_K, 0)_K$, (4.8) implies

$$H'_\mu(z)_{KK}(\hat{z}_K - z_K^*) = (\hat{H}'(z_K) - H'_\mu(z)_{KK})(z_K^* - z_K) + \hat{r}(z_K), \tag{4.10}$$

where we let $\hat{r}(z_K) := \hat{H}(z_K^*) - \hat{H}(z_K) - \hat{H}'(z_K)(z_K^* - z_K)$. Since $\hat{H}$ is continuous differentiable so that $\|\hat{r}(z_K)\|/\|z_K - z_K^*\| \to 0$ as $\|z_K - z_K^*\| \to 0$, it follows from (4.9) tending to zero entrywise as $\mu \to 0$ and $\|z - z^*\| \leq \mu/\delta_1$ that the right-hand side of (4.10) tends to zero componentwise faster than μ. Also, as $(z,\mu) \to (z^*,0)$, $H'_\mu(z)_{KK}$ converges entrywise to the matrix

$$\begin{bmatrix} 0 & 0 & I & 0 \\ 0 & I & 0 & 0 \\ M^*_{II} & M^*_{IJ} & -I & 0 \\ M^*_{JI} & M^*_{JJ} & 0 & -I \end{bmatrix},$$

where $M^* := F'(x^*)$ (see (4.9)), which is nonsingular due to M^*_{II} being nonsingular. Thus, by taking δ_1 smaller if necessary, we can further assume that $\|(H'_\mu(z)_{KK})^{-1}\|$ is bounded, so then (4.10) yields $\|\hat{z}_K - z^*_K\|/\mu \to 0$ uniformly in z as $\mu \to 0$. Also $\hat{z}_{K^c} = z^*_{K^c} = 0$ trivially.

If in addition g and F have the stated properties, then we obtain that $\|\hat{r}(z_K)\| = O(\|z_K - z^*_K\|^2) = O(\mu^2)$, $1 - E_{ii} = O(\mu)$ for all $i \in I$, $E_{ii} = O(\mu)$ for all $i \in J$, and $\|M_{KK} - \hat{M}_{KK}\| = O(\mu)$ for all z sufficiently near z^* and all $\mu \in \Re_{++}$ sufficiently small, from which we deduce that $\|\hat{z}_K - z^*_K\| = O(\mu^2)$. [Here $b = O(a)$ means $b \le \kappa a$ for some constant $\kappa \in \Re_{++}$ independent of (z,μ).]
■

We note that g and F satisfy the additional assumptions in Lemma 4.2 when g is chosen by (1.5) or (1.6) and F is twice continuously differentiable on $\Re^n$. Also, the fact that (4.9) goes to zero entrywise superlinearly in $\|z_K - z^*_K\|$, which we used in the proof, is known in the case $K = N$ and g is twice continuously differentiable and satisfies $\int_{-\infty}^{\infty} \tau g''(\tau)d\tau < \infty$ [2, 7, 10]. The next lemma shows that if $\|z - z^*\|$ is in the order of μ and coincides with z^* in the degenerate components, then $(z,\mu) \in \mathcal{N}_\beta$.

Lemma 4.3 *Fix any $\beta \in \Re_{++}$ and $g \in \mathcal{CM}$ with $g(0) \le \beta/\sqrt{2n}$, and any $z^* = (x^*,y^*) \in S$. Let I^*, J^*, K^* be given by (4.3) and $\kappa := .9/\max\{\sqrt{2n}, \sqrt{2}(\|F'(x^*)\| + 1)\}$. There exists $\delta_3 \in \Re_{++}$ such that for all $z = (x,y) \in \Re^{2n}$ and $\mu \in (0,\delta_3]$ satisfying $\|z - z^*\| \le \kappa\beta\mu$ and $x_{(K^*)^c} = y_{(K^*)^c} = 0$, we have $(z,\mu) \in \mathcal{N}_\beta$.*

Proof. Let $\hat{\beta} := \beta/\sqrt{2n}$. First, since $g(\tau) - \tau \to 0$ monotonically as $\tau \to \infty$, there exists $\tau_1 \in \Re$ such that $0 < g(\tau) - \tau \le .1\hat{\beta}$ for all $\tau \ge \tau_1 - .9\hat{\beta}$. Then, for any $a \ge \tau_1$ and $|b| \le .9\hat{\beta}$, we have $0 < g(a-b) - (a-b) \le .1\hat{\beta}$, implying in the case $b > 0$ that $-.9\hat{\beta} \le -b < g(a-b) - a \le .1\hat{\beta} - b \le .1\hat{\beta}$ and in the case $b \le 0$ that $0 \le -b \le g(a-b) - a \le .1\hat{\beta} - b \le \hat{\beta}$. Thus, in either case $|a - g(a-b)| \le \hat{\beta}$. Second, since $g(\tau) \to 0$ monotonically as $\tau \to -\infty$, there exists $\tau_2 \in \Re$ such that $0 < g(-\tau) \le .1\hat{\beta}$ for all $\tau \ge \tau_2 - .9\hat{\beta}$. Then, for any $b \ge \tau_2$ and $|a| \le .9\hat{\beta}$, we have $0 < g(a-b) \le .1\hat{\beta}$, implying in the case $a > 0$ that $-.9\hat{\beta} \le -a < g(a-b) - a \le .1\hat{\beta} - a \le .1\hat{\beta}$ and in the case $a \le 0$ that $0 \le -a < g(a-b) - a \le .1\hat{\beta} - a \le \hat{\beta}$. Thus, in either case $|a - g(a-b)| \le \hat{\beta}$.

Choose any $\delta_3 \in \Re_{++}$ sufficiently small so that for all $z = (x,y) \in \Re^{2n}$ and $\mu \in (0,\delta_3]$ satisfying $\|z - z^*\| \le \kappa\beta\mu$, we have $x_i/\mu \ge \tau_1$ for all $i \in I^*$

and $y_i/\mu \geq \tau_2$ for all $i \in J^*$. Then for $i \in I^*$, we have $x_i/\mu \geq \tau_1$ and $|y_i/\mu| = |y_i - y_i^*|/\mu \leq \kappa\beta \leq .9\hat{\beta}$ so the first fact above yields

$$|x_i/\mu - g((x_i - y_i)/\mu)| \leq \hat{\beta},$$

and, for $i \in J^*$, we have $y_i/\mu \geq \tau_2$ and $|x_i/\mu| = |x_i - x_i^*|/\mu \leq \kappa\beta \leq .9\hat{\beta}$ so the second fact above yields the same. Suppose in addition that $x_{(K^*)^c} = y_{(K^*)^c} = 0$. Then for $i \in (K^*)^c$, we have $|x_i/\mu - g((x_i - y_i)/\mu)| = |g(0)| \leq \hat{\beta}$. Thus $\|x - G((x-y)/\mu)\| \leq \sqrt{n}\hat{\beta} = \beta/\sqrt{2}$. Finally, we have

$$\begin{aligned} \|F(x) - y\| &= \|F(x) - F(x^*) + y^* - y\| \\ &= \|M(x - x^*) + y^* - y\| \\ &\leq \|M\|\|x - x^*\| + \|y^* - y\| \\ &\leq (\|M\| + 1)\kappa\beta\mu \\ &\leq .9(\|M\| + 1)\beta\mu/(\sqrt{2}(\|F'(x^*)\| + 1)), \end{aligned}$$

where $M := \int_0^1 F'(x^* + \tau(x - x^*))d\tau$. Since M converges entrywise to $F'(x^*)$ as $\|z - z^*\| \to 0$, by taking δ_3 smaller if necessary, we can assume that $.9(\|M\| + 1) \leq \|F'(x^*)\| + 1$. Thus $\|F(x) - y\|/\mu \leq \beta/\sqrt{2}$. Then, by (1.4), $\|H_\mu(z)\|/\mu \leq \beta$. ∎

By combining Lemmas 4.1, 4.2 and 4.3, we obtain the following local superlinear convergence result for Algorithm 2.1 with $\hat{z}^t$ chosen according to (4.1)–(4.2).

Proposition 4.1 *Fix any $\beta \in \Re_{++}$ and $g \in \mathcal{CM}$ with $g(0) \leq \beta/\sqrt{2n}$, and any $z^* = (x^*, y^*) \in S$ such that $F'(x^*)_{I^*I^*}$ is nonsingular and (4.4) holds, where I^*, J^*, K^* are given by (4.3). Assume $H'_\mu(z)$ is nonsingular for all $(z, \mu) \in \mathcal{N}_\beta$ with $\mu \leq \mu^0 \in \Re_{++}$. Let $\{(z^t, \mu^t)\}_{t=0,1\ldots}$ be generated by Algorithm 2.1 with $\hat{z}^t$ chosen by (4.1)–(4.2) and π chosen such that $\chi(\mu)/\pi(\mu) \leq \kappa\beta$ for all $\mu \in \Re_{++}$ sufficiently small, where χ is the function given in Lemma 4.2 (via Lemma 4.1) and κ is the constant defined in Lemma 4.3. There exists $\delta \in \Re_{++}$ such that if $\|z^{\bar{t}} - z^*\| + \mu^{\bar{t}} \leq \delta$ for some $\bar{t}$, then $\{z^t\}$ converges to z^* and $\mu^{t+1} \leq \pi(\mu^t)$ for all $t \geq \bar{t}$.*

Proof. By Proposition 3.1, $\{(z^t, \mu^t)\}_{t=0,1\ldots}$ is well defined and satisfies $(z^t, \mu^t) \in \mathcal{N}_\beta$ for all t. Then, by (4.1) and Lemmas 4.1 and 4.2, there exists $\delta_2 \in \Re_{++}$ which, together with χ, has the property that

$$\|z^t - z^*\| \leq \mu^t/\delta_2,\ \|\hat{z}^t - z^*\| \leq \chi(\mu^t) \ \text{ whenever } \|z^t - z^*\| + \mu^t \leq \delta_2,\ K^t = K^*. \tag{4.11}$$

Also, by Lemma 4.3 and $\hat{z}^t_{(K^t)^c} = 0$, there exists $\delta_3 \in \Re_{++}$ such that

$$(\hat{z}^t, \pi(\mu^t)) \in \mathcal{N}_\beta \ \text{ whenever } \|\hat{z}^t - z^*\| \leq \kappa\beta\pi(\mu^t),\ \pi(\mu^t) \leq \delta_3,\ K^t = K^*.$$

Since $\chi(\mu) \le \kappa\beta\pi(\mu) \le \kappa\beta\delta_3$ for all μ sufficiently small, the above two facts imply that there exists $\delta \in (0, \delta_2]$ such that

$$\chi(\mu^t) \le \kappa\beta\pi(\mu^t),\ (\hat{z}^t, \pi(\mu^t)) \in \mathcal{N}_\beta \ \text{ whenever } \|z^t - z^*\| + \mu^t \le \delta,\ K^t = K^*, \tag{4.12}$$

in which case our choice of α^t in Algorithm 2.1 yields $\alpha^t = 1$, $z^{t+1} = \hat{z}^t$ and $\mu^{t+1} \le \pi(\mu^t)$. By taking δ smaller if necessary (depending on $\min_{i\in I^*} x_i^*$, $\min_{i\in J^*} y_i^*$ and δ_2 only), we can also assume that whenever $\|z^t - z^*\| + \mu^t \le \delta$, we have $x_i^t > \sqrt{\mu^t}$ for all $i \in I^*$, $y_i^t > \sqrt{\mu^t}$ for all $i \in J^*$, and from $\|z^t - z^*\| \le \mu^t/\delta_2$ (see (4.11)) that $|x_i^t| \le \mu^t/\delta_2 < \sqrt{\mu^t}$ and $|y_i^t| \le \mu^t/\delta_2 < \sqrt{\mu^t}$ for all $i \in (K^*)^c$, in which case (4.2) yields $K^t = K^*$. Then (4.11) and (4.12) and $\delta \le \delta_2$ yield

$$\|z^{t+1} - z^*\| \le \kappa\beta\pi(\mu^t),\ \mu^{t+1} \le \pi(\mu^t) \ \text{ whenever } \|z^t - z^*\| + \mu^t \le \delta. \tag{4.13}$$

By taking δ smaller if necessary, we can assume that $\max\{\kappa\beta, 1\}\pi(\mu)/\mu < 1/2$ for all $\mu \in (0, \delta]$, so $\kappa\beta\pi(\mu) < \mu/2 \le \delta/2$. Then, if $\|z^{\bar{t}} - z^*\| \le \delta/2$ and $\mu^{\bar{t}} \le \delta/2$, a simple induction argument using (4.13) yields $\|z^t - z^*\| \le \kappa\beta\pi(\mu^t) \le \delta/2$ and $\mu^{t+1} \le \pi(\mu^t) < \mu^t/2$ for all $t \ge \bar{t}$. This shows that $\{\mu^t\} \to 0$ superlinearly and $\{z^t\} \to z^*$. ∎

Proposition 4.1 shows that the rate of local superlinear convergence depends on the rate at which $\pi(\mu) \to 0$ as $\mu \to 0$. If g and F satisfy the additional assumptions in Lemma 4.2 so that χ can be chosen such that $\limsup_{\mu\to 0} \chi(\mu)/\mu^2 < \infty$, then any π having the additional property $\lim_{\mu\to 0} \pi(\mu)/\mu^2 = \infty$ can be used (e.g., $\pi(\mu) = \mu^2|\ln(\mu)|$). This allows Algorithm 2.1 to achieve a convergence rate as close to quadratic as desired.

5 PRELIMINARY COMPUTATIONAL EXPERIENCE

To gain some understanding of the numerical behavior/performance of Algorithm 2.1 with $\hat{z}^t$ chosen as described in Proposition 4.1, we implemented this method in Matlab and ran it on a set of test problems. In this section we describe the implementation and report on our preliminary numerical experience with it.

In our Matlab implementation of Algorithm 2.1, we set, for a given $x^0 \in \Re^n$, $y^0 = F(x^0)$, $\mu^0 = \|H_0(z^0)\|/4$ and $\beta = 1.5\|H_{\mu^0}(z^0)\|/\mu^0$. Also, we choose $\pi(\mu) = \min\{(1-\sigma)\mu, \mu^{1.8}\}$, $\psi = .7$, and $\sigma = \min\{.1, \hat{\sigma}\}$, where $\hat{\sigma} := \beta/(\beta + \sqrt{n}g(0))$. Finally, we choose $\hat{z}^t$ by (4.1)–(4.2) whenever $\mu^t < 0.1$ (otherwise we set $\alpha^t = 0$) and choose $\nu^t = 0$ (corresponding to α^t being either 0 or 1). The above parameter choices, though reasonable, were made without much fine-tuning and can conceivably be improved. To further decrease μ^t, we replace σ by $\hat{\sigma}$ in (2.3) whenever (2.2) with this replacement is satisfied by $\alpha = 0$.

In our tests, we choose x^0 to be the vector of 1s multiplied by either .1 or 1 or 10, and we terminate the method whenever $\mu^t \le 10^{-7}$. We choose g given by either (1.5) or (1.6). For g given by (1.6), due to Matlab treating $g'(\tau)$ as zero for large $|\tau|$, we replace $g(\tau)$ by its linear approximation $g(30) + g'(30)(\tau - 30)$ for $\tau > 30$ and analogously for $\tau < -30$. The performance of the method on a set of six test problems is summarized in Table 1. The test problems are (see [16, 28]):

Kojima-Shindo problem: Here $n = 4$ and F is not a P_0-function on $\Re^n_+$.

Nash-Cournot problem: Here $n = 10$ and F is defined on $\Re^n_+$ only.

Lemke problem: Here $n = 30$ and $F(x) = Mx + q$, where the (i, j)th entry of $M \in \Re^{n\times n}$ equals 2 (respectively, 1 and 0) if $j > i$ (respectively, $j = i$ and $j < i$) for all i and j (so M is positive semidefinite), and every entry of q equals -1.

Harker-Pang problem: Here $n = 30$ and $F(x) = Mx + q$, where $M = AA^T + B + D$ and every entry of $A \in \Re^{n\times n}$ and of skew-symmetric $B \in \Re^{n\times n}$ is uniformly generated from $(-5, 5)$ and every diagonal entry of the diagonal $B \in \Re^{n\times n}$ is uniformly generated from $(0, 0.3)$ (so M is positive definite), and every entry of $q \in \Re^n$ is uniformly generated from $(-500, 0)$.

Skew Symmetric LCP: Here $n = 30$ and $F(x) = Mx+q$, where $M = A-A^T$ and every entry of $A \in \Re^{n\times n}$ is uniformly generated from $[-5, 5]$ (so M is skew-symmetric), and $q = -M\bar{x} + \bar{y}$, where each entry of $\bar{x}$ has equal probability of being 0 or being uniformly generated from $[0, 5]$ and each entry of $\bar{y}$ is 0 if the corresponding entry of $\bar{x}$ is nonzero and otherwise has equal probability of being 0 or being uniformly generated from $[0, 5]$ (so $(\bar{x}, \bar{y})$ is typically not strictly complementary).

Symmetric Positive Semi-Definite LCP: Here $n = 30$ and $F(x) = Mx+q$, where $M = AA^T$ and every entry of $A \in \Re^{n\times 20}$ is uniformly generated from $[-1, 1]$ (so M is symmetric positive semi-definite), and $q = -M\bar{x} + \bar{y}$, where each entry of $\bar{x}$ has equal probability of being 0 or being uniformly generated from $[0, 10]$ and each entry of $\bar{y}$ is 0 if the corresponding entry of $\bar{x}$ is nonzero and otherwise has equal probability of being 0 or being uniformly generated from $[0, 10]$ (so $(\bar{x}, \bar{y})$ is typically not strictly complementary).

As can be seen from Table 1, the performance varies, depending on the problem and the starting point x^0, but is relatively insensitive to the choice of g given by either (1.5) or (1.6). The number of iterations, between 8 and 49, is acceptable for the level of accuracy achieved. In [28], a similar set of test problems were solved using a non-continuation Newton-type method in fewer iterations, but with less accuracy. [This method also solves an $n \times n$ system of linear equations at each iteration, so its work per iteration is similar to that for Algorithm 2.1.] In general, we found the use of $\hat{z}^t$ to significantly accelerate the local convergence of Algorithm 2.1, enabling μ^t to decrease very rapidly in the last few iterations before termination. Moreover, in cases where $\{z^t\}$ converged to a non-strictly complementary solution $z^* = (x^*, y^*)$, the index sets I^t, J^t, K^t correctly identified I^*, J^*, K^* in the later iterations. However, we caution that the effectiveness of the acceleration depends on the nonsingularity of $F'(x^*)_{I^*I^*}$ and (4.4). This acceleration mechanism was ineffective (i.e., nfast=0) on the fifth problem, for which (4.4) likely fails, and the number of iterations is correspondingly higher. The same occurs on the the sixth problem if the rank of M

Problem			g of (1.5)	g of (1.6)
Name	n	x^0	iter/nf/nfast/resid	iter/nf/nfast/resid
Kojima-Shindo	4	(.1,...,.1)	$10/16/3/1{\cdot}10^{-20}$	$10/16/3/9{\cdot}10^{-16}$
		(1,...,1)	$9/14/4/4{\cdot}10^{-16}$	$9/17/4/9{\cdot}10^{-30}$
		(10,...,10)	$23/96/4/6{\cdot}10^{-29}$	$26/136/3/9{\cdot}10^{-16}$
Nash-Cournot	10	(.1,...,.1)	$31/68/4/1{\cdot}10^{-14}$	$24/40/4/4{\cdot}10^{-14}$
		(1,...,1)	$24/39/4/4{\cdot}10^{-14}$	$18/28/3/2{\cdot}10^{-14}$
		(10,...,10)	$26/102/4/4{\cdot}10^{-14}$	$25/116/3/1{\cdot}10^{-14}$
Lemke	30	(.1,...,.1)	$8/12/4/9{\cdot}10^{-22}$	$8/12/4/3{\cdot}10^{-41}$
		(1,...,1)	$14/18/4/6{\cdot}10^{-22}$	$11/15/4/8{\cdot}10^{-41}$
		(10,...,10)	$21/25/4/3{\cdot}10^{-21}$	$15/18/3/7{\cdot}10^{-29}$
Harker-Pang	30	(.1,...,.1)	$29/40/2/9{\cdot}10^{-13}$	$28/44/3/1{\cdot}10^{-12}$
		(1,...,1)	$28/40/2/9{\cdot}10^{-13}$	$26/37/3/6{\cdot}10^{-13}$
		(10,...,10)	$25/38/4/1{\cdot}10^{-12}$	$23/36/3/1{\cdot}10^{-12}$
SkewSymLCP	30	(.1,...,.1)	$49/109/0/2{\cdot}10^{-8}$	$47/139/0/6{\cdot}10^{-9}$
		(1,...,1)	$48/115/0/2{\cdot}10^{-8}$	$45/112/0/2{\cdot}10^{-8}$
		(10,...,10)	$43/111/0/2{\cdot}10^{-8}$	$43/156/0/2{\cdot}10^{-8}$
SymPSDLCP	30	(.1,...,.1)	$23/30/2/1{\cdot}10^{-7}$	$27/38/1/1{\cdot}10^{-7}$
		(1,...,1)	$22/29/2/1{\cdot}10^{-7}$	$22/29/2/7{\cdot}10^{-8}$
		(10,...,10)	$17/24/2/1{\cdot}10^{-7}$	$17/24/2/5{\cdot}10^{-8}$

Table 1.1 Performance of Algorithm 2.1 on six test problems, as indicated by the number of iterations to termination (iter), the number of F-evaluations (nf), the number of iterations in which $\alpha^t \neq 0$ (nfast), and the residual $\|H_0(x, F(x))\|$ upon termination (resid).

is changed from 20 to 10, with a resulting increase in the number of iterations of at least threefold.

6 CONCLUSION AND EXTENSIONS

In this paper we have studied the global (linear) convergence and local superlinear convergence of a non-interior continuation method, based on the Chen-Mangasarian class of smoothing functions, for complementarity problems. The method uses an inexpensive active-set strategy to attain local superlinear convergence under conditions milder than those for other methods.

There is a number of directions in which our work can be extended. One direction is extension to smoothing functions outside the Chen-Mangasarian class, such as one suggested by Kanzow mentioned in Section 1. Another direction is extension to interior-point path-following methods, which we are presently studying. A third direction is a more comprehensive computational study that considers detailed implementation issues such as problem preprocessing, nonmonotone line search, and uses a broader set of test problems such as those from the MCPLIB and NETLIB libraries. We hope to study these and other extensions presently or in the future.

References

[1] Burke, J. and Xu, S., The global linear convergence of a non-interior path-following algorithm for linear complementarity problems, Preprint, Department of Mathematics, University of Washington, Seattle, Washington (December 1996); *Math. Oper. Res.*, to appear.

[2] Chen, B. and Chen, X., A global linear and local quadratic continuation smoothing method for variational inequalities with box constraints, Report, Department of Management and Systems, Washington State University, Pullman, Washington (March 1997).

[3] Chen, B. and Chen, X., A global and local superlinear continuation-smoothing method for P_0+R_0 and monotone NCP, Report, Department of Management and Systems, Washington State University, Pullman, Washington (May 1997); *SIAM J. Optim.*, to appear.

[4] Chen, B. and Harker, P. T., A non-interior-point continuation method for linear complementarity problems, *SIAM J. Matrix Anal. Appl.*, 14 (1993), 1168-1190.

[5] Chen, B. and Harker, P. T., A continuation method for monotone variational inequalities, *Math. Programming*, 69 (1995), 237-253.

[6] Chen, B. and Harker, P. T., Smooth approximations to nonlinear complementarity problems, *SIAM J. Optim.*, 7 (1997), 403-420.

[7] Chen, B. and Xiu, N., A global linear and local quadratic non-interior continuation method for nonlinear complementarity problems based on Chen-Mangasarian smoothing function, Report, Department of Management and Systems, Washington State University, Pullman, Washington (February 1997); *SIAM J. Optim.*, to appear.

[8] Chen, C. and Mangasarian, O. L., Smoothing methods for convex inequalities and linear complementarity problems, *Math. Programming*, 71 (1995), 51-69.

[9] Chen, C. and Mangasarian, O. L., A class of smoothing functions for nonlinear and mixed complementarity problems, *Comput. Optim. Appl.*, 5 (1996), 97-138.

[10] Chen, X., Qi, L., and Sun, D., Global and superlinear convergence of the smoothing Newton method and its application to general box constrained variational inequalities, *Math. Comput.*, 67 (1998), 519-540.

[11] Chen, X. and Ye, Y., On homotopy-smoothing methods for variational inequalities, Report, School of Mathematics, University of New South Wales, Sydney, Australia (December 1996); *SIAM J. Control Optim.*, to appear.

[12] Cottle, R. W., Giannessi, F., and Lions, J.-L., Eds., *Variational Inequalities and Complementarity Problems: Theory and Applications*, John Wiley & Sons, New York, New York, 1980.

[13] Cottle, R. W., Pang, J.-S., and Stone, R. E., *The Linear Complementarity Problem*, Academic Press, New York, New York, 1992.

[14] Ferris, M. C. and Pang, J.-S., editors, *Complementarity and Variational Problems: State of the Art*, SIAM Publishing, Philadelphia, Pennsylvania, 1997.

[15] Gabriel, S. A. and Moré, J. J., Smoothing of mixed complementarity problems, in *Complementarity and Variational Problems: State of the Art*, edited by M. C. Ferris and J.-S. Pang, SIAM Publishing, Philadelphia, Pennsylvania, 1997, 105-116.

[16] Harker, P., and Pang, J.-S., Finite-dimensional variational inequality and nonlinear complementarity problems: a survey of theory, algorithms and applications, *Math. Programming*, 48 (1990), 161–220.

[17] Horn, R. A. and Johnson, C. R., *Matrix Analysis*, Cambridge University Press, Cambridge, United Kingdom, 1985.

[18] Hotta, K. and Yoshise, A., Global convergence of a class of non-interior-point algorithms using Chen-Harker-Kanzow functions for nonlinear complementarity problems, Discussion Paper 708, Institute of Policy and Planning Sciences, University of Tsukuba, Tsukuba, Japan (December 1996).

[19] Jiang, H., Smoothed Fischer-Burmeister equation methods for the complementarity problem, Report, Department of Mathematics, University of Melbourne, Parkville, Australia (June 1997).

[20] Kanzow, C., Some noninterior continuation methods for linear complementarity problems, *SIAM J. Matrix Anal. Appl.*, 17 (1996), 851-868.

[21] Kanzow, C., A new approach to continuation methods for complementarity problems with uniform *P*-functions, *Oper. Res. Letters*, 20 (1997), 85-92.

[22] Kanzow, C. and Jiang, H., A continuation method for (strongly) monotone variational inequalities, Preprint, Institute of Applied Mathematics, University of Hamburg, Hamburg, Germany (October 1994); *Math. Programming*, to appear.

[23] Kanzow, C., Yamashita, N., and Fukushima, M., New NCP-functions and their properties, *J. Optim. Theory Appl.*, 94 (1997), 115-135.

[24] Kojima, M., Megiddo, N., and Mizuno, S., A general framework of continuation methods for complementarity problems, *Math. Oper. Res.*, 18 (1993), 945–963.

[25] Kojima, M., Megiddo, N., and Noma T., Homotopy continuation methods for nonlinear complementarity problems, *Math. Oper. Res.*, 16 (1991), 754–774.

[26] Kojima, M., Megiddo, N., Noma, T. and Yoshise, A., *A Unified Approach to Interior Point Algorithms for Linear Complementarity Problems*, Lecture Notes in Computer Science 538, Springer-Verlag, Berlin, 1991.

[27] Kojima, M., Noma, T., and Yoshise, A., Global convergence in infeasible-interior-point algorithms, *Math. Programming*, 65 (1994), 43–72.

[28] Luo, Z.-Q. and Tseng, P., A new class of merit functions for the nonlinear complementarity problem, in *Complementarity and Variational Problems: State of the Art*, edited by M. C. Ferris and J.-S. Pang, SIAM Publishing, Philadelphia, Pennsylvania, 1997, 204–225.

[29] Moré, J. J. and Rheinboldt, W. C., On P- and S-functions and related classes of n-dimensional nonlinear mappings, *Linear Algebra Appl.*, 6 (1973), 45-68.

[30] Pang, J.-S., Complementarity problems, in *Handbook on Global Optimization*, edited by R. Horst and P. Pardalos, Kluwer Academic Publishers, Norwell, Massachusetts, 1995, 271–338.

[31] Qi, L. and Sun, D., Improving the convergence of non-interior point algorithms for nonlinear complementarity problems, Preprint, School of Mathematics, University of New South Wales, Sydney, Australia (May 1997); *Math. Comput.*, to appear.

[32] Qi, L., Sun, D. and Zhou, G., A new look at smoothing Newton methods for nonlinear complementarity problems and box constrained variational inequalities, Preprint, School of Mathematics, University of New South Wales, Sydney, Australia (June 1997).

[33] Smale, S., Algorithms for solving equations, in *Proceedings of the International Congress of Mathematicians*, edited by A. M. Gleason, American Mathematical Society, Providence, Rhode Island (1987), 172-195.

[34] Tseng, P., Simplified analysis of an $O(nL)$-iteration infeasible predictor-corrector path following method for monotone LCP, in *Recent Trends in Optimization Theory and Applications*, edited by R. P. Agarwal, World Scientific, Singapore, 1995, 423-434.

[35] Tseng, P., An infeasible path-following method for monotone complementarity problems, *SIAM J. Optim.*, 7 (1997), 386–402.

[36] Wright, S. J. and Ralph, D., A superlinear infeasible-interior-point algorithm for monotone complementarity problems, *Math. Oper. Res.*, 21 (1996), 815-838.

[37] Xu, S., The global linear convergence an infeasible non-interior path-following algorithm for complementarity problems with uniform P-functions, Preprint, Department of Mathematics, University of Washington, Seattle, Washington (December 1996).

[38] Xu, S., The global linear convergence and complexity of a non-interior path-following algorithm for monotone LCP based on Chen-Harker-Kanzow-Smale smoothing functions, Preprint, Department of Mathematics, University of Washington, Seattle, Washington (February 1997).

Reformulation: Nonsmooth, Piecewise Smooth,
Semismooth and Smoothing Methods, pp. 405–420
Edited by M. Fukushima and L. Qi

A New Merit Function and a Descent Method for Semidefinite Complementarity Problems [1]

Nobuo Yamashita[†] and Masao Fukushima[†]

Abstract Recently, Tseng extended several merit functions for the nonlinear complementarity problem to the semidefinite complementarity problem (SDCP) and investigated various properties of those functions. In this paper, we propose a new merit function for the SDCP based on the squared Fischer-Burmeister function and show that it has some favorable properties. Particularly, we give conditions under which the function provides a global error bound for the SDCP and conditions under which it has bounded level sets. We also present a derivative-free method for solving the SDCP and prove its global convergence under suitable assumptions.

Key Words semidefinite complementarity problem, merit function, error bound, bounded level sets, derivative-free descent method.

1 INTRODUCTION

In the last several years, the semidefinite programming problem (SDP) has been studied intensively; for a survey, see [18]. One of the reasons why many researchers pay much attention to it is that SDP can be applied to some important problems such as linear programming problems, control problems and combinatorial optimization problems. Particularly, it is well known that the SDP relaxation for some combinatorial optimization problems can provide a better bound for the original problem than the LP and Lagrangian relaxations.

[1]The work of the authors was supported in part by the Scientific Research Grant-in-Aid from the Ministry of Education, Science, Sports and Culture, Japan. The work of the first author was also supported by the Research Fellowships of the Japan Society for the Promotion of Science for Young Scientists.

[†]Department of Applied Mathematics and Physics, Graduate School of Informatics, Kyoto University, Kyoto 606-8501, Japan, nobuo@kuamp.kyoto-u.ac.jp, fuku@kuamp.kyoto-u.ac.jp

As a natural generalization of SDP, the semidefinite complementarity problem (SDCP) is defined as follows [16, 17]. Let $\mathcal{X}$ denote the space of $n \times n$ block-diagonal matrices with a certain fixed block structure. Then SDCP is to find a matrix $x \in \mathcal{S}$ such that

$$\text{[SDCP]} \qquad \langle x, F(x) \rangle = 0,\ x \in \mathcal{K},\ F(x) \in \mathcal{K},$$

where $\mathcal{S}$ denotes the subspace comprising those $x \in \mathcal{X}$ that are symmetric, $\mathcal{K} \subset \mathcal{S}$ denotes the closed convex cone comprising those elements of $\mathcal{S}$ that are positive semidefinite, F is a mapping from $\mathcal{S}$ into itself, and $\langle \cdot, \cdot \rangle$ is the inner product defined by

$$\langle x, y \rangle := \mathrm{tr}[xy],$$

where $x, y \in \mathcal{S}$ and $\mathrm{tr}[\cdot]$ denotes the matrix trace. When the set $\mathcal{S}$ is restricted to be the space of diagonal matrices, this problem reduces to the nonlinear complementarity problem (NCP) [7, 14]. When $F : \mathcal{S} \to \mathcal{S}$ is affine, the SDCP is called the semidefinite linear complementarity problem [11].

Since SDCP is considered an extension of NCP, it is natural to expect that the SDCP can be solved by applying the solution methods that have been developed for NCP. In particular, various reformulations of NCP as a minimization problem have recently been proposed and studied vigorously [2, 3, 4, 5, 9, 10, 12, 13]. In this paper, we study a reformulation of SDCP to a minimization problem. We call the objective function of an equivalent minimization problem a merit function for the SDCP. A pioneering work in the study of merit functions for SDCP has recently been done by Tseng [17]. He showed that some well-known merit functions for NCP such as the gap function [1], the regularized gap function [4], the implicit Lagrangian [13], the squared Fischer-Burmeister function [3] and Luo and Tseng's class of functions [12] can be extended to SDCP. The gap function and the regularized gap function constitute an equivalent minimization problem on $\mathcal{K}$, whereas the implicit Lagrangian, the squared Fischer-Burmeister function and Luo and Tseng's class of functions constitute an equivalent minimization problem on $\mathcal{S}$ [17]. Tseng [17] also gave sufficient conditions under which any stationary point of these merit functions is a solution of the SDCP. Specifically, such a sufficient condition is the strong monotonicity of F for the regularized gap function and the implicit Lagrangian, while it is the monotonicity of F for the squared Fischer-Burmeister function and Luo and Tseng's class of functions. Some of those conditions have recently been refined by Qi and Chen [15]. Moreover, Tseng [17] showed that the regularized gap function and the implicit Lagrangian extended to SDCP provide error bounds for the SDCP if F is strongly monotone and Lipschitz continuous. However, it has yet to be answered whether or not there exist merit functions that provide error bounds for the SDCP under the strong monotonicity of F only, or when a merit function has bounded level sets. Error bounds play an important role in analyzing the convergence rate of iterative methods for solving the SDCP. The boundedness of level sets of a merit function ensures that the sequence generated by a descent method has at least one accumulation point. For NCP, Kanzow et al. [10] proposed a merit function that provides a global error bound under the strong monotonicity of F only and has bounded level sets under the monotonicity of F and the strict feasibility of the NCP. Hence, it is natural to expect that there exists a merit function for the SDCP that enjoys such desirable properties. The purpose of the paper is to propose such a merit function and investigate its properties.

In this paper, we propose the following merit function for the SDCP:

$$f(x) := \psi(\langle x, F(x)\rangle) + \phi(x, F(x)), \tag{1.1}$$

where $\psi : \Re \to \Re$ and $\phi : \mathcal{S} \times \mathcal{S} \to \Re$ are defined by

$$\psi(t) := \frac{1}{4}\max\{0, t\}^4 \tag{1.2}$$

and

$$\phi(a, b) := \frac{1}{2}\|(a^2 + b^2)^{\frac{1}{2}} - a - b\|^2, \tag{1.3}$$

respectively, and the norm $\|\cdot\|$ is defined by

$$\|x\| := \sqrt{\langle x, x\rangle} = \sqrt{\mathrm{tr}[x^2]}.$$

The function ϕ is the squared Fischer-Burmeister function extended to the space $\mathcal{S}$. As shown later, the function f inherits various favorable properties of the squared Fischer-Burmeister function. Moreover, we show that f provides an error bound for the SDCP under strong monotonicity of F only and has bounded level sets under suitable conditions. Note that the function f does not reduce to an existing merit function for NCP when the set $\mathcal{S}$ is restricted to be the space of diagonal matrices. Hence it is considered a new merit function for NCP as well. Note also that it is considered a modification of the function of Luo and Tseng's class of functions. In fact, if the second term $\phi(x, F(x))$ of $f(x)$ is replaced with $\frac{1}{2}\|[(x^2 + F(x)^2)^{\frac{1}{2}} - x - F(x)]_+\|^2$, where $[\cdot]_+$ denotes the orthogonal projection onto $\mathcal{K}$, then f becomes the function $f_{LT} : \mathcal{S} \to R$ defined by

$$f_{LT}(x) := \psi(\langle x, F(x)\rangle) + \frac{1}{2}\|[(x^2 + F(x)^2)^{\frac{1}{2}} - x - F(x)]_+\|^2,$$

which is a function of Luo and Tseng's class [12]. Tseng [17] showed a condition for a stationary point of f_{LT} to be a solution of the SDCP, but did not investigate properties of f_{LT} concerning error bounds and the boundedness of level sets. As the function f_{LT} is similar to f, f_{LT} in fact enjoys the same properties as those of f which will be shown later in this paper. Nevertheless, we focus on and analyze the function f because of the following two reasons. One is that f is simpler to deal with than f_{LT} since ϕ does not involve the projection $[\cdot]_+$ onto $\mathcal{K}$. The other is that f_{LT} does not seem to enjoy any particular properties f does not have as long as SDCP is concerned. For NCP, f_{LT} is convex under some conditions [12], but f is not convex in general under the same conditions. However, this advantage of f_{LT} over f does not carry over to SDCP. In fact,

Qi and Chen [15] gave a counterexample which shows that f_{LT} for SDCP is not convex under the conditions given in [12].

The paper is organized as follows. In Section 2, we mention some basic results that will be used in the subsequent sections. In Section 3, we show that the function f defined by (1.1) serves as a merit function for SDCP. We also give conditions under which any stationary point of f is a solution of the SDCP. In Section 4, we consider conditions under which f provides an error bound for SDCP and conditions under which the level sets of f are bounded. We propose a derivative-free descent method for solving SDCP and prove global convergence of the method in Section 5. In Section 6, we conclude the paper with some remarks.

2 BASIC PROPERTIES

In this section, we mention some basic results that will play an important role in the subsequent analyses.

First, we define some concepts that will frequently be used in the paper. We basically adopt the notations used in [17].

Definition 2.1 **(a)** *For any matrix $x \in \mathcal{S}$, $[x]_+$ and $[x]_-$ denote the orthogonal projections of x onto $\mathcal{K}$ and $-\mathcal{K}$, respectively.*

(b) *For any $n \times n$ matrix x, $\mathrm{sym}[x]$ is defined by*

$$\mathrm{sym}[x] := x + x^T,$$

where x^T denotes the transpose of x.

(c) *For any $c \in \mathcal{K}$, the set $\mathcal{S}_c \subset \mathcal{S}$ denotes the subspace comprising those $x \in \mathcal{S}$ whose nullspace contains the nullspace of c.*

(d) *For any $c \in \mathcal{K}$, the linear mapping $L_c : \mathcal{S}_c \to \mathcal{S}_c$ is defined by*

$$L_c[x] := cx + xc.$$

(e) *For any subspace $\mathcal{T} \subseteq \mathcal{S}$, a mapping $G : \mathcal{T} \to \mathcal{T}$ is said to be positive definite on $\mathcal{T}$ if*

$$\langle x, G[x]\rangle > 0 \quad \text{for any } x \in \mathcal{T} \ (x \neq 0).$$

With these definitions, the following properties hold.

Lemma 2.1 *[17]*

(a) *For any $x \in \mathcal{S}$, we have*

$$x = [x]_+ + [x]_-, \quad [x]_+[x]_- = 0.$$

(b) *For any $a, b \in \mathcal{S}$, let $c := (a^2 + b^2)^{\frac{1}{2}}$. Then $a, b \in \mathcal{S}_c$.*

(c) *For any $c \in \mathcal{K}$, the linear mapping L_c is positive definite and hence invertible on $\mathcal{S}_c$. Specifically, for any $x \in \mathcal{S}_c$, $L_c^{-1}[x]$ is the unique matrix $y \in \mathcal{S}_c$ satisfying*

$$cy + yc = x.$$

(d) *For any $c \in \mathcal{K}$ and $x, y \in \mathcal{S}_c$, we have*

$$\langle y, L_c[x] \rangle = \langle L_c[y], x \rangle, \quad xL_c^{-1}[c] = L_c^{-1}[c]x = \frac{1}{2}x$$

and

$$xL_c^{-1}[x] = 0 \;\Rightarrow\; x = 0.$$

(e) *For any $a \in \mathcal{K}$ and $b \in \mathcal{S}$, if $a^2 - b^2 \in \mathcal{K}$, then $a - b \in \mathcal{K}$.*

(f) *For any $a \in \mathcal{K}$ and $b \in \mathcal{S}$, we have $\langle a, b \rangle \leq \langle a, [b]_+ \rangle$.* ∎

Parts (a)–(f) in Lemma 2.1 are given in [17] as (1.12), (6.7), the argument after (6.2), Lemma 6.1 (b), (6.3) and (1.11), respectively.

The proposed function f consists of the function ψ defined by (1.2) and the squared Fischer-Burmeister function ϕ defined by (1.3). We show some properties of ψ and ϕ, which will be used in Section 3 where we consider conditions under which any stationary point of f is a solution of the SDCP. The following lemma is concerned with the function ψ. The proof is easy and hence omitted.

Lemma 2.2 *The function ψ is differentiable and $t\nabla\psi(t) \geq 0$ for all $t \in \Re$. Moreover, if $t\nabla\psi(t) = 0$, then $t \leq 0$.* ∎

Next we give some important properties of the function ϕ. Note that the term "differentiable" means "Fréchet differentiable".

Lemma 2.3 **(a)** *$\phi(a, b) \geq 0$ for all $(a, b) \in \mathcal{S} \times \mathcal{S}$. $\phi(a, b) = 0$ and $(a, b) \in \mathcal{S} \times \mathcal{S}$ if and only if $a \in \mathcal{K}$, $b \in \mathcal{K}$ and $\langle a, b \rangle = 0$.*

(b) *ϕ is differentiable on $\mathcal{S} \times \mathcal{S}$ and*

$$\begin{aligned} \nabla_a\phi(a, b) &= \operatorname{sym}[L_c^{-1}[c - a - b](a - c)] \in \mathcal{S}, \\ \nabla_b\phi(a, b) &= \operatorname{sym}[L_c^{-1}[c - a - b](b - c)] \in \mathcal{S}, \end{aligned}$$

where $c := (a^2 + b^2)^{\frac{1}{2}} \in \mathcal{K}$.

(c) *For every $(a, b) \in \mathcal{S} \times \mathcal{S}$, we have*

$$\langle \nabla_a\phi(a, b), \nabla_b\phi(a, b) \rangle \geq \|(c - a - b)g\|^2,$$

where $c := (a^2 + b^2)^{\frac{1}{2}} \in \mathcal{K}$ and $g := L_c^{-1}[c - a - b] \in \mathcal{S}$.

(d) *For every $(a, b) \in \mathcal{S} \times \mathcal{S}$, we have*

$$\langle a, \nabla_a\phi(a, b) \rangle + \langle b, \nabla_b\phi(a, b) \rangle = \|c - a - b\|^2,$$

where $c := (a^2 + b^2)^{\frac{1}{2}} \in \mathcal{K}$.

Proof. Parts (a), (b) and (c) are proven in [17, Lemma 6.3]. So we prove (d) only. By (b) we have

$$\begin{aligned}
&\langle a, \nabla_a \phi(a,b)\rangle + \langle b, \nabla_b \phi(a,b)\rangle \\
&= \operatorname{tr}\left[L_c^{-1}[c-a-b]((a-c)a + a(a-c) + (b-c)b + b(b-c))\right] \\
&= \operatorname{tr}\left[L_c^{-1}[c-a-b]L_c[c-a-b]\right] \\
&= \|c-a-b\|^2,
\end{aligned}$$

where the first equality follows from the fact that $\operatorname{tr}[xyz] = \operatorname{tr}[yzx] = \operatorname{tr}[zxy]$ for any $x, y, z \in \mathcal{S}$, the second equality follows from the definitions of c and L_c, and the last equality follows from Lemma 2.1 (d). ∎

In the following, we present some lemmas that will be useful in characterizing the asymptotic behavior of the merit function f.

Lemma 2.4 *For any* $a \in \mathcal{S}$, *we have*

$$\|[a+b]_+\| \geq \|[a]_+\| \quad \textit{for all } b \in \mathcal{K}.$$

Proof. For $a \in \mathcal{S}$, let $\lambda_1 \leq \lambda_2 \leq \cdots \leq \lambda_n$ be its eigenvalues. For every $b \in \mathcal{K}$, there exists an orthogonal matrix p such that pbp^T is diagonal. Let $\mu_1, \mu_2, \cdots, \mu_n$ be the eigenvalues of b. Then $\mu_i \geq 0$ for all i, and b is written as

$$b = \sum_{i=1}^{n} \mu_i p_i p_i^T,$$

where p_i is the ith column of p. Thus we have

$$a + b = a + \sum_{i=1}^{n} \mu_i p_i p_i^T.$$

Let $\hat{\lambda}_1 \leq \hat{\lambda}_2 \leq \cdots \leq \hat{\lambda}_n$ be the eigenvalues of $a+b$. Then, by [6, Theorem 8.1.5], the inequality $\hat{\lambda}_i \geq \lambda_i$ holds for every i. Hence we have

$$\|[a+b]_+\|^2 = \operatorname{tr}\left[[a+b]_+^2\right] = \sum_{i:\hat{\lambda}_i > 0} \hat{\lambda}_i^2 \geq \sum_{i:\lambda_i > 0} \lambda_i^2 = \operatorname{tr}\left[[a]_+^2\right] = \|[a]_+\|^2,$$

which shows the desired inequality. ∎

Lemma 2.5 *For every* $(a,b) \in \mathcal{S} \times \mathcal{S}$, *we have*

$$2\|(a^2+b^2)^{\frac{1}{2}} - a - b\|^2 \geq 2\|[(a^2+b^2)^{\frac{1}{2}} - a - b]_+\|^2 \geq \|[-a]_+\|^2 + \|[-b]_+\|^2.$$

Proof. The first inequality follows from the fact that

$$\|x\|^2 = \|[x]_+\|^2 + \|[x]_-\|^2$$

holds for any $x \in \mathcal{S}$. Next we prove the second inequality. Since $(a^2+b^2)^{\frac{1}{2}} - a \in \mathcal{K}$ by Lemma 2.1 (e), we have from Lemma 2.4

$$\|[(a^2+b^2)^{\frac{1}{2}} - a - b]_+\|^2 \geq \|[-b]_+\|^2.$$

In a similar way, we obtain the inequality

$$\|[(a^2+b^2)^{\frac{1}{2}} - a - b]_+\|^2 \geq \|[-a]_+\|^2.$$

Adding the above two inequalities yields the desired inequality. ∎

Now we give a key lemma for the boundedness of level sets of f, which will be used in Section 4. This result is an extension of [9, Lemma 2.6] which states the growth behavior of Fischer-Burmeister function on the space $\Re^2$.

Lemma 2.6 *For any sequence $\{(a^k,b^k)\} \subseteq \mathcal{S} \times \mathcal{S}$, let $\lambda_1^k \leq \lambda_2^k \leq \cdots \leq \lambda_n^k$ and $\mu_1^k \leq \mu_2^k \leq \cdots \leq \mu_n^k$ denote the eigenvalues of a^k and b^k, respectively. Then the following statements hold.*

(a) *If $\lambda_1^k \to -\infty$, then $\phi(a^k,b^k) \to \infty$.*

(b) *Suppose that $\{\lambda_1^k\}$ and $\{\mu_1^k\}$ are bounded below. If $\lambda_n^k \to \infty$, then for arbitrary positive definite matrices $g, h \in \mathcal{S}$, we have*

$$\langle g, a^k \rangle + \langle h, b^k \rangle \to \infty.$$

Proof. Part (a) follows directly from Lemma 2.5 and the definition (1.3) of ϕ. To prove (b), let p^k be an orthogonal matrix such that $p^k a^k (p^k)^T = \mathrm{diag}[\lambda_1^k, \cdots, \lambda_n^k]$ and let $g^k := p^k g (p^k)^T$. Then we have

$$\begin{aligned}
\langle g, a^k \rangle &= \mathrm{tr}\left[g a^k\right] \\
&= \mathrm{tr}\left[p^k g (p^k)^T p^k a^k (p^k)^T\right] \\
&= \mathrm{tr}\left[g^k \mathrm{diag}[\lambda_1^k, \cdots, \lambda_n^k]\right] \\
&= \sum_{i=1}^n \lambda_i^k g_{ii}^k. \qquad (2.4)
\end{aligned}$$

Note that g^k is positive definite and hence its diagonal elements g_{ii}^k are all positive. Moreover, we claim that, for each i, the sequence $\{g_{ii}^k\}$ is bounded away from zero. To see this, suppose that there exists an index i and a subsequence $\{g_{ii}^k\}_{k \in \kappa}$ converging to zero. Since the sequence $\{p^k\}$ of orthogonal matrices is bounded, it follows that, by taking a further subsequence if necessary, the sequence $\{p^k\}_{k \in \kappa}$ is convergent to a limit, say p^*, so that $\{g^k\}_{k \in \kappa}$ converges to $g^* = p^* g (p^*)^T$. Since p^* is also an orthogonal matrix, g^* must be positive definite. However this contradicts the hypothesis that g_{ii}^* is zero. Hence $\{g_{ii}^k\}$ is bounded away from zero for each i. Since $\{\lambda_i^k\}$ is bounded below for every i and $\lambda_n^k \to \infty$, it then follows from (2.4) that

$$\langle g, a^k \rangle \to \infty.$$

In a similar way, we can show that $\{\langle h, b^k\rangle\}$ is bounded below. Hence (b) is proved. ∎

Finally we recall some concepts about a mapping $F : \mathcal{S} \to \mathcal{S}$.

Definition 2.2 **(a)** *F is monotone if*

$$\langle x - y, F(x) - F(y)\rangle \geq 0 \quad \textit{for all } x, y \in \mathcal{S}.$$

(b) *F is strongly monotone with modulus $\mu > 0$ if*

$$\langle x - y, F(x) - F(y)\rangle \geq \mu\|x - y\|^2 \quad \textit{for all } x, y \in \mathcal{S}.$$

(c) *$\nabla F(x)$ denotes the Jacobian of F at $x \in S$, viewed as a linear mapping from $\mathcal{S}$ to $\mathcal{S}$.*

(d) *$\nabla F(x)$ is positive semidefinite if*

$$\langle y, \nabla F(x)y\rangle \geq 0 \qquad \textit{for all } y \in \mathcal{S}.$$

Note that if F is monotone, then $\nabla F(x)$ is positive semidefinite for all $x \in \mathcal{S}$.

3 EQUIVALENT UNCONSTRAINED MINIMIZATION

In this section, we show the equivalence between the SDCP and the minimization problem

$$\min_{x \in \mathcal{S}} \; f(x), \tag{3.5}$$

where f is defined by (1.1). Moreover, we give conditions under which any stationary point of problem (3.5) is a solution of the SDCP.

Theorem 3.1 *The function f is nonnegative on $\mathcal{S}$. $f(x) = 0$ and $x \in \mathcal{S}$ if and only if x is a solution of the SDCP.*

Proof. The theorem is evident from the definition of f and Lemma 2.3 (a). ∎

This theorem indicates that we can solve the SDCP by finding a global optimum of the minimization problem (3.5). However, the merit function f is not convex in general. So it is important to know when any stationary point of f is a global minimum of f. As shown in the next theorem, the function f has this property if F is monotone. Under the same condition, Tseng [17] showed that any stationary point of the squared Fischer-Burmeister function and the function f_{LT} is a solution of the SDCP. The proof given below is similar to that of [17, Proposition 6.1].

Theorem 3.2 *If F is differentiable, then so is f and the gradient of f is given by*

$$\begin{aligned} \nabla f(x) \;&=\; \nabla\psi(\langle x, F(x)\rangle)(\nabla F(x)x + F(x)) \\ &\quad + \nabla_a\phi(x, F(x)) + \nabla F(x)\nabla_b\phi(x, F(x)). \end{aligned} \tag{3.6}$$

Moreover, if F is monotone, then any stationary point of f is a solution of the SDCP.

Proof. If F is differentiable, then from the definition of f, Lemma 2.2, Lemma 2.3 (b) and the chain rule, f is also differentiable and (3.6) holds.

Next we prove the last part of the theorem. Define $d(x)$ by

$$d(x) := -\nabla\psi(\langle x, F(x)\rangle)x - \nabla_b\phi(x, F(x)).$$

Note that $d(x) \in \mathcal{S}$, because $\nabla_b\phi(x, F(x))$ belongs to $\mathcal{S}$ by Lemma 2.3 (b). Put $\alpha := \nabla\psi(\langle x, F(x)\rangle)$. Note that $\alpha \geq 0$ from the definition of ψ. Then we have

$$\begin{aligned}
&\langle d(x), \nabla f(x)\rangle \\
&= -\langle \alpha x + \nabla_b\phi(x, F(x)), \nabla F(x)(\alpha x + \nabla_b\phi(x, F(x))) \\
&\quad +\alpha F(x) + \nabla_a\phi(x, F(x))\rangle \\
&\leq -\langle \alpha x + \nabla_b\phi(x, F(x)), \alpha F(x) + \nabla_a\phi(x, F(x))\rangle \\
&= -\alpha^2\langle x, F(x)\rangle - \alpha(\langle x, \nabla_a\phi(x, F(x))\rangle + \langle F(x), \nabla_b\phi(x, F(x))\rangle) \\
&\quad -\langle \nabla_a\phi(x, F(x)), \nabla_b\phi(x, F(x))\rangle \\
&\leq -\alpha^2\langle x, F(x)\rangle - \langle \nabla_a\phi(x, F(x)), \nabla_b\phi(x, F(x))\rangle \\
&\leq -\langle \nabla_a\phi(x, F(x)), \nabla_b\phi(x, F(x))\rangle
\end{aligned} \tag{3.7}$$

where the first inequality follows from the monotonicity of F, the second inequality follows from Lemma 2.3 (d), and the third inequality follows from the fact that $\alpha \geq 0$ and $\alpha\langle x, F(x)\rangle \geq 0$ by Lemma 2.2. Moreover, by Lemma 2.3 (c), we have

$$\langle \nabla_a\phi(x, F(x)), \nabla_b\phi(x, F(x))\rangle \geq \|(c - a - b)g\|^2, \tag{3.8}$$

where $a := x$, $b := F(x)$, $c := (x^2 + F(x)^2)^{\frac{1}{2}}$ and $g := L_c^{-1}[c - a - b]$. Since $\nabla f(x) = 0$ at a stationary point x, it follows from (3.7) and (3.8) that

$$\|(c - a - b)g\|^2 \leq 0,$$

which means $(c-a-b)g = (c-a-b)L_c^{-1}[c-a-b] = 0$. Since $c-a-b \in \mathcal{S}_c$ by Lemma 2.1 (b), we have $c - a - b = 0$ by Lemma 2.1 (d). Hence x is a solution of the SDCP by Lemma 2.3 (a). ∎

For the NCP, similar results have been proved under weaker conditions. In fact, Facchinei and Soares [2] showed that any stationary point of the squared Fischer-Burmeister function is a solution of the NCP when the function F involved in the NCP is a P_0 function. Tseng raised an interesting question as to how the P_0 property on $\Re^n$ can be extended to the space $\mathcal{S}$. As an answer to this question, Qi and Chen [15] recently proposed the following condition on the mapping $\nabla F(x) : \mathcal{S} \to \mathcal{S}$.

Condition 1. For any $y \in \mathcal{S}$ with $y \neq 0$ and for any nonsingular matrix $p \in \Re^{n\times n}$, there exists an index i such that $(py)_i \neq 0$ and $((py)\nabla F(x)(py)^T)_{ii} \geq 0$.

Qi and Chen [15] called $\nabla F(x)$ with this property a P_0 matrix and showed that any stationary point x of the squared Fischer-Burmeister function is a solution of the SDCP if $\nabla F(x)$ satisfies Condition 1.

We can also prove that, under Condition 1, any stationary point of the function f defined by (1.1) is a solution of the SDCP. We only state the result without proof since the proof is similar to [15, Theorem 3.3].

Theorem 3.3 *Let $x \in \mathcal{S}$ be a stationary point of f. If $\nabla F(x)$ satisfies Condition 1, then x is a solution of the SDCP.* ∎

4 ERROR BOUND AND BOUNDED LEVEL SETS

In this section we study conditions under which level sets of f are bounded. Moreover, we give conditions under which f provides a global error bound for the SDCP.

First we establish a theorem about the boundedness of level sets of f.

Theorem 4.1 *Suppose that $F : \mathcal{S} \to \mathcal{S}$ is differentiable and monotone and that the SDCP is strictly feasible, i.e., there exists a positive definite matrix $\hat{x} \in \mathcal{S}$ such that $F(\hat{x})$ is positive definite. Then the level set $\mathcal{L}(\gamma) := \{x \in \mathcal{S} \mid f(x) \leq \gamma\}$ is nonempty and bounded for all $\gamma \geq 0$.*

Proof. First we prove the boundedness of level sets. Assume that there exists an unbounded sequence $\{x^k\} \subseteq \mathcal{L}(\gamma)$ for some $\gamma \geq 0$. It can be seen that the sequences of the minimum eigenvalues of $\{x^k\}$ and $\{F(x^k)\}$ are bounded below. In fact, if not, it follows from Lemma 2.6 (a) that $f(x^k) \to \infty$, which contradicts $\{x^k\} \subseteq \mathcal{L}(\gamma)$. Hence, the unboundedness of $\{x^k\}$ implies that the sequence of the maximum eigenvalues of $\{x^k\}$ tends to infinity. Let $\hat{x}$ be a strictly feasible solution of the SDCP. Since F is monotone, we have

$$\langle x^k, F(\hat{x})\rangle + \langle \hat{x}, F(x^k)\rangle \leq \langle x^k, F(x^k)\rangle + \langle \hat{x}, F(\hat{x})\rangle. \tag{4.9}$$

It follows from Lemma 2.6 (b) and the positive definiteness of $\hat{x}$ and $F(\hat{x})$ that the left-hand side of (4.9) tends to infinity. Hence, $\langle x^k, F(x^k)\rangle \to \infty$. Then we have $\psi(\langle x^k, F(x^k)\rangle) \to \infty$, which yields $f(x^k) \to \infty$. This contradicts $\{x^k\} \subseteq \mathcal{L}(\gamma)$.

Next we prove that $\mathcal{L}(\gamma)$ is nonempty for all $\gamma \geq 0$. Choose an arbitrary $\bar{x} \in \mathcal{S}$ and let $\bar{\gamma} := f(\bar{x})$. Then $\mathcal{L}(\bar{\gamma})$ is nonempty and, as shown above, bounded. Thus $\mathcal{L}(\bar{\gamma})$ contains a local minimizer $\tilde{x}$ of f, which satisfies the stationarity condition $\nabla f(\tilde{x}) = 0$. It then follows from Theorems 3.2 and 3.1 that $f(\tilde{x}) = 0$. Hence, $\mathcal{L}(0)$ is nonempty. Since $\mathcal{L}(0) \subseteq \mathcal{L}(\gamma)$ for all $\gamma \geq 0$, $\mathcal{L}(\gamma)$ is nonempty for all $\gamma \geq 0$. ∎

In a way similar to the above proof, we can prove the boundedness of level sets of a function of Luo and Tseng's class [12], provided it satisfies the properties stated in Lemma 2.6. Specifically, the function f_{LT} has bounded level sets when F is monotone and the SDCP is strictly feasible. On the other hand, the implicit Lagrangian and the squared Fischer-Burmeister function do not necessarily have bounded level sets under the same conditions, because those functions may be bounded above even if $\langle x, F(x)\rangle \to \infty$. Thus, the term $\psi(\langle x, F(x)\rangle)$ involved in f and f_{LT} plays an important role in ensuring the boundedness of level sets.

Combining Theorem 4.1 and Theorem 3.2, we can easily show the existence of a solution and the boundedness of the solution set of the SDCP.

Corollary 4.1 *Suppose that the assumptions in Theorem 4.1 are satisfied. Then the SDCP has a nonempty bounded solution set.* ∎

We note that Shida and Shindo [16] show the result stated in Corollary 4.1 under slightly weaker conditions. But the above proof is quite different from that of [16].

Next we show that f provides a global error bound for strongly monotone SDCPs. As in the case of the boundedness of level sets, the term $\psi(\langle x, F(x)\rangle)$ is the key to the error bound result. Note that not only the strong monotonicity of F but also the Lipschitz continuity of F is required for the implicit Lagrangian to provide a global error bound for the SDCP [17].

Theorem 4.2 *Let F be strongly monotone with modules μ. Then there exists a constant $\tau > 0$ such that*

$$\tau\|x-x^*\|^2 \le \max\{0, \langle x, F(x)\rangle\} + \|[-x]_+\| + \|[-F(x)]_+\| \quad \text{for all } x \in \mathcal{S}, \tag{4.10}$$

where x^* is the unique solution of the SDCP. Moreover, there exists a constant $\tau' > 0$ such that

$$\tau'\|x - x^*\|^2 \le f(x)^{\frac{1}{4}} + f(x)^{\frac{1}{2}} \quad \text{for all } x \in \mathcal{S}. \tag{4.11}$$

Proof. Since F is strongly monotone, we have

$$\begin{aligned}
\mu\|x - x^*\|^2 &\le \langle x - x^*, F(x) - F(x^*)\rangle \\
&= \langle x, F(x)\rangle + \langle x^*, -F(x)\rangle + \langle F(x^*), -x\rangle \\
&\le \max\{0, \langle x, F(x)\rangle\} + \langle x^*, [-F(x)]_+\rangle + \langle F(x^*), [-x]_+\rangle \\
&\le \max\{0, \langle x, F(x)\rangle\} + \|x^*\|\|[-F(x)]_+\| + \|F(x^*)\|\|[-x]_+\| \\
&\le \max\{1, \|x^*\|, \|F(x^*)\|\}\left(\max\{0, \langle x, F(x)\rangle\}\right. \\
&\quad \left. + \|[-F(x)]_+\| + \|[-x]_+\|\right),
\end{aligned}$$

where the second inequality follows from Lemma 2.1 (f). Setting

$$\tau := \mu / \max\{1, \|x^*\|, \|F(x^*)\|\}$$

then yields the inequality (4.10).

By the definition of the function ψ and Lemma 2.5, we have

$$\max\{0, \langle x, F(x)\rangle\} = \sqrt{2}\psi(\langle x, F(x)\rangle)^{\frac{1}{4}} \le \sqrt{2} f(x)^{\frac{1}{4}}$$

and

$$\|[-F(x)]_+\| + \|[-x]_+\| \le 2\sqrt{2}\phi(x, F(x))^{\frac{1}{2}} \le 2\sqrt{2}f(x)^{\frac{1}{2}},$$

which yield the desired inequality (4.11). ∎

From this theorem, we obtain another result about the boundedness of level sets of f.

Corollary 4.2 *If F is differentiable and strongly monotone, then the level set $\mathcal{L}(\gamma) := \{x \in \mathcal{S} \mid f(x) \le \gamma\}$ is nonempty and bounded for all $\gamma \ge 0$.* ∎

5 DESCENT METHOD

In this section, we propose a descent method for solving the minimization problem (3.5) and prove its global convergence. The proposed method uses the following matrix $d(x) \in \mathcal{S}$ as a search direction:

$$d(x) := -\nabla\psi(\langle x, F(x)\rangle)x - \nabla_b\phi(x, F(x)). \tag{5.12}$$

Since $d(x)$ does not involve the Jacobian $\nabla F(x)$, the method is considered an extension of the derivative-free methods developed for NCP [8, 19]. The proposed method will turn out to be particularly useful for problems in which the Jacobian of F is costly to evaluate.

The next lemma shows that, when $\nabla F(x)$ is positive semidefinite, the matrix $d(x)$ is a descent direction of f whenever x is not a solution of the problem.

Lemma 5.1 *Suppose that x is not a solution of the SDCP. If $\nabla F(x)$ is positive semidefinite, then the matrix $d(x) \in \mathcal{S}$ defined by (5.12) satisfies the descent condition $\langle \nabla f(x), d(x)\rangle < 0$.*

Proof. As shown in the proof of Theorem 3.2, we have

$$\langle \nabla f(x), d(x)\rangle \le -\|(c-a-b)g\|^2, \tag{5.13}$$

where $a := x, b := F(x), c := (a^2 + b^2)^{\frac{1}{2}}$ and $g := L_c^{-1}[c-a-b]$. Hence, we have

$$\langle \nabla f(x), d(x)\rangle \le 0.$$

Moreover, if $\langle \nabla f(x), d(x)\rangle = 0$, then (5.13) implies $(c-a-b)g = (c-a-b)L_c^{-1}[c-a-b] = 0$. It then follows from Lemma 2.1 (d) that $c-a-b=0$. Hence x is a solution of the SDCP by Lemma 2.3 (a). However this contradicts the assumption. Consequently, we have $\langle \nabla f(x), d(x)\rangle < 0$. ∎

Now we state the descent method that uses the search direction $d(x)$.

Algorithm 5.1

(S0) Choose $x^0 \in \mathcal{S}, \varepsilon \ge 0, \sigma \in (0, 1/2), \beta \in (0,1)$ and set $k := 0$.

(S1) If $f(x^k) \le \varepsilon$, then stop.

(S2) Let $d(x^k) := -\nabla\psi(\langle x^k, F(x^k)\rangle)x^k - \nabla_b\phi(x^k, F(x^k))$.

(S3) Compute a steplength $t_k := \beta^{m_k}$, where m_k is the smallest nonnegative integer m satisfying the Armijo-type condition

$$f(x^k + \beta^m d(x^k)) \le (1 - \sigma\beta^{2m})f(x^k).$$

(S4) Set $x^{k+1} := x^k + t_k d(x^k), k := k + 1$, and go to S1.

Note that the generated sequence $\{x^k\}$ lies in $\mathcal{S}$ because $d(x^k) \in \mathcal{S}$ for every k.

Next we show the global convergence of Algorithm 5.1. Here we suppose that $\varepsilon = 0$ so that the algorithm generally generates an infinite sequence, i.e., x^k is not a solution of the SDCP for every k.

Theorem 5.1 *Suppose that F is continuously differentiable. Suppose either that F is monotone and the SDCP is strictly feasible or that F is strongly monotone. Then the sequence generated by Algorithm 5.1 has at least one accumulation point and any accumulation point is a solution of the SDCP.*

Proof. First, we claim that, whenever x^k is not a solution, there exists a nonnegative integer m_k in S3 of the algorithm. Assume the contrary. Then for any positive integer m, we have

$$f(x^k + \beta^m d(x^k)) - f(x^k) > -\sigma\beta^{2m}f(x^k).$$

Dividing the both sides by β^m and letting $m \to \infty$ yield

$$\langle \nabla f(x^k), d(x^k)\rangle \ge 0.$$

Since $\nabla F(x)$ is positive semidefinite under the given hypotheses, this contradicts Lemma 5.1. Hence we can find an integer m_k in S3.

Next we show that the sequence generated by the algorithm has at least one accumulation point. By the descent property of Algorithm 5.1, the sequence $\{f(x^k)\}$ is monotonically decreasing. Hence, by Theorem 4.1 or Corollary 4.2, the generated sequence is bounded, and hence it has at least one accumulation point.

The last part of the theorem can be proven in a way similar to that of [8, Proposition 11]. For completeness, we give a proof. Let x^* be an arbitrary accumulation point of the generated sequence $\{x^k\}$ and $\{x^k\}_{k\in\kappa}$ be a subsequence converging to x^*. Since $d(\cdot)$ is continuous, $\{d(x^k)\}_{k\in\kappa}$ converges to the limit $d(x^*)$. First consider the case where there exists a constant $\bar{\beta}$ such that $\beta^{m_k} \ge \bar{\beta} > 0$ for all $k \in \kappa$. Then since

$$f(x^{k+1}) \le (1 - \sigma\bar{\beta}^2)f(x^k)$$

for all $k \in \kappa$ and the entire sequence $\{f(x^k)\}$ is decreasing, we have $f(x^*) = 0$. This implies that x^* is a solution of the SDCP. Next consider the case where there exists a further subsequence such that $\beta^{m_k} \to 0$. Note that by Armijo's rule in S3, we have

$$f(x^k + \beta^{m_k-1}d(x^k)) - f(x^k) > -\sigma\beta^{2(m_k-1)}f(x^k).$$

Dividing both side by β^{m_k-1} and passing to the limit on the subsequence, we get

$$\langle \nabla f(x^*), d(x^*)\rangle \geq 0.$$

In view of Lemma 5.1, we deduce that x^* is a solution of the SDCP. ∎

6 CONCLUDING REMARKS

In this paper, we have proposed a new merit function for the SDCP. We have shown that the function enjoys a number of favorable properties. In particular, it provides a global error bound under strong monotonicity of F only and it has bounded level sets when F is monotone and the SDCP is strictly feasible.

Finally, we point out that the function f is a new merit function, even if we regard it as a merit function for NCP. For NCP, f is written as

$$f(x) := \psi(\langle x, F(x)\rangle) + \frac{1}{2}\sum_{i=1}^{n}\left(\sqrt{x_i^2 + F_i(x)^2} - x_i - F_i(x)\right)^2,$$

where F is a function from $\Re^n$ into $\Re^n$, and the inner product is simply defined as $\langle x, y\rangle := x^T y$ for $x, y \in \Re^n$. As shown in this paper, f constitutes an unconstrained differentiable minimization problem equivalent to the given NCP. Any stationary point of f is a solution of the NCP if F is monotone. Moreover, f provides a global error bound for the NCP under the monotonicity of F only, and f has bounded level sets if F is monotone and the NCP is strictly feasible. Note that, for NCP, when the first term $\psi(\langle x, F(x)\rangle)$ of f is replaced by $\sum_{i=1}^{n}\psi(x_i F_i(x))$, we can show that the above mentioned properties of f are preserved under the P property of F instead of the monotonicity of F. Hence, f retains all desirable properties of the following merit function f_{KYF} for the NCP proposed by Kanzow et al. [10].

$$f_{KYF}(x) := \sum_{i=1}^{n}\psi(x_i F_i(x)) + \frac{1}{2}\sum_{i=1}^{n}\left[\sqrt{x_i^2 + F_i(x)^2} - x_i - F_i(x)\right]_+^2.$$

Moreover, there are at least two advantages of f over the function f_{KYF}. One is that f is twice differentiable at every nondegenerate solution of the NCP, while f_{KYF} is not twice differentiable at any solution of the NCP. The other is that f behaves like the squared Fischer-Burmeister function near each solution of the NCP. Thanks to this property, we may construct a hybrid method for NCP by combining a steepest descent method for minimizing f and Newton's method for the system of equations derived from the Fischer-Burmeister function. In this method, the steepest descent method guarantees the global convergence and Newton's method ensures a rapid local convergence property.

References

[1] A. Auslender, Optimisation: Méthodes Numériques, Masson: Paris, 1976.

[2] F. Facchinei and J. Soares, "A new merit function for nonlinear complementarity problems and a related algorithm," SIAM Journal on Optimization, Vol.7, pp.225–247, 1997.

[3] A. Fischer, "An NCP-function and its use for the solution of complementarity problems," in D.-Z. Du, L. Qi and R.S. Womersley (Eds.), Recent Advances in Nonsmooth Optimization, World Scientific Publishers: Singapore, pp. 88–105, 1995.

[4] M. Fukushima, "Equivalent differentiable optimization problems and descent methods for asymmetric variational inequality problems," Mathematical Programming, Vol. 53, pp. 99–110, 1992.

[5] M. Fukushima, "Merit functions for variational inequality and complementarity problems," in G. Di Pillo and F. Giannessi (Eds.), Nonlinear Optimization and Applications, Plenum Press: New York, pp. 155–170, 1996.

[6] G.H. Golub and C.F. Van Loan, Matrix Computation, The Johns Hopkins University Press: Baltimore, 1989.

[7] P.T. Harker and J.-S. Pang, "Finite-dimensional variational inequality and nonlinear complementarity problem: A survey of theory, algorithms and applications," Mathematical Programming, Vol. 48, pp. 161–220, 1990.

[8] H. Jiang, "Unconstrained minimization approaches to nonlinear complementarity problems," Vol. 9, pp. 169–181, 1996.

[9] C. Kanzow, "Global convergence properties of some iterative methods for linear complementarity problems," SIAM Journal on Optimization, Vol.6, pp. 326–341, 1996.

[10] C. Kanzow, N. Yamashita and M. Fukushima, "New NCP-functions and their properties," Journal of Optimization Theory and Application, Vol. 94, pp. 115–135, 1997.

[11] M. Kojima, S. Shindoh and S. Hara, "Interior-point methods for the monotone semidefinite linear complementarity problems in symmetric matrices," SIAM Journal on Optimization, Vol. 7, pp. 86–125, 1997.

[12] Z.-Q. Luo and P. Tseng, "A new class of merit functions for the nonlinear complementarity problem," in M.C. Ferris and J.-S. Pang (Eds.), Complementarity and Variational Problems: State of the Art, SIAM: Philadelphia, pp. 204–225, 1997.

[13] O.L. Mangasarian and M.V. Solodov, "Nonlinear complementarity as unconstrained and constrained minimization," Mathematical Programming, Vol. 62, pp. 277–297, 1993.

[14] J.-S. Pang, "Complementarity problems," in R. Horst and P. Pardalos (Eds.), Handbook of Global Optimization, Kluwer Academic Publishers: Boston, Massachusetts, pp. 271–338, 1994.

[15] H.D. Qi and X. Chen, "On stationary points of merit functions for semidefinite complementarity problems," working paper, Institute of Computational Mathematics and Scientific/Engineering Computing, Chinese Academy of Sciences, Beijing, China, 1997.

[16] M. Shida and S. Shindoh, "Monotone semidefinite complementarity problems", Research Report 312, Department of Mathematical and Computing Sciences, Tokyo Institute of Technology, Tokyo, Japan, 1996.

[17] P. Tseng, "Merit functions for semi-definite complementarity problems", Mathematical Programming, to appear.

[18] L. Vandenberghe and S. Boyd, "Semidefinite Programming," SIAM Review, Vol. 38, pp. 49–95, 1996.

[19] N. Yamashita and M. Fukushima, "On stationary points of the implicit Lagrangian for nonlinear complementarity problems," Journal of Optimization Theory and Applications, Vol. 84, pp. 653–663, 1995.

Reformulation: Nonsmooth, Piecewise Smooth, Semismooth and Smoothing Methods, pp. 421–441
Edited by M. Fukushima and L. Qi

Numerical Experiments for a Class of Squared Smoothing Newton Methods for Box Constrained Variational Inequality Problems [1]

Guanglu Zhou[†], Defeng Sun[†] and Liqun Qi[†]

Abstract In this paper we present a class of squared smoothing Newton methods for the box constrained variational inequality problem. This class of squared smoothing Newton methods is a regularized version of the class of smoothing Newton methods proposed in [25]. We tested all the test problem collections of GAMSLIB and MCPLIB with all available starting points. Numerical results indicate that these squared smoothing Newton methods are extremely robust and promising.

Key Words variational inequality problem, smoothing approximation, smoothing Newton method, regularization method, convergence.

1 INTRODUCTION

Consider the box constrained variational inequality problem (BVIP for short): Find $y^* \in X = \{y \in \Re^n | \ a \leq y \leq b\}$, where $a \in \{\Re \cup \{-\infty\}\}^n$, $b \in \{\Re \cup \{\infty\}\}^n$ and $a < b$, such that

$$(y - y^*)^T F(y^*) \geq 0 \quad \text{for all } y \in X, \tag{1.1}$$

where $F : D \to \Re^n$ is a continuously differentiable function on some open set D, which contains X. When $X = \Re^n_+$, BVIP reduces to the nonlinear complementarity problem: Find $y^* \in \Re^n_+$ such that

$$F(y^*) \in \Re^n_+ \quad \text{and} \quad F(y^*)^T y^* = 0. \tag{1.2}$$

[1]This work is supported by the Australian Research Council.
[†]School of Mathematics, The University of New South Wales, Sydney 2052, Australia. E-mail: zhou@maths.unsw.edu.au, sun@maths.unsw.edu.au, L.Qi@unsw.edu.au

Let Π_X be the projection operator on X. It is well known that solving BVIP is equivalent to solving the following Robinson's normal equation

$$E(x) := F(\Pi_X(x)) + x - \Pi_X(x) = 0 \tag{1.3}$$

in the sense that if $x^* \in \Re^n$ is a solution of (1.3) then $y^* := \Pi_X(x^*)$ is a solution of (1.1), and conversely if y^* is a solution of (1.1) then $x^* := y^* - F(y^*)$ is a solution of (1.3) [27]. Let $N := \{1, 2, \cdots, n\}$ and

$$\begin{aligned} I_\infty &= \{i \in N \mid a_i = -\infty \text{ and } b_i = +\infty\}, \\ I_{ab} &= \{i \in N \mid a_i > -\infty \text{ and } b_i < +\infty\}, \\ I_a &= \{i \in N \mid a_i > -\infty \text{ and } b_i = +\infty\}, \\ I_b &= \{i \in N \mid a_i = -\infty \text{ and } b_i < +\infty\}. \end{aligned}$$

Define

$$W(x) := F(\Pi_X(x)) + x - \Pi_X(x) + \alpha T(x), \tag{1.4}$$

where $\alpha \geq 0$ and $T : \Re^n \to \Re^n$ is defined by

$$T_i(x) = \begin{cases} 0 & \text{if } i \in I_\infty \\ 0 & \text{if } i \in I_{ab} \\ [(\Pi_X(x))_i - a_i][F_i(\Pi_X(x))]_+ & \text{if } i \in I_a \\ [(\Pi_X(x))_i - b_i][-F_i(\Pi_X(x))]_+ & \text{if } i \in I_b \end{cases}, \quad i \in N.$$

Properties of $W(x)$ have been studied in [30] in the case that $a_i = 0$ and $b_i = +\infty$ for all $i \in N$.

We can easily prove the following lemma.

Lemma 1.1 $E(x) = 0$ *if and only if* $W(x) = 0$.

By using the Gabriel-Moré smoothing function for $\Pi_X(\cdot)$, we can construct approximations for $W(\cdot)$:

$$G(u, x) := M(u, x) + \alpha S(u, x), \quad (u, x) \in \Re^n \times \Re^n, \tag{1.5}$$

where $M(u, x) := F(p(u, x)) + x - p(u, x)$ and $S : \Re^{2n} \to \Re^n$ is defined by

$$S_i(u, x) = \begin{cases} 0 & \text{if } i \in I_\infty \\ 0 & \text{if } i \in I_{ab} \\ [p_i(u, x) - a_i][p_i(u, F(p(u, x)) + a) - a_i] & \text{if } i \in I_a \\ [p_i(u, x) - b_i][b_i - p_i(u, F(p(u, x)) + b)] & \text{if } i \in I_b \end{cases}, \quad i \in N,$$

where $p(u, x)$ was defined in [25] and will be reviewed in the next section. We note that for any $(u, x) \in \Re^n \times \Re^n$, $p(u, x) \in X$ [25]. So we can assume that F has definition on X only in order that $G(\cdot)$ has definition on $\Re^n \times \Re^n$. This is a very nice feature.

Recently, smoothing Newton methods have attracted a lot of attention in the literature partially due to their superior numerical performance [1], e.g., see [2, 3, 5, 6, 7, 8, 9, 10, 16, 23, 25, 31] and references therein. Among them the first globally and superlinearly (quadratically) convergent smoothing Newton method was proposed by Chen, Qi and Sun in [9]. The result of [9] has been further investigated by Chen and Ye [10]. But they all assumed that F had definition on the whole space $\Re^n$. Qi, Sun and Zhou in [25] avoided this requirement by making use of the mapping $M(\cdot)$ and used one smoothing approximation function instead of using an infinite sequence of those functions.

Regularization methods for solving monotone complementarity problems have been studied by several authors [4, 12, 20, 26, 29]. Facchinei and Kanzow [12] replaced the monotonicity assumption by a P_0-function condition and showed that many properties of regularization methods still hold for this larger class of problems. Sun [29] proposed a regularization smoothing Newton method for solving nonlinear complementarity problem under the assumption that F is a P_0-function and obtained some stronger results for monotone complementarity problems. H.-D. Qi [20] proposed a regularized smoothing Newton method for the nonlinear complementarity problem and the box constrained variational inequality problem by using the developments on regularization methods and smoothing Newton methods. The global convergence of this method was proved under the assumption that F is a P_0-function and the solution set of the problem (1.1) is nonempty and bounded.

In this paper we propose a class of squared smoothing Newton methods for the box constrained variational inequality problem and present the numerical results of this class of methods. This class of squared smoothing Newton methods is a regularized version of the class of smoothing Newton methods proposed in [25]. In the next section we will give some definitions. This class of squared smoothing Newton methods will be proposed in section 3. In section 4 we will report numerical results of these methods. We then make some final remarks in section 5.

To ease our discussion, we introduce some notation here: If $u \in \Re^n$, $\mathrm{diag}(u)$ is the diagonal matrix whose i-th diagonal element is u_i. For a continuously differentiable function $\Phi : \Re^m \to \Re^m$, we denote the Jacobian of Φ at $x \in \Re^m$ by $\Phi'(x)$, whereas the transposed Jacobian as $\nabla\Phi(x)$. $\|\cdot\|$ denotes the Euclidean norm. If X is a subset in $\Re^n$, we denote by $intX$ the interior of X. If V is an $m \times m$ matrix with entries V_{jk}, $j,k = 1,\ldots,m$, and $\mathcal{J}$ and $\mathcal{K}$ are index sets such that $\mathcal{J},\mathcal{K} \subseteq \{1,\ldots,m\}$, we denote by $V_{\mathcal{J}\mathcal{K}}$ the $|\mathcal{J}| \times |\mathcal{K}|$ sub-matrix of V consisting of entries V_{jk}, $j \in \mathcal{J}$, $k \in \mathcal{K}$. If $V_{\mathcal{J}\mathcal{J}}$ is nonsingular, we denote by $V/V_{\mathcal{J}\mathcal{J}}$ the Schur-complement of $V_{\mathcal{J}\mathcal{J}}$ in V, i.e., $V/V_{\mathcal{J}\mathcal{J}} := V_{\mathcal{K}\mathcal{K}} - V_{\mathcal{K}\mathcal{J}}V_{\mathcal{J}\mathcal{J}}^{-1}V_{\mathcal{J}\mathcal{K}}$, where $\mathcal{K} = \{1,\ldots,m\}\backslash\mathcal{J}$.

2 SOME PRELIMINARIES

We first restate some definitions.

Definition 2.1 *A matrix $V \in \Re^{n\times n}$ is called a*

(a) P_0-matrix if, for every $x \in \Re^n$ with $x \neq 0$, there is an index $i_0 = i_0(x)$ with

$$x_{i_0} \neq 0 \quad \textit{and} \quad x_{i_0}[Vx]_{i_0} \geq 0;$$

(b) P-matrix if, for every $x \in \Re^n$ *with* $x \neq 0$*, it holds that*

$$\max_i x_i[Vx]_i > 0.$$

Definition 2.2 *A function* $F : D \to \Re^n$*,* $D \subseteq \Re^n$*, is called a*

(a) P_0*-function on D if, for all* $x, y \in D$ *with* $x \neq y$*, there is an index* $i_0 = i_0(x, y)$ *with*

$$x_{i_0} \neq y_{i_0} \quad \text{and} \quad (x_{i_0} - y_{i_0})[F_{i_0}(x) - F_{i_0}(y)] \geq 0;$$

(b) P-function on D if, for all $x, y \in D$ *with* $x \neq y$*, it holds that*

$$\max_i (x_i - y_i)[F_i(x) - F_i(y)] > 0;$$

(c) uniform P-function on D if there is a constant $\mu > 0$ *such that*

$$\max_i (x_i - y_i)[F_i(x) - F_i(y)] \geq \mu \|x - y\|^2$$

holds for all $x, y \in D$*.*

Definition 2.3 *A function* $F : D \to \Re^n$*,* $D \subseteq \Re^n$*, is called a*

(a) monotone function on D if, for all $x, y \in D$ *with* $x \neq y$*,*

$$(x - y)^T[F(x) - F(y)] \geq 0;$$

(b) strictly monotone function on D if, for all $x, y \in D$ *with* $x \neq y$*,*

$$(x - y)^T[F(x) - F(y)] > 0.$$

It is known that every uniform P-function is P-function and every P-function is a P_0-function. Moreover, the Jacobian of a continuously differentiable P_0-function (uniform P-function) at a point is a P_0-matrix (P-matrix).

We now restate the definition of $p(u, x)$, $(u, x) \in \Re^n \times \Re^n$, given in [25]. For each $i \in N$, $p_i(u, x) = q(u_i, a_i, b_i, x_i)$ and for any $(\mu, c, d, w) \in \Re \times \{\Re \cup \{-\infty\}\} \times \{\Re \cup \{\infty\}\} \times \Re$ with $c \leq d$, $q(\mu, c, d, w)$ is defined by

$$q(\mu, c, d, w) = \begin{cases} \phi(|\mu|, c, d, w) & \text{if } \mu \neq 0 \\ \Pi_{[c,d]\cap\Re}(w) & \text{if } \mu = 0 \end{cases}, \tag{2.1}$$

and $\phi(\mu, c, d, w)$, $(\mu, w) \in \Re_{++} \times \Re$ is a Gabriel-Moré smoothing approximation function [14]. The definition of $\phi(\cdot)$ is as follows: Let $\rho : \Re \to \Re_+$ be a density function, i.e., $\rho(s) \geq 0$ and $\int_{-\infty}^{\infty} \rho(s)ds = 1$, with a bounded absolute mean, that is

$$\kappa := \int_{-\infty}^{\infty} |s|\rho(s)ds < \infty. \tag{2.2}$$

For any three numbers $c \in \Re \cup \{-\infty\}$, $d \in \Re \cup \{\infty\}$ with $c \leq d$ and $e \in \Re$, the median function mid$(\cdot)$ is defined by

$$\text{mid}(c, d, e) = \Pi_{[c,d]\cap\Re}(e) = \begin{cases} c & \text{if } e < c \\ e & \text{if } c \leq e \leq d \\ d & \text{if } d < e \end{cases}.$$

Then the Gabriel-Moré smoothing function $\phi(\mu, c, d, w)$ for $\Pi_{[c,d]\cap\Re}(w)$ [14] is defined by

$$\phi(\mu, c, d, w) = \int_{-\infty}^{\infty} \text{mid}(c, d, w - \mu s)\rho(s)ds, \quad (\mu, w) \in \Re_{++} \times \Re. \tag{2.3}$$

If $c = -\infty$ and/or $d = \infty$, the value of ϕ takes the limit of ϕ as $c \to -\infty$ and/or $d \to \infty$, correspondingly. For example, if c is finite and $d = \infty$, then

$$\phi(\mu, c, \infty, w) = \lim_{d' \to \infty} \phi(u, c, d', w), \quad (u, w) \in \Re_{++} \times \Re.$$

For the sake of convenience, let $\phi_{cd} : \Re_{++} \times \Re \to \Re$ be defined by

$$\phi_{cd}(\mu, w) := \phi(\mu, c, d, w), \quad (\mu, w) \in \Re_{++} \times \Re \tag{2.4}$$

and for any given $\mu \in \Re_{++}$, let $\phi_{\mu cd} : \Re \to \Re$ be defined by

$$\phi_{\mu cd}(w) := \phi(\mu, c, d, w), \quad w \in \Re. \tag{2.5}$$

Lemma 2.1 *[14, Lemma 2.3] For any given $\mu > 0$, the mapping $\phi_{\mu cd}(\cdot)$ is continuously differentiable with*

$$\phi'_{\mu cd}(w) = \int_{(w-d)/\mu}^{(w-c)/\mu} \rho(s)ds,$$

where $\phi_{\mu cd}(\cdot)$ is defined by (2.5). In particular, $\phi'_{\mu cd}(w) \in [0, 1]$.

Lemma 2.2 *[25, Lemma 2.2] The mapping $\phi_{cd}(\cdot)$ defined by (2.4) is Lipschitz continuous on $\Re_{++} \times \Re$ with Lipschitz constant $L := 2\max\{1, \kappa\}$.*

Let $q_{cd} : \Re^2 \to \Re$ be defined by

$$q_{cd}(\mu, w) = q(\mu, c, d, w), \quad (\mu, w) \in \Re^2, \tag{2.6}$$

where $q(\mu, c, d, w)$ is defined by (2.1).

Lemma 2.3 *[25, Lemma 2.3] The mapping $q_{cd}(\cdot)$ is globally Lipschitz continuous on $\Re^2$ with the same Lipschitz constant as in Lemma 2.2.*

Some most often used Gabriel-Moré smoothing functions like the neural networks smoothing function, the Chen-Harker-Kanzow-Smale smoothing function and the uniform smoothing function are discussed in [25].

In this paper, unless otherwise stated, we always assume that $c \in \Re \cup \{-\infty\}$, $d \in \Re \cup \{\infty\}$ and $c \leq d$. By Lemma 2.2 of [14], for any $(\mu, w) \in \Re_{++} \times \Re$,

$$\phi(\mu, c, d, w) \in [c, d] \cap \Re,$$

and so, for any $(u,x) \in \Re^n \times \Re^n$,

$$p(u,x) \in X. \tag{2.7}$$

Then the mapping $G(\cdot)$ defined in (1.5) is well defined on $\Re^{2n}$ while $F(\cdot)$ is only required to have definition on X, the feasible region.

From Lemma 2.1, for any given $\mu \in \Re_{++}$, $\phi_{\mu cd}(\cdot)$ is continuously differentiable at any $w \in \Re$. Moreover, for several most often used Gabriel-Moré smoothing functions it can be verified that $\phi_{cd}(\cdot)$ is also continuously differentiable at any $(\mu, w) \in \Re_{++} \times \Re$. In this paper, we are interested in smoothing functions with this property, which we make as an assumption.

Assumption 2.1 *The function $\phi_{cd}(\cdot)$ is continuously differentiable at any $(\mu, w) \in \Re_{++} \times \Re$.*

Let $z := (u,x) \in \Re^n \times \Re^n$. For some $\lambda \geq 0$, define $\bar{G} : \Re^{2n} \to \Re^n$ by

$$\bar{G}_i(z) = G_i(z) + \lambda u_i p_i(u,x), \; i \in N,$$

where $G(\cdot)$ is defined in (1.5). Define $H : \Re^{2n} \to \Re^{2n}$ by

$$H(z) := \begin{pmatrix} u \\ \bar{G}(z) \end{pmatrix}. \tag{2.8}$$

Then H is continuously differentiable at any $z = (u,x) \in \Re^n_{++} \times \Re^n$ if Assumption 2.1 is satisfied.

For any $u \in \Re^n_{++}$ and $x \in \Re^n$, define $c(u,x), d(u,x) \in \Re^n$ by

$$c_i(u,x) = \partial p_i(u,x)/\partial x_i, \; d_i(u,x) = \partial p_i(u,x)/\partial u_i, \; i \in N.$$

For any $u \in \Re^n_{++}$ and $x \in \Re^n$, define $D^U(u,x)$, $C^X(u,x)$, $P^N(u,x)$, $P^U(u,x)$, $P^X(u,x) \in \Re^n$ by

$$D_i^U(u,x) = \begin{cases} 0 & \text{if } i \in I_\infty \cup I_{ab} \\ d_i(u, F(p(u,x)) + a) & \text{if } i \in I_a \\ d_i(u, F(p(u,x)) + b) & \text{if } i \in I_b \end{cases},$$

$$C_i^X(u,x) = \begin{cases} 0 & \text{if } i \in I_\infty \cup I_{ab} \\ c_i(u, F(p(u,x)) + a) & \text{if } i \in I_a \\ c_i(u, F(p(u,x)) + b) & \text{if } i \in I_b \end{cases},$$

$$P_i^N(u,x) = \begin{cases} 0 & \text{if } i \in I_\infty \cup I_{ab} \\ p_i(u,x) - a_i & \text{if } i \in I_a \\ b_i - p_i(u,x) & \text{if } i \in I_b \end{cases},$$

$$P_i^U(u,x) = \begin{cases} 0 & \text{if } i \in I_\infty \cup I_{ab} \\ d_i(u,x)[p_i(u, F(p(u,x)) + a) - a_i] & \text{if } i \in I_a \\ d_i(u,x)[b_i - p_i(u, F(p(u,x)) + b)] & \text{if } i \in I_b \end{cases}$$

and

$$P_i^X(u,x) = \begin{cases} 0 & \text{if } i \in I_\infty \cup I_{ab} \\ c_i(u,x)[p_i(u,F(p(u,x))+a) - a_i] & \text{if } i \in I_a \\ c_i(u,x)[b_i - p_i(u,F(p(u,x))+b)] & \text{if } i \in I_b \end{cases}, \quad i \in N.$$

Theorem 2.1 *Suppose that Assumption 2.1 holds for a chosen smoothing function*
$\phi(\mu,c,d,w)$, $(\mu,w) \in \Re_{++} \times \Re$. *Then*

(i) *The mapping $H(\cdot)$ is continuously differentiable at any $z = (u,x) \in \Re^n_{++} \times \Re^n$ and*

$$H'(z) = \begin{pmatrix} I & 0 \\ \bar{G}'_u(z) & \bar{G}'_x(z) \end{pmatrix}, \tag{2.9}$$

where

$$\begin{aligned} \bar{G}'_u(z) &= [F'(p(z)) - I + \lambda \mathrm{diag}(u)]\mathrm{diag}(d(u,x)) + \lambda \mathrm{diag}(p(z)) \\ &\quad + \alpha \mathrm{diag}(P^U(u,x)) + \alpha \mathrm{diag}(P^N(u,x))\mathrm{diag}(D^U(u,x)) \\ &\quad + \alpha \mathrm{diag}(P^N(u,x))\mathrm{diag}(C^X(u,x))F'(p(z))\mathrm{diag}(d(u,x)), \end{aligned}$$

$$\begin{aligned} \bar{G}'_x(z) &= \{[I + \alpha \mathrm{diag}(P^N(u,x))\mathrm{diag}(C^X(u,x))]F'(p(z)) \\ &\quad + \lambda \mathrm{diag}(u)\}\, \mathrm{diag}(c(u,x)) \\ &\quad + I - \mathrm{diag}(c(u,x)) + \alpha \mathrm{diag}(P^X(u,x)) \end{aligned}$$

and for each $i \in N$, $c_i(u,x) \in [0,1]$.

(ii) *If $\lambda > 0$ and for some $z \in \Re^n_{++} \times \Re^n$, $F'(p(z))$ is a P_0-matrix, then $H'(z)$ is nonsingular.*

Proof. (i) Since Assumption 2.1 is satisfied for $\phi(\cdot)$, from the definition, we know that $H(\cdot)$ is continuously differentiable at any $z = (u,x) \in \Re^n_{++} \times \Re^n$. By direct computation we have (2.9). From Lemma 2.1 and the definition of $p_i(\cdot)$, $c_i(u,x) \in [0,1]$, $i \in N$.
(ii) Suppose that $\lambda > 0$ and for some $z \in \Re^n_{++} \times \Re^n$, $F'(p(z))$ is a P_0-matrix. From (i) and the definition of $C^X(u,x), P^N(u,x)$ and $P^X(u,x)$, we have $C_i^X(u,x) \in [0,1], P_i^N(u,x) \geq 0$ and $P_i^X(u,x) \geq 0$, for $i \in N$. Then $Q = I + \alpha \mathrm{diag}(P^N(u,x))\mathrm{diag}(C^X(u,x))$ is a positive diagonal matrix. So $QF'(p(z))$ is a P_0-matrix and $QF'(p(z)) + \lambda \mathrm{diag}(u)$ is a P-matrix. From [7, Lemma 2] we have that $\bar{G}'_x(z)$ is nonsingular. It then follows from (2.9) that $H'(z)$ is also nonsingular. ∎

In order to design high-order convergent Newton methods we need the concept of semismoothness. Semismoothness was originally introduced by Mifflin [19] for functionals. Convex functions, smooth functions, and piecewise linear functions are examples of semismooth functions. The composition of semismooth functions is still a semismooth function [19]. In [24], Qi and J. Sun extended the definition of semismooth functions to $\Phi : \Re^{m_1} \to \Re^{m_2}$. A locally Lipschitz continuous vector valued function $\Phi : \Re^{m_1} \to \Re^{m_2}$ has a generalized Jacobian $\partial\Phi(x)$ as in Clarke [11]. Φ is said to be *semismooth* at $x \in \Re^{m_1}$, if

$$\lim_{\substack{V \in \partial\Phi(x+th') \\ h' \to h,\ t \downarrow 0}} \{Vh'\} \tag{2.10}$$

exists for any $h \in \Re^{m_1}$. It has been proved in [24] that Φ is semismooth at x if and only if all its component functions are. Also, $\Phi'(x;h)$, the directional derivative of Φ at x in the direction h, exists and equals the limit in (2.10) for any $h \in \Re^{m_1}$ if Φ is semismooth at x.

Lemma 2.4 *[24] Suppose that $\Phi : \Re^{m_1} \to \Re^{m_2}$ is a locally Lipschitzian function and semismooth at x. Then*

(i) *for any $V \in \partial\Phi(x+h)$, $h \to 0$,*

$$Vh - \Phi'(x;h) = o(\|h\|);$$

(ii) *for any $h \to 0$,*

$$\Phi(x+h) - \Phi(x) - \Phi'(x;h) = o(\|h\|).$$

The following lemma is extracted from Theorem 2.3 of [24].

Lemma 2.5 *Suppose that $\Phi : \Re^{m_1} \to \Re^{m_2}$ is a locally Lipschitzian function. Then the following two statements are equivalent:*

(i) *$\Phi(\cdot)$ is semismooth at x.*

(ii) *Φ is directionally differentiable at x, and for any $V \in \partial\Phi(x+h)$, $h \to 0$,*

$$Vh - \Phi'(x;h) = o(\|h\|).$$

A stronger notion than semismoothness is strong semismoothness. $\Phi(\cdot)$ is said to be *strongly semismooth* at x if Φ is semismooth at x and for any $V \in \partial\Phi(x+h)$, $h \to 0$,

$$Vh - \Phi'(x;h) = O(\|h\|^2).$$

(Note that in [22] and [24] different names for strong semismoothness are used.) A function Φ is said to be a (strongly) semismooth function if it is (strongly) semismooth everywhere.

Recall that from Lemma 2.3 the function $q_{cd}(\cdot)$ defined by (2.6) is globally Lipschitz continuous on $\Re^2$. Then, from Lemma 2.5 and the definition of strong semismoothness, we can prove in the above mentioned three usual cases [25] that $q_{cd}(\cdot)$ is strongly semismooth at $x \in \Re^2$, i.e., for any $V \in \partial q_{cd}(x+h)$, $h \to 0$,

$$Vh - q'_{cd}(x;h) = O(\|h\|^2). \tag{2.11}$$

3 A CLASS OF SQUARED SMOOTHING NEWTON METHODS

Choose $\bar{u} \in \Re^n_{++}$ and $\gamma \in (0,1)$ such that $\gamma\|\bar{u}\| < 1$. Let $\bar{z} := (\bar{u}, 0) \in \Re^n \times \Re^n$. Define the merit function $\psi : \Re^{2n} \to \Re_+$ by

$$\psi(z) := \|H(z)\|^2$$

and define $\beta : \Re^{2n} \to \Re_+$ by

$$\beta(z) := \gamma \min\{1, \psi(z)\}.$$

Let

$$\Omega := \{z = (u, x) \in \Re^n \times \Re^n | \; u \geq \beta(z)\bar{u}\}.$$

Then, because for any $z \in \Re^{2n}$, $\beta(z) \leq \gamma < 1$, it follows that for any $x \in \Re^n$,

$$(\bar{u}, x) \in \Omega.$$

Proposition 3.1 *The following relations hold:*

(i) $H(z) = 0 \Longleftrightarrow \beta(z) = 0 \Longleftrightarrow H(z) = \beta(z)\bar{z}$.

(ii) $H(z) = 0 \Longrightarrow u = 0$ *and* $y = \Pi_X(x)$ *is a solution of (1.1).*

(iii) *If* $x = y - F(y)$*, where* y *is a solution of (1.1), then* $H(0, x) = 0$.

The proof of this proposition is similar to that of Proposition 4.1 in [25], so we omit it.

Algorithm 3.1

Step 0. *Choose constants* $\delta \in (0,1)$, $\sigma \in (0, 1/2)$, $\alpha \geq 0$ *and* $\lambda \geq 0$. *Let* $u^0 := \bar{u}$, $x^0 \in \Re^n$ *be an arbitrary point and* $k := 0$.

Step 1. *If* $H(z^k) = 0$ *then stop. Otherwise, let* $\beta_k := \beta(z^k)$.

Step 2. *Compute* $\Delta z^k := (\Delta u^k, \Delta x^k) \in \Re^n \times \Re^n$ *by*

$$H(z^k) + H'(z^k)\Delta z^k = \beta_k \bar{z}. \tag{3.1}$$

Step 3. *Let* l_k *be the smallest nonnegative integer* l *satisfying*

$$\psi(z^k + \delta^l \Delta z^k) \leq [1 - 2\sigma(1 - \gamma\|\bar{u}\|)\delta^l]\psi(z^k). \tag{3.2}$$

Define $z^{k+1} := z^k + \delta^{l_k}\Delta z^k$.

Step 4. *Replace* k *by* $k+1$ *and go to Step 1.*

Remark. Algorithm 3.1 is actually the smoothing Newton method proposed in [25] for the case that $\alpha = 0$ and $\lambda = 0$. When $\lambda > 0$, Algorithm 3.1 has better properties than the original version of the smoothing Newton method given in [25]. The parameter α is introduced in order to improve the numerical performance.

Lemma 3.1 *Suppose that F is a P_0-function on X, $\lambda > 0$ and that $\bar{u}, \tilde{u} \in \Re^n$ are two positive vectors such that $\bar{u} \geq \tilde{u} > 0$. Then for any sequence $\{z^k = (u^k, x^k)\}$ such that $\tilde{u} \leq u^k \leq \bar{u}$ and $\|x^k\| \to +\infty$ we have*

$$\lim_{k\to\infty} \psi(z^k) = +\infty. \tag{3.3}$$

Proof. For the sake of contradiction, suppose that there exists a sequence $\{z^k = (u^k, x^k) \in \Re^n \times \Re^n\}$ such that $\tilde{u} \leq u^k \leq \bar{u}$, $\|x^k\| \to \infty$ and $\psi(z^k)$ is bounded. It is easy to prove that

$$|\mathrm{mid}(a_i, b_i, x_i^k)| \to \infty \Longrightarrow |x_i^k| \to \infty \ \text{ and } \ |x_i^k - \mathrm{mid}(a_i, b_i, x_i^k)| \to 0, \quad i \in N. \tag{3.4}$$

From Lemma 2.3 and the definition of $p(\cdot)$, there exists a constant $L' > 0$ such that

$$|p_i(u^k, x^k) - \mathrm{mid}(a_i, b_i, x_i^k)| \leq L'|u_i^k|, \quad i \in N. \tag{3.5}$$

From (3.4) and (3.5) we have

$$|p_i(u^k, x^k)| \to \infty \Longrightarrow \{|x_i^k - p_i(u^k, x^k)|\} \text{ is bounded.} \tag{3.6}$$

Define the index set J by $J := \{i \in N \mid \{p_i(u^k, x^k)\} \text{ is unbounded}\}$. Then it follows that $J \neq \emptyset$ because otherwise $\|\bar{G}(z^k)\| = \|F(p(z^k)) + x^k - p(z^k) + \alpha S(z^k) + \lambda \mathrm{diag}(u^k)p(z^k)\| \to \infty$. Let $\bar{z}^k = (\bar{u}^k, \bar{x}^k) \in \Re^n \times \Re^n$ be defined by

$$\bar{u}_i^k = \begin{cases} u_i^k & \text{if } i \notin J \\ 0 & \text{if } i \in J \end{cases}$$

and

$$\bar{x}_i^k = \begin{cases} x_i^k & \text{if } i \notin J \\ 0 & \text{if } i \in J \end{cases}, \quad i \in N.$$

Then

$$p_i(\bar{z}^k) = \begin{cases} p_i(z^k) & \text{if } i \notin J \\ \mathrm{mid}(a_i, b_i, 0) & \text{if } i \in J \end{cases}, \quad i \in N.$$

Hence $\{\|p(\bar{z}^k)\|\}$ is bounded. Because F is a P_0-function on X, we have

$$\begin{aligned} 0 &\leq \max_{i\in N}[p_i(z^k) - p_i(\bar{z}^k)][F_i(p(z^k)) - F_i(p(\bar{z}^k))] \\ &= \max_{i\in J}[p_i(z^k) - p_i(\bar{z}^k)][F_i(p(z^k)) - F_i(p(\bar{z}^k))] \\ &= [p_i(z^k) - p_i(\bar{z}^k)][F_i(p(z^k)) - F_i(p(\bar{z}^k))], \end{aligned} \tag{3.7}$$

where $i \in J$ is one of indices for which the maximum is attained, without loss of generality, assumed to be independent of k. Since $i \in J$, we have

$$|p_i(z^k)| \to \infty.$$

From (3.7) and the boundedness of $\{F_i(p(\bar{z}^k))\}$, we have that $F_i(p(z^k))$ does not tend to $-\infty$ if $p_i(z^k) \to +\infty$ and $F_i(p(z^k))$ does not tend to $+\infty$ if $p_i(z^k) \to -\infty$.

We now consider two cases.

Case 1: $p_i(z^k) \to +\infty$.

In this case, we have that $i \in I_\infty \cup I_a$. Since $S_i(z^k) \geq 0$, from (3.6) we have

$$\bar{G}_i(z^k) = F_i(p(z^k)) + x_i^k - p_i(z^k) + \alpha S_i(z^k) + \lambda u_i^k p_i(z^k) \to +\infty.$$

Case 2: $p_i(z^k) \to -\infty$.

In this case, $i \in I_\infty \cup I_b$. Since $S_i(z^k) \leq 0$, from (3.6) we have

$$\bar{G}_i(z^k) = F_i(p(z^k)) + x_i^k - p_i(z^k) + \alpha S_i(z^k) + \lambda u_i^k p_i(z^k) \to -\infty.$$

In either case we get $\psi(z^k) \to +\infty$, which is a contradiction. So we complete our proof. ∎

Remark. Lemma 3.1 is not true if $\tilde{u} = 0$ even if F is strictly monotone. To see this, we may consider the function $F(x) = e^x - 1$, $x \in \Re$. This function was provided by H.-D. Qi [21]. Suppose that $X = \Re$. Then $\psi(z) = u^2 + F(x)^2$ and when $x \to -\infty$ and $u = 0$, $\psi(z) \to 1$. This clearly shows that ψ may have unbounded level sets. However, if the solution set of (1.1) is bounded, we can prove the global convergence of our methods (see Theorem 3.1) under the assumption that F is a P_0-function on X.

Assumption 3.1 **(i)** *F is a P_0-function on X.*

(ii) *The solution set of the problem (1.1) is nonempty and bounded.*

Theorem 3.1 *Suppose that Assumptions 2.1 and 3.1 are satisfied and $\lambda > 0$. Then the infinite sequence $\{z^k\}$ generated by Algorithm 3.1 is bounded and each accumulation point $\bar{z}$ of $\{z^k\}$ is a solution of $H(z) = 0$.*

Proof. By combining Lemma 3.1 and the proof of Theorem 4.1 [29] and Theorem 4.6 [20], we can prove this theorem. We omit the details. ∎

Theorem 3.2 *Suppose that Assumptions 2.1 and 3.1 are satisfied, $\lambda > 0$ and z^* is an accumulation point of the infinite sequence $\{z^k\}$ generated by Algorithm 3.1. Suppose that H is semismooth at z^* and that all $V \in \partial H(z^*)$ are nonsingular. Then the whole sequence $\{z^k\}$ converges to z^*,*

$$\|z^{k+1} - z^*\| = o(\|z^k - z^*\|) \tag{3.8}$$

and

$$u_i^{k+1} = o(u_i^k), \ i \in N. \tag{3.9}$$

Furthermore, if H is strongly semismooth at z^*, then

$$\|z^{k+1} - z^*\| = O(\|z^k - z^*\|^2) \tag{3.10}$$

and

$$u_i^{k+1} = O\left((u_i^k)^2\right), \ i \in N. \tag{3.11}$$

Proof. See [25, Theorem 7.1] for a similar proof. ∎

Next, we study under what conditions all the matrices $V \in \partial H(z^*)$ are nonsingular at a solution point $z^* = (u^*, x^*) \in \Re^n \times \Re^n$ of $H(z) = 0$. Apparently, $u^* = 0$ and x^* is a solution of (1.3). For convenience of handling notation we denote

$$\mathcal{I} := \{i |\ a_i < x_i^* < b_i \ \&\ F_i(\Pi_X(x^*)) = 0, \ i \in N\},$$

$$\begin{aligned} \mathcal{J} : &= \{i |\ x_i^* = a_i \ \&\ F_i(\Pi_X(x^*)) = 0, \ i \in N\} \\ &\cup \{i |\ x_i^* = b_i \ \&\ F_i(\Pi_X(x^*)) = 0, \ i \in N\} \end{aligned}$$

and

$$\begin{aligned} \mathcal{K} : &= \{i |\ x_i^* < a_i \ \&\ F_i(\Pi_X(x^*)) > 0, \ i \in N\} \\ &\cup \{i |\ x_i^* > b_i \ \&\ F_i(\Pi_X(x^*)) < 0, \ i \in N\}. \end{aligned}$$

Then

$$\mathcal{I} \cup \mathcal{J} \cup \mathcal{K} = N.$$

By rearrangement we assume that $\nabla F(\Pi_X(x^*))$ can be rewritten as

$$\nabla F(\Pi_X(x^*)) = \begin{pmatrix} \nabla F(\Pi_X(x^*))_{\mathcal{I}\mathcal{I}} & \nabla F(\Pi_X(x^*))_{\mathcal{I}\mathcal{J}} & \nabla F(\Pi_X(x^*))_{\mathcal{I}\mathcal{K}} \\ \nabla F(\Pi_X(x^*))_{\mathcal{J}\mathcal{I}} & \nabla F(\Pi_X(x^*))_{\mathcal{J}\mathcal{J}} & \nabla F(\Pi_X(x^*))_{\mathcal{J}\mathcal{K}} \\ \nabla F(\Pi_X(x^*))_{\mathcal{K}\mathcal{I}} & \nabla F(\Pi_X(x^*))_{\mathcal{K}\mathcal{J}} & \nabla F(\Pi_X(x^*))_{\mathcal{K}\mathcal{K}} \end{pmatrix}.$$

BVIP is said to be R-regular at x^* if $\nabla F(\Pi_X(x^*))_{\mathcal{I}\mathcal{I}}$ is nonsingular and its Schur-complement in the matrix

$$\begin{pmatrix} \nabla F(\Pi_X(x^*))_{\mathcal{I}\mathcal{I}} & \nabla F(\Pi_X(x^*))_{\mathcal{I}\mathcal{J}} \\ \nabla F(\Pi_X(x^*))_{\mathcal{J}\mathcal{I}} & \nabla F(\Pi_X(x^*))_{\mathcal{J}\mathcal{J}} \end{pmatrix}$$

is a P-matrix, see [28].

Proposition 3.2 *Suppose that $z^* = (u^*, x^*) \in \Re^n \times \Re^n$ is a solution of $H(z) = 0$. If BVIP is R-regular at x^*, then all $V \in \partial H(z^*)$ are nonsingular.*

Proof. It is easy to see that for any $V \in \partial H(z^*)$ there exists a matrix $W = (W_u, W_x) \in \partial \bar{G}(z^*)$ with $W_u, W_x \in \Re^{n \times n}$ such that

$$V = \begin{pmatrix} I & 0 \\ W_u & W_x \end{pmatrix}.$$

Hence, proving that V is nonsingular is equivalent to proving that W_x is nonsingular. For any $U = (U_u, U_x) \in \partial p(z^*)$, by the definition of p, we have

$$U \in \partial p_1(z^*) \times \partial p_2(z^*) \times \cdots \times \partial p_n(z^*) = \partial p(z^*).$$

Then for each $i \in N$, the ith row of U, $U_i \in \partial p_i(z^*)$. Apparently, from the definition of p and Lemma 2.1,

$$U_x = \text{diag}\{(u_x)_i,\ i \in N\},$$

where $(u_x)_i$ is defined by

$$\begin{cases} (u_x)_i = 1 & \text{if } i \in \mathcal{I} \\ (u_x)_i \in [0,1] & \text{if } i \in \mathcal{J} \\ (u_x)_i = 0 & \text{if } i \in \mathcal{K} \end{cases}.$$

Define $P^N(z^*) \in \Re^n$ by

$$P_i^N(z^*) = \begin{cases} 0 & \text{if } \ i \in I_\infty \cup I_{ab} \\ p_i(z^*) - a_i & \text{if } \ i \in I_a \\ b_i - p_i(z^*) & \text{if } \ i \in I_b \end{cases}, \ i \in N.$$

For any $W = (W_u, W_x) \in \partial\bar{G}(z^*)$ with $W_u, W_x \in \Re^{n\times n}$ there exist $U = (U_u, U_x) \in \partial p(z^*)$ and $C^X(z^*) \in \Re^n$ such that

$$\begin{aligned} W_x &= F'(p(z^*))U_x + I - U_x \\ &\quad +\alpha\text{diag}(P^N(z^*))\text{diag}(C^X(z^*))F'(p(z^*))U_x \\ &\quad +\alpha\text{diag}(P^X(z^*)), \end{aligned}$$

where

$$P_i^X(z^*) = \begin{cases} 0 & \text{if } \ i \in I_\infty \cup I_{ab} \\ (u_x)_i[(\Pi_X(F(p(z^*)) + a)_i - a_i] & \text{if } \ i \in I_a \\ (u_x)_i[b_i - (\Pi_X(F(p(z^*)) + b)_i] & \text{if } \ i \in I_b \end{cases}, \ i \in N$$

and

$$C_i^X(z^*) = \begin{cases} 0 & \text{if } \ i \in I_\infty \cup I_{ab} \\ A_i & \text{if } \ i \in I_a \\ B_i & \text{if } \ i \in I_b \end{cases}, \quad i \in N,$$

where

$$A_i = \begin{cases} 1 & \text{if } \ F_i(\Pi_X(x^*)) > 0 \\ 0 & \text{if } \ F_i(\Pi_X(x^*)) < 0 \\ \epsilon_1 \in [0,1] & \text{if } \ F_i(\Pi_X(x^*)) = 0 \end{cases}$$

and

$$B_i = \begin{cases} 0 & \text{if } \ F_i(\Pi_X(x^*)) > 0 \\ 1 & \text{if } \ F_i(\Pi_X(x^*)) < 0 \\ \epsilon_2 \in [0,1] & \text{if } \ F_i(\Pi_X(x^*)) = 0 \end{cases}.$$

Let $D = \alpha\text{diag}(P^N(z^*))\text{diag}(C^X(z^*))$. We have that D is a nonnegative diagonal matrix. By inspecting the structure of P^X we have that $P_i^X = 0$, for all $i \in N$. Then we have

$$W_x = (I + D)F'(p(z^*))U_x + I - U_x.$$

Let $Q = W_x^T(I+D)^{-1}$. Then

$$Q = U_x \nabla F(p(z^*)) + (I - U_x)(I+D)^{-1}.$$

Thus, for each $i \in \mathcal{J}$, there exists $\lambda_i \in [0,1]$ such that

$$Q_i = \begin{cases} \nabla F(p(z^*))_i & \text{if } i \in \mathcal{I} \\ \lambda_i \nabla F(p(z^*))_i + (1-\lambda_i)(1+D_{ii})^{-1}e_i & \text{if } i \in \mathcal{J} \\ (1+D_{ii})^{-1}e_i & \text{if } i \in \mathcal{K} \end{cases},$$

where e_i is the ith unit row vector of $\Re^n$ and $\nabla F(p(z^*))_i$ is the ith row vector of $\nabla F(p(z^*))$, $i \in N$. Then, by [13, Proposition 3.2] we can prove that Q, and so W_x, is nonsingular under the assumption of R-regularity (note that $p(z^*) = \Pi_X(x^*))$. Hence, any $V \in \partial H(z^*)$ is nonsingular. So, we complete our proof. ∎

4 NUMERICAL RESULTS

Algorithm 3.1 was implemented in MATLAB and was run on a SUN Sparc Server 3002 for all test problems with all available starting points from the test problem collections GAMSLIB and MCPLIB [1] (note that there are three starting points in `ehl_kost` with the same data and so we only list the results for `ehl_kost` with the first starting point in Table 1.2). Throughout the computational experiments, unless otherwise stated, we chose the Chen-Harker-Kanzow-Smale smoothing function and used the following parameters:

$$\delta = 0.5, \sigma = 10^{-4}, \bar{u} = 0.2e, \gamma = \min\{10^{-5}, 0.2/\|\bar{u}\|\} \text{ and } \lambda = 0.05,$$

where e is the vector of all ones.

To improve the numerical behaviour of Algorithm 3.1, we replaced the standard (monotone) Armijo-rule by a nonmonotone line search as described in Grippo, Lampariello and Lucidi [15], i.e., we computed the smallest nonnegative integer l such that

$$z^k + \delta^l \Delta z^k \in \Omega \tag{4.1}$$

and

$$\psi(z^k + \delta^l \Delta z^k) \le \mathcal{W}_k - 2\sigma(1 - \gamma\|\bar{u}\|)\delta^l \psi(z^k), \tag{4.2}$$

where $\mathcal{W}_k$ is given by

$$\mathcal{W}_k = \max_{j=k-m_k,\ldots,k} \psi(z^j)$$

and where, for given nonnegative integers m and s, we set

$$m_k = 0$$

if $k \le s$ at the kth iteration, whereas we set

$$m_k := \min\{m_{k-1} + 1, m\}$$

at all other iterations. In our implementation, we use

$$m = 8 \quad \text{and} \quad s = 2.$$

We terminated our iteration when one of the following conditions was satisfied

$$k > 3000, R(x^k) := \|p(z^k) - \Pi_X[p(z^k) - F(p(z^k))]\|_\infty \leq 10^{-6} \text{ or } ls > 80,$$

where ls was the number of line search at each step.

Using this algorithmic environment, we made some preliminary test runs using different values of the parameter α. In view of these preliminary experiments, it seems that α should be large if the iteration point is far away from a solution and α should be reduced if the iteration point is getting closer to a solution of the problem. This motivated us to use a dynamic choice of α for our test runs. More precisely, we updated α using the following rules:

(a) Set $\alpha = 10^4$ at the beginning of each iteration.

(b) If $R(x^k) < 10$, then set $\alpha = 100$.

(c) If $R(x^k) < 10^{-2}$ or $k \geq 80$, then set $\alpha = 10^{-3}$.

(d) If $R(x^k) < 10^{-3}$, then set $\alpha = 10^{-6}$.

The numerical results which we obtained are summarized in Tables 1.1-1.3. In these tables, **Dim** denotes the number of the variables in the problem, **Start. point** denotes the starting point, **Iter** denotes the number of iterations, which is also equal to the number of Jacobian evaluations for the function F, **NF** denotes the number of function evaluations for the function F and $R(x^k)$ denotes the value of $R(x)$ at the final iteration.

The results reported in Tables 1.1-1.3 show that the squared smoothing Newton methods are extremely promising and robust. The algorithm was able to solve almost all the problems. There are just three problems with superscript triple-asterisk which our algorithm was not able to solve because the steplength was getting too small. They are **gemmge**, **vonthmcp** and **hydroc20**. However we can solve these problems if we change some of the parameters. The results reported for these three problems were obtained by letting $\alpha = 10^{-10}$ and $\lambda = 10^{-6}$ while keeping other parameters unchanged.

Lastly, it is deserved to point out that domain violation phenomenon does not occur during our computation because $p(u, x) \in intX$ for all $(u, x) \in \Re^n_{++} \times \Re^n$. This is a very nice feature of our methods.

5 CONCLUSIONS

In this paper we present a regularized version of a class of squared smoothing Newton methods, originally proposed in [25], for the box constrained variational

Table 1.1 Numerical results for the problems from GAMSLIB

Problem	Dim	Iter	NF	$R(x^k)$
cafemge	101	29	30	8.0×10^{-7}
cammcp	242	8	9	9.1×10^{-8}
cammge	128	15	16	1.2×10^{-7}
cirimge	9	3	4	4.5×10^{-10}
co2mge	208	36	140	4.1×10^{-8}
dmcmge	170	13	21	3.4×10^{-13}
ers82mcp	232	6	8	1.9×10^{-9}
etamge	114	16	84	4.8×10^{-9}
finmge	153	10	11	7.8×10^{-10}
gemmcp	262	1	2	1.9×10^{-7}
gemmge***	178	15	18	6.6×10^{-7}
hansmcp	43	9	18	6.2×10^{-8}
hansmge	43	72	75	2.4×10^{-10}
harkmcp	32	14	18	6.5×10^{-13}
harmge	11	11	48	8.7×10^{-8}
kehomge	9	17	35	4.0×10^{-8}
kormcp	78	5	6	4.0×10^{-10}
mr5mcp	350	9	11	2.6×10^{-9}
nsmge	212	13	14	7.5×10^{-10}
oligomcp	6	7	11	1.1×10^{-7}
sammge	23	4	5	6.1×10^{-8}
scarfmcp	18	10	14	4.9×10^{-9}
scarfmge	18	14	19	1.5×10^{-10}
shovmge	51	88	90	2.3×10^{-9}
threemge	9	9	10	1.6×10^{-10}
transmcp	11	5	24	4.6×10^{-10}
two3mcp	6	6	7	1.3×10^{-8}
unstmge	5	8	9	1.3×10^{-8}
vonthmcp***	125	2609	19066	5.9×10^{-9}
vonthmge	80	94	516	1.8×10^{-8}
wallmcp	6	4	5	1.1×10^{-8}

inequality problem. As can be seen from the numerical results, these methods are fairly robust and promising. The global convergence of these methods were proved under the assumption that F is a P_0-function on X and the solution set of the problem (1.1) is nonempty and bounded. This assumption may be the weakest one known in the literature.

In Algorithm 3.1 we always assume that the iteration matrix $H'(z)$ is nonsingular. This is guaranteed by assuming that F is a P_0-function on X. In this paper, We have not discussed how to handle the case that $H'(z)$ is singular. By introducing a gradient direction if necessary, Kanzow and Pieper [18] described a strategy for handling the singularity issue of the iteration matrices for the smoothing Newton method proposed in [9]. Whether or not the idea introduced in [18] is applicable to our method is an interesting question.

Acknowledgements

The authors would like to thank Robert S. Womersley for his help in preparing the computation and two referees for their helpful comments.

References

[1] S.C. Billups, S.P. Dirkse and M.C. Ferris, "A comparison of algorithms for large-scale mixed complementarity problems", *Computational Optimization and Applications, 7* (1997), 3–25.

[2] B. Chen and X. Chen, "A global linear and local quadratic continuation smoothing method for variational inequalities with box constraints", Preprint, Department of Management and Systems, Washington State University, Pullman, March 1997.

[3] B. Chen and X. Chen, "A global and local superlinear continuation-smoothing method for $P_0 + R_0$ and monotone NCP", *SIAM J. Optimization*, to appear.

[4] B. Chen, X. Chen and C. Kanzow, "A penalized Fischer-Burmeister NCP-function: Theoretical investigation and numerical results", AMR 97/28, Applied Mathematics Report, School of Mathematics, the University of New South Wales, Sydney 2052, Australia, September 1997.

[5] B. Chen and P.T. Harker, "A non-interior-point continuation method for linear complementarity problems", *SIAM Journal on Matrix Analysis and Applications, 14* (1993), 1168–1190.

[6] B. Chen and P.T. Harker, "Smooth approximations to nonlinear complementarity problems", *SIAM Journal on Optimization, 7* (1997), 403–420.

[7] B. Chen, P.T. Harker and M. C. Pınar, "Continuation methods for nonlinear complementarity problems via normal maps", Preprint, Department of Management and Systems, Washington State University, Pullman, November 1995.

[8] C. Chen and O.L. Mangasarian, "A class of smoothing functions for nonlinear and mixed complementarity problems", *Computational Optimization and Applications, 5* (1996), 97–138.

[9] X. Chen, L. Qi, and D. Sun, "Global and superlinear convergence of the smoothing Newton method and its application to general box constrained variational inequalities", *Mathematics of Computation, 67* (1998), 519-540.

[10] X. Chen and Y. Ye, "On homotopy-smoothing methods for variational inequalities", *SIAM J. Control Optim.*, to appear.

[11] F.H. Clarke, *Optimization and Nonsmooth Analysis*, Wiley, New York, 1983.

[12] F. Facchinei and C. Kanzow, "Beyond monotonicity in regularization methods for nonlinear complementarity problems", DIS Working Paper 11-97, Università di Roman "La Sapienza", Roma, Italy, May 1997.

[13] F. Facchinei and J. Soares, "A new merit function for nonlinear complementarity problems and a related algorithm", *SIAM Journal on Optimization, 7* (1997), 225–247.

[14] S.A. Gabriel and J.J. Moré, "Smoothing of mixed complementarity problems", in *Complementarity and Variational Problems: State of the Art*, M.C. Ferris and J.S. Pang, eds., SIAM, Philadelphia, Pennsylvania, pp. 105–116, 1997.

[15] L. Grippo, F. Lampariello and S. Lucidi, "A nonmonotone line search technique for Newton's method", *SIAM Journal on Numerical Analysis, 23* (1986), pp. 707-716.

[16] K. Hotta and A. Yoshise "Global convergence of a class of non-interior-point algorithms using Chen-Harker-Kanzow functions for nonlinear complementarity problems", Discussion Paper Series No. 708, Institute of Policy and Planning Sciences, University of Tsukuba, Tsukuba, Ibaraki 305, Japan, December 1996.

[17] H. Jiang, "Smoothed Fischer-Burmeister equation methods for the nonlinear complementarity problem", Preprint, Department of Mathematics, the University of Melbourne, Victoria 3052, Australia, June 1997.

[18] C. Kanzow and H. Pieper, "Jacobian smoothing methods for general complenmentarity problems", Technical Report 97-08, Computer Sciences Department, University of Wisconsin, Madison, WI, October 1997.

[19] R. Mifflin, "Semismooth and semiconvex functions in constrained optimization", *SIAM Journal Control and Optimization, 15* (1977), 957–972.

[20] H.-D. Qi, "A regularized smoothing Newton method for box constrained variational inequality problems with P_0 functions", Preprint, Institute of Computational Mathematics and Scientific/Engineering Computing, Chinese Academy of Sciences, July 1997.

[21] H.-D. Qi, Private communication, August 1997.

[22] L. Qi, "Convergence analysis of some algorithms for solving nonsmooth equations", *Mathematics of Operations Research, 18* (1993), 227–244.

[23] L. Qi and D. Sun, "Improving the convergence of non-interior point algorithms for nonlinear complementarity problems", *Mathematics of Computation*, to appear.

[24] L. Qi and J. Sun, "A nonsmooth version of Newton's method", *Mathematical Programming, 58* (1993), 353–367.

[25] L. Qi, D. Sun and G. Zhou, "A new look at smoothing Newton methods for nonlinear complementarity problems and box constrained variational

inequalities", AMR 97/13, Applied Mathematics Report, School of Mathematics, the University of New South Wales, Sydney 2052, Australia, June 1997.

[26] G. Ravindran and M.S. Gowda, "Regularization of P_0 functions in box variational inequality problem", Preprint, Department of Mathematics and Statistics, University of Maryland Baltimore County, Baltimore, Maryland 2150, USA, September 1997.

[27] S. M. Robinson, "Normal maps induced by linear transformation", *Mathematics of Operations Research, 17* (1992), 691–714.

[28] S.M. Robinson, "Generalized equations", in *Mathematical Programming: The State of the Art*, A. Bachem, M. Grötschel and B. Korte, eds., Springer-Verlag, Berlin, pp. 346–347, 1983.

[29] D. Sun, "A regularization Newton method for solving nonlinear complementarity problems", *Applied Mathematics and Optimization*, to appear.

[30] D. Sun and L. Qi, "On NCP-functions", AMR 97/32, Applied Mathematics Report, School of Mathematics, the University of New South Wales, Sydney 2052, Australia, October 1997.

[31] S. Xu, "The global linear convergence of an infeasible non-interior path-following algorithm for complementarity problems with uniform P-functions", Preprint, Department of Mathematics, University of Washington, Seattle, WA 98195, December 1996.

Table 1.2 Numerical results for the problems from MCPLIB

Problem	Dim	Start. point	Iter	NF	$R(x^k)$
bertsekas	15	(1)	8	9	4.1×10^{-8}
bertsekas	15	(2)	10	11	5.5×10^{-9}
bertsekas	15	(3)	60	87	3.8×10^{-9}
billups	1	(1)	9	19	7.3×10^{-10}
bert_oc	5000	(1)	11	30	8.4×10^{-7}
bratu	5625	(1)	43	156	3.4×10^{-10}
choi	13	(1)	5	6	1.2×10^{-10}
colvdual	20	(1)	7	8	1.8×10^{-14}
colvdual	20	(2)	8	10	5.5×10^{-9}
colvnlp	15	(1)	7	8	1.1×10^{-14}
colvnlp	15	(2)	8	10	1.8×10^{-9}
cycle	1	(1)	7	12	0
ehl_k40	41	(1)	11	13	9.2×10^{-8}
ehl_k60	61	(1)	13	15	1.5×10^{-8}
ehl_k80	81	(1)	13	15	4.6×10^{-13}
ehl_kost	101	(1)	14	17	1.3×10^{-12}
explcp	16	(1)	6	8	1.0×10^{-7}
freebert	15	(1)	9	10	2.3×10^{-8}
freebert	15	(2)	25	70	4.7×10^{-7}
freebert	15	(3)	9	10	2.4×10^{-8}
freebert	15	(4)	9	10	5.5×10^{-8}
freebert	15	(5)	88	987	1.9×10^{-7}
freebert	15	(6)	9	10	5.6×10^{-8}
gafni	5	(1)	5	10	1.4×10^{-9}
gafni	5	(2)	5	8	1.7×10^{-10}
gafni	5	(3)	6	15	1.5×10^{-11}
hanskoop	14	(1)	22	35	1.5×10^{-7}
hanskoop	14	(2)	20	31	1.5×10^{-12}
hanskoop	14	(3)	21	37	1.0×10^{-7}
hanskoop	14	(4)	17	29	5.1×10^{-11}
hanskoop	14	(5)	9	11	2.6×10^{-8}
hydroc06	29	(1)	7	10	2.5×10^{-7}
hydroc20***	99	(1)	8	10	3.1×10^{-7}
jel	6	(1)	6	7	1.3×10^{-8}
josephy	4	(1)	10	30	6.1×10^{-10}
josephy	4	(2)	6	7	1.6×10^{-7}
josephy	4	(3)	17	18	1.9×10^{-12}
josephy	4	(4)	5	6	1.2×10^{-8}
josephy	4	(5)	4	6	9.9×10^{-7}
josephy	4	(6)	6	7	9.2×10^{-10}
kojshin	4	(1)	6	10	2.6×10^{-9}
kojshin	4	(2)	7	8	5.5×10^{-8}
kojshin	4	(3)	18	19	7.9×10^{-14}
kojshin	4	(4)	5	7	1.3×10^{-7}
kojshin	4	(5)	6	9	3.5×10^{-7}

Table 1.3 (continued) Numerical results for the problems from MCPLIB

Problem	Dim	Start. point	Iter	NF	$R(x^k)$
kojshin	4	(6)	6	9	2.5×10^{-7}
mathinum	3	(1)	4	5	3.7×10^{-9}
mathinum	3	(2)	5	6	4.1×10^{-12}
mathinum	3	(3)	9	13	2.6×10^{-8}
mathinum	3	(4)	5	6	1.3×10^{-9}
mathisum	4	(1)	4	5	1.6×10^{-9}
mathisum	4	(2)	5	6	2.3×10^{-9}
mathisum	4	(3)	10	11	5.0×10^{-10}
mathisum	4	(4)	4	5	7.6×10^{-9}
methan08	31	(1)	5	6	6.6×10^{-13}
nash	10	(1)	6	7	4.0×10^{-9}
nash	10	(2)	9	10	1.4×10^{-7}
obstacle	2500	(1)	7	8	5.5×10^{-13}
obstacle	2500	(2)	9	13	5.8×10^{-13}
opt_cont31	1024	(1)	11	16	4.1×10^{-10}
opt_cont127	4096	(1)	12	24	1.6×10^{-8}
opt_cont255	8193	(1)	14	33	3.1×10^{-8}
opt_cont511	16384	(1)	16	54	5.6×10^{-8}
pgvon105	105	(1)	71	200	6.3×10^{-7}
pgvon105	105	(2)	13	34	1.0×10^{-7}
pgvon105	105	(3)	13	34	1.0×10^{-7}
pgvon106	106	(1)	40	175	9.9×10^{-7}
pies	42	(1)	36	340	4.5×10^{-13}
powell	16	(1)	87	304	1.0×10^{-7}
powell	16	(2)	20	22	5.0×10^{-7}
powell	16	(3)	25	35	8.0×10^{-7}
powell	16	(4)	18	20	4.2×10^{-7}
powell_mcp	8	(1)	6	7	6.5×10^{-12}
powell_mcp	8	(2)	7	8	1.8×10^{-12}
powell_mcp	8	(3)	8	9	4.7×10^{-8}
powell_mcp	8	(4)	7	8	1.2×10^{-7}
scarfanum	13	(1)	9	18	5.4×10^{-10}
scarfanum	13	(2)	9	16	3.0×10^{-10}
scarfanum	13	(3)	8	9	6.3×10^{-9}
scarfasum	14	(1)	15	20	2.8×10^{-9}
scarfasum	14	(2)	20	43	1.6×10^{-10}
scarfasum	14	(3)	15	28	2.1×10^{-10}
scarfbnum	39	(1)	18	76	5.2×10^{-12}
scarfbnum	39	(2)	26	154	4.3×10^{-14}
scarfbsum	40	(1)	27	259	2.0×10^{-12}
scarfbsum	40	(2)	31	283	5.1×10^{-7}
sppe	27	(1)	60	386	1.9×10^{-7}
sppe	27	(2)	10	11	7.3×10^{-13}
tobin	42	(1)	18	86	2.5×10^{-13}
tobin	42	(2)	11	29	2.7×10^{-10}

Applied Optimization

1. D.-Z. Du and D.F. Hsu (eds.): *Combinatorial Network Theory.* 1996
ISBN 0-7923-3777-8
2. M.J. Panik: *Linear Programming: Mathematics, Theory and Algorithms.* 1996
ISBN 0-7923-3782-4
3. R.B. Kearfott and V. Kreinovich (eds.): *Applications of Interval Computations.* 1996 ISBN 0-7923-3847-2
4. N. Hritonenko and Y. Yatsenko: *Modeling and Optimimization of the Lifetime of Technology.* 1996 ISBN 0-7923-4014-0
5. T. Terlaky (ed.): *Interior Point Methods of Mathematical Programming.* 1996
ISBN 0-7923-4201-1
6. B. Jansen: *Interior Point Techniques in Optimization.* Complementarity, Sensitivity and Algorithms. 1997 ISBN 0-7923-4430-8
7. A. Migdalas, P.M. Pardalos and S. Storøy (eds.): *Parallel Computing in Optimization.* 1997 ISBN 0-7923-4583-5
8. F.A. Lootsma: *Fuzzy Logic for Planning and Decision Making.* 1997
ISBN 0-7923-4681-5
9. J.A. dos Santos Gromicho: *Quasiconvex Optimization and Location Theory.* 1998
ISBN 0-7923-4694-7
10. V. Kreinovich, A. Lakeyev, J. Rohn and P. Kahl: *Computational Complexity and Feasibility of Data Processing and Interval Computations.* 1998
ISBN 0-7923-4865-6
11. J. Gil-Aluja: *The Interactive Management of Human Resources in Uncertainty.* 1998
ISBN 0-7923-4886-9
12. C. Zopounidis and A.I. Dimitras: *Multicriteria Decision Aid Methods for the Prediction of Business Failure.* 1998 ISBN 0-7923-4900-8
13. F. Giannessi, S. Komlósi and T. Rapcsák (eds.): *New Trends in Mathematical Programming.* Homage to Steven Vajda. 1998 ISBN 0-7923-5036-7
14. Ya-xiang Yuan (ed.): *Advances in Nonlinear Programming.* Proceedings of the '96 International Conference on Nonlinear Programming. 1998 ISBN 0-7923-5053-7
15. W.W. Hager and P.M. Pardalos: *Optimal Control.* Theory, Algorithms, and Applications. 1998 ISBN 0-7923-5067-7
16. Gang Yu (ed.): *Industrial Applications of Combinatorial Optimization.* 1998
ISBN 0-7923-5073-1
17. D. Braha and O. Maimon (eds.): *A Mathematical Theory of Design: Foundations, Algorithms and Applications.* 1998 ISBN 0-7923-5079-0

Applied Optimization

18. O. Maimon, E. Khmelnitsky and K. Kogan: *Optimal Flow Control in Manufacturing*. Production Planning and Scheduling. 1998 ISBN 0-7923-5106-1
19. C. Zopounidis and P.M. Pardalos (eds.): *Managing in Uncertainty: Theory and Practice*. 1998 ISBN 0-7923-5110-X
20. A.S. Belenky: *Operations Research in Transportation Systems:* Ideas and Schemes of Optimization Methods for Strategic Planning and Operations Management. 1998 ISBN 0-7923-5157-6
21. J. Gil-Aluja: *Investment in Uncertainty*. 1999 ISBN 0-7923-5296-3
22. M. Fukushima and L. Qi (eds.): *Reformulation: Nonsmooth, Piecewise Smooth, Semismooth and Smooting Methods*. 1999 ISBN 0-7923-5320-X
23. M. Patriksson: *Nonlinear Programming and Variational Inequality Problems*. A Unified Approach. 1999 ISBN 0-7923-5455-9

KLUWER ACADEMIC PUBLISHERS – DORDRECHT / BOSTON / LONDON